Achieving sustainable cultivation of cassava

Volume 1: Cultivation techniques

It is widely recognised that agriculture is a significant contributor to global warming and climate change. Agriculture needs to reduce its environmental impact and adapt to current climate change whilst still feeding a growing population, i.e. become more 'climate-smart'. Burleigh Dodds Science Publishing is playing its part in achieving this by bringing together key research on making the production of the world's most important crops and livestock products more sustainable.

Based on extensive research, our publications specifically target the challenge of climate-smart agriculture. In this way we are using 'smart publishing' to help achieve climate-smart agriculture.

Burleigh Dodds Science Publishing is an independent and innovative publisher delivering high quality customer-focused agricultural science content in both print and online formats for the academic and research communities. Our aim is to build a foundation of knowledge on which researchers can build to meet the challenge of climate-smart agriculture.

For more information about Burleigh Dodds Science Publishing simply call us on +44 (0) 1223 839365, email info@bdspublishing.com or alternatively please visit our website at www.bdspublishing.com.

Related titles:

Achieving sustainable cultivation of cassava Volume 2: Genetics, breeding, pests and diseases
Print (ISBN 978-1-78676-004-3); Online (ISBN 978-1-78676-006-7, 978-1-78676-007-4)

Achieving sustainable cultivation of grain legumes Volume 1: Advances in breeding and cultivation techniques
Print (ISBN 978-1-78676-136-1); Online (ISBN 978-1-78676-138-5, 978-1-78676-139-2)

Achieving sustainable cultivation of grain legumes Volume 2: Improving cultivation of particular grain legumes
Print (ISBN 978-1-78676-140-8); Online (ISBN 978-1-78676-142-2, 978-1-78676-143-9)

Chapters are available individually from our online bookshop: https://shop.bdspublishing.com

climate•SMART•publishing
IN AGRICULTURAL SCIENCE

BURLEIGH DODDS SERIES IN AGRICULTURAL SCIENCE

NUMBER 20

Achieving sustainable cultivation of cassava

Volume 1: Cultivation techniques

Edited by Dr Clair Hershey, formerly International Center for Tropical Agriculture (CIAT), Colombia

burleigh dodds
SCIENCE PUBLISHING

Published by Burleigh Dodds Science Publishing Limited
82 High Street, Sawston, Cambridge CB22 3HJ, UK
www.bdspublishing.com

Burleigh Dodds Science Publishing, 1518 Walnut Street, Suite 900, Philadelphia, PA 19102-3406, USA

First published 2017 by Burleigh Dodds Science Publishing Limited

Library of Congress Control Number: 2017936826

British Library Cataloguing in Publication Data
A catalogue record for this book is available from the British Library

ISBN 978-1-78676-000-5 (print)
ISBN 978-1-78676-002-9 (online)
ISBN 978-1-78676-003-6 (online)
ISSN 2059-6936 (print)
ISSN 2059-6944 (online)

Typeset by Deanta Global Publishing Services, Chennai, India
Printed by Lightning Source

Contents

Series list x

Introduction xiv

Part 1 The cassava plant and its uses

1 The emergence of cassava as a global crop 3
Doyle McKey, University of Montpellier, France; and Marc Delêtre, Trinity College Dublin, Ireland
1 Introduction 3
2 Origins and early cultivation of cassava in the Americas 4
3 Introduction of cassava in the Old World 15
4 Conclusion 24
5 Where to look for further information 25
6 Acknowledgements 25
7 References 26

2 Use and nutritional value of cassava roots and leaves as a traditional food 33
Aurélie Bechoff, Natural Resources Institute, University of Greenwich, UK
1 Introduction 33
2 Diversity in traditional cassava products 35
3 Products from cassava roots 35
4 Products from cassava leaves 40
5 Chemical composition of cassava roots 43
6 Chemical composition of cassava leaves 44
7 Anti-nutrients in roots and leaves 46
8 Case study: biofortified cassava roots made into gari and fufu in Nigeria 47
9 Future trends 50
10 Conclusion 51
11 Where to look for further information 52
12 References 53

3 Cassava for industrial uses 57
Kuakoon Piyachomkwan and Sittichoke Wanlapatit, National Center for Genetic Engineering and Biotechnology, Thailand; and Klanarong Sriroth, Kasetsart University, Thailand
1 Introduction 57
2 Cassava root composition 58
3 Cassava chips 59
4 Cassava starch: production and properties 68
5 Cassava starch: products 76
6 Cassava starch: industrial applications 84
7 Summary 86
8 Where to look for further information 86
9 References 87

4 New uses and processes for cassava 89
Keith Tomlins and Ben Bennett, Natural Resources Institute, University of Greenwich, UK
1 Introduction 89
2 Current and developing cassava production and use 89
3 New products from cassava: the case of High Quality Cassava Flour (HQCF) in Africa 92
4 Summary and future trends 96
5 Conclusion 97
6 Where to look for further information 97
7 Acknowledgements 98
8 References 98

Part 2 Current cassava cultivation and opportunities for improvement

5 Cassava cultivation in Asia 101
Tin Maung Aye, CIAT, Vietnam
1 Introduction 101
2 Cassava production in Asia: an overview 102
3 Varietal improvement 104
4 Other key research areas for developing cassava production in Asia 110
5 Case study: developing sustainable cassava production in Cambodia 113
6 Summary 118
7 Where to look for further information 119
8 References 120

6 Cassava cultivation in sub-Saharan Africa 123
Dunstan S. C. Spencer, Enterprise Development Services Ltd, Sierra Leone; and Chuma Ezedinma, UNIDO, Nigeria
1 Introduction 123
2 Cassava's contribution to incomes in SSA 130
3 Cassava's contribution to food and nutrition security 133
4 Drivers of change in cassava production in SSA 140
5 Future trends and conclusion 145
6 References 146

7 Cassava cultivation in Latin America 149
Olivier F. Vilpoux, Denilson de Oliveira Guilherme and Marney Pascoli Cereda, Catholic University of Campo Grande, Brazil
1 Introduction 150
2 Global and Latin American approaches to cassava production 150
3 Cassava competitiveness in Latin America 154
4 Boosting cassava production in Latin America 158
5 Improving the Latin American starch industry 162
6 Brazilian cassava flour (farinha) 164
7 Consumption of boiled and fried cassava 165
8 Future trends: cassava for animal feed 171
9 Where to look for further information 172
10 References 172

8 Drivers of change for cassava's multiple roles in sustainable development 175
Clair H. Hershey, formerly CIAT, Colombia
1 Introduction 175
2 The development of cassava as a crop 177
3 The role cassava can play in development 179
4 Constraints on cassava cultivation and production 180
5 Addressing constraints: genetic diversity and breeding 182
6 Addressing constraints: crop and soil management 185
7 Addressing constraints: processing, market and gender issues 187
8 Addressing constraints: policy issues 189
9 A new research paradigm for cassava 191
10 Cassava and climate change 192
11 Conclusions 195
12 Where to look for further information 195
13 References 196

9 Targeting smallholder farmers to adopt improved cassava technologies: challenges and opportunities 199
Kolawole Adebayo, Federal University of Agriculture, Nigeria
1 Introduction 199
2 Research on improving cassava production and understanding smallholder farmers and their decisions 201
3 Research on smallholders' support systems 203
4 Case studies 205
5 Summary 209
6 Future trends in research 209
7 Where to look for further information 209
8 References 210

10 GCP21: a global cassava partnership for the 21st century 213
Claude M. Fauquet and Joe Tohme, International Center for Tropical Agriculture (CIAT), Colombia
1 Introduction 213
2 GCP21 as a platform for the science of cassava production 214
3 Making connections in cassava research 215
4 Looking for gaps in R&D from basic science to final products 217
5 Improving global access to information and technology 219
6 Conclusion 220
7 Where to look for further information 220
8 References 221

Part 3 Sustainable intensification of cassava cultivation

11 Integrated crop management for cassava cultivation in Asia 225
Tin Maung Aye, CIAT, Vietnam; and Reinhardt Howeler, CIAT Emeritus - formerly CIAT, Thailand
1 Introduction 225
2 Implementing GAP in cassava pre-harvesting 227
3 Implementing GAP for cassava crop health 235

4 Implementing GAP in cassava post-harvesting 244
5 Adoption of GAP by farmers 245
6 Conclusion 249
7 Future trends 250
8 References 250

12 Seed systems management in cassava cultivation 255
James George and Sarojini Amma Sunitha, ICAR-Central Tuber Crops Research Institute, India
1 Introduction 255
2 General description of cassava seed 256
3 Formal and informal seed systems 257
4 Quality and standards for planting material 258
5 Production of quality planting material 262
6 Rapid multiplication techniques 269
7 Formal seed system 272
8 Case study: mini-stake-based rapid multiplication at ICAR-CTCRI, India 274
9 An integrated approach for the future 280
10 Future trends and conclusion 281
11 Where to look for further information 281
12 References 281

13 Cassava cultivation and soil productivity 285
Reinhardt Howeler, CIAT Emeritus - formerly CIAT, Thailand
1 Introduction 285
2 Nutrient removal by the cassava crop 287
3 The effect of cassava on soil productivity 290
4 Nutrient losses in eroded sediments and runoff 293
5 Diagnosis of nutritional problems 294
6 Conclusions 298
7 References 298

14 Addressing nutritional disorders in cassava cultivation 301
Reinhardt Howeler, CIAT Emeritus - formerly CIAT, Thailand
1 Introduction 301
2 Nitrogen and phosphorus deficiencies 302
3 Potassium, calcium, magnesium and sulphur deficiencies 306
4 Micronutrients and soil pH 308
5 Short- and long-term nitrogen, phosphorus and potassium (NPK) requirements of cassava: an overview 311
6 Soil and nutrient management for cassava cultivation in Latin America 311
7 Soil and nutrient management for cassava cultivation in sub-Saharan Africa 315
8 Soil and nutrient management for cassava cultivation in Asia 317
9 Conclusions 324
10 References 324

15 Nutrient sources and their application in cassava cultivation 331
Reinhardt Howeler, CIAT Emeritus - formerly CIAT, Thailand
1 Introduction 331
2 Organic and inorganic nutrient sources 331
3 When and where to apply chemical fertilizers, manures and soil amendments 335
4 Determination of fertilizer rates and nitrogen, phosphorus and potassium (NPK) balance 336
5 Site-specific nutrient management 338
6 Fertilizer requirements for balanced nutrition 342
7 Genetic variation in nutrient use efficiency (NUE) and tolerance to low soil fertility 347
8 Conclusions 352
9 References 352

16 Intercropping and crop rotations in cassava cultivation: a production systems approach 355
Thomas W. Kuyper, Wageningen University, The Netherlands; and Samuel Adjei-Nsiah, International Institute of Tropical Agriculture (IITA), Ghana
1 Introduction 355
2 Advantages of cassava as a crop 356
3 Crop rotations in cassava 358
4 Cassava intercropping 359
5 Yield increases through cassava intercropping 359
6 Achieving yield increases in cassava intercropping 361
7 Soil fertility enhancement through cassava intercropping 363
8 Farmer experimentation during cassava cropping 364
9 Cassava functional traits 365
10 Future trends and conclusion 368
11 Where to look for further information 369
12 References 369

17 Value chain approaches to mechanization in cassava cultivation and harvesting in Africa 375
George Marechera and Grace Muinga, African Agricultural Technology Foundation (AATF), Kenya
1 Introduction 375
2 The benefits of mechanization 376
3 Obstacles to mechanization by smallholders 377
4 The AATF cassava mechanization experience 380
5 Mechanized production options for cassava 383
6 Assessing the effectiveness of mechanization 388
7 Business models for giving famers access to farm machinery 392
8 Conclusions 396
9 References 397

Index 399

Series list

Title	Series number
Achieving sustainable cultivation of maize - Vol 1 From improved varieties to local applications Edited by: Dr Dave Watson, CGIAR Maize Research Program Manager, CIMMYT, Mexico	001
Achieving sustainable cultivation of maize - Vol 2 Cultivation techniques, pest and disease control Edited by: Dr Dave Watson, CGIAR Maize Research Program Manager, CIMMYT, Mexico	002
Achieving sustainable cultivation of rice - Vol 1 Breeding for higher yield and quality Edited by: Prof. Takuji Sasaki, Tokyo University of Agriculture, Japan	003
Achieving sustainable cultivation of rice - Vol 2 Cultivation, pest and disease management Edited by: Prof. Takuji Sasaki, Tokyo University of Agriculture, Japan	004
Achieving sustainable cultivation of wheat - Vol 1 Breeding, quality traits, pests and diseases Edited by: Prof. Peter Langridge, The University of Adelaide, Australia	005
Achieving sustainable cultivation of wheat - Vol 2 Cultivation techniques Edited by: Prof. Peter Langridge, The University of Adelaide, Australia	006
Achieving sustainable cultivation of tomatoes Edited by: Dr Autar Mattoo, USDA-ARS, USA & Prof. Avtar Handa, Purdue University, USA	007
Achieving sustainable production of milk - Vol 1 Milk composition, genetics and breeding Edited by: Dr Nico van Belzen, International Dairy Federation (IDF), Belgium	008
Achieving sustainable production of milk - Vol 2 Safety, quality and sustainability Edited by: Dr Nico van Belzen, International Dairy Federation (IDF), Belgium	009
Achieving sustainable production of milk - Vol 3 Dairy herd management and welfare Edited by: Prof. John Webster, University of Bristol, UK	010
Ensuring safety and quality in the production of beef - Vol 1 Safety Edited by: Prof. Gary Acuff, Texas A&M University, USA & Prof.James Dickson, Iowa State University, USA	011
Ensuring safety and quality in the production of beef - Vol 2 Quality Edited by: Prof. Michael Dikeman, Kansas State University, USA	012
Achieving sustainable production of poultry meat - Vol 1 Safety, quality and sustainability Edited by: Prof. Steven C. Ricke, University of Arkansas, USA	013
Achieving sustainable production of poultry meat - Vol 2 Breeding and nutrition Edited by: Prof. Todd Applegate, University of Georgia, USA	014
Achieving sustainable production of poultry meat - Vol 3 Health and welfare Edited by: Prof. Todd Applegate, University of Georgia, USA	015

Achieving sustainable production of eggs - Vol 1 016
Safety and quality
Edited by: Prof. Julie Roberts, University of New England, Australia

Achieving sustainable production of eggs - Vol 2 017
Animal welfare and sustainability
Edited by: Prof. Julie Roberts, University of New England, Australia

Achieving sustainable cultivation of apples 018
Edited by: Dr Kate Evans, Washington State University, USA

Integrated disease management of wheat and barley 019
Edited by: Prof. Richard Oliver, Curtin University, Australia

Achieving sustainable cultivation of cassava - Vol 1 020
Cultivation techniques
Edited by: Dr Clair Hershey, formerly International Center for Tropical Agriculture (CIAT), Colombia

Achieving sustainable cultivation of cassava - Vol 2 021
Genetics, breeding, pests and diseases
Edited by: Dr Clair Hershey, formerly International Center for Tropical Agriculture (CIAT), Colombia

Achieving sustainable production of sheep 022
Edited by: Prof. Johan Greyling, University of the Free State, South Africa

Achieving sustainable production of pig meat - Vol 1 023
Safety, quality and sustainability
Edited by: Prof. Alan Mathew, Purdue University, USA

Achieving sustainable production of pig meat - Vol 2 024
Animal breeding and nutrition
Edited by: Prof. Julian Wiseman, University of Nottingham, UK

Achieving sustainable production of pig meat - Vol 3 025
Animal health and welfare
Edited by: Prof. Julian Wiseman, University of Nottingham, UK

Achieving sustainable cultivation of potatoes - Vol 1 026
Breeding, nutritional and sensory quality
Edited by: Prof. Gefu Wang-Pruski, Dalhousie University, Canada

Achieving sustainable cultivation of oil palm - Vol 1 027
Introduction, breeding and cultivation techniques
Edited by: Prof. Alain Rival, Center for International Cooperation in Agricultural Research for Development (CIRAD), France

Achieving sustainable cultivation of oil palm - Vol 2 028
Diseases, pests, quality and sustainability
Edited by: Prof. Alain Rival, Center for International Cooperation in Agricultural Research for Development (CIRAD), France

Achieving sustainable cultivation of soybeans - Vol 1 029
Breeding and cultivation techniques
Edited by: Prof. Henry Nguyen, University of Missouri, USA

Achieving sustainable cultivation of soybeans - Vol 2 030
Diseases, pests, food and non-food uses
Edited by: Prof. Henry Nguyen, University of Missouri, USA

Achieving sustainable cultivation of sorghum - Vol 1 031
Genetics, breeding and production techniques
Edited by: Prof. Bill Rooney, Texas A&M University, USA

Achieving sustainable cultivation of sorghum - Vol 2 032
Sorghum utilisation around the world
Edited by: Prof. Bill Rooney, Texas A&M University, USA

Achieving sustainable cultivation of potatoes - Vol 2 033
Production and storage, crop protection and sustainability
Edited by: Dr Stuart Wale, Potato Dynamics Ltd, UK

Achieving sustainable cultivation of mangoes 034
Edited by: Prof. Víctor Galán Saúco, Instituto Canario de Investigaciones Agrarias (ICIA), Spain & Dr Ping Lu, Charles Darwin University, Australia

Achieving sustainable cultivation of grain legumes - Vol 1 035
Advances in breeding and cultivation techniques
Edited by: Dr Shoba Sivasankar et al., CGIAR Research Program on Grain Legumes, ICRISAT, India

Achieving sustainable cultivation of grain legumes - Vol 2 036
Improving cultivation of particular grain legumes
Edited by: Dr Shoba Sivasankar et al., CGIAR Research Program on Grain Legumes, ICRISAT, India

Achieving sustainable cultivation of sugarcane - Vol 1 037
Cultivation techniques, quality and sustainability
Edited by: Prof. Philippe Rott, University of Florida, USA

Achieving sustainable cultivation of sugarcane - Vol 2 038
Breeding, pests and diseases
Edited by: Prof. Philippe Rott, University of Florida, USA

Achieving sustainable cultivation of coffee 039
Breeding and quality traits
Edited by: Dr Philippe Lashermes, Institut de Recherche pour le Développement (IRD), France

Achieving sustainable cultivation of bananas - Vol 1 040
Cultivation techniques
Edited by: Prof. Gert Kema, Wageningen University, The Netherlands & Prof. André Drenth, University of Queensland, Australia

Global Tea Science 041
Current status and future needs
Edited by: Dr V. S. Sharma, Formerly UPASI Tea Research Institute, India & Dr M. T. Kumudini Gunasekare, Coordinating Secretariat for Science Technology and Innovation (COSTI), Sri Lanka

Integrated weed management 042
Edited by: Emeritus Prof. Rob Zimdahl, Colorado State University, USA

Achieving sustainable cultivation of cocoa - Vol 1 043
Genetics, breeding, cultivation and quality
Edited by: Prof. Pathmanathan Umaharan, Cocoa Research Centre – The University of the West Indies, Trinidad and Tobago

Achieving sustainable cultivation of cocoa - Vol 2 044
Diseases, pests and sustainability
Edited by: Prof. Pathmanathan Umaharan, Cocoa Research Centre – The University of the West Indies, Trinidad and Tobago

Water management for sustainable agriculture 045
Edited by: Prof. Theib Oweis, Formerly ICARDA, Lebanon

Improving organic animal farming 046
Edited by: Dr Mette Vaarst, Aarhus University, Denmark & Dr Stephen Roderick, Duchy College, Cornwall, UK

Improving organic crop cultivation 047
Edited by: Prof. Ulrich Köpke, University of Bonn, Germany

Managing soil health for sustainable agriculture - Vol 1 048
Fundamentals
Edited by: Dr Don Reicosky, USDA-ARS, USA

Managing soil health for sustainable agriculture - Vol 2 049
Monitoring and management
Edited by: Dr Don Reicosky, USDA-ARS, USA

Rice insect pests and their management 050
E. A. Heinrichs, Francis E. Nwilene, Michael J. Stout, Buyung A. R. Hadi & Thais Freitas

Improving grassland and pasture management in temperate agriculture 051
Edited by: Prof. Athole Marshall & Dr Rosemary Collins, University of Aberystwyth, UK

Precision agriculture for sustainability 052
Edited by: Dr John Stafford, Silsoe Solutions, UK

Introduction

1 The roles, challenges and opportunities for cassava in development

For the last century agricultural scientists have embraced and met the challenges of providing technologies that gave farmers the capacity to stay ahead of the growing food demands from increasing populations. Hunger and malnutrition still exist -- and cause far too much misery and livelihood risk -- but UN statistics show clear progress in reducing food deficits at the global level. Agricultural science has ever more tools at its disposal, but at the same time an expanding panorama of expectations from its users. While the Green Revolution of the 1960s and beyond made landmark progress by focusing mainly on productivity of cereal grains under high input conditions, the world rightly demands more attention to environmental impact of agriculture (water quality, soil erosion, genetic resources), quality and nutrition of food, dwindling inputs for expanding productivity, and social justice issues such as equity and opportunity for women. Agricultural sciences are slowly adapting to these new realities, but not quickly enough, and with too few resources committed to success, especially in developing countries. This is the broader context within which we consider the cassava story as a key target for sustainable improvement.

Although cassava is little-known outside the tropics, globally it is the fourth most important calorie source after wheat, rice and maize. It is especially renowned as a crop that supports farmers in more marginal conditions and with poor access to inputs, due to its ability to yield well under low-fertility soils and with periodic drought where more sensitive crops would be devastated. Although it was domesticated from its wild ancestors in the Americas, over half of global production now comes from Africa where it has a leading role in food security. Cassava is largely an energy crop. Early in the growth cycle, beginning at about 2-3 months, some of the roots convert to starch storage organs. From a physiological standpoint, this is a mechanism to store energy through periods of stress, and for recovery when the stress is alleviated. In agriculture, farmers typically aim to minimize stress so that as much of the root starch as possible can be recovered for human use. Although cassava is often referred to as a *subsistence crop*, this is in fact only a small part of present-day production. There is a growing dichotomy between cassava as a low-input, low-yield crop of the poor, and cassava as a vibrant and highly flexible crop driven by industrial demand. Even in Africa where it is nearly entirely a crop of small farmers, mainly for human food, a large majority of production is marketed for off-farm use. In Asia, most of production is already for industrial uses, and with the exception of a few countries, especially Indonesia, India and the Philippines, human consumption is relatively low. Even in industrial markets, most of production is on small farms. Nonetheless, large-scale industrial systems are emerging around the globe, driven by robust demand for cassava products and by technologies that facilitate production and processing at scale. Some facets of cassava production are not easy to industrialize, i.e. to mechanize or implement at large scale, for several reasons. These constraints are gradually being overcome. Planting and harvesting can be mechanized. But there are still many aspects of production that favor small farmer management, and cassava production is likely to be dominated by small farmers for decades to come. Just as the crop is grown primarily by poor farmers, consumers also tend to be from the lower income strata. For governments and development agencies who want to differentially benefit the poor, cassava is an ideal target for sustainable improvement.

2 Science progress and the needs for supporting cassava's future roles

Cassava's features, especially its minimal production in developed countries, have often meant that research investment has long been far less than for most important food crops. This situation was reversed to some degree when two International Agricultural Research Centers – the International Center for Tropical Agriculture (CIAT) in Cali, Colombia and the International Institute for Tropical Agriculture (IITA) in Ibadan, Nigeria were founded in the late 1960s, and were mandated to work on cassava on a global scale (historically, CIAT in the Americas and Asia, and IITA in Africa). This investment motivated considerable parallel investment by national research programs around the world. Through the 1980s, most countries with significant cassava production had developed human and physical capital to improve the crop and the livelihoods of people who depended on it. However, by the 1990s, and especially in Latin America, many countries discontinued or severely cut funding to cassava research. A sort of *development fatigue* had set in for many traditional donors, and there was a rising belief that that the private sector would take on much of the responsibility for crop improvement research. In other crops, and especially maize and soybeans, research was more and more taken on by the private sector, which was able to get a return on investment through the sale of technologies such as seed and chemicals. But cassava provided much less opportunity for profit from the private sector, and therefore lagged in technology development. By early into the 21st century, some of the key donor agencies for agricultural development began to recognize the need and the potential to invest in cassava, especially the Bill and Melinda Gates Foundation. In addition, the CGIAR -- parent organization of CIAT and IITA -- renewed its interest in cassava and recognized the synergies that could be achieved by joint work on key starchy, vegetatively propagated crops. Thus, in 2011, the Root, Tuber and Banana CGIAR Research Program (RTB-CRP) began operation, involving four CGIAR centers and five key crops (cassava, potato, sweet potato, bananas/plantains and yams). At about the same time, some of first major private sector investment in cassava research was initiated to develop amylose-free (*waxy*) cassava varieties in Thailand and Colombia.

While cassava still receives far less research funding in proportion to its value, compared to other major food crops, there has been key interest in recent years from donors, governments, and a limited number of private sector organizations to better fund cassava research. Cassava has been reasonably well-positioned in the molecular revolution, in terms of development of breeding techniques and information. However, the practical applications remain largely a future hope, and impact to date from breeding has been though more conventional approaches.

3 Bringing together the latest research and development information, and expected outcomes of this book

There is no recent comprehensive review of cassava research and development. With the explosion of information in general, and specifically about crop research, scientists have a difficult challenge to stay broadly informed about any crop, even one in which they may be specialized. These two volumes bring together global experts across the spectrum of cassava production and utilization topics, to distill and analyze information toward the broader context of achieving sustainable cassava cultivation. Apart from the broad chapter contents, the reader is provided extensive reference lists for further consultation and in-depth learning. This book will enable single-source consultation of a wide range of topics relevant to cassava R&D well into the future.

4 Section and chapter overviews

These two volumes present a comprehensive review of the history, the current strategies and the future potential to further transform cassava value chains toward sustainable systems. Each volume divides into three complementary sections.

Volume 1: Part 1 describes the cassava plant and its uses. **Chapter 1** takes an ethno-biological journey through the crop's early history, informed especially by practices still carried out today in some of the crop's most traditional growing areas of South America's Amazon basin. **Chapter 2** describes the high diversity of food products from cassava, especially in Africa, where over 90% of cassava is destined for human food use. Many of the processes were brought from the crop's homeland in the America centuries ago, and expanded and modified to meet local needs. **Chapter 3** reviews the opposite end of the spectrum – cassava's diverse industrial uses, and especially starch. Much of SE Asia's production is destined for industrial uses, but Africa and Latin America are also advancing in these more sophisticated, value-added markets. **Chapter 4** shows the potential and some examples of the transformation of traditional products and markets into higher value markets in Africa, especially for high quality cassava flour (HQCF).

Part 2 of Volume 1 begins within a broad overview of cassava production, processing and use across Asia, Africa and Latin America, **Chapter 5, 6 and 7,** respectively. Although the crop was domesticated in the Americas, only about 18% of current production comes from this region, and over half from Africa. These chapters compare and contrast the commonalities and differences among the regions, with lessons from each that can support sustainable development goals. **Chapter 8** draws on the regional overviews to highlight some of the global challenges and opportunities for sustainable cassava development, and the drivers of research and policy for setting priorities. Clearly there is no single strategy that applies globally, and every strategy fits into the context of an evolving social, economic, agronomic and environmental environment, among others. **Chapter 9** discusses the need to target and involve small-holder farmers in the development of cassava technologies, with case studies from Africa. **Chapter 10** presents the Global Cassava Partnership for the 21st Century (GCP21) as a global support partnership for cassava research and development. GCP21 is a not-for-profit international alliance of 45 organizations, aiming to fill gaps in cassava R&D in order to unlock the potential of cassava for improving food security and income, especially for the poor.

Part 3 of Volume 1 includes a comprehensive coverage of production practices for sustainable intensification of cassava production. **Chapter 11** describes the full range of production practices that growers should take into account to improve yields, profitability and sustainability – land preparation, plant populations, cropping systems, weed management, pest and disease management, and harvesting. The chapter draws heavily on experiences from Asia, but with relevance around the world. **Chapter 12** reviews the critical management of *seed* in cassava – normally through the vegetative reproduction through stem pieces. Currently nearly all seed is managed under informal systems, but this is slowly changing. Practices and systems developed in India inform the needs, challenges and experiences globally to develop improved seed systems. **Chapters 13, 14 and 15** give extensive coverage to best practices for managing cassava nutrition to achieve high and sustainable yields, through fertility management and soil conservation. As the demands increase for higher yield through sustainable practices, effective long-term soil fertility maintenance is a core strategy. The chapters draw especially on very extensive research and production experience in Asia and Latin America, where fertilizer use is far more common than in Africa. The chapters cover the relationship between soil fertility and crop productivity, how to diagnose nutritional needs and disorders, and best practices to achieve sustainable productivity through nutrient application. **Chapter 16** describes rotation and intercropping in cassava cultivation, especially common in Africa. There is a growing body of evidence on the sustainability and income advantages of diversified crop systems. Because of cassava's long period in the field, and its slow early growth, there is a wide range of alternative options for managing multiple crops that are complementary to cassava's growth and development. Finally, **Chapter 17** of Volume 1 reviews the principles of mechanization for all aspects of cassava production, which until now has been limited, especially in Africa. The chapter uses a case study of the Cassava Mechanization and Agro-Processing Project of the Africa Agricultural Technology Foundation to illustrate principles, challenges and opportunities for cassava mechanization.

Volume 2 of *Achieving Sustainable Cassava Cultivation* covers genetic resources, breeding, and pests and diseases. **Part 1** focuses on genetic resources and breeding tools. **Chapter 1** reviews knowledge on cassava's unique growth and development features as a perennial crop managed as an annual, its vegetative propagation, and the fact that it has no phasic development as is the case for the grain and grain legume crops. This understanding is key to improving the crop through both management and breeding. **Chapter 2** describes the cassava's genetic resources, especially the *ex situ* collections managed by the International Agricultural Research Centers (CIAT and IITA), a fundamental resource for the crop's genetic improvement. **Chapter 3** delves into the genetic basis for cassava breeding, and provides novel and innovative strategies to move beyond the current plateau for improving cassava yield.In particular, the chapter describes in detail the rationale and the possible strategies to exploit heterosis and to make cassava breeding more efficient through the use of inbreeding. **Chapter 4** brings the reader up to date on the fast-moving repertoire of molecular knowledge that supports cassava breeding. While cassava lagged behind other major crops initially in the molecular revolution, it has been catching up fast in recent years and molecular techniques are poised to bring major benefits to the growers and consumers of this crop. **Chapter 5** provides the technical background and describes specific molecular tools for making cassava breeding more efficient. The use of marker assisted selection (MAS) and genomic selection (GS) are covered in some detail. **Chapter 6** reviews the current status and the potential for improving cassava through genetic engineering, or genetic modification through targeting the insertion of genes directly into the crop's genome to achieve novel traits. In spite of rapid scientific advances, the regulatory environment limits impact at the field level while governments and the public assess and absorb the potential risks and benefits.

Part 2 of Volume 2 looks at specific breeding goals for root yield and quality, and progress toward reaching them. **Chapter 7** uses a case study for cassava breeding programs in Thailand, which has had one of the world's most successful cassava improvement efforts through the collaboration of two local centers (The Department of Agriculture and Kasetsart University), with additional support from International Centers (CIAT) and the Thai cassava industry. The fact that breeding efforts were aimed exclusively at industrial markets allowed the Thai programs to focus on yield and starch content, and to gain valuable experience in breeding for these two traits. **Chapter 8** reviews a comprehensive program in West Africa to improve the nutritional content of cassava, for pro-vitamin A. The program supported by the HarvestPlus initiative demonstrates the many components of the value chain that need to be considered – from breeding through measuring nutritional impact at the household level. This case study is illustrative of the need for cassava improvement programs to be fully integrated with processors and consumers to develop successful products. **Chapter 9** reveals the importance and the complexity of breeding for the fine-tuned consumer preferences, especially in Africa where a plethora of different products and their regional variations present major challenges to breeders with regard to identifying priority traits and their prioritization for breeding. The chapter uses examples from Africa to sort through some of these complexities and guide breeders in their planning.

Part 3 of Volume 2 covers pest and disease management in cassava, including weeds. These are areas that have received insufficient attention in the past, due to a widespread belief that cassava is a rugged crop that will produce reasonably well without any need to manage pests and diseases. However, as production practices have intensified, and especially as insects, mites and pathogens have moved with ease around the world, it is now well-understood that sustainable production is possible only with good management of these problems. **Chapter 10** describes the major diseases affecting cassava and their integrated management. It is noteworthy that two of Africa's most devastating problems – cassava mosaic disease and cassava brown streak disease – have not been found in the Americas where the crop and nearly all other of its pests and disease have co-evolved. Asia, once nearly free of serious disease problems, is recently experiencing new challenges, for example with witches broom disease and Sri Lankan cassava mosaic disease. **Chapter 11** reviews the arthropod pest complex of cassava and the importance of integrated pest management strategies. Because of cassava's long growth cycle, along with other factors, pesticide applications are usually not economically or ecologically sustainable. Cassava entomologists have a long success in biological control of major pests, especially the cassava mealybug, once one of Africa's most devastating pest problems and now brought under control in most regions through a parasitic wasp introduced from cassava's homeland in the Americas. **Chapter 12** discusses integrated weed management for cassava, with a focus on Africa. Currently, the vast majority of weed control is done manually in Africa. However, the demands for more efficient and effective weed control are creating the need for research on new options, both mechanical and chemical, as well as refined crop management such as intercropping and supporting early-vigor and shading through fertilization, plant spacing and variety selection. In Africa, most weeding is done by women, and new options for better weed management can have broad gender implications for the continent.

Part 1

The cassava plant and its uses

Chapter 1

The emergence of cassava as a global crop

Doyle McKey, University of Montpellier, France; and Marc Delêtre, Trinity College Dublin, Ireland

1 Introduction
2 Origins and early cultivation of cassava in the Americas
3 Introduction of cassava in the Old World
4 Conclusion
5 Where to look for further information
6 Acknowledgements
7 References

1 Introduction

Among tropical crops, cassava[1] is unique in its ability to produce economically viable yields with little input of materials and labour, even in marginal ecological conditions (McKey and Beckerman, 1993; Burns et al., 2010), and plays a crucial role in efforts to achieve food security in the tropics. We trace the origins of cassava as a crop, focusing on the coevolution of the key adaptations – biological adaptations of the plant and cultural adaptations of its cultivators – that led to cassava's success and its diffusion throughout the Neotropics. We then trace the diffusion and early cultivation of cassava in areas where it was introduced following the Columbian exchange. In these areas, cassava diffusion was often accompanied by a process of cultural reappropriation that we consider to be a form of 'cultural domestication'. This complicated plant – the world's only staple crop that can be dangerously toxic if not correctly prepared – was introduced by Europeans throughout the tropical world, but without the Amerindian folk knowledge with which it had coevolved. We identify two key co-adaptations of the crop and the native Amazonians who shepherded its domestication and diffusion throughout the Neotropics: (1) a mixed clonal/sexual reproductive system, resulting from the match between the plant's traits and cultivators' use of space over time, and (2) built-in chemical defences against herbivores and pathogens and practices to manage these defences safely. Neither of these themes has received the attention they both deserve from cassava breeders, agronomists and other stakeholders engaged in designing paths to food security. We discuss examples where initial mismatches between the plant's ecological traits and farming practices were partly overcome by farmers' ingenuity, which

1. An argument can be made on linguistic grounds for the use of 'manioc' rather than 'cassava' as the English name of this plant (Gade, 2002). For consistency with the rest of the book, we use 'cassava' here.

http://dx.doi.org/10.19103/AS.2016.0014.04

led to many technical innovations. Although socio-economic conditions, farming practices, even the plant itself, have all changed in myriad ways since the crop's domestication, some things are timeless. As we try to design a sustainable future in a changing world, cassava will likely play an ever more prominent role as other crops fail (Brown et al., 2016), and it may be wise to draw some lessons from the past.

2 Origins and early cultivation of cassava in the Americas

Unravelling the domestication history of a crop plant is a three-faceted problem: (1) one must determine the crop's botanical origin, that is, the wild ancestor(s) from which it is derived; (2) one must determine the region where the wild ancestors were first cultivated; and (3) one must determine the environmental conditions that prevailed there at the time. For cassava, all of these aspects have long been debated but some major issues are still unresolved (Isendahl, 2011; Clement et al., 2016). Several reasons explain the continuing uncertainty. First, cassava is a thoroughly domesticated plant that differs in numerous traits from all its extant wild relatives (McKey et al., 2012), and there exists no incipiently domesticated population that might offer direct evidence of the identity of cassava's wild ancestor (Clement, 2014). Secondly, the archaeological record of consumption of cassava has many gaps. Remains of its tuberous roots are rarely preserved, except in arid environments. Starch grains of *Manihot* are distinctive, but like roots require particular conditions for their preservation. Phytoliths[2] in contrast can be preserved in a great range of environments, but cassava appears to have only one type of phytolith that is distinctive (Chandler-Ezell et al., 2006). Mean size of starch grains of modern cassava roots is larger than for wild *Manihot* spp., but there is some overlap (Perry, 2002). Distinguishing the cultivated plant from its wild relatives based on these remains would require at the very least larger samples of these microfossils than are likely to be available in most archaeological deposits. Finally, plant distributions have shifted with climatic fluctuations, so that the current range of a wild ancestor may differ from its range in the past, and locations where ancient cultivation is documented may have presented environments that differed greatly from those found in the region today.

These three questions, what wild plant was domesticated, where this happened, and the kind of environment that prevailed when domestication began, are linked to fascinating questions about the process itself: How did domestication happen? What was the chain of events that led from management of wild progenitors by pre-agricultural peoples to cultivation? Which plant traits changed under domestication, and how did the forces of selection – natural and artificial – bring about these transformations? For cassava, these questions too have long been debated, and gaps in our knowledge still remain.

2.1 Cassava's wild ancestors

Cassava was long thought to be a 'compilospecies', the product of complex hybridization and introgression involving multiple wild *Manihot* spp. (Rogers and Appan, 1973).

2. Phytoliths are microscopic silica bodies that precipitate in and around cells in many plants. When plant tissues decompose, the silica is deposited in the soil, forming a record of past vegetation.

However, genetic evidence now points clearly to populations of a single wild taxon as the closest wild relative, and sole ancestor, of the crop (Olsen and Schaal, 1999). According to current classification, cassava is even considered conspecific with this wild relative, *M. esculenta* Crantz subsp. *flabellifolia* (Pohl) Ciferri (Allem, 1994). Allem (1994) considered several other named *Manihot* spp. to be synonymous with this taxon, which, in his broad definition, comprises populations found along the southern, eastern and northern rims of Amazonia. Current genetic evidence specifically identifies southwestern Amazonia as the cradle of domesticated cassava (Olsen and Schaal, 1999; Léotard et al., 2009). Plants by the wild taxon, especially those from populations of southwestern Brazil, are quite similar to cassava in gross morphology. Long considered feral derivatives of cassava, they attracted little attention until molecular studies by Olsen and Schaal (1999) identified them as ancestral and not derived.

Understanding the ecology of cassava's wild ancestor turns out to be key to understanding several crucial features of the crop's ecology, as well as understanding how the crop was transformed during domestication. We have studied the ecology of cassava's wild ancestor in Rondônia (Ménard et al., 2013), within the region where genetic evidence indicates domestication took place. Here, in the shifting mosaic of forest and savanna environments, ecological succession to forest is frequently interrupted by disturbances

M. e. flabellifolia (French Guiana)

Figure 1 Individuals of the close wild relative of cassava (*Manihot esculenta* subsp. *flabellifolia*, known locally as *M. tristis*) in coastal savannas of French Guiana. Left: shrub habit in early succession. Right: vine habit in secondary forest. The white arrow marks the viny stem (photos: Doyle McKey).

such as fire (or nowadays forest clearing), creating habitats propitious to cassava's wild ancestor. Its tuberous roots store underground reserves, enabling rapid regrowth after fire or vegetation clearing. Where disturbance is infrequent and vegetation cover becomes denser, the plant expresses an astonishing plasticity of growth form (Fig. 1). From a much-branched shrub in open environments, cassava's ancestor turns into a vine capable of climbing into the canopy of low forest as the vegetation closes and grows higher. Unless a new disturbance occurs, the plant eventually dies out, leaving a bank of seeds that remain dormant in the soil.

As we will see further down, the biology of seed dispersal, dormancy and germination is crucial to understanding cassava's ecology and evolution. The story begins with the seed and seedling biology of cassava's wild ancestor. After pollination and fruit maturation, seed dispersal occurs in two stages. First, in a process aptly named 'ballistic autochory', the fibres of the woody capsule dry and twist, building up tension that culminates in the explosive dehiscence of the capsule, projecting each of its three seeds up to several metres around the plant. Each seed bears an oily caruncle, which plays a key role in the second stage of dispersal. The caruncle functions as an elaiosome – a food body rich in lipids and other nutrients – that attracts ants (Elias and McKey, 2000). Ants carry the seeds to their nests, where the elaiosome is fed to the brood. Seeds are then discarded in refuse piles near the nest, where they become buried at varying depths in the soil. Thus protected, seeds can lie dormant for years, 'waiting' for a disturbance to germinate and establish. Seeds of many light-loving plants use increased light level, or a change in the quality of light when vegetation is removed, as a cue for germination. Buried in the soil, however, the seeds of cassava's wild ancestor are unable to detect light but respond to an increase in soil temperature as a signal of favourable conditions for germination. At temperatures typical of vegetation-shaded soils in the lowland tropics (25°C), seeds remain dormant, but when soil temperatures reach 35°C or more, signalling the removal of vegetation cover by a disturbance, seed germination is triggered if ample water is available (Pujol et al., 2002).

A picture thus emerges of cassava's wild ancestor as a plant adapted to dynamic forest-savanna ecotones, thriving in open areas but capable of surviving long periods in closed vegetation by adopting viny life forms and having seeds capable of prolonged dormancy, in effect 'waiting' for the next disturbance. This picture holds throughout the large range of this plant. In the coastal savannas and inselberg rock-savannas of French Guiana, populations of cassava's wild relative are known as *M. tristis*, which Allem (1994) considered as synonymous with *M. esculenta* subsp. *flabellifolia*. Although plants in French Guiana less often grow to become large vines – in keeping with the lower-stature vegetation that characterizes the plant's habitats there – their traits and their ecology are similar to those of populations in the region where cassava was domesticated. Landscapes of southwestern Amazonia appear to have been more open, drier and fire-prone at the time of cassava's domestication there than they are today (Arroyo-Kalin, 2010). They may thus have been more similar to those of present-day French Guianan coastal savannas. Together, populations in the two regions give us a good idea of the ecology of cassava's ancestor when domestication began.

2.2 From wild to domesticated

How did pre-agricultural peoples transform cassava's wild ancestor into the crop? Numerous records of *Manihot* spp. microfossils from sites scattered from Amazonia to Mesoamerica indicate that various wild relatives of cassava were consumed by pre-agricultural peoples

(Arroyo-Kalin, 2010). However, these archaeological records give little insight into how plants or landscapes were managed. Little appears to be known about how wild *Manihot* spp. are used by present-day Amerindians. In a review of the use of wild plants in tropical South America, Lévi-Strauss (1952) noted few instances of wild roots and tubers used as food, and ascribed the paucity of observations to the fact that they are collected nowadays only in times of scarcity, as a substitute for domesticates. Interestingly, in the one case he cites, not only were roots of wild *Manihot* grated for flour, but their oily seeds were also reported to be eaten.

In the absence of more substantial data, researchers have speculated about the practices of foraging peoples that ultimately led to cultivation. The wild ancestor of cassava is fire-friendly. As documented in some other parts of the world (Daniau et al., 2010), foragers in southwestern Amazonia may have burned landscapes to encourage the plant and to enhance other wild resources. Observing the plant's ability to take root from pieces of cut stems, they may have started planting stem fragments to augment density. Arroyo-Kalin (2012) proposed another scenario, drawing upon Anderson's (1952) classic 'dump-heap' or 'midden' model of domestication: After roots were separated and processed, stem fragments discarded onto dump heaps could have taken root, giving hunter-gatherers an opportunity to observe and learn about one important property of the cassava plant: its capacity to propagate clonally. By fostering the plants' growth in this enriched soil, gatherers would have gradually become planters and cultivators.

Like most other perennial domesticated plants (or those whose wild ancestors were perennial), cassava is propagated vegetatively. This mode of reproduction has several agronomic advantages, the most important of which is the possibility to 'freeze' and multiply interesting genotypes endlessly. Because most perennial plants are highly outcrossed, it is much more difficult to bring (initially rare) favoured genotypes to high frequency via sexual reproduction in perennial plants than in annual plants, which – with some conspicuous exceptions such as maize and rye – are frequently selfers (Zohary, 1984). However, many perennials have some ability to sprout from vegetative parts. Taking advantage of this capacity would have enabled cultivators to bypass seedling and sapling stages, limiting mortality during these often vulnerable juvenile stages, and greatly shortening the time from planting to harvest.

We do not know how foraging peoples began planting and cultivating stem fragments of cassava's wild ancestor. In any event, once they did, this greatly altered selection on the plant's traits. Cultivators likely favoured plants with large tuberous roots. Selection also acted on stems, the plant part used in propagation. Cassava's wild ancestor produces adventitious roots from cambium exposed where the stem is cut. Also, the trailing stems of large viny individuals sometimes take root where they come into contact with the soil. However, stems of domesticated cassava have a much greater capacity to produce adventitious roots when the stem is planted. Roots originate primarily from meristematic tissue in large, paired protuberances located on either side of the leaf insertion at each node.[3] These protuberances are absent in the wild ancestor. Internal anatomy and chemical composition of stems also changed during domestication (Ménard et al., 2013). These changes were likely driven by selection on the new function of stems as propagules. Stems became thicker, producing cuttings more resistant to desiccation. Cell walls became thinner and cell contents more voluminous. With starch concentrations

3. This is particularly true when cassava stakes are planted obliquely, as they usually are. Stakes planted vertically may root mostly from the cut end.

three times greater than in stems of the wild ancestor, stems of domesticated cassava yield cuttings that provide the young plant with a large reserve of energy. Interestingly, stems of domesticated cassava are also much more brittle than those of the wild ancestor. Stem flexibility is important in the wild ancestor, ensuring its success as a liana. Domesticated cassava, however, is cultivated as a shrub, and traits important only in the lianescent phase are no longer maintained by selection (Ménard et al., 2013). More than a simple side effect of other changes, stem brittleness may also confer a positive advantage in the domesticate. Before iron tools, the simplest way to prepare stem 'cuttings' would have been to break the stems, and stem brittleness could be a further adaptation to vegetative propagation, driven or reinforced by early farmers' preference for plants whose stems broke easily and cleanly.

2.3 Mixed clonal/sexual reproductive system of cassava

Numerous traits thus appear to have evolved during cassava domestication (see also Mondolot et al., 2008; Pujol et al., 2008). This fact is difficult to square with the well-known limits to evolution in purely clonal populations, and with the frequent notion that evolution mostly proceeds by somatic mutations in vegetatively propagated crops (see review in McKey et al. [2012]). Clonal propagation keeps genotypes in a fixed state, and in the absence of recombination, mutation is the only mechanism by which genetic diversity can be enhanced. Fixation of somatic mutations is a slow process, however, and could not explain alone the diversity of newly evolved traits in the domesticated cassava. This suggests the action of numerous cycles of recombination (needed to generate variation) and selection (to fix certain traits and maintain agronomic performance).

Several studies have shown how sexual reproduction comes into play to drive evolution in populations of domesticated cassava, and how the mixed clonal/sexual reproductive system that characterizes the crop's traditional management by Amerindian farmers permits its continuing adaptation.

Whereas several clonally propagated crops have lost their ability to reproduce sexually, for example, bananas (*Musa sapientum* var. *paradisiaca*) and some yams (*Dioscorea* spp.), domesticated cassava has largely conserved its wild ancestor's sexual reproductive biology. The fact that these plants have time to flower and fruit may be explained by an agronomic advantage peculiar to cassava. While its roots are highly perishable once harvested (more about this below), they have a great capacity for 'live storage' in the soil: mature plants can be left in the ground in fields for long periods (up to three years; Jones, 1959) and their roots can be harvested as needed ('piecemeal harvesting'). Many plants thus have time to flower and fruit before being uprooted. Flowers of domesticated cassava are pollinated by insects and the plant is a preferential outcrosser. Its seeds are dispersed in two stages and buried by ants. In the Guianas, the same ant species disperses and buries domesticated cassava seeds in forest slash-and-burn parcels (Elias and McKey, 2000) and seeds of the wild ancestor in savannas (Renard et al., 2011). Once buried, seeds of domesticated cassava, like those of the wild ancestor, remain dormant in the soil until a disturbance removes vegetation cover (waiting sometimes up to 50 years; Elias et al., 2000a). For domesticated cassava, the disturbance is usually the re-opening of a slash-and-burn parcel after a fallow period. Thus, at about the same time that farmers plant their clonally propagated stem cuttings, large numbers of seedlings, each a unique genotype, appear in the newly opened parcel. These 'volunteer' seedlings (Fig. 2) are closely observed by cultivators, who allow

Figure 2 Cassava volunteer seedlings in slash-and-burn fields. Left: A young seedling still bearing foliaceous cotyledons (white arrow). Right: A larger plant issued from a seed. Basal internodes (i.e. segments of the stem between successive leaf scars) of plants issued from seed are very short. Because these internodes remain unchanged in length as the plant grows, even large plants issued from seeds can be distinguished from those issued from stem cuttings (photos: Marc Delêtre).

them to grow. Those that survive to maturity are harvested and their roots processed, like those of the clonally propagated landraces. If the cultivator is satisfied with the yield and the characteristics of the root, the stem is divided into cuttings that are soon replanted and multiplied. After a few repetitions of such trials, descendants of plants issued from seed join the stocks of clonal landraces (each of which originated in the same way at some time in the past). Although volunteers presenting novel combinations of morphological characters are sometimes multiplied as new landraces, most end up assimilated into already named landraces they phenotypically resemble (Elias et al., 2000a; Duputié et al., 2009). A landrace is thus constituted of a multitude of clones, some common, some rare, which although phenotypically similar (Elias et al., 2001a) may have diverse genetic backgrounds (Elias et al., 2001b).

Cassava's mixed clonal/sexual system under this kind of dynamic management maintains both agronomic performance in the present and the diversity that is necessary for continued adaptation in the future. This kind of cassava plantation is simultaneously a production farm and an agricultural experimental station. Most of the parcel is planted with reliable, high-performance clones, but new genotypes continuously appear, sifted by both natural and artificial selection to favour the fastest-growing plants (Pujol et al., 2005b; Pujol and McKey, 2006), and ensuring the system's resilience. Passage through the seed stage also confers another important advantage: many of the systemic viral pathogens that accumulate in clones do not cross the placental barrier, so that plants issued from seeds begin their life free of infection. Cassava varieties also accumulate deleterious mutations with time, the 'Müller's ratchet' (Felsenstein, 1974) of genetic load. Recombination through sexual reproduction permits stirring the genetic pool to produce new combinations of alleles, thereby increasing diversity, but also reducing the deleterious effects of the 'genetic load' of mutations in clonal lineages. For small-scale farmers, the mixed reproductive system of cassava means they can maintain those genotypes which they like most, while still being able to select new ones issued from sexual reproduction. Genetic load and pathogen load may explain why some cultivators consider plants issued from seed as being 'young' and vigorous.

The wild ancestor's adaptations to forest-savanna ecotones – seed dispersal, burial and dormancy – pre-adapted it to the field-and-fallow cycles of swidden farming systems, which mimic natural cycles of disturbances in unmanaged forests. Grafted onto these 'wild' ecological traits, new cultural traits – the clonal propagation of cassava by farmers but also their keen observation and integration of sexually produced seedlings – gave rise to a system that ensured the resilience of cassava cultivation as it spread over a wide range of environments.

This mixed clonal/sexual reproductive system, based on the recruitment by cultivators of 'volunteer' plants issued from true seeds as new clones, is widespread in Amazonia today, not only among Amerindians (McKey and Beckerman, 1993; Elias et al., 2001a) but also among 'caboclo' farmers of mixed ethnic and cultural heritage (Fraser, 2010). What evidence do we have that the integration of volunteer seedlings has characterized cassava cultivation since the plant's domestication? Something close to a 'smoking gun' is provided by the most surprising change under domestication that can be inferred by comparing cassava and its wild ancestor: a change in the functional morphology of seedlings (Pujol et al., 2005a). That traits of seedlings evolved during domestication shows that they continued to play an important role in the crop's life cycle, for otherwise selection would not have acted on them.

How seedling traits changed during domestication also makes ecological sense. Seeds of cassava's wild ancestor are usually buried in the soil following dispersal by ants. 'Wild' cassava seeds are characterized by hypogeal germination, that is, the hypocotyl does not elongate during germination and cotyledons remain buried in the soil. Enclosed in the testa, they serve as stored reserves. The epicotyl grows to the surface, and the seedling's first photosynthetic organs are its (very small) first true leaves. In contrast, seeds of domesticated cassava exhibit epigeal germination. The hypocotyl elongates, pushing the cotyledons above the soil. The cotyledons emerge from the testa and expand into foliaceous photosynthetic organs, much larger than the first true leaves of the wild ancestor (Pujol et al., 2005b). These differences are truly remarkable, because seedling functional morphology is usually evolutionarily conservative, and usually invariant within a genus (Garwood, 1996). In contrast, comparison of cassava and its closest wild relatives indicate that domestication has produced a radically different germination type in the domesticated plant in less than 10 000 years.

What selective pressures drove these changes? Seedling morphology in the wild ancestor confers stress tolerance to the plant: if the above-ground parts of the young seedling are destroyed by fire, drought or herbivores, the plant still has underground reserves (the storage cotyledons) and underground meristems (along the buried part of the epicotyl) that give it a second chance. However, this tolerance to stress comes at the cost of low initial growth rate: its photosynthetic surface is initially very small. In contrast, the domesticate has sacrificed stress tolerance to maximize initial growth rate. If its above-ground parts are destroyed, it has neither underground reserves nor meristems; but if they are not destroyed, its large foliaceous cotyledons enable it to grow rapidly. In crop fields, where disturbance (clearing and burning of the fallow vegetation) is followed predictably by the beginning of the rainy season, and where the soil is enriched by ashes from the burn, the risk of stress is lower than in wild environments, and the balance of selective forces is tipped towards maximizing the opportunities for rapid growth (Pujol et al.,

2005a). Another selective force acts in the same direction: slower-growing seedlings are more likely to be selectively weeded by farmers (Pujol et al., 2005b).

The integration of volunteer seedlings, and the mixed clonal/sexual reproductive system that results, thus appear to have characterized the traditional management of populations of cassava in its continent of origin since domestication began. This feature of cassava's reproductive biology, perfectly suited to field/fallow cycles – and in fact at least partially a result of such cycles – tests new genotypes in each generation. The evolutionary flexibility conferred by this system may account in large part for the resilience of cassava cultivation. Where cultivation practices diminish or eliminate the role of sexual reproduction, there may be a price to pay – sometimes perceptible only after a long lapse of time – for the loss of the crop's evolutionary potential.

2.4 Cyanogenesis and domestication

Cassava is unique in another way: it is probably the sole widely cultivated crop that can be deadly poisonous if not correctly prepared. Both cassava and its wild relatives produce cyanogenic glucosides. In intact cells, the glucosides and the enzymes (glucosidases) that cleave them are stored in different compartments. When the plant's cellular structure is destroyed (e.g. by a vertebrate herbivore, a phytophagous insect or a plant-pathogenic bacteria or fungus), the two come in contact, releasing highly toxic hydrocyanic acid. Many domesticated crop plants possess antinutritional compounds, including cyanogenic glucosides, which function as chemical defences against the plants' enemies, for example, sorghum (*Sorghum bicolor* [L.] Moench), almonds (*Prunus dulcis* [Mill.] D.A. Webb, Rosaceae), Lima beans (*Phaseolus lunatus* L., Fabaceae), white clover (*Trifolium repens* L., Fabaceae), rubber tree (*Hevea brasiliensis* Müll. Arg., Euphorbiaceae) and some yams (*Dioscorea* spp.) (Jones, 1998). Cassava, however, is exceptional in producing high concentrations of a highly toxic substance. Given the large quantities in which it is typically consumed, cassava would be dangerously toxic were these compounds not removed by processing.

All cassava varieties are cyanogenic, but the concentration of cyanogens varies greatly among varieties, particularly in the tuberous roots, the part usually consumed by humans. Cassava varieties are classed as 'sweet' or 'bitter' depending on the cyanogen content of their roots. Varieties classed as 'sweet' are considered to be non-toxic; their roots can be eaten boiled, roasted or even raw, and usually have HCN concentrations lower than 100 mg HCN × kg^{-1} fresh weight. Varieties classed as 'bitter' are toxic, and can be consumed only after cyanide has been removed by various procedures. Simple cooking does not detoxify bitter cassava; cooking denatures the glucosidase, but does not destroy the glucoside. Ingested glucosides can be cleaved by glucosidases present in human guts, releasing HCN in sometimes lethal quantities. HCN concentrations in root flesh of bitter manioc can be greater than 500 mg HCN × kg^{-1} fresh weight in the most toxic cultivars (Dufour, 1988) and concentrations of up to four times that are known (Burns et al., 2012). Environmental conditions also affect cyanogen concentration, which generally increases when the plant experiences stressful conditions; a drought-stressed plant will have higher cyanogen concentrations than the same plant under favourable conditions (McKey and Beckerman, 1993; Burns et al., 2012).

The toxicity of cassava and cassava products has two facets: Acute toxicity results from ingestion of large doses of HCN (e.g. eating boiled roots of bitter cassava); chronic toxicity results from long-term ingestion of smaller amounts of HCN (e.g. residual quantities in improperly prepared cassava products). Both acute and chronic cassava toxicity are almost unheard of in the Americas, but both occur in areas where the crop was introduced since the Columbian exchange (see Section 3.4).

The variation in toxicity among cassava varieties has long fascinated researchers. Why does this variation exist? What are the advantages and disadvantages of bitter and of sweet varieties, and how do these help explain the patterns in their respective ecological and geographic distributions? Much evidence points to the role of cyanogens in chemical

Figure 3 Processing of bitter cassava by Makushi Amerindians in southern Guyana. The roots are rasped on a wooden plank (a) bearing large numbers of embedded small metal triangles (b). Grating destroys cellular compartmentation, allowing enzymatic hydrolysis of the cyanogenic glucosides to release HCN. The thoroughly grated pulp is placed in the *tipiti* or manioc sleeve press (c), a woven cylinder open at the top and with a closed loop at top and bottom. The top loop is placed around a rafter of the house, while a pole serving as a lever is placed into the bottom loop. The juice containing the cyanide is squeezed out by the person sitting on the lever. Once removed from the *tipiti* (d), the detoxified pulp can be made into various storable products (photos: Doyle McKey).

defence, in plants in general (Gleadow and Woodrow, 2002) and in cassava (reviewed in McKey et al., 2010). As in wild plants, natural selection acts on cassava's chemical defences. However, humans also exert artificial selection on toxicity, and understanding the variation in sweet and bitter cassava ultimately hinges on understanding how the two sources of selection interact.

This question was reviewed by McKey and Beckerman (1993) and McKey et al. (2010). In nature, the strength of natural selection on toxicity should vary with the probability of attack by herbivores or pathogens, and with the cost incurred by the plant when such attacks occur. Under farmers' management, artificial selection against toxicity prevails if cassava roots are eaten raw, roasted or boiled. But when cassava is detoxified (e.g. by procedures involving grating, maceration or fermentation; see Fig. 3–5), artificial selection against toxicity is neutralized. Natural selection for increased toxicity can act unchecked, and can even be reinforced by artificial selection as increased toxicity coevolves with greater efficiency of detoxification procedures, or as cultivators acquire a cultural preference for the taste of products made from highly cyanogenic varieties (Wilson and Dufour, 2002). In the great diversity of natural and cultural environments in Amazonia, divergent selection on toxicity has produced varieties with roots sweeter than those of any *Manihot* known in the wild, but also varieties with roots far more toxic than those of any wild congener (McKey and Beckerman, 1993). Bitter cassava is a staple crop in central and

Figure 4 Processing of bitter cassava in Gabon, Central Africa. The preferred way to prepare cassava is usually in the form of cassava sticks. Peeled roots are soaked for several days in water to initiate fermentation (a), roots are then crushed, fibres are discarded and the pulp is worked on a 'cassava plank' (b). The paste is then shaped into cassava sticks which are wrapped in Marantaceae leaves and steam cooked (c) (photos: Marc Delêtre).

eastern Amazonia; sweet cassava is widespread in the Neotropics. In Amerindian groups that grow both kinds of varieties, sweet cassava is often grown as a minor vegetable crop in home gardens, where soils are richer and proximity to human settlements discourages mammalian herbivores. In contrast, bitter cassava is usually grown as a staple crop in slash-and-burn fields further away from settlements and thus more vulnerable to herbivores, and often on poorer soils (McKey and Beckerman, 1993). Reality is of course more nuanced than this general picture. Not all herbivores or pathogens are dissuaded by cyanogens, and most organisms have some capacity to detoxify them. However, cyanogens persist as widespread and effective chemical defences because they are expensive to detoxify, costing the herbivore a molecule of sulphur-containing amino acid for each HCN molecule detoxified (Gleadow and Woodrow, 2002; McKey et al., 2010). Likewise, bitter cassava is not restricted to poor soils, being also cultivated on enriched anthropogenic *terra preta* soils (Fraser et al., 2012). Finally, patterns that hold in Amazonia are less marked in areas further from the crop's centre of origin, for example, in Brazil's Atlantic Forest region (Peroni and Hanazaki, 2002).

Cassava processing serves not only the purpose of detoxification but also that of yielding a storable product. These two functions are inseparable. In fact, processing is *required* if cassava is to be stored. Like seeds, most roots and tubers can be stored after harvest, sometimes for up to several months. Cassava roots, however, have limited shelf life and rapidly suffer post-harvest physiological deterioration (Wenham, 1995). Any roots not immediately consumed must be rapidly converted into a storable product. It is ironic that the perishability of fresh roots, a trait perceived today as a major agronomic problem in cassava, may have been crucial in setting the conditions for the evolution of a key adaptation of this plant. By making it necessary to process the crop if a surplus is to be stored, the perishability of fresh roots provided frequent opportunities for selection in favour of increased toxicity. Processing techniques, selection practices and the plant itself have thus coevolved to produce a built-in chemical defence that can be removed after harvest. It can be argued that this trait, more than any other, has led to cassava's success as a major tropical crop.

Defence and storage provide us with the beginning of an answer as to why the toxicity of cassava is so variable, but a host of related additional questions remain: Were the roots of the wild ancestor sweet or bitter? If they were bitter, why was a highly toxic plant cultivated and domesticated? Did sweet and bitter varieties have distinct origins as domesticates, or did the differences evolve after domestication?

Cyanogenesis appears to be a universal trait in the genus *Manihot*, but not all species have bitter roots. Of the seven species for which reports of HCN concentrations in fleshy roots were available 26 years ago, two species would be classed as sweet (62 and 97 mg $\times$ kg^{-1}, respectively) and five as bitter (ranging from 125 to 430 mg $\times$ kg^{-1}) (McKey and Beckerman, 1993). Since then, cassava's closest wild relative, *M. esculenta* subsp. *flabellifolia*, has been added to the list. In populations from southwestern Amazonia, where domestication originated, starchy roots of this taxon are reported to contain a mean of 244 mg HCN $\times$ kg^{-1} fresh weight (Perrut-Lima et al., 2014). It thus appears that roots of the wild ancestor would have been dangerously toxic and required processing. This should not be surprising. The underground storage organs of most wild plants have some kind of defence, often chemical, and foraging peoples often employ complex procedures to detoxify them (Lévi-Strauss, 1952; Johns and Kubo, 1988). The wild ancestors of many domesticated plants are toxic (Johns, 1990). What distinguishes cassava, however, is that dangerously toxic concentrations were maintained, and likely even increased, during domestication.

Numerous studies suggest that cassava was domesticated only once, and that differentiation into sweet and bitter varieties evolved afterwards, but which came first is still highly debated. The recent study by Perrut-Lima et al. (2014) clears up one question about cassava toxicity and domestication: cassava's wild ancestor was bitter, not sweet. Other uncertainties persist, however. Arroyo-Kalin (2010) argued that selection first acted to reduce toxicity (sweet came first, bitter came later). He posited that cassava was first cultivated in anthropogenic dump-heap environments, where rich soils favoured plants with low levels of defences. Only later, when cassava had spread to central Amazonia, did intensive cultivation on poor soils, the need to produce a storable surplus for large sedentary populations and other factors combine to favour highly toxic varieties. Mühlen et al. (2013) interpret their data on genetic differentiation among sweet and bitter cassava varieties as favouring this scenario. However, in the dendrogram they present, it is notable that the base of the tree – indeed, its three most basal branches, which comprise over one-third of all individuals sampled – consists almost entirely of bitter individuals. The mystery therefore persists; although the sample studied by Mühlen et al. (2013) is the largest ever used to study this question, it does not include traditional landraces from the crop's area of origin in southwestern Amazonia.

The study by Mühlen et al. (2013) nevertheless allows two interesting conclusions. First, although sweet and bitter varieties diverged from a common domesticated ancestor, once they arose they diffused largely independently. The tree shows deep phylogenetic splits between sweet and bitter lineages, with large branches that are sometimes entirely sweet or entirely bitter. Although information on the location of the individuals is lacking, at least some of these 'radiations' appear to include individuals from widely spread localities. The hypothesis of an independent diffusion of sweet and bitter varieties matches with field observations which show that, at least in Amazonia, the two kinds are managed as distinct crops, with sweet cassava often grown in home gardens and bitter cassava grown in slash-and-burn fields (McKey and Beckerman, 1993). Interestingly, this pattern is not marked in the Atlantic Forest region of Brazil (Peroni and Hanazaki, 2002), where cassava cultivation does not have the long history it does in Amazonia. Secondly, the dendrogram suggests that the direction of evolutionary change in toxicity has varied over time and space: lineages that are predominantly bitter or sweet have occasionally given rise to varieties of the other category. Although Mühlen et al. (2013) did not specify the methods they used to assign ancestral character states, it appears from inspection of the tree that transitions from bitter to sweet are more frequent than the converse. It would be interesting to assess the relative roles of gene flow (hybridization) and selection in these cases.

3 Introduction of cassava in the Old World

Following their conquest of South America, Europeans introduced cassava into their colonies throughout the tropics, first into Africa (Carter et al., 1992), then into tropical Asia and Oceania (Lebot, 2009). These diasporic populations of cassava then embarked on their own complex history of interactions with local biotic and cultural environments, a history that has never been fully explored. One theme in this history is common to many plants – including crop plants – that have been introduced by humans to areas outside their region of origin: the initial escape from natural enemies that were present in the plant's area of origin (McKey and Beckerman, 1993; Wood and Lenné, 2011; Ellen and Soselisa,

2012b). Already better defended than most crop plants against herbivores and pathogens in its Neotropical homeland, cassava long enjoyed even greater (relative) freedom from pests and diseases in its areas of introduction. As noted by Ellen and Soselisa (2012b: 32), 'The absence of natural competitors and pathogens, whether viral, bacterial or fungal, gives cassava the characteristics of an invasive species in many diasporic contexts.' Africa and Asia, and especially the latter, harbour lower numbers of arthropod pests (Bellotti, 2001), viral diseases (Calvert and Thrush, 2001), and nematode and fungal pathogens (Hillocks and Wydra, 2001) than are present in the Neotropics. In Vanuatu, cassava is so far completely free of significant pests and diseases (Sardos et al., 2008). However, herbivores of cassava have been introduced from the Neotropics (Fabres et al., 1981; Yaninek and Herren, 1988). For some of these, their own natural enemies were later introduced in turn (Herren et al., 1987). Also, new herbivores and pathogens appear over time in areas of crop introduction (Strong, 1979), and new enemies of cassava evolved to exploit this abundant new resource. These include the Cassava Mosaic and Cassava Brown Streak diseases in Africa (Legg and Thresh, 2003), both prominent threats to regional food security, and the witches' broom disease in Asia (Graziosi et al., 2016).

Cassava agronomy has been relatively well studied from a comparative perspective between the crop's area of origin and regions where the crop was introduced. Cassava toxicity as a public health problem has also been well studied (Burns et al., 2010). However, as noted by Delêtre (2010) in Africa and Ellen and Soselisa (2012b) in Indonesia, in contrast to studies in the crop's Neotropical homeland, few studies in areas of introduction have examined how these agricultural dimensions of cassava are articulated with cultural dimensions. This section of our chapter focuses on what we term the 'cultural domestication' of diasporic populations of this complex crop.

What do we mean by the term 'cultural domestication'? The process of domestication consists not only of evolutionary change in the crop plant but also cultural evolution of the practices required for the crop's adaptive management. When a crop is introduced into a new continent, the practices that have coevolved with it in its area of origin are rarely transmitted in unaltered form. For some crops, the consequences can be dramatic. Where maize was adopted as a staple crop outside Meso-America, for example, the micronutrient deficiency disease pellagra became frequent. Pellagra was unknown in the crop's area of origin, where maize is traditionally prepared mixed with lime, in a process called 'nixtamalization' that increases the biological availability of niacin. Its benefits being unknown, this process was not transmitted into the areas of introduction. The harmful effects caused by absence of such processing led to cultural changes to mitigate these effects. These changes can be viewed as a form of 'cultural domestication'. In most areas, the harmful effects of maize have been countered by vitamin supplements and improved economic conditions that allow people to have more diverse diets. Still, maize has never been fully 'culturally domesticated' in some areas of introduction. Nixtamalization (and other processes that accomplish the same function) has never been adopted in maize's areas of introduction, and nutritional deficiency diseases are still frequent in some areas where maize dominates diets, notably in parts of Africa and Asia (Ranum et al., 2014).

Like maize, cassava is a highly productive plant, a good traveller but with some cumbersome baggage: its successful exploitation requires some technical mastery. Earlier, we argued that Amerindian cultivators evolved cultural adaptations centring on two key aspects: (1) managing the crop's chemical-defence polymorphism and potential toxicity and (2) avoiding the consequences of a purely clonal reproductive system via practices that ensure the continued role of sexual reproduction in the crop's biology. But what was the fate of these cultural adaptations in cassava's areas of introduction? Did imperfect transmission lead to maladaptive mismatches between the plant and farming practices? Did such mismatches favour in turn modifications in the plant, in practices, or both? Have new coadaptations of plant and practices evolved? Or do mismatches persist, as in the case of maize?

3.1 History of introduction and diffusion

Many studies have shown that the 'environment' in which cultivated plants evolve is complex, and that much can be learned from approaching the question in terms of people – plant interactions (Anderson, 1960; Salick, 1995). History, in particular, is an important factor to consider, as the vectors and modes of cassava spread in its areas of introduction have had a considerable influence on the processes of reappropriation of the new crop by local populations (what we have termed 'cultural domestication'). Examining the plant's historical trajectories, and the diversity of the cultures into which it was introduced, sets the stage for our analysis of the fate of the key biocultural coadaptations of cassava and its cultivators.

3.1.1 Africa and Madagascar

Africa was the first continent into which cassava was introduced from its South American homeland. Cassava was introduced into Africa on repeated occasions; first, along the Atlantic coast in the sixteenth century, in São Tomé, Congo and West Africa. Then, in the eighteenth century, cassava was brought into East Africa (Jones, 1959). How many times cassava was introduced, and how many varieties these introductions represented, is not known, but given the technical difficulty at the time of transporting cassava stem cuttings overseas, the Portuguese most likely selected a large range of local landraces in order to evaluate them in their African colonies (Pickersgill, 1998).

Introduction of a crop plant into a new region can be accompanied by a reduction in genetic diversity, as only a subset of overall diversity is introduced. However, the tremendous present-day diversity of African cassava (Lefèvre and Charrier, 1993; Pickersgill, 1998, Fregene et al., 2003; Mkumbira et al., 2003; Manu-Aduening et al., 2005; Kizito et al., 2005, 2007; Delêtre, 2010) gives little evidence of such a 'bottleneck'. In fact, based on neutral markers, levels of genetic diversity of cassava in Africa today appear similar to those encountered in Amazonia (Beeching et al., 1993; Fregene et al., 2000, 2003). The origins and true extent of this diversity are still unresolved questions.

Despite an initially small founding population and further reductions of the initial 'pool' of cassava landraces due to genetic drift and the Cassava Mosaic and Cassava Brown Streak disease pandemics (Patil and Fauquet, 2009; Legg et al., 2011), any loss of genetic diversity appears to have been balanced by processes of restoration. Multiple introductions certainly contributed to mitigate, or even cancel the initial 'bottleneck' effect. Introduction of plants into a single area from different source areas in the crop's homeland can spatially compress genetic diversity, favouring gene flow in smallholders' farms between crop types that would never have encountered each other in the continent of origin. The impressive diversity of beans in East Africa (Martin and Adams, 1987a,b; Pickersgill, 1998) and of watermelons in Brazil (Romão, 2000) probably resulted in this way. In East Africa, beans of Mesoamerican and Andean origin may be found in the same field, favouring gene flow and producing novel types not known elsewhere (Pickersgill, 1998). Similarly, spatial compression of cassava diversity could have occurred in areas where the Portuguese were particularly active in the sixteenth and seventeenth centuries, for example, on Fernando Pó (now Bioko), on the islands of São Tomé and Príncipe, in Sierra Leone, or around Luanda and the mouth of the Congo River (Jones, 1959). Like beans, cassava may have undergone secondary diversification in Africa (Lefèvre and Charrier, 1993; Pickersgill, 1998), and geographic compression of diversity may have been a contributing factor.

Another part of the explanation for the diversity of cassava in Africa is surely the great cultural diversity of African peoples, and those groups that adopted cassava each reinvented cassava farming in their own particular way (Fresco, 1986; Richards, 1985). Cultural diversity probably also accounts for much of the strong disparities observed in varietal richness of cassava when comparing different regions (e.g. Fregene et al., 2003) or even different villages in a single region (Fregene et al., 2003; Delêtre, 2010). However, the biocultural mechanisms underlying the secondary diversification of cassava in Africa, and the influence of cultural variation in farming practices on cassava diversity, are still unclear.

Compounding the cultural diversity factor is the historical context of cassava introduction into Africa, in particular the socio-economic and political circumstances surrounding cassava's adoption by African societies. Variation in the context of early exchanges between European traders and African populations appears to have affected how, and how quickly, different societies adopted cassava (Delêtre, 2010). Although cassava was first introduced to Africa in the second half of the sixteenth century, its diffusion moved in fits and starts and took almost 400 years to complete. In Central Africa, the diffusion of cassava was initially carried out mainly – or solely – by African populations (Jones, 1959). Through a phenomenon of cultural impregnation facilitated by the strong ties between Portuguese settlers and the kingdom of Kongo, cassava spread rapidly along the Congo River and was firmly adopted in most Central African countries by the end of the eighteenth century (Jones, 1959). In West Africa, however, the crop remained largely restricted to coastal areas until late in the eighteenth century. Around the same time, cassava was reintroduced along Africa's east coast, in Madagascar and in Mozambique (Jones, 1959), but like in West Africa the crop did not begin to spread until the nineteenth and twentieth centuries. Cultural and environmental barriers seem to have considerably impeded its initial adoption as a crop (Carter et al., 1992). Cassava diffusion continued through colonial times, and the crop reached its present geographical limits (between around 30°N and 30°S) in the second half of the twentieth century.

In many parts of Africa, the adoption of cassava accompanied dramatic socio-economic transformations of African societies following the onset of formal colonial rule in the late nineteenth to early twentieth centuries (Jones, 1959; Adam, 1980; Geffray, 1989; Delêtre, 2010). Promoting cassava as a staple obeyed both a political and an economic rationale: securing food for populations in the short term, while, in the long term, transforming the regional agriculture into cashcropping with the development of cacao plantations (Sautter, 1966; Guyer, 1978; Dounias, 1993). The adoption of cassava, which requires less labour than traditional regional staples such as yams or plantains, left farmers available for labour on cash crops such as cotton, coffee and cocoa (Carter et al., 1992; Bahuchet and Philippson, 1998; Cloarec-Heiss and Nougayrol, 1998), and colonial authorities were a direct agent of the spread of cassava farming, encouraging, sometimes forcibly, the adoption of cassava as a subsistence crop in most of West and East Africa, and in parts of Congo, Cameroon and Gabon (Jones, 1959). Studies of the history of cassava in Africa are still scarce (summarized in Delêtre, 2010) but give us some information about how African populations perceived this plant, and how these perceptions may have affected the crop's adoption and spread. Was it eagerly adopted or was it resented as an imposition? Perceptions and attitudes likely varied among societies, and this likely affected its early diffusion, but today the crop is culturally integrated and valued throughout Africa.

Today, in Africa, as for most other crops, smallholder farmers are the main agents of cassava diffusion, through the social exchange of stem cuttings. As in other crops (e.g. peanuts [Cogels, 2002] and sorghum [Labeyrie et al., 2014; Westengen et al., 2014]), the social rules that govern these exchanges play an important role in shaping regional dynamics of cassava diversity. In Gabon, for example, village-level diversity of cassava was shown to be related to kinship systems (Delêtre et al., 2011). This is because kinship systems define marriage prohibitions and matrimonial networks, which in turn channel the circulation of crop landraces. Throughout the country, marriages are traditionally exogamous (i.e. between different clans) and residence is always virilocal (i.e. the woman moves into her husband's village). In southern Gabon, where societies are predominantly matrilineal, the young bride brings cuttings from the cassava varieties grown by her mother to her husband's village, thus contributing to enrich varietal diversity in the community. Over 65 genetically distinct landraces can be present in a single village. In patrilineal societies of northern Gabon, however, the bride moves empty-handed into her husband's village, and receives cassava cuttings from her mother-in-law (already resident in the village). Not surprisingly, the number of varieties per village is much lower than in southern Gabon (Delêtre et al., 2011). Because they structure social networks of exchange between kin (related by descent) and affines (related by marriage), marriages thus favour or limit exchanges of seeds between farmer communities and modify patterns of crop diversity at the regional level. As in South America (Chernela, 1986; Emperaire et al., 1998; Pinton and Emperaire, 2001; Kawa et al., 2013), exchanges of crop landraces play a key role in on-farm dynamics of cassava diversity in Africa. That cassava became such an intrinsic part of core cultural rules in African societies is a testimony of the degree of cultural reappropriation of the crop in Africa.

3.1.2 Tropical Asia and Oceania

The history of cassava's introduction and diffusion in other parts of the Old World is less richly documented than in Africa. Early centres of introduction in Asia were India, Java and the Philippines (Onwueme, 2002). Cassava was initially used primarily as an emergency crop by resource-poor farmers on marginal land, but began spreading in the nineteenth century. In Southeast Asia, it became increasingly significant through the early twentieth century (Ellen and Soselisa, 2012a,b). Although throughout tropical Asia the most fertile soils are usually planted in rice (Onwueme, 2002), by 1970 cassava had surpassed in importance many other traditional starch staples, particularly in those areas with low rainfall and poor soils. The studies by Ellen and Soselisa (2012a,b), who compared a group specialized in the growing and processing of bitter cassava and another group in which sweet varieties predominate in a multi-crop system, are among the few in the region that focus on the integration of cultural and agricultural dimensions. They show patterns highly reminiscent of those frequently observed in Africa and in the Neotropics. In Oceania, the articulation of cultural and agricultural dimensions of cassava cultivation has been best studied in Vanuatu (Sardos et al., 2008), where only sweet varieties are cultivated. Cassava's introduction into Oceania is relatively recent. The crop reached New Caledonia in the 1850s and Vanuatu in 1859 (Sardos et al., 2008). Although culturally less valued than the region's traditional starchy staples – yams, taro and sweet potato – cassava is increasingly planted throughout Oceania (Onyueme, 2002).

3.2 Cultural domestication in the cassava diaspora

We treat here the cultural re-appropriation in the Old World of the two aspects of domestication that we emphasized in discussing the biocultural co-evolution of cassava and farmers' practices in its neotropical homeland: the crop's chemical defence polymorphism and its mixed clonal/sexual reproductive system.

3.2.1 Management of cassava's chemical-defence polymorphism

As in the crop's Neotropical homeland, bitter and sweet varieties of cassava are both widespread in Africa, and the ecological contrasts that have developed between the two types (as in Indonesia [Ellen and Soselisa, 2012b]) largely mirror the patterns seen in the Neotropics (McKey and Beckerman, 1993). As in the Americas, bitter cassava is generally more resistant to pests and diseases than sweet varieties (Poubom et al., 2005; Ellen and Soselisa, 2012b) and their roots can be left in 'live storage' in the ground for much longer periods than those of sweet cassava (see McKey and Beckerman, 1993 and McKey et al., 2010 for review). Similarly, as in the Neotropics, where cassava is the main staple, bitter varieties generally predominate and are usually preferred in Africa (Nweke et al., 1994, 2001), whereas sweet varieties tend to be more important only in complex cropping systems where cassava is a secondary crop (Delêtre, 2010; Ellen and Soselisa, 2012b). Chiwona-Karltun et al. (1998, 2000) offer a particularly fascinating and multi-faceted analysis of why farmers in Malawi prefer bitter cassava, suggesting notably that bitterness confers upon cassava roots a protection against opportunistic theft, as their detoxification requires time-consuming processing. The parallels between the Neotropics and Africa in the ecological and cultural niches occupied by sweet and bitter varieties are striking, and may be partly explained by the transmission across continents of general knowledge about the strengths and susceptibilities of each group of varieties. However, across the diversity of ecological and cultural contexts to which cassava was exposed during its diffusion in Africa and Asia, the roles of sweet and bitter varieties must have been frequently re-adjusted, if not reinvented.

The parallels between the two continents are not perfect, however. One difference in particular may have had important consequences. Whereas in South America (and in some places in Indonesia [Ellen and Soselisa, 2012b]) the two crops are usually grown separately (either in separate patches within the same field, in separate fields, or with sweet cassava in home gardens and bitter cassava in slash-and-burn fields) in places where both types are found, in Africa bitter and sweet cassava are frequently grown in mixtures in the same field (Delêtre, 2010; Bradbury et al., 2013). The more frequent sympatry of sweet and bitter varieties in African fields may favour more frequent hybridization, and may explain why the two groups of varieties are not genetically differentiated in Africa in contrast to South America (Bradbury et al., 2013; Mühlen et al., 2013).

The blurring of the genetic divide between bitter and sweet cassava in Africa could result in difficulties for farmers to efficiently discriminate between the two kinds and manage their toxicity. At the local scale, farmers can often distinguish sweet varieties and bitter varieties by morphological traits that are artificially maintained in association with each variety through clonal propagation (see Ellen and Soselisa, 2012b; Delêtre et al., 2016). For cassava as a whole, however, no morphological traits are universally associated with cyanogen content (Rogers, 1965; Rogers and Appan, 1973; Boster, 1984; Nye, 1991), and hybridization can break the artificial correlations farmers once used to distinguish and manage the two categories. Varieties regarded as 'safe' for consumption with minimal processing may be found upon harvest to be unexpectedly bitter, while 'bitter' varieties may turn out to have reduced cyanogen content, with unknown consequences for their resistance in the field and capacity for live storage.

The jury is still out on precisely what these results mean in terms of cyanogen levels and the management of toxicity. Cassava landraces are often categorized into 'bitter' or 'sweet' based on their processing requirements rather than from empirical testing of their toxicity (Nye, 1991). Cyanogen contents were not measured in the study by Bradbury et al. (2013), and where farmers place the threshold of what they consider safe to eat without elaborate processing may vary among communities, and even among individuals. Furthermore, factors other than genotype affect both cyanogen content and taste. Our vision of cassava's cyanogen content as a continuum is based largely on diasporic populations in Africa affected by post-introduction hybridization, and systematic comparative studies of cyanogen content of bitter and sweet varieties grown by Amazonian Amerindians might reveal a more strongly bimodal distribution of cyanogen content.

The blurring of bitter and sweet cassava in Africa probably interacted with a much more important factor. It is likely that the processing methods initially applied to cassava in Africa were not as effective in detoxifying the crop as those developed over centuries by Amazonian Amerindians. This is suggested by patterns in the occurrence of chronic cassava toxicity. Despite the widespread prevalence of potentially toxic bitter varieties in both Africa and the Neotropics, chronic toxicity is virtually unknown in the crop's area of origin but abundantly documented in Africa and other areas of introduction (Tylleskär, 1994). Chronic cassava toxicity is rare in parts of Africa where there has been a long history of cultivation of bitter cassava. In those regions, detoxification methods are often comparable in effectiveness to those employed by Amerindians (Dufour, 2007). In contrast, the prevalence of chronic cassava toxicity is high in groups that have relatively recently adopted bitter cassava as a staple crop, suggesting that detoxification methods have improved over time in areas of introduction, but after an initial period of cultural adjustments. Epidemics of Konzo, a paralytic disease associated with chronic cassava toxicity, have also been linked with political instability in Central and East Africa, where refugee populations forced to escape war zones lack the time or access to water to detoxify cassava roots and bypass or shorten some processing steps (Cliff et al., 2011). The patterns observed underscore our contention that toxicity is a problem whose solution lies not so much in breeding low-cyanogen or acyanogenic plants as it does in providing farmers with the knowledge to effectively detoxify the plant, and the protein resources (or the money to buy them) necessary for efficient physiological detoxification of residual dietary cyanide (McKey et al., 2010).

3.2.2 Mixed clonal/sexual systems in the cassava diaspora

The mixed clonal/sexual system that characterizes cassava under 'traditional' cultivation throughout Amazonia hinges ultimately on farmers' weeding behaviour: whether or not they are interested in volunteer seedlings, letting them grow and incorporating some of them as new clones. Are smallholder farmers in Africa, Asia and Oceania also interested in volunteer seedlings? We do not know the circumstances of the transmission of cassava farming techniques to African farmers. However, it is unlikely that those who first introduced the crop into Africa transmitted Amerindian knowledge intact, particularly knowledge about a practice that is not of obvious agronomic interest. As clonal propagation provides a straightforward and abundant source of planting material, farmers have little need for cassava seedlings, unless driven by necessity, such as following the loss of the original varieties owing to drought or epiphytotics (e.g. Jennings, 1963, 1970). Farmers will only start experimenting with cassava seedlings if they develop a curiosity for them. Farmers working with cassava for the first time may have lacked experience with crops with a similar ecology that could have served as a model for learning to come to grips with cassava. However, if farmers did not at least occasionally incorporate volunteer seedlings, this would signify the loss of a key mechanism to compensate for the inevitable reduction of diversity in a purely clonal system, with dire consequences for the stability of yield and resilience of the cropping system over medium or long time frames. Gibson et al. (2000) have highlighted the same potential problem for sweet potato in Africa.

Do farmers of diasporic populations of cassava incorporate volunteer seedlings? What we know so far indicates that the answer to this question varies greatly from one place to another, and sometimes even among farmers in a single village. What we know also suggests one or two twists in the basic story.

The most richly documented case study is from Gabon. Delêtre (2010) studied cassava diversity and management practices in ten villages, representing ten different ethnic groups. In only one village did farmers' practices towards volunteer seedlings resemble those widespread in Amazonia: farmers showed great interest in and knowledge about cassava seedlings as a source of diversity, and reported that they regularly tested seedlings as new clones. Interestingly, this ethnic group was among those who established early relations with Portuguese traders, and among the first populations in Gabon to get acquainted with cassava – probably as early as 1700 – and the main actors of the onset of cassava farming in the region, long before other populations discovered the crop. Some of them travelled willingly to Brazil (Patterson, 1975), favouring cultural exchanges between Africans and Amerindians in the Americas. Africans likely learned during these travels the ways to

Figure 5 Preparation of cassava flour ('farinha' or 'gari') in Africa. 'Gari' is popular in West Africa, where African slaves returning from Brazil (the 'Brasilias'; see Verger [1968]) introduced the technique after 1750, greatly hastening the crop's adoption in the region (Jones 1957). After the roots are grated (a), the pulp is left to dry in the sun, then roasted on cooking plates (b). The flour can then be stored in tight containers for over a year (photos: Marc Delêtre).

detoxify cassava roots and prepare cassava flour, which they know under its Portuguese name, '*farinha*' (Fig. 5).

In most of the villages studied, however, farmers treated cassava seedlings as weeds and systematically removed them from their fields. Cassava seedlings have low value as a resource, as they produce only a single taproot.[4] Seedlings are also sometimes the object of mistrust: owing to their novel traits, or trait combinations, they do not fit into the categories whose properties are known. In one village in which farmers favoured sweet varieties, volunteer seedlings were discarded because farmers observed they were sometimes bitter, even when phenotypically indistinguishable from a sweet variety (Delêtre et al., 2016).

This widespread lack of interest in seedlings contrasts with high levels of intravarietal diversity, and genetic studies showed that the set of plants ascribed by a farmer to a single landrace often included a few individuals with unique recombinant genotypes: as in Amazonia (Elias et al., 2000a), a landrace was often a polyphyletic group, with some rare genotypes resulting from independent, relatively recent events of sexual reproduction (Delêtre et al., 2016). In other words, as in Amazonia, volunteer seedlings are recruited as planting material – although mostly inadvertently. An explanation of the apparent discrepancy between farmers' reported behaviour and the genetic patterns detected was suggested by observing what farmers actually do in their fields: they remove seedlings, but this removal is unwittingly selective. Seedlings whose phenotypes conspicuously diverge from those of any of their familiar landraces are removed, whereas plants that resemble one landrace are thought to be resprouts from planted cuttings or stem fragments left over after harvest, and are thus trusted. Some of these plants undoubtedly are landrace clones, but many are true volunteer seedlings with unique recombinant genotypes. This selective incorporation of seedlings based on 'perceptual indistinctiveness' (Delêtre et al., 2016) – in contrast to the process of selection based on perceptual distinctiveness (Gibson, 2009) – injects a small but regular dose of recombination, maintaining a mixed clonal/sexual reproductive system. Although this system generates little visually conspicuous diversity, its impact on traits that farmers find difficult to predict from morphology (e.g. variation in disease resistance or toxicity) is perhaps not negligible.

Such accidental incorporation of volunteer seedlings might be quite widespread in Africa. It could help explain the great genetic diversity of cassava in Africa despite the frequent lack of interest in cassava seedlings that we have observed in Gabon and in Cameroon (McKey, unpublished field observations) and that others have noted elsewhere (Fresco, 1986; de Waal et al., 1997; Manu-Aduening et al., 2005). Selection based on perceptual indistinctiveness could also help explain the apparent mismatch between high neutral genetic diversity of cassava in Africa and low phenotypic diversity. For example, to name just two conspicuous traits, the diversity we have observed in stem architecture and in leaf morphology appears much lower in African cassava fields than in Amerindian fields in Guyana (Elias et al., 2001b) and Peru (Boster, 1983).

Interest in volunteer seedlings and their incorporation as new clones also appear to vary among sites in other regions where cassava has been introduced. In Indonesia, Ellen and Soselisa (2012b) consider that 'there is little evidence for incorporation of volunteer seedlings, either deliberate or inadvertent'. In Vanuatu, however, farmers often propagate

4. In contrast, Amerindian farmers in Guyana recognize that a plant from seed is 'handicapped' and that its worth cannot be judged until it is propagated as cuttings, which can produce starch storage roots from adventitious roots produced at each buried node (Elias et al., 2000).

volunteer seedlings, and the resulting mixed/clonal sexual system contributes to the surprisingly high genetic diversity of cassava in the archipelago (Sardos et al., 2008). Before cassava's introduction, farmers in Oceania had experience with crops that resemble this crop in some features of their ecology, for example, *Abelmoschus manihot* L. (Medik.) ('island cabbage') and *Ipomoea batatas* L. (Lam.) (sweet potato), both of which are clonally propagated but also produce seeds that germinate in farmers' fields. In Oceania, the latter is known to have a mixed clonal/sexual system in homegardens (Roullier et al., 2013). Cultivators' familiarity with such crops may have created some 'habits of mind' (Ross et al., 2005), facilitating the transposition to cassava of knowledge and practices acquired while managing similar indigenous plants.

Far from the crop's area of origin, other wild relatives have contributed to cassava's diversification. *Manihot glaziovii* Muell. Arg., the Ceara rubber tree, has been introduced throughout the Old World tropics. In Africa, it was introduced near the end of the nineteenth century in an attempt to develop rubber production (Jones, 1959). Although the experiment failed, *M. glaziovii* is now naturalized in Africa (Halsey et al., 2008), and spontaneous *M. esculenta* × *M. glaziovii* hybrids have been reported in Madagascar (Cours, 1951) and Côte d'Ivoire (Lefèvre and Kouakou, 1987; Lefèvre and Charrier, 1993). There is also genetic evidence that some African cultivars are descendants of spontaneous *M. esculenta* × *M. glaziovii* hybrids (Beeching et al., 1993). Incorporation could have occurred when, on a former rubber plantation, farmers opened a new cassava farm, triggering the germination of seeds from *M. glaziovii* and favouring interspecific gene flow. In East Africa, interspecific hybridization with *M. glaziovii* was also used to breed for resistance to the Cassava Mosaic Disease (CMD), starting as early as 1937 (Bueno, 1985). These East African interspecific hybrids became the main source of resistance to CMD in the IITA cassava breeding programme (Legg et al., 2014), and many widely adopted cultivars resulting from this programme show widespread introgression by *M. glaziovii* (Bredeson et al., 2016).

In Indonesia, *M. glaziovii* has been used in an interesting farmer-created innovation. In the Mukibat system, which takes its name from the Javan farmer who invented it, *M. glaziovii* scions are grafted onto a cassava rootstock, greatly increasing the storage root mass produced (Bruijn and Dharmaputra, 1974).

4 Conclusion

Coevolution of cassava's traits and farmers' management practices have created a remarkably diverse crop which shows extraordinary variation in its ecology and in its biocultural interactions with farmers, both within and between cassava's centre of origin and areas of introduction. In this chapter, we have focused on two dimensions of this variation that we believe have been largely overlooked, and should be more widely appreciated by researchers and stakeholders interested in the contribution of cassava to food security and human well-being: the management of a chemically defended but potentially toxic crop, and the creation of diversity through farmers' capitalization (conscious or not) of cassava's greatest asset, its mixed clonal/sexual reproductive system. Global change adds new challenges to the management of toxicity (McKey et al., 2010; Brown et al., 2016) and may make it even more important that cassava landrace populations in farmers' fields maintain their capacity to generate genetic diversity. Future cassava breeding programmes and extension plans might reap great profit from taking more effectively into account the biocultural variation in this fascinating and complex crop.

5 Where to look for further information

The key reference on the closest wild relatives of cassava is provided in the paper by Olsen and Schaal (1999) and its conclusions are followed here. Still, the origin of cassava remains underexplored compared to crops such as maize. Recent developments have revealed unexpected twists in the story of that crop's ancestry (e.g. Hufford et al., 2013; Piperno et al., 2015), and it would not be surprising if the ancestry of cassava and its evolution under domestication also turn out to be more complicated than what we presently know.

McKey et al. (2010) and McKey et al. (2012) provide an entry into the literature on the evolutionary ecology and ethnobiology of cassava and other vegetatively propagated crop plants, focussing on the mixed clonal/sexual reproductive systems that characterize many of them (e.g. cassava under cultivation by Native American farmers). For a detailed comparison of the traits of stems of cassava and its closest wild relatives, not covered by these earlier reviews, see Ménard et al. (2013).

A definitive analytical history of cassava's introduction and diffusion in the Old World, treating cultural as well as agronomic dimensions, has yet to be written. Jones (1959), Carter et al. (1992) and Delêtre (2010) are key references for Africa. A recent master's thesis (Leestma, 2015), discovered after completion of this chapter, offers additional information. For Asia and the Pacific, the most informative references we have found are the studies by Ellen and Soselisa (2012a, b) and Sardos et al. (2008).

In discussing chronic cassava toxicity in Africa, we have emphasized the role of imperfect knowledge about detoxification, but the problem is much too complex to be explained by any single factor. How poverty, drought and war join with cyanogen intake to produce one disease caused by chronic cassava toxicity, konzo, is masterfully reviewed by Nzwalo and Cliff (2011). The review by Burns et al. (2010) offers a good discussion of how cyanogenesis and other complexities of cassava will affect its future role in food security.

6 Acknowledgements

Our research on cassava has benefited from funding by several sources. The programme 'Future of Tropical Forest Peoples' (European Commission, DG VIII) and several grants from France's Ministry of Research funded several PhD students who worked on the project. DM received grants from the French agencies Bureau des Ressources Génétiques, Mission pour la Création du Parc de la Guyane, Contrat Plan État Région Guyane, and through the programmes 'Tropical Ecosystems' of the Ministry of Ecology and Sustainable Development and 'Impact of Biotechnologies on Agroecosystems' of the Ministry of Research and Higher Education. Research was also supported by the award to DM of the 2006 Terra Ficaria Prize (Fondation Yves Rocher/Institut de France). MD was funded by a doctoral grant from IRCSET and a Marie Curie IEF grant under the People Programme (Marie Skłodowska-Curie Actions) of the European Union's Seventh Framework Programme FP7/2007–2013/(REA grant agreement N°623764). Our greatest gratitude is to the Laboratoire Universitaire des Traditions Orales et Dynamiques Contemporaines (LUTO-DC) and the Université Omar Bongo (UOB), Libreville, who hosted the project in Gabon, and to the farmers who received us and shared their knowledge with us in French Guiana, Guyana, Gabon and Vanuatu.

7 References

Adam, M. (1980), 'Manioc, rente foncière et situation des femmes dans les environs de Brazzaville'. *Cah. Etud. Afr.*, 20(77), 5–48.

Allem, A. C. (1994), 'The origin of *Manihot esculenta* Crantz (Euphorbiaceae)'. *Genet. Resour. Crop Evol.*, 41(3), 133–50.

Anderson, E. (1952), *Plants, Man and Life*. Little, Brown and Company, Boston.

Anderson, E. (1960), 'The evolution of domestication'. In S. Tax (ed.), *Evolution after Darwin*, Vol. II. *The Evolution of Man: Man, Culture and Society*, University of Chicago Press, Chicago, pp. 67–84.

Arroyo-Kalin, M. (2010), 'The Amazonian Formative: crop domestication and anthropogenic soils'. *Diversity*, 2(4), 473–504.

Arroyo-Kalin, M. (2012), 'Slash-burn-and-churn: Landscape history and crop cultivation in pre-Columbian Amazonia'. *Quatern. Int.*, 249, 4–18.

Bahuchet, S. and Philippson, G. (1998), 'Les plantes d'origine américaine en Afrique bantoue: une approche linguistique'. In M. Chastanet (ed.), *Plantes et Paysages d'Afrique – Une Histoire à Explorer*, Karthala, Paris, pp. 87–116.

Beeching, J. R., Marmey, P., Gavalda, M. C., Noirot, M., Haysom, H. R., Hughes, M. A. and Charrier, A. (1993), 'An assessment of genetic diversity within a collection of cassava (*Manihot esculenta* Crantz) germplasm using molecular markers'. *Ann. Bot.*, 72(6), 515–20.

Boster, J. S. (1983), 'A comparison of the diversity of Jivaroan gardens with that of the tropical forest'. *Hum. Ecol.*, 11(1), 47–68.

Boster, J. S. (1984), 'Inferring decision making from preferences and behaviour: An analysis of Aguaruna Jívaro manioc selection'. *Hum. Ecol.*, 12(4), 3443–58.

Bradbury, E. J., Duputié, A., Delêtre, M., Roullier, C., Narváez-Trujillo, A., Manu-Aduening, J. A., Emshwiller, E. and McKey, D. (2013), 'Geographic differences in patterns of genetic differentiation among bitter and sweet cassava (*Manihot esculenta*: Euphorbiaceae)'. *Am. J. B.*, 100(5), 857–66.

Bredeson, J. V., Lyons, J. B., Prochnik, S. E., Wu, G. A., Ha, C. M., Edsinger-Gonzales, E., Grimwood, J., Schmutz, J., Rabbi, I. Y., Egesi, C., Nauluvula, P., Lebot, V., Ndunguru, J ., Mkamilo, G., Bart, R. S., Setter, T. L., Gleadow, R. M., Kulakow, P., Ferguson, M. E., Rounsley, S. and Rokhsar, D. S. (2016), 'Sequencing wild and cultivated cassava and related species reveals extensive interspecific hybridization and genetic diversity'. *Nat. Biotechnol.*, 34(5), 562–70.

Brown, A. L., Cavagnaro, T. R., Gleadow, R. and Miller, R. E. (2016), Interactive effects of temperature and drought on cassava growth and toxicity: Implications for food security? *Glob. Change Biol.*, 22: 3461–73. doi:10.1111/gcb.13380.

Bruijn, G. D. and Dharmaputra, T. S. (1974), 'The Mukibat system, a high-yielding method of cassava production in Indonesia'. *Neth. J. Agric. Sci.*, 22(2), 89–100.

Bueno, A. (1985), 'Hybridization and breeding methodologies appropriate to cassava'. In C. Hershey (ed.), *Cassava Breeding: A Multidisciplinary Review, Proceedings of a Workshop held in the Philippines, 4–7 March 1985*, pp. 51–66.

Burns, A., Gleadow, R., Cliff, J., Zacarias, A. and Cavagnaro, T. (2010), 'Cassava: The drought, war and famine crop in a changing world'. *Sustainability*, 2(11), 3572–3607.

Burns, A. E., Gleadow, R. M., Zacarias, A. M., Cuambe, C. E., Miller, R. E. and Cavagnaro, T. R. (2012), 'Variations in the chemical composition of cassava (*Manihot esculenta* Crantz) leaves and roots as affected by genotypic and environmental variation'. *J. Agric. Food Chem.*, 60(19), 4946–56.

Carter, S. E., Fresco, L. O., Jones, P. G., Fairbairn, J. N. (1992), *An Atlas of Cassava in Africa: Historical, Agroecological and Demographic Aspects of Crop Distribution*, CIAT, Cali, Colombia.

Chandler-Ezell, K., Pearsall, D. M. and Zeidler, J. A. (2006), 'Root and tuber phytoliths and starch grains document manioc (*Manihot esculenta*) arrowroot (*Maranta arundinacea*) and llerén (*Calathea* sp.) at the Real Alto site, Ecuador'. *Econ. Bot.*, 60(2), 103–20.

Chernela, J. M. (1987), 'Os cultivares de mandioca na área do Uaupés (Tukâno)'. In B. Ribeiro (ed.), *SUMA: Etnológica Brasilieira*, FINEP, Petrópolis, pp. 151–8.

Chiwona-Karltun, L., Mkumbira, J., Saka, J., Bovin, M., Mahungu, N. M. and Rosling, H. (1998), 'The importance of being bitter: A qualitative study on cassava cultivar preference in Malawi'. *Ecol. Food Nutri.*, 37(3), 219–45.

Chiwona-Karltun, L., Tylleskär, T., Mkumbira, J., Gebre-Medhin, M. and Rosling, H. (2000), 'Low dietary cyanogen exposure from frequent consumption of potentially toxic cassava in Malawi'. *Int. J. Food Sci. Nutr.*, 51(1), 33–43.

Clement, C. R. (2014), 'Crop domestication in the Amazon.' In *Encyclopaedia of the History of Science, Technology, and Medicine in Non-western Cultures*, Springer, Dordrecht, pp. 1–7. DOI 10.1007/978-94-007-3934-5_9876-1.

Clement, C. R., Rodrigues, D. P., Alves-Pereira, A., Mühlen, G. S., Cristo-Araújo, M. D., Moreira, P. A., Lins, J. and Reis, V. M. (2016), 'Crop domestication in the upper Madeira River basin'. *Bol. Mus. Para. Emílio Goeldi. Cienc. Hum.*, 11(1), 193–205.

Cliff, J., Muquingue, H., Nhassico, D., Nzwalo, H. and Bradbury, J. H. (2011), 'Konzo and continuing cyanide intoxication from cassava in Mozambique'. *Food Chem. Toxicol.*, 49(3), 631–5.

Cloarec-Heiss, F. and Nougayrol, P. (1998), 'Des noms et des routes: la diffusion des plantes américaines en Afrique centrale (RCA-Tchad)'. In M. Chastanet (ed.), *Plantes et Paysages d'Afrique – Une Histoire à Explorer*, Karthala, Paris, pp. 117–64.

Cogels, S. (2002), *Les Ntumu du sud-Cameroun forestier: Une société de non-spécialistes. Système de production, stratégies d'usage des ressources et enjeux du changement*. PhD dissertation. Université Libre de Bruxelles, Brussels, Belgium.

Cours, G. (1951), 'Le manioc à Madagascar'. *Mémoire de l'institut scientifique de Madagascar, série B*, 3, 203–400.

Daniau, A. L., d'Errico, F. and Goñi, M. F. S. (2010), 'Testing the hypothesis of fire use for ecosystem management by Neanderthal and Upper Palaeolithic modern human populations'. *PLOs ONE*, 5(2), e9157.

Delêtre, M. (2010), *The ins and outs of manioc diversity in Gabon, Central Africa: A pluridisciplinary approach to the dynamics of genetic diversity of* Manihot esculenta *Crantz (Euphorbiaceae)*. PhD dissertation. Trinity College, University of Dublin, Ireland.

Delêtre, M., Hodkinson, T. R. and McKey, D. (2016), 'Perceptual selection and the unconscious selection of 'volunteer' seedlings in clonally propagated crops: An example with African cassava (*Manihot esculenta* Crantz) using ethnobotany and population genetics'. *Genet. Resour. Crop Evol.* DOI: 10.1007/s10722-016-0390-3.

Delêtre, M., McKey, D. and Hodkinson, T. R. (2011), 'Marriage exchanges, seed exchanges, and the dynamics of manioc diversity'. *Proc. Natl. Acad. Sci. U.S.A.*, 108(45), 18249–54.

De Waal, D., Chinjinga, F. R., Johansson, L., Kanju, E. F. and Nathaniels, N. (1997), 'Village-based cassava breeding in Tanzania'. In L. van Veldhuizen, A. Waters-Bayer and R. Ramirez (eds), *Farmers' Research in Practice: Lessons from the Field*, International Technology Publications, London, pp. 83–8.

Dounias, E. (1993), *Dynamique et gestion différentielles du système de production à dominante agricole des Mvae du sud Cameroun forestier*. PhD dissertation. Université des Sciences et Techniques du Languedoc, Montpellier, France.

Dufour, D. L. (1988), 'Cyanide content of cassava (*Manihot esculenta*) cultivars used by Tukanoan Indians in northwest Amazonia'. *Econ. Bot.*, 42(2), 255–66.

Dufour, D. L. (2007), '"Bitter" cassava: toxicity and detoxification'. In R. Ortiz and N. Nassar (eds), *Proceedings of the First International Meeting on Cassava Breeding, Biotechnology and*

Ecology. University of Brasilia, Brasilia, pp. 171–84. http://www.geneconserve.pro.br/siteantigo/meeting2/proceedings.pdf#page=171.
Duputié, A., Massol, F., David, P., Haxaire, C. and McKey, D. (2009b), 'Traditional Amerindian cultivators combine directional and ideotypic selection for sustainable management of cassava genetic diversity'. *J. Evol. Biol.*, 22(6), 1317–25.
Elias, M. and McKey, D. (2000), 'The unmanaged reproductive ecology of domesticated plants in traditional agroecosystems: An example involving cassava and a call for data'. *Acta Oecol.*, 21(3), 223–30.
Elias, M., McKey, D., Panaud, O., Anstett, M. C. and Robert, T. (2001a), 'Traditional management of cassava morphological and genetic diversity by the Makushi Amerindians (Guyana, South America): Perspectives for on-farm conservation of crop genetic resources'. *Euphytica*, 120(1), 143–57.
Elias, M., Panaud, O. and Robert, T. (2000b), 'Assessment of genetic variability in a traditional cassava (*Manihot esculenta* Crantz) farming system, using AFLP markers'. *Heredity*, 85(3), 219–30.
Elias, M., Penet, L., Vindry, P., McKey, D., Panaud, O. and Robert, T. (2001b), 'Unmanaged sexual reproduction and the dynamics of genetic diversity of a vegetatively propagated crop plant, cassava (*Manihot esculenta* Crantz), in a traditional farming system'. *Mol. Ecol.*, 10(8), 1895–1907.
Elias, M., Rival, L. and McKey, D. (2000a), 'Perception and management of cassava (*Manihot esculenta* Crantz) diversity among the Makushi Amerindians of Guyana (South America)'. *J. Ethnobiol.*, 20(1), 239–65.
Ellen, R. and Soselisa, H. L. (2012a), 'Cassava diversity and toxicity in relation to environmental degradation. A feature of food security in the Moluccas, Indonesia'. In A.-K. Hornidge and C. Antweiler (eds), *Environmental Uncertainty and Local Knowledge: Southeast Asia as a Laboratory of Global Ecological Change*, Transcript Verlag, Bielefeld, Germany, pp. 215–42.
Ellen, R. F. and Soselisa, H. L. (2012b). 'A comparative study of the socio-ecological concomitants of cassava (*Manihot esculenta* Crantz) diversity, local knowledge and management in eastern Indonesia'. *Ethnobot. Res. Appl.*, 10, 015–035. DOI: http://dx.doi.org/10.17348/era.10.0.015-035.
Emperaire, L., Pinton, F. and Second, G. (1998), 'Gestion dynamique de la diversité variétale du manioc en Amazonie du Nord-Ouest'. *Nat. Sci., Sociétés*, 6(2), 27–42.
Fabres, G., Kiyindou, A. and Epouna-Mouinga, S. (1981), 'Première quantification du phénomène de gradation des populations de *Phenacoccus manihoti* Matile-Ferrero (Hom. Pseudococcidae) en République Populaire du Congo'. *Agronomie*, 1(6), 483–6.
Felsenstein, J. (1974), 'The evolutionary advantage of recombination'. *Genetics*, 78(2), 737–56.
Fraser, J. A. (2010), 'Caboclo horticulture and Amazonian dark earths along the middle Madeira River, Brazil'. *Hum. Ecol.*, 38(5), 651–62.
Fraser, J. A., Alves-Pereira, A., Junqueira, A. B., Peroni, N. and Clement, C. R. (2012), 'Convergent adaptations: bitter manioc cultivation systems in fertile anthropogenic dark earths and floodplain soils in Central Amazonia'. *PLoS One*, 7(8), e43636.
Fregene, M., Bernal, A., Duque, M., Dixon, A. and Tohme, J. (2000), 'AFLP analysis of African cassava (*Manihot esculenta* Crantz) germplasm resistant to the cassava mosaic disease (CMD)'. *Theor. Appl. Genet.*, 100(5), 678–85.
Fregene, M. A., Suarez, M., Mkumbira, J., Kulembeka, H., Ndedya, E., Kulaya, A., Mitchel, S., Gullberg, U., Rosling, H., Dixon, A. G. O., Dean, R. and Kresovich, S. (2003), 'Simple sequence repeat marker diversity in cassava landraces: genetic diversity and differentiation in an asexually propagated crop'. *Theor. Appl. Genet.*, 107(6), 1083–93.
Fresco, L. O. (1986), *Cassava in Shifting Cultivation – A Systems Approach to Agricultural Technology Development in Africa*, Royal Tropical Institute, Amsterdam, The Netherlands.
Gade, D. W. (2002), 'Names for *Manihot esculenta*: Geographical variations and lexical clarification'. *J. Latin Am. Geo.*, 1(1), 55–74.
Garwood, N. C. (1996) 'Functional morphology of tropical tree seedlings'. In M. D. Swaine (ed.), *The Ecology of Tropical Forest Tree Seedlings*, UNESCO and Parthenon, Paris and Carnforth (UK), pp. 59–129.

Geffray, C. (1989), 'Les hommes au travail, les femmes au grenier. La société makhuwa (Erati) des années trente à 1956'. *Cah. Sci. Hum.*, 25(3), 313–24.
Gibson, R. W. (2009), 'A review of perceptual distinctiveness in landraces including an analysis of how its roles have been overlooked in plant breeding for low-input farming systems'. *Econ. Bot.*, 63(3), 242–5.
Gibson, R. W., Jeremiah, S. C., Aritua, V., Msabaha, R. P., Mpembe, I. and Ndunguru, J. (2000), Sweet potato virus disease in sub-Saharan Africa: Evidence that neglect of seedlings in the traditional farming system hinders the development of superior resistant landraces. *J. Phytopathol.*, 148(7–8), 441–7.
Gleadow, R. M. and Woodrow, I. E. (2002), 'Mini-Review: constraints on effectiveness of cyanogenic glycosides in herbivore defense'. *J. Chem. Ecol.*, 28(7), 1301–13.
Graziosi, I., Minato, N., Alvarez, E., Ngo, D. T., Hoat, T. X., Aye, T. M., Pardo, J. M., Wongtiem, P. and Wyckhuys, K. A. (2016), 'Emerging pests and diseases of South-east Asian cassava: A comprehensive evaluation of geographic priorities, management options and research needs'. *Pest Manag. Sci.*, 72(6), 1071–89.
Guyer, J. I. (1978), 'The food economy and French colonial rule in central Cameroon'. *J. Afr. Hist.* 19(4), 577–97.
Halsey, M. E., Olsen, K. M., Taylor, N. J. and Chavarriaga-Aguirre, P. (2008), 'Reproductive biology of cassava (*Manihot esculenta* Crantz) and isolation of experimental field trials'. *Crop Sci.*, 48(1), 49–58.
Herren, H. R., Neuenschwander, P., Hennessey, R. D. and Hammond, W. N. O. (1987), 'Introduction and dispersal of *Epidinocarsis lopezi* (Hym., Encyrtidae), an exotic parasitoid of the cassava mealybug, *Phenacoccus manihoti* (Hom., Pseudococcidae), in Africa'. *Agric. Ecosyst. Environ.*, 19(2), 131–44.
Hufford, M. B., Lubinksy, P., Pyhäjärvi, T., Devengenzo, M. T., Ellstrand, N. C and Ross-Ibarra, J. (2013), 'The genomic signature of crop-wild introgression in maize'. *PLoS Genet.*, 9(5), e1003477.
Isendahl, C. (2011), The domestication and early spread of manioc (*Manihot esculenta* Crantz): A brief synthesis.' *Lat. Am. Ant.*, 22(4), 452–68.
Jennings, D. L. (1963), 'Variation in pollen and ovule fertility in varieties of cassava and the effect of interspecific crossing on fertility'. *Euphytica*, 12(1), 69–76.
Jennings, D. L. (1970), Cassava in Africa. *Field Crop Abstr.*, 23(3), 271–8.
Johns, T. (1990), *With Bitter Herbs they Shall Eat it: Chemical Ecology and the Origins of Human Diet and Medicine*, University of Arizona Press, Tucson.
Johns, T. and Kubo, I. (1988), 'A survey of traditional methods employed for the detoxification of plant foods'. *J. Ethnobiol.*, 8(1), 81–129.
Jones, D. A. (1998), 'Why are so many food plants cyanogenic?' *Phytochemistry,* 47(2), 155–62.
Jones, W. O. (1957), 'Manioc: An example of innovation in African economies'. *Econ. Devel. Cult. Change*, 5(2), 97–117.
Jones, W. O. (1959), *Manioc in Africa,* Stanford University Press, Stanford, CA.
Kawa, N. C., McCarty, C. and Clement, C. R. (2013), 'Manioc varietal diversity, social networks, and distribution constraints in rural Amazonia'. *Curr. Anthropol.*, 54(6), 764–70.
Kizito, B. E., Bua, A., Fregene, M., Egwang, T., Gullberg, U. and Westerbergh, A. (2005), 'The effect of cassava mosaic disease on the genetic diversity of cassava in Uganda'. *Euphytica*, 146(1–2), 45–54.
Kizito, B. E., Chiwona-Karltun, L., Egwang, T., Fregene, M. and Westerbergh, A. (2007), 'Genetic diversity and variety composition of cassava on small-scale farms in Uganda: an interdisciplinary study using genetic markers and farmer interviews'. *Genetica*, 130(3), 301–18.
Labeyrie, V., Deu, M., Barnaud, A., Calatayud, C., Buiron, M., Wambugu, P., Manel, S., Glaszmann, J.-C. and Leclerc. C. (2014), 'Influence of ethnolinguistic diversity on the sorghum genetic patterns in subsistence farming systems in Eastern Kenya'. *PLoS One*, 9(3), e92178.
Lebot, V. (2009), *Tropical Root and Tuber Crops: Cassava, Sweet Potato, Yams and Aroids* (No. 17,. CABI, Wallingford, UK.

Leestma, J. C. (2015), *A Tropical Flour: Manioc in the Afro-Brazilian World, 1500-1800* Master's thesis, University of New Mexico, Albuquerque, NM. http://dspace.unm.edu/handle/1928/27939 (consulted December 26, 2016).

Lefèvre, F. and Charrier, A. (1993), 'Isozyme diversity within African *Manihot* germplasm'. *Euphytica*, 66(1–2), 73–80.

Lefèvre, F. and Kouakou, B. (1987), 'Prospections de *Manihot glaziovii* Muell. Arg. en Côte d'Ivoire'. ORSTOM. http://horizon.documentation.ird.fr/exl-doc/pleins_textes/doc34-03/25215.pdf.

Legg, J. P. and Thresh, J. M. (2003), 'Cassava virus diseases in Africa. In *Proceedings of the First International Conference on Plant Virology in Sub-Saharan Africa (4–8 June, Ibadan, Nigeria)*, pp. 517–52. IITA, Ibadan, Nigeria.

Legg, J. P., Jeremiah, S. C., Obiero, H. M., Maruthi, M. N., Ndyetabula, I., Okao-Okuja, G., Bouwmeester, H., Bigirimana, S., Tata-Hangy, W., Gashaka, G.,Mkamilo, G., Alicai, T. and Lava Kumar, P. (2011), 'Comparing the regional epidemiology of the cassava mosaic and cassava brown streak pandemics in Africa'. *Virus Res.*, 159(2), 161–70.

Legg, J., Somado, E. A., Barker, I., Beach, L., Ceballos, H., Cuellar, W., Elkhoury, W., Gerling, D., Helsen, J., Hershey, C., Jarvis, A., Kulakov, P., Kumar, L., Lorenzen, J., Lynam, J., McMahon, M., Maruthi, G., Miano, D., Mtunda, K., Natwuruhunga, P., Okogbenin, E., Pezo, P., Terry, E., Thiele, G., Thresh, M., Wadsworth, J., Walsh, S., Winter, S., Tohme, J. and Fauquet, C. (2014), 'A global alliance declaring war on cassava viruses in Africa'. *Food Secur.*, 6(2), 231–48.

Léotard, G., Duputié, A., Kjellberg, F., Douzery, E. J., Debain, C., De Granville, J. J. and McKey, D. (2009), 'Phylogeography and the origin of cassava: new insights from the northern rim of the Amazonian basin'. *Mol. Phylogenet. Evol.*, 53(1), 329–34.

Lévi-Strauss, C. (1952), 'The use of wild plants in tropical South America'. *Econ. Bot.*, 6(3), 252–70.

Manu-Aduening, J. A., Lamboll, R. I., Dankyi, A. A. and Gibson, R. W. (2005), 'Cassava diversity in Ghanaian farming systems'. *Euphytica*, 144(3), 331–40.

Martin, G. B. and Adams, M. W. (1987a), 'Landraces of *Phaseolus vulgaris* in northern Malawi. I. Regional variation'. *Econ. Bot.*, 41(2), 190–203.

Martin, G. B. and Adams, M. W. (1987b) 'Landraces of *Phaseolus vulgaris* in northern Malawi. II. Generation and maintenance of variability'. *Econ. Bot.*, 41(2), 204–15.

McKey, D. and Beckerman, S. (1993), 'Chemical ecology, plant evolution and traditional manioc cultivation systems'. In C. Hladik, A. Hladik, O. F. Linares, H. Pagezy, A. Semple and M. Hadley (eds), *Tropical Forests, People and Food. Biocultural Interactions and Applications to Development*, Parthenon, Carnforth. UNESCO, Paris, pp. 83–112.

McKey, D., Elias, M., Pujol, B. and Duputié, A. (2010), 'The evolutionary ecology of clonally propagated domesticated plants'. *New Phytol.*, 186(2), 318–32.

McKey, D., Elias, M., Pujol, B. and Duputié, A. (2012), 'Ecological approaches to crop domestication'. In P. Gepts, R. Bettinger, S. B. Brush, T. Famula, P. E. McGuire and C. O. Qualset (eds), *Biodiversity in Agriculture. Domestication, Evolution, and Sustainability*, Cambridge University Press, Cambridge, UK, pp. 377–406.

Ménard, L., McKey, D., Mühlen, G. S., Clair, B. and Rowe, N. P. (2013), 'The evolutionary fate of phenotypic plasticity and functional traits under domestication in manioc: Changes in stem biomechanics and the appearance of stem brittleness'. *PLOs ONE*, 8(9), e74727.

Mkumbira, J., Chiwona-Karltun, L., Lagercrantz, U., Mahungu, N. M., Saka, J., Mhone, A., Bokanga, M., Brimer, L., Gullberg, U. and Rosling, H. (2003), 'Classification of cassava into "bitter" and "cool" in Malawi: From farmers' perception to characterisation by molecular markers'. *Euphytica*, 132(1), 7–22.

Mondolot, L., Marlas, A., Barbeau, D., Gargadennec, A., Pujol, B. and McKey, D. (2008), 'Domestication and defence: foliar tannins and C/N ratios in cassava and a close wild relative'. *Acta Oecol.*, 34(2), 147–54.

Nweke, F., Spencer, D. and Lynam, J. (2001), *The Cassava Transformation: Africa's Best Kept Secret*, Michigan State University, East Lansing.

Nweke, F. I., Dixon, A. G. O., Asiedu, R. and Folayan, S. A. (1994), '*Cassava Varietal Needs of Farmers and the Potential for Production Growth in Africa*'. Working Paper No. 10. Collaborative Study of Cassava in Africa. IITA, Ibadan.

Nye, M. (1991), 'The mis-measure of manioc (*Manihot esculenta*, Euphorbiaceae)'. *Econ. Bot.*, 45(1), 47–57.

Nzwalo, H. and Cliff, J. (2011), 'Konzo: from poverty, cassava, and cyanogen intake to toxico-nutritional neurological disease'. *PLoS Negl Trop Dis,*, 5(6), e1051.

Olsen, K. M. and Schaal, B. A. (1999), 'Evidence on the origin of cassava: Phylogeography of *Manihot esculenta*'. *Proc. Natl. Acad. Sci. U.S.A.*, 96(10), 5586–91.

Onwueme, I. C. (2002), 'Cassava in Asia and the Pacific'. In R. J. Hillocks, J. M. Thresh and A. C. Bellotti (eds), *Cassava: Biology, Production and Utilization*, CABI, Wallingford, UK, pp. 55–65.

Patil, B. L. and Fauquet, C. M. (2009), 'Cassava mosaic geminiviruses: Actual knowledge and perspectives'. *Mol. Plant Pathol.*, 10(5), 685–701.

Patterson, K. D. (1975), *The Northern Gabon Coast to 1875*, Clarendon Press, Oxford.

Pautasso, M., Aistara, G., Barnaud, A., Caillon, S., Clouvel, P., Coomes, O., Delêtre, M., Demeulenaere, E., De Santis, P., Doering, T., Eloy, L., Emperaire, L., Garine, E., Goldringer, I., Jarvis, D., Joly, H., Leclerc, C., Louafi, S., Martin, P., Massol, F., McGuire, S., McKey, D., Padoch, C., Soler, C., Thomas, M. and Tramontini, S. (2012), 'Seed exchange networks for agrobiodiversity conservation. A review'. *Agron. Sustain. Dev.*, 33(1), 151–75.

Peroni, N. and Hanazaki, N. (2002). Current and lost diversity of cultivated varieties, especially cassava, under swidden cultivation systems in the Brazilian Atlantic Forest. *Agric. Ecosyst. Environ.*, 92(2), 171–83.

Perrut-Lima, P., Mühlen, G. S. and Carvalho, C. R. (2014), 'Cyanogenic glycoside content of *Manihot esculenta* subsp. *flabellifolia* in south-central Rondônia, Brazil, in the center of domestication of *M. esculenta* subsp. *esculenta*'. *Genet. Resour. Crop Evol.*, 61(6), 1035–8.

Perry, L. (2002), 'Starch granule size and the domestication of manioc (*Manihot esculenta*) and sweet potato (*Ipomoea batatas*)'. *Econ. Bot.*, 56(4), 335–49.

Pickersgill, B. (1998), 'Crop introductions and the development of secondary areas of diversity'. In H. D. V. Prendergast, N. L. Etkin, D. R. Harris and P. J. Houghton (eds), *Plants for Food and Medicine*, Royal Botanic Gardens, Kew, pp. 93–105.

Pinton, F. and Emperaire, L. (2001), 'Le manioc en Amazonie brésilienne: Diversité variétale et marché'. *Genet. Sel. Evol.*, 33(S1), S491–512.

Piperno, D. R., Holst, I., Winter, K. and McMillan, O. (2015), 'Teosinte before domestication: Experimental study of growth and phenotypic variability in late Pleistocene and early Holocene environments'. *Quaternary International*, 363, 65–77.

Poubom, C. F. N., Awah, E. T., Tchuanyo, M. and Tengoua, F. (2005), 'Farmers' perceptions of cassava pests and indigenous control methods in Cameroon'. *Int. J. Pest Manage.*, 51(2), 157–64.

Pujol, B. and McKey, D. (2006), 'Size asymmetry in intraspecific competition and the density-dependence of inbreeding depression in a natural plant population: a case study in cassava (*Manihot esculenta* Crantz, Euphorbiaceae)'. *J. Evol. Biol.*, 19(1), 85–96.

Pujol, B., Salager, J. L., Beltran, M., Bousquet, S. and McKey, D. (2008), 'Photosynthesis and leaf structure in domesticated cassava (Euphorbiaceae) and a close wild relative: have leaf photosynthetic parameters evolved under domestication?'. *Biotropica*, 40(3), 305–12.

Pujol, B., Gigot, G., Laurent, G., Pinheiro-Kluppel, M., Elias, M., Hossaert-McKey, M. and McKey, D. (2002), 'Germination ecology of cassava (*Manihot esculenta* Crantz, Euphorbiaceae) in traditional agroecosystems: Seed and seedling biology of a vegetatively propagated domesticated plant'. *Econ. Bot.*, 56(4), 366–79.

Pujol, B., Mühlen, G., Garwood, N., Horoszowski, Y., Douzery, E. J. and McKey, D. (2005a), 'Evolution under domestication: contrasting functional morphology of seedlings in domesticated cassava and its closest wild relatives'. *New Phytol.*, 166(1), 305–18.

Pujol, B., Patrice, D. and McKey, D. (2005b), 'Microevolution in agricultural environments: how a traditional Amerindian farming practice favours heterozygosity in cassava (*Manihot esculenta* Crantz, Euphorbiaceae)'. *Ecol. Lett.* 8(2), 138–47.

Ranum, P., Peña-Rosas, J. P. and Garcia-Casal, M. N. (2014), 'Global maize production, utilization, and consumption'. *Ann. N. Y. Acad. Sci.*, 1312(1), 105–12.

Renard, D., Schatz, B. and McKey, D. B. (2010), 'Ant nest architecture and seed burial depth: implications for seed fate and germination success in a myrmecochorous savanna shrub'. *Ecoscience*, 17(2), 194–202.
Richards, P. (1985), *Indigenous Agricultural Revolution – Ecology and Food Production in West Africa*, Unwin Hyman, London.
Rogers, D. J. (1965), 'Some botanical and ethnological considerations of *Manihot esculenta*'. *Econ. Bot.*, 19(4), 369–77.
Rogers, D. J. and Appan, S. G. (1973), *Manihot, Manihotoides (Euphorbiaceae).* Hafner Press, New York.
Romão, R. L. (2000), 'Northeast Brazil: a secondary centre of diversity for watermelon (*Citrullus lanatus*)'. *Genet. Resour. Crop Evol.*, 47(2), 207–13.
Ross, N., Medin, D., Coley, J. D. and Atran, S. (2002), 'Cultural and experiential differences in the development of folkbiological induction'. *Cognitive Dev.*, 18(1), 25–47.
Roullier, C., Kambouo, R., Paofa, J., McKey, D. and Lebot, V. (2013), 'On the origin of sweet potato (*Ipomoea batatas* (L.) Lam.) genetic diversity in New Guinea, a secondary centre of diversity'. *Heredity*, 110(6), 594–604.
Salick, J. (1995), 'Toward an integration of evolutionary ecology and economic botany: personal perspectives on plant/people interactions'. *Ann. Missouri Bot. Gard.*, 82(1), 25–33.
Santos Mühlen, G., Alves-Pereira, A., Clement, C. R., Losada Valle, T. (2013), 'Genetic diversity and differentiation of Brazilian bitter and sweet manioc varieties (*Manihot esculenta* Crantz, Euphorbiaceae) based on SSR molecular markers'. *Tipití*, 11(2), 66–73.
Sardos, J., McKey, D., Duval, M. F., Malapa, R., Noyer, J. L. and Lebot, V. (2008), 'Evolution of cassava (*Manihot esculenta* Crantz) after recent introduction into a South Pacific island system: The contribution of sex to the diversification of a clonally propagated crop'. *Genome*, 51(11), 912–21.
Sautter, G. (1966), *De l'Atlantique au Fleuve Congo: une Géographie du Sous-peuplement. République du Congo, République gabonaise*. La Haye, Mouton, Paris.
Strong Jr, D. R. (1979), 'Biogeographic dynamics of insect-host plant communities'. *Annu. Rev. Entomol.*, 24(1), 89–119.
Thomas, M., Dawson, J. C., Goldringer, I. and Bonneuil, C. (2012), 'Seed exchanges, a key to analyze crop diversity dynamics in farmer-led on-farm conservation'. *Genet. Resour. Crop Evol.*, 58(3), 321–38.
Tylleskär, T. (1994), 'The association between cassava and the paralytic disease konzo'. *Acta Hortic.*, 375(33), 321–31.
Verger, P. (1968), *Flux et reflux de la traite des nègres entre le golfe de Bénin et Bahia de todos os santos du XVIIe au XIXe siècle*. Mouton, Paris.
Wenham, J. E. (1995), *'Post-harvest deterioration of cassava: a biotechnology perspective'*. Plant Production and Protection Paper FAO No 130. FAO Plant Production and Protection, Rome Italy.
Westengen, O. T., Okongo, M. A., Onek, L., Berg, T., Upadhyaya, H., et al. (2014) Ethnolinguistic structuring of sorghum genetic diversity in Africa and the role of local seed systems. *Proc. Natl. Acad. Sci. U.S.A.*, 111(39), 14100–5.
Wilson, W. M. and Dufour, D. L. (2002), 'Why "bitter" cassava? Productivity of "bitter" and "sweet" cassava in a Tukanoan Indian settlement in the northwest Amazon'. *Econ. Bot.* 56(1), 49–57.
Wood, D. and Lenné, J. M. (2011), '4 Crop introduction and agrobiodiversity management'. In J. M. Lenné and D. Wood (eds), *Agrobiodiversity Management for Food Security: A Critical Review*, CABI, Wallingford, UK, pp. 53–63.
Yaninek, J. S. and Herren, H. R. (1988), 'Introduction and spread of the cassava green mite, *Mononychellus tanajoa* (Bondar) (Acari: Tetranychidae), an exotic pest in Africa and the search for appropriate control methods: a review'. *Bull. Entomol. Res.*, 78(1), 1–13.
Zohary, D. (1984), 'Modes of evolution under domestication. 'In W. F. Grant (ed.), *Plant Biosystematics*, Academic Press, Toronto, Canada, pp. 579–96.

Chapter 2

Use and nutritional value of cassava roots and leaves as a traditional food

Aurélie Bechoff, Natural Resources Institute, University of Greenwich, UK

1 Introduction
2 Diversity in traditional cassava products
3 Products from cassava roots
4 Products from cassava leaves
5 Chemical composition of cassava roots
6 Chemical composition of cassava leaves
7 Anti-nutrients in roots and leaves
8 Case study: biofortified cassava roots made into gari and fufu in Nigeria
9 Future trends
10 Conclusion
11 Where to look for further information
12 References

1 Introduction

Cassava is second after rice in importance as a source of carbohydrates in developing and tropical countries, and the fifth most important staple crop globally (FAOSTAT 2013). In developing countries, over half a billion people consume cassava as food and rely on it as an important source of nutrition and income. Cassava's main commercial product is the long tuberous starchy root. Size varies widely, but averages about 5–7 cm diameter and 20–40 cm long. The root has a mostly dark brown, but sometimes light brown or white, peel and generally a white or cream interior flesh. Cassava leaves are alternate palmate and smooth leaves with lobes between 7 cm and 15 cm long that are also edible (Lancaster and Brooks 1983) (Fig. 1).

http://dx.doi.org/10.19103/AS.2016.0014.01

Figure 1 (a) Cassava leaves and plant and (b) cassava roots a day after harvest. Source: Bechoff, A. 2013, Nigeria.

In Asia, with some exceptions such as Indonesia, cassava is mostly used in non-food applications. Most production is destined for industrial starch, animal feed and ethanol, and only a small proportion is for food uses. In Africa, on the other hand, cassava is used mostly for human food and is therefore an important crop for food security.

Cassava was domesticated in America, and introduced to Africa and Asia later in history. Cassava has since been fully integrated as part of the local cuisine in those two continents. Both the starchy and fresh leaves can be used for human consumption to prepare a variety of dishes.

Cassava has many advantages as a food crop in in the tropics: it is drought tolerant, and can yield relatively well in poor soils, though it will also respond well to good management. It does not have a defined maturity period, and roots can be *stored* in the soil for long periods, allowing flexibility of harvest on an as-needed basis for food or for sale. Cassava is a cheap source of carbohydrates and therefore of energy. Under relatively harsh climate conditions and with little input, cassava can often produce more energy per hectare than any other crop. The versatility of cassava has made it become a source of livelihood to about 300 million people in sub-Saharan Africa, with many of them being poor smallholder farmers. With the new challenges coming up in the twenty-first century such as global warming and increase in world population, cassava's role in food security is likely to increase in the years to come.

Although the cassava crop has many qualities, it also presents a number of weaknesses or challenges:

- Cassava roots and leaves contain cyanogenic glucosides, which release the highly toxic hydrogen cyanide (HCN) on hydrolysis. Cultivars with low levels – below approximately 100 ppm – are said to have low cyanogenic potential (CNP) and are typically referred to as *sweet*. These roots may be eaten directly after boiling or frying. Cultivars with higher CNP are referred to as *bitter* (although CNP is not a precisely related to level of bitterness), and require exhaustive processing in order to reduce these compounds to safe levels. In all places where cultivars with high

CNP are traditionally grown, people have developed effective ways to reduce toxicity through processing techniques and poisoning is normally rare.

- Cassava roots have a short *shelf life* of 2–3 days after harvest. Growers and consumers overcome this constraint by quickly processing or consuming the roots (Blagbrough et al., 2010; Sánchez et al., 2010).
- Cassava roots are composed mostly of starch and contain very few proteins and other micronutrients of importance for a balanced diet. High consumption of cassava roots without other complementary foods can lead to hidden hunger, that is, deficiency of micronutrients such as vitamins and minerals. Cassava leaves, on the other hand, contain some proteins and vitamins that could compensate the lack of the nutrient in roots, but the leaves are often underutilized for human food.

2 Diversity in traditional cassava products

The main constraint of cassava for direct use is its cyanogenic potential. Both roots and leaves can contain high amounts of cyanogenic compounds that can potentially be harmful if the products are not correctly processed. This feature of cassava has resulted in the development of numerous processed products. The processing method and the number of processing steps are linked to the level of cyanide content (*sweet* or *bitter* varieties). We have focused on traditional African products in this review; however, it should be noted that traditional products of cassava can also be found in other parts of the world (i.e. in Central/South America where many of the African products have an ancient origin).

Cassava roots also have a very short shelf life at ambient temperature under tropical conditions (24–48 h) (Westby 2002), which requires rapid consumption or processing after harvest.

3 Products from cassava roots

Products can be divided into two categories:

1. Non-fermented products that are processed to a stable or consumable form in a single step. This applies mainly to sweet cassava (e.g. boiling or frying) and to the intermediate cassava products of either sweet or bitter cassava, such as dried cassava chunks, flour and starch;
2. Fermented products that are made by several steps of processing. This applies mostly to the bitter-type varieties that require cyanide detoxification. Roots and leaves contain enzymes (i.e. β-glucosidase) that are released from the cell vacuoles during processing and cleave the cyanogenic compounds when put into contact through processing such as grating or pounding. Subsequent operations (i.e. fermentation; cooking or roasting) are important to ensure that cyanide is fully volatilized from the product. There are two different types of root fermentation: a solid-state fermentation (directly on the mash) or a submerged fermentation or soaking of the whole roots.

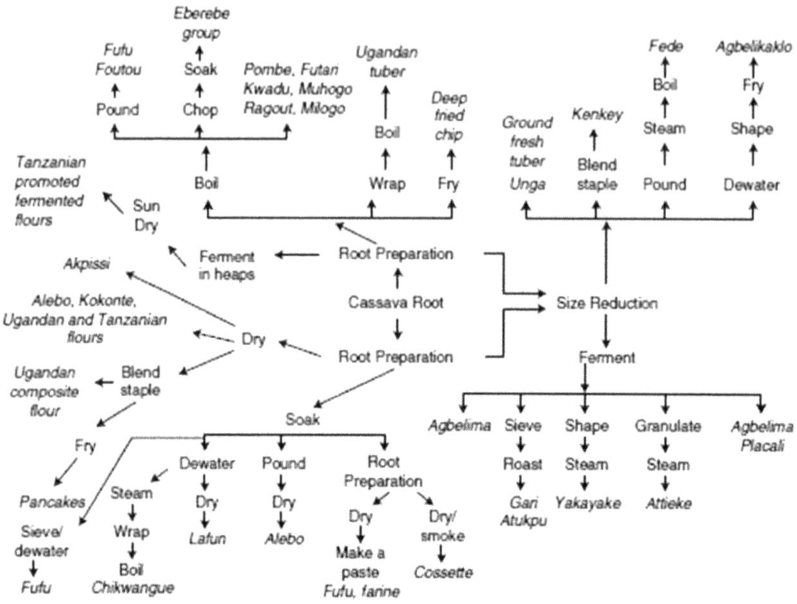

Figure 2 Interrelationship of cassava products from 233 villages in 6 African countries. Source: Westby, 2002.

Figure 2 presents examples of the many different types of products that can be encountered in Africa. Cassava products are made through several steps of processing. Cassava roots undergo peeling, followed by a size reduction (e.g. grating or chopping), boiling or soaking. This is followed by fermenting and/or drying. A large variety of products can be prepared using a combination of different steps.

3.1 Non-fermented products made in a single-step process

Sweet varieties of cassava (that may contain low levels of cyanogenic compounds) are used for processes that generally involve only a single step because of the low initial cyanogenic compound that can be reduced easily to a safe level for consumption. High dry matter content varieties are generally preferred because the texture is firmer and less moist.

Boiling or frying are the most common processes. After peeling, roots can be processed as whole or as chunks or slices. Cooked or boiled cassava as whole roots or pieces of different varieties have diverse tastes (e.g. variable sweetness) and texture (e.g. different levels of stickiness or softness) (Charoenkul et al., 2006; Hongbété et al., 2011).

3.2 Non-fermented products made into industrial products

Intermediate or industrial products are in most cases non-fermented because the quality can be better controlled on non-fermented products rather than on spontaneously fermented ones. The quality of intermediate products has to comply with quality standards because these are industrial products.

In Asia, starch, also called tapioca, is one of the most common products derived from cassava roots. The product is odourless, bland in taste and not sour, with low particle size (<250 µm) and a minimum of 95% starch (ARS 2012). Cassava starch, also called tapioca, is used as a thickener and a stabilizer in various industries (soups, baby food, confectionery, glue and adhesives, pharmaceuticals, detergent soaps, textile etc.)

The use of dried chips is a traditional intermediate product for industry and also for animal feeding in Asia and Latin America, but becoming increasingly popular in Africa as well. Cassava roots can be chopped as irregular chunks or uniform slices and dried. Pellets of regular shape and size may be produced using a pelleting machine that compresses and dries the products. It is essential that the chips or pellets are well dried, ideally to not more than 14% moisture, for a good storability (Falade and Akingbala 2010). Dried cassava can also be ground into flour for use in different food and non-food applications.

The Natural Resources Institute (NRI) of the University of Greenwich, UK, in collaboration with other local partners, has been promoting for many years the use of High Quality Cassava Flour (HQCF), especially in Africa where there is scope in the market for improved cassava processed products. HQCF is a flour of industrial quality that can be used in multiple applications such as a partial substitute for wheat flour in the baking industry or substitute for maize flour, maize or cassava starch in various industries. Good-quality roots are selected based on high dry matter content, absence of rots, root form and age (10–12 months). Roots are processed to meet specific quality criteria. For instance, roots are grated rather than chipped, to minimize cyanogenic residues. The mash is then pressed using a hydraulic press instead of a manual press as used in more rudimentary processing. Pressed wet cake is dried with a maximum loading and the final moisture content is between 10% and 12%. The milled and sieved flour should have a particle size <0.25 mm. The roots should be processed in less than 24 h to meet HQCF quality standards. HQCF can be used to substitute other types of flours, including maize or wheat, and also starch in the food industry for products such as bread, biscuits and confectionary products.

3.3 Products fermented by water submersion

Fermentation by submersion means that whole cassava or pieces of cassava are immersed in water and left to ferment for several days.

Fermentation has been used ancestrally as a means of food preservation and it is a process with a long history in cassava. Fermentation of cassava is a process by which specific and appreciated product flavours are developed, and the multiplication of lactic acid bacteria that acidifies the medium limits the proliferation of undesirable and pathogenic microorganisms (e.g. *E. coli*).

3.3.1 Fufu

Fufu refers to different products depending on the country where it is found. In Ghana, fufu is a non-fermented paste made out of roots from sweet varieties that are boiled and pounded. In Nigeria, fufu refers to a fermented dough that is made of cassava (whole or slices) steeped in water and fermented for up to three days, sieved and then cooked into a homogeneous paste. Fufu is one of the staple foods in both these countries. Fufu develops its typical taste and aroma during fermentation. A smooth texture, a characteristic aroma and a homogeneous colour are important quality criteria for the consumers (Falade and

Figure 3 Dried fufu from Nigeria exported to UK, Belgium, the Netherlands, United States, Canada and Australia.

Akingbala 2010; Oyewole and Ogundele, 2004; Tomlins et al., 2007). Fufu can also be dried to increase its shelf life and to reduce transportation costs, especially for export (Fig. 3).

3.3.2 Lafun

Lafun is a fibrous flour popular in Nigeria. When reconstituted into a paste it looks similar to fufu. In the traditional lafun preparation, fresh cassava roots are cut into chunks and steeped for 3–4 days or until the roots become soft. Chunks are then sun-dried for several days (often on the side of the road) and then milled into flour. Unlike fufu, lafun is not sieved and therefore the fibres are still present, which gives it a coarser texture (Cassava Biz 2005; Oyewole and Afolami 2001).

3.3.3 Chikwangue and 'bâton de manioc'

Chikwangue, also called cassava bread, and similar products such as 'bâton de manioc' (Fig. 4) are common forms of cassava consumption in Central Africa, in mostly French-speaking countries (i.e. Cameroon, Gabon, Congo and DRC). Chikwangue and 'bâton de manioc' are prepared by peeling and chopping roots into pieces that are left covered in water for about 2 days. Afterwards central fibres are removed, the pieces washed with clear water, squeezed and made into a smooth paste. The paste is then wrapped in young leaves (*Megaphrynium macrostachyum*, or other species of *Marantaceae* (arrowroot family)) bound with raffia grass and boiled or steamed. Sometimes the product is pre-gelatinized

Figure 4 'Bâton de manioc' ready for consumption. Source: Bechoff, A., Cameroon, 2015.

in the initial cooking and later it is fully cooked (O'Brien et al., 1992; Agbor-Egbe and Mbome 2006; Treche et al., 1995).

3.4 Solid-state fermented products

As opposed to submerged fermentation, solid-state fermentation is conducted directly on the product without addition of water. Cassava roots are generally processed into a mash that is then fermented for variable periods (1–5 days). In most cases, fermentation is a spontaneous one where the indigenous floras present in the mash are responsible for inducing the fermentation. A drying step sometimes is added after the fermentation in order to obtain a product with a longer shelf life.

3.4.1 Gari and kpokpogari

Gari is a slightly acidic and semolina-like food product that is very popular in West Africa (e.g. Benin and Nigeria). In Nigeria, gari is the main staple food and therefore plays an important role in food security. Gari is made by peeling (washing), grating, fermenting, pressing and roasting cassava. Gari can equally be eaten dry or reconstituted into a paste (using boiling water) called eba. There are different types of gari based on the duration of fermentation, the size of the granule, addition of red palm oil, etc. Levels of quality are based on the physicochemical characteristics of the roots, the hygiene practices and degree of coarseness (Achinewhu et al., 1998). Ability of gari to swell when reconstituted into eba by the addition of boiling water may also be an important quality criterion (Blanshard et al., 1993; Irtwange and Achimba 2009; Oduro et al., 2000; Owuamanam et al., 2010).

Kpokpogari is a variant product from gari that is popular in the Niger Delta area of Nigeria. During processing of kpokpogari, the starch portion of the product is separated from the fibrous portion by sifting with addition of water. The fibrous part is collected and after fermentation roasted in a hot pan. During the roasting step, the mash is constantly stirred until an intermediately dried product is obtained (kpokpogari) with granules that are larger than common gari (Sanni et al., 2005).

3.4.2 Attieke

Attieke is a semolina-like product popular in Côte d'Ivoire, similar to gari and kpokpogari. The initial steps of preparation are similar to gari but the latter steps involve sun drying and steaming instead of roasting. Attieke has a slightly sour taste and is similar to couscous in appearance (Djeni et al., 2011).

3.4.3 Agbelima

Agbelima is a fermented dough commonly consumed in West Africa (i.e. Ghana, Togo and Benin). Processing steps are peeling, grating and fermenting similarly to other solid fermented products. However, as opposed to gari and fufu products that undergo spontaneous fermentation, the preparation of agbelima involves addition of kudeme, a traditional fermentation inoculum, which aids the breaking down of the texture through the activity of a number of tissue-degrading enzymes. The product is mostly consumed in a wet form. A dried form (flour) was however developed by FRI (Food Research Institute) in Ghana, but the local consumption is still marginal and the flour is mostly promoted for export (Dziedzoave 1996; Dziedzoave et al., 1999, 2000).

In summary, the products made from cassava roots are of different types (non-fermented or fermented; dried or wet) and the choice of the process is generally based on the cyanogenic potential of the roots (either sweet or bitter varieties). Food products made using single step are boiled or fried whole roots or pieces from sweet varieties whilst products made using several steps are fufu, lafun and chikwangue, and gari, agbelima and attieke from bitter varieties mostly. Products are often regional and the preparation varies from one ethnical group to another. In addition, the type of products made are climate related: for example, wet products such as baton and fufu are mostly produced in tropical humid climates whilst products such as lafun that require sun drying are preferably made under drier and more Sahelian climates.

4 Products from cassava leaves

Cassava is mostly grown for its starchy roots. However, the stem, leaves and petioles are also edible (Lancaster and Brooks 1983; Latif and Müller 2015). The consumption of cassava leaves is not as frequent as that of cassava roots, but they are nutritious and their consumption together with the roots could add valuable proteins and vitamins to the diet. Cassava leaves are consumed mostly in Central Africa (Congo, DRC, Angola, Gabon, Central Africa, Liberia, etc.), South and South-east Asia (Indonesia, Malaysia and Sri Lanka) and Brazil (Latif and Müller 2015).

4.1 General process

The number of recipes for cassava leaves is not as considerable as for cassava roots. The basis of most recipes is similar: the first step of cassava leaves preparation generally involves chopping, grinding or pounding (i.e. for about 15 minutes) followed by cooking. Cooking duration is variable (15–60 minutes). Boiled cassava leaves have the appearance of cooked spinach. Depending on how much water is removed, the cooked product may be a thick paste or a thinner stew. There are a number of variants, based mostly on the different ingredients added to the leaves.

4.2 Different recipes

Cassava leaves play an important part in the diet in Congo, accounting for about 60% of the vegetable consumption. Cassava is considered as 'all sufficient' in Congo – a complete meal – because people get 'bread' from the roots and 'meat' from the leaves. Cooked cassava leaves are also called 'old man's meat' because little chewing is required (Achidi et al., 2005; Latif and Müller 2015). Sakasaka is a popular dish in both Congo and Sierra Leone; leaves are sometimes blanched before the pounding step (Achidi et al., 2005). Eggplants and onions can be added during pounding. During cooking, palm oil or groundnut paste may be added. Cooking time is about 60 minutes and by the end, the water is mostly evaporated. However, another variant of this dish uses bicarbonate and requires less water; cooking time is also reduced because of the bicarbonate properties (Achidi et al., 2005). Sakasaka is served as a side with meat or fish and 'baton de manioc'/ chikwangue or rice. Sometimes cassava leaves are sun-dried before pounding, as reported in Angola and DRC (Lancaster and Brooks 1983).

Although Nigeria is the main producer of cassava in the world, cassava leaves are not commonly consumed. In some areas of Nigeria, however (eastern parts: Benue State), some tribes regularly consume cassava leaves. Sweet cassava varieties called 'red sweet cassava' are selected and the youngest soft leaves from the fresh top are collected. The petiole of the leaves is reddish in nature, unlike the more common green petiole. Leaves are washed with water, squeezed and prepared in the form of egusi (melon seed). Cooking is 10–15 minutes. The dish is served with pounded yam, maize flour dough or a cassava root dish such as akpu and eba.

Researchers at IITA in Nigeria, working with HarvestPlus, are promoting the consumption of cassava leaves and have adapted recipes for amaranth leaves to cassava leaves. The condiment recipe given below is based on a typical Nigerian Yoruba recipe (see also Fig. 5).

Figure 5 Ingredients for cassava leave recipes developed in Nigeria. Source: Bechoff, A., 2013. From left to right: cassava leaf; knorr cubes; red hot pepper; dried fish; salt; onions; crayfish; and palm oil.

Ingredients for 500 g of ground leaves: Palm oil (350 g); red peppers (5 pieces) (60 g); knorr cubes (4); dried fish (213 g without bones); crayfish (160 g); salt (4 g); onions (3 medium size) sliced (260 g)

Nigerian recipe: Fry the onions in palm oil, add pepper juice. Add dried fish (without bones) and pounded crayfish. Add knorr cubes. The sauce is ready after about 10–15 minutes.

Boil the thawed ground leaves for about 10–15 minutes. Add the sauce and cook for a further 1–2 minutes. The quantity of leaves/sauce was selected by local cooks according to the Yoruba traditional practices. The ratio thawed ground leaves/sauce was 1.3/1.0.

Table 1 Proximate, vitamin and mineral composition of cassava roots and leaves. Source: Montagnac et al., 2009[a]; USDA food composition cassava, 2015[b]; Latif & Müller, 2015[c]

	Proximate composition (for 100 g)	Cassava roots[a,b]	Cassava leaves[a,] (dry weight basis[c])
Macronutrients	Food energy (kJ)	526–611	209–251
	Dry matter (g)	30–40	10–30
	Protein (g)	0.3–3.5	1.0–10.0 (17.7–38.1)
	Lipid (g)	0.03–0.5	0.2–2.9 (3.5–7.3)
	Carbohydrate (g)	35–38	7.0–18.3 (31.9–64.7)
	Dietary fibre (g)	0.1–3.7	0.5–10.0
	Ash (g)	0.4–1.7	0.7–14.5 (4–16.1)
Vitamins	B1: Thiamine (mg)	0.03–0.28	0.06–0.31
	B2: Riboflavin (mg)	0.03–0.05	0.21–0.74
	B3: Niacin (mg)	0.6–1.1	1.3–2.8
	C: Ascorbic acid (mg) (vitamin C)	15–50	60–370
	A: Carotenoids (μg)	5–35	8300–11800
Minerals	Calcium (mg)	16–176	37–708 (430–1140)
	Phosphorus (mg)	6–152	27–211 (180–420)
	Iron (mg)	0.27–14.0	0.4–8.3 (15–27)
	Potassium (g)	0.25–0.27	0.35 (1.4–2.3)
	Magnesium (g)	0.03	0.12 (0.26–0.37)
	Copper (ppm)	2.0	3.00 (40–70)
	Zinc (ppm)	14	71.00 (120–210)
	Sodium (ppm)	76	51.0 (380–1200)
	Manganese (ppm)	3	72.0 (160–250)

Note: Values are indicated on a fresh weight basis except those between brackets[c].

Table 1 summarizes the proximate, vitamin and mineral composition of cassava roots and leaves.

The chemical constituents comprise the macronutrients (carbohydrate, fat, lipid, protein and fibres) and the micronutrients (vitamins and minerals). Cassava chemical composition is determined by both genes and the environment.

5 Chemical composition of cassava roots

5.1 Macronutrients

Carbohydrate content is high (35–38% fresh weight) – a result of the high starch content of cassava. For this reason cassava roots are considered mostly as an energy food (Montagnac et al., 2009; Burns et al., 2012). Cassava contains more carbohydrates than potatoes and cereals (i.e. wheat, rice, maize and sorghum). The main simple sugars are sucrose, glucose, fructose and maltose. Total sugars range from 0.2% to 18.8%, with 3.7% on average and reducing sugars between 0 and 15.7 with 1.2% on average (Sánchez et al., 2009). In sweet cassava varieties, up to 17% of carbohydrates were sucrose and there were small amounts of dextrose and fructose (Montagnac et al., 2009). Bitter varieties were shown to have higher total sugar levels than sweet ones in a study by Padonou et al. (2005). Higher sugar levels in some bitter varieties could mask the bitter taste of the cyanogenic glucosides.

Of the carbohydrates in cassava roots, starch occupies about 85% on a dry weight basis (Sánchez et al., 2009; Sayre et al., 2011). Starch properties are related to the amylose/amylopectin ratio. Amylose content is 15–20% on average but high-amylose mutants with up to 30% amylose have been found (Ceballos et al., 2007). Because of the high proportion of starch, there is a strong positive correlation between starch content and dry matter content in roots, on a fresh basis. Starch quality and content are influenced by both genetics and growing conditions linked to the environment (soil water availability; temperature; soil type and soil nutrient status) (Sánchez et al., 2009).

Fibre content – quite low in roots – ranges from 0.1% to 3.7%, and is influenced especially by root age.

Protein and lipid contents (up to 3.5% and 0.5%, respectively) are considered low in cassava compared to seed products such as the grains. Lipid content of cassava was lower than that found in maize and sorghum; however, it was higher than that found in potato and comparable to rice (Montagnac et al., 2009). The protein content was much lower than that of maize and sorghum, which have about 10% protein. In cassava roots, half of the protein content consists of whole proteins, whereas the other half is amino acids. Some essential amino acids such as methionine, cysteine and tryptophan are present in very low quantities; nonetheless, other compounds such as arginine, glutamate and aspartate are present in substantial amounts (Montagnac et al., 2009). Chavez et al. (2000) believed that there is space for improvement in protein in cassava roots through breeding. A screening of cassava varieties (Chávez et al., 2005) from various parts of the world revealed that many varieties from Central and South America have higher protein contents in the roots, and this was believed to result from natural introgression of genes from wild relatives. The diversity of protein content found in nature for cassava roots gives hope that there are ways for higher protein cassava varieties to be developed either using conventional breeding or through genetic engineering. Nonetheless, it should be recognized that cassava storage roots evolved primarily as a means to store energy (starch) for regrowth after a stress period (primarily drought), and they are not well suited to accumulate high levels of proteins. Additionally, farmers and consumers also would have focused their selection of varieties on the basis of starch yield and quality rather than protein.

5.2 Minerals and vitamins

Cassava roots contain calcium, iron, potassium, magnesium, copper, zinc and manganese in levels comparable to those of many legumes (Montagnac et al., 2009). Chavez et al. (2000) reported average levels of 590 mg/kg of calcium on a dry matter basis.

Levels for iron and zinc of 9.6 mg/kg and 6.4 mg/kg, respectively, were reported in roots (Chavez et al., 2000). These levels were considered low to meet nutritional requirements according to Burns et al. (2012), and there is a need to improve these essential minerals in cassava. To date, there have not been concentrated attempts to improve iron and zinc through conventional breeding techniques so the potential for improvement is unknown. Limited content of these minerals in most soils where cassava is grown may limit success. Improvement of mineral content using transgenics or gene editing could also provide a more successful route.

Vitamin C (ascorbic acid) content in cassava roots is relatively high – from 15 to 45 mg/100 g in the edible portions (Montagnac et al., 2009). On the other hand, low amounts of the B vitamins (thiamin, riboflavin and niacin) were reported. Vitamins C and B are labile and their content was reported to significantly lower during processing.

Products from cassava roots are often staple foods and therefore critical to food security in sub-Saharan Africa, especially by providing the carbohydrate component of the diet. Mineral and vitamin levels in cassava roots are overall lower than in most cereals, legumes and other root and tuber crops. In addition, these levels are further reduced during processing. Cassava roots are high sources of energy but have low protein and fat, and lack essential minerals and vitamins (Montagnac et al., 2009). Products from cassava roots therefore require either genetic improvement and/or addition of other dishes that can complement these deficiencies (e.g. sauce, meat, fish and vegetables – including cassava leaves). The strategy (genetic improvement or diet diversity) will depend on the particular nutritional component and the relative access to genetic diversity versus access to diverse diets. In many cases, a combined, complementary approach can be used.

6 Chemical composition of cassava leaves

Leaves contain less carbohydrate than roots but have more nutrients that are essential to tackle hidden hunger, such as vitamins and minerals.

The nutrient composition of leaves depends on a number of factors: variety of cassava, age of the plant, age of the leaves (young to mature) and size of the leaves (Montagnac et al., 2009; Latif and Müller 2015).

6.1 Macronutrients

The carbohydrates in cassava leaves are mainly starch, with amylose content varying from 19% to 24% (Montagnac et al., 2009), levels similar but slightly higher compared to the roots. Leaves contain about three times less carbohydrate than roots but more protein, lipid, fibres, minerals and vitamins (Montagnac et al., 2009).

Although Montagnac et al. (2009) reported values of protein levels of up to 10%, Chavez et al. (2000) mentioned that levels could reach 30% (DW basis) and showed that leaf protein could be an important nutritional supplement to the roots that had very low protein content. It was reported that cassava leaves have an essential amino acid content higher than soya bean protein. The total protein content of cassava leaves is 5 to 10 times to that of roots and is equivalent to the protein content of egg (10.9 g/100 g) (Montagnac et al., 2009; Latif and Müller 2015). However, in terms of amino acid balance, methionine and isoleucine are deficient whilst lysine levels are higher when compared to egg. Cassava leaves could be a valuable source of lysine to complement lysine-deficient diets such as rice, wheat, sorghum and millet (Lancaster and Brooks 1983). However, processing can significantly reduce these levels by denaturing amino acids, as was reported by Lancaster and Brooks (1983).

6.2 Minerals and vitamins

Cassava leaves are rich in minerals (potassium, phosphorus, magnesium and calcium) and vitamins (B1, B2, C and A) (Montagnac et al., 2009).

Leaves contain about 100 times more calcium than the roots and 2–3 times more phosphorus. Leaves are also concentrated in other minerals such as potassium, magnesium, copper, zinc, iron and manganese. However, the levels of manganese, zinc, iron, copper and sodium are lower than other crops according to Latif and Müller (2015). Montagnac et al. (2009) nonetheless reported that iron and zinc levels are equivalent to those found in sweet potato leaves. Calcium is comparable to the levels found in broccoli and magnesium content is more than in broccoli but lower than in sweet potato leaves. Overall Montagnac et al. (2009) stated that the mineral content of cassava leaves was similar to that most other leaves.

The vitamin content of cassava leaves is high: Vitamins B1 (thiamin) and B3 (niacin) levels are more than that of legumes and leafy legumes with the exception of soya bean. Compared to roots levels of B1 and B3 in cassava leaves are 4 to 5 times more. In addition, B1 and B3 levels surpass some of the contents in animal products such as fresh egg, cheese and whole milk. Vitamin B2 (riboflavin) content surpasses levels of all of the above, including soya bean and is about ten times more than that of roots (Montagnac et al., 2009).

The vitamin A (carotenoids) content of cassava leaves was superior to that of other legumes and leafy legumes and the vitamin C content was also high compared to other vegetables. Levels of carotenoids (vitamin A) are about a 1000 times more in leaves compared to roots (Montagnac et al., 2009). Vitamin E is low in both leaves and roots. The processing techniques can lead to large reductions in vitamins, in particular vitamin C that is very labile. In addition, storage of cassava leaves can also lead to reduction in vitamin C (Lancaster and Brooks 1983). Vitamin B was reduced by half while boiling and only slightly reduced during steaming (Lancaster and Brooks 1983). Vitamin A or carotenoids can also be affected, slightly in boiling and steaming and more drying followed by storage. Leaves are used as a by-product, used for animal feeding or often simply discarded. They are an important part of nutrient recycling when allowed to remain in the field. Although leaves are nutritious, their use is as not as common as the roots. Reasons might be palatability and lack of knowledge of their nutritional properties, but also the higher levels of anti-nutrients compared to the roots, which requires exhaustive processing to reduce them to a safe level.

7 Anti-nutrients in roots and leaves

The main nutritional constraint of both roots and leaves is the need for detoxification due to the presence of cyanogenic compounds. It should be noted that cassava leaves also contain other anti-nutrients such as high fibre content, tannin, polyphenols and phytate that reduce nutrient bioavailability (Latif and Müller 2015). Methionine availability in particular is negatively influenced by the presence of tannins (Latif and Müller 2015).

Non-protein nitrogen components in cassava roots are nitrite, nitrate and cyanogens. Of all the anti-nutrients, cyanogens are the most toxic compounds present in the cassava leaves and roots. These can be found under three different forms: cyanogenic glucosides (95% linamarin and 5% lotaustralin), cyanohydrin and free cyanide. Cyanogenic compounds can cause an upper moto-neuron disease called konzo. The disease results in cerebral damage and epidemic paralysis that is irreversible. Ruminant animals are less vulnerable to these compounds because these animals have an ability to detoxify the leaves in their digestive system.

Cyanogenic potential in the roots varies between 14 ppm and 3274 ppm, with 340 ppm on average (Sánchez et al., 2009). The concentration of cyanogenic compounds in the leaves is 5–20 times that of the roots, and leaves also have 200 times more linamarase activity (Bokanga 1994).

Figure 6 describes the different steps of cassava detoxification. Cyanogen glucoside (95% linamarin and 5% lotaustralin) hydrolysis occurs in cassava mash in presence of linamarase (alpha-β-glucosidase), an enzyme naturally present in cassava. Linamarase releases acetone cyanohydrin and glucose. Subsequently, acetone cyanohydrin either can be spontaneously cleaved into acetone and hydrogen cyanide or an enzyme (lyase) can act as a catalyst to the reaction. While hydrogen cyanide is readily removed during cassava processing, the presence of residual linamarin and its breakdown product (acetone cyanohydrin) in cassava-based food products has been a cause for concern because of their possible effects on health. Acetone cyanohydrin is stable under acidic conditions and unstable under alkaline conditions.

The hydrolytic enzymes (linamarase and hydroxynitrile lyase) are situated in the cell walls of cassava and are released when the cell matrix is disrupted, for example, by

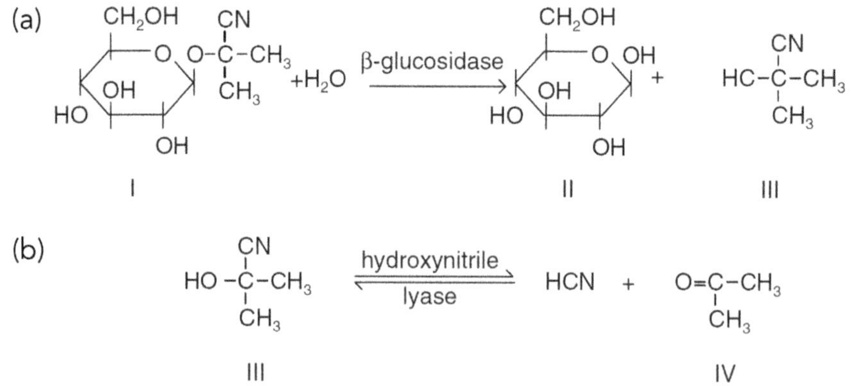

Figure 6 (a) Enzymatic breakdown of linamarin (compound I); glucose (compound II); acetone cyanohydrin (compound III), (b) cyanide (HCN) and acetone (compound IV) (Westby 2002).

processes such as grating roots or pounding leaves (Latif and Müller 2015), allowing the substrate (cyanogens) to be catalyzed. Thus the initial grating step is important to help detoxify, that is, to reduce cyanogens. In addition, heating can help cyanide (gas form) to be liberated from the product matrix. By reducing the amount of residual cyanide, heating and physical cell disruption are important to make cassava products safer.

8 Case study: biofortified cassava roots made into gari and fufu in Nigeria

8.1 Project outline

Vitamin A deficiency is a major public health issue in developing countries and especially in sub-Saharan Africa. In Nigeria, it is estimated that 20% of pregnant women and about 30% of children under five are vitamin A deficient (Mainasara 2011). HarvestPlus is currently developing and promoting biofortified cassava (yellow cassava) (Fig. 7) that contains

Figure 7 Peeled roots of yellow biofortified cassava and leaves. Source: Bechoff, A., 2013, Nigeria.

provitamin A carotenoids that could contribute to reducing vitamin A deficiency. Varieties of yellow cassava contain between 5 ppm and 25 ppm (Ceballos et al., 2013) of total carotenoids, whilst non-biofortified (white) varieties have only 0.05–0.35 ppm (Table 1).

The HarvestPlus-funded project described here (2012–14), led by the NRI, UK investigated the retention of carotenoids and consumer acceptability of yellow varieties of cassava in Nigeria. These yellow varieties were processed into gari and fufu using the traditional methods applied to the local (white) varieties. (Fig. 8 and 9)

Figure 8 Cassava fufu made from a yellow biofortified variety. Source: Bechoff, A., 2013, Nigeria.

Figure 9 Gari and eba from yellow biofortified cassava. Source: Bechoff, A., 2013, Nigeria.

8.2 Project outcomes

In its raw state, yellow-fleshed cassava mainly contains provitamin A carotenoids. These have been reported to be trans-β-carotene, 13-cis and 9-cis-β-carotene (Fig. 10) (Rodriguez-Amaya and Kimura 2004; Chavez et al., 2007; Thakkar et al., 2009).

Provitamin A carotenoids are chemically unstable molecules that are affected by external factors such as temperature, light and oxygen. (Rodriguez-Amaya and Kimura 2004). Hence provitamin A levels in yellow-fleshed cassava can be affected by processes such as gari and fufu production. Therefore, measuring the retention of provitamin A during/after processing (i.e. gari and fufu) is important to ensure that the biofortified food retains provitamin A levels that could contribute to a significant part of the vitamin A intake of the people who consume it.

Bechoff et al. (2015) demonstrated that carotenoids in gari degraded quickly during storage at ambient temperature: carotenoid content, and hence vitamin A activity of the gari products, decreased markedly with storage time and temperature (Table 2).

This work advises a limited storage period of gari products from yellow cassava at ambient temperature (25 days maximum). Gari, eba and fufu made from yellow cassava were well accepted by the local population, but the main issue was the limited shelf life of gari from biofortified cassava. The research overall showed that yellow cassava products have a significant carotenoid content, were well accepted by the local population and could therefore contribute to tackling vitamin A deficiency. These project results have implications for the promotion of yellow biofortified cassava to farmers, processors and different stakeholders in developing countries.

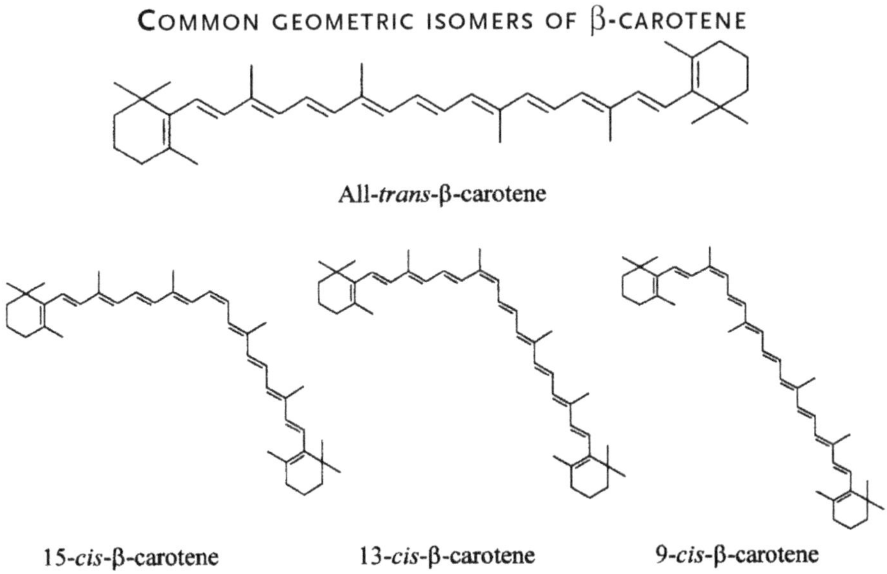

Figure 10 Isomers of β-carotene. Source: Rodriguez-Amaya and Kimura (2004).

Table 2 Estimation of vitamin A activity (μg retinol equivalent) for a 100 g portion of gari made from yellow biofortified cassava (BG) and from white cassava with palm oil (RPG) (μg RAE[a]) during storage and maintained at constant humidity (a_w = 0.5 for NaBr)

RAE[a] estimate		5:1[b]		12:1[b]	
T (°C)	Storage days	BG	RPG	BG	RPG
	Initial	301±7	203±6	126±3	85±2
19	24	256±11	209±11	107±5	87±4
	49	205±5	161±24	85±2	67±10
	60	200±7	175±13	83±3	73±5
	80	156±9	140±18	65±4	58±8
26	18	245±23	181±8	102±10	75±3
	24	250±26	190±11	104±11	79±4
	49	150±29	135±7	63±12	56±3
	60	130±20	154±9	54±8	64±4
	80	99±15	130±12	41±6	54±5
33	10	201±14	148±16	84±6	61±6
	18	204±12	200±6	85±5	83±2
	24	172±15	189±8	72±6	79±3
	49	87 ±7	137±10	36±3	57±4
	60	58±2	123±6	24±1	51±2
	80	41±3	93±4	17±1	39±2
40	10	148±26	164±5	62±11	68±2
	18	140±7	183±5	58±3	76±2
	24	99±3	157±9	41±1	65±4
	31	68±2	144±7	28±1	60±3
	49	29±2	99±3	12±1	41±1
	60	21±1	80±7	9±0.4	33±3
	80	12±0.2	42±1	5±0.1	18±0.3

Mean of triplicate determinations ± standard deviation. [a]Retinol Activity Equivalent. [b]RAE was calculated for a bioconversion factor of 5:1 (La Frano et al., 2013) estimate = (All-trans-β-carotene content/5 + minor β-carotene content/10) × unit (g) or for a bioconversion factor of 12:1 (National Academy of Sciences/Institute of Medicine 2001). Classical estimate = {All-trans- β-carotene content/12 + minor β-carotene content/24} × unit (g). Minor compounds are epoxy and cis β-carotene that are estimated to possess half of trans β-carotene activity.

9 Future trends

The presence of anti-nutrients, especially cyanogenic compounds, is a major factor for the safe consumption of cassava roots and leaves. Adequate processing is required to detoxify these products to an acceptable and safe level. However, extensive processing can also result in loss of nutritional components such as vitamins (i.e. vitamin A, B and C), denaturation of the proteins and loss of the minerals due to leaching. Retention studies are needed to better understand effects of specific processes and storage conditions. On the one hand, research should look at the preservation of nutritive value in cassava roots and leaves; on the other hand, it should consider reaching a safe level for detoxification of cyanogenic compounds.

Cassava roots are poor in micronutrients such as vitamin A, zinc and iron and also in protein. Breeders have been working on improving the nutritional value of cassava roots using mainly conventional breeding, but also genetic engineering. Thanks to breeding efforts, carotenoid content in the cassava roots have reached a level that was beyond initial expectations (up to 25 ppm). There is still considerable debate among nutritionists and breeders with regard to best approaches to improve mineral and protein nutrition in deficit populations – whether to focus on diet diversity or biofortification through breeding. Both approaches have merit and challenges, and some combination seems prudent.

Cassava leaves are more nutritious but their use is still unexploited in places where cassava is mostly grown for its roots. Leaves may be underutilized because of their high level of toxicity compared to roots, but there might also be a lack of awareness in local populations who consume roots about the nutritional content and palatability of leaves. There is potential to promote cassava leaves; however, research on the types of dishes suiting the local population, consumer acceptance of cassava leaves and modifying food habits is required. Education on the uses of cassava leaves (including farming management of the leaf harvest to optimize and balance income and nutrition through combined leaf and root yields) should be undertaken as part of the promotion of cassava leaves.

10 Conclusion

Cassava is an important staple crop in developing countries, and contributes to food security of millions of people, especially poor smallholder farmers in sub-Saharan Africa. Products from cassava roots are incredibly diverse and vary from country to country, and among ethnic groups. The types of products are linked to tradition, convenience, access to processing resources and, very importantly, the need for detoxification of cassava roots and leaves that contain cyanogenic glucosides. Roots of sweet varieties (low cyanogen content) require little processing; a single step such as boiling or frying is sufficient to bring the level under the safe level of 10 ppm recommended by the Food Agriculture Organization. On the other hand, bitter varieties (average to high cyanogen content) of roots require more extensive processing. Grinding, pounding and heating are among the steps that help detoxify the cassava by enzymatic catalysis of cyanogens (releasing naturally present enzymes in the cassava) and liberation of cyanide. Products can be fermented by submersion in water or as a mash (solid-state fermentation). Some products are dried (i.e. gari, attieke and flour) and others are processed as 'wet' (fufu, agbelima, chikwangue, etc.). Processes for cassava leaves are not as varied as for roots; the general steps are grinding or pounding, followed by cooking. The variants in leaf processing are mostly from the different ingredients added to the leaves and the duration of cooking. Roots and leaves are good sources of potassium, magnesium, and vitamins B and C. Whilst roots mainly consist of carbohydrate (starch), leaves have considerably more protein (i.e. especially high lysine content), and also carotenoids that could be complementary from a nutritional point of view. Both roots and leaves however are deficient in methionine, an important amino acid for the diet. In addition, levels of zinc and iron are not high enough for roots and leaves to be a substantial source of these essential minerals. Efforts have been made to develop and promote biofortified varieties that contain higher carotenoid contents in the roots (giving them a yellow colour), and there are hopes that these varieties could help tackle vitamin A deficiency.

Nutritional content of cassava varieties can be improved over current levels by 'breeding'. Researchers also need to look at selection in relation to the consumer taste and acceptance in order to promote more nutritional varieties (e.g. biofortified cassava containing carotenoids). Cassava's nutritional drawbacks have led national and international research institutes such as CGIAR, NGOs and the private sector to look for appropriate strategies to solve these challenges using different approaches. Examples are the BioCassavaPlus programme that is looking at improving the poor nutritional content by developing nutritionally improved varieties using transgenic strategies. Targets for improvement are low cyanogenic potential, pest and disease resistance. PPD resistance, higher and better protein content, and enhanced micronutrient content (i.e. iron, zinc and carotenoids) (Sayre et al., 2011). Other approaches use conventional breeding techniques to improve micronutrients of interest in the cassava roots (i.e. HarvestPlus) (Bouis and Welch 2013). Breeding for nutrition should focus on the people who are food insecure, and on the specific nutrients for which they have the most critical deficits. Education, promotion and effective scaling out of appropriate technologies need to be combined and optimized in order to increase adoption of these varieties by nutritionally vulnerable people.

Understanding potential market demand is paramount. For instance, are there new or improved forms of products from cassava roots and leaves that would be suitable in particular for the growing urban populations in sub-Saharan Africa? This might include dried forms of some of the wet products that could be prepared more quickly than the traditional products (e.g. fufu). The demand for competing products to cassava for starch, such as potato, sweet potato, maize and rice should also be considered. In addition, exported cassava of better quality could compete with locally grown cassava, and provide income for diversified diets as compared to direct consumption and high dependence on cassava. These points should be taken into account for the development of cassava products. Cassava leaves also have potential to be used to increase protein and vitamin intake but are still very much underexploited. For instance, the development of cassava leaf protein concentrate could be explored (Latif and Müller 2015). This could help tackle protein and vitamin A malnutrition in some areas where currently only the roots are used for food. Cassava has the potential for impact on nutrition and health through diverse and nutritionally improved products from roots and leaves, and through generation of income from the sales of cassava that can be used indirectly to improve nutrition and health. In order to achieve this, consumers' tastes, nutritional requirements and market demand, all need to be met.

11 Where to look for further information

Research Programme on Roots, Tubers and Bananas. Consultative Group for International Agricultural Research (CGIAR) http://www.rtb.cgiar.org

Natural Resources Institute website and newsletter http://www.nri.org [NRI awarded Queen's Anniversary Prize for ground-breaking work on cassava in 2015]

International Society for Tropical Root Crops http://www.istrc.org/

12 References

Achinewhu, S. C., Barber, L. I. and Ijeoma, I. O. (1998). Physicochemical properties and garification (gari yield) of selected cassava cultivars in Rivers State, Nigeria. *Plant Foods Hum. Nutr.*, 52, 133–40.

Agbor-Egbe, T. and Mbome, I. L. (2006). The effects of processing techniques in reducing cyanogen levels during the production of some Cameroonian cassava foods. *J. Food Compos. Anal.*, 19, 354–63.

ARS:846 (2012). Food grade cassava starch specifications CD-ARS 846. First edition 2012. ARSO-201. African Organization for Standardization. https://law.resource.org/pub/ars/ibr/ars.846.2012.pdf.

Bechoff, A., Chijioke, U., Tomlins, K. I., Govinden, P., Ilona, P., Westby, A., and Boy, E. (2015). Carotenoid stability during storage of yellow gari made from biofortified cassava or with palm oil. *J. Food Comp. Ana.*, 44, 36–44.

Blagbrough, I. S., Bayoumi, S. A., Rowan, M. G. and Beeching, J. R. (2010). Cassava: an appraisal of its phytochemistry and its biotechnological prospects. *Phytochemistry*, 71(17), 1940–51.

Blanshard, A. F., Dahniya, M. T., Poulter, N. H. and Taylor, A. J. (1993). Quality of cassava foods in Sierra Leone. *J. Sci. Food Agric.*, 64(4), 425–32.

Bokanga, M. (1994). Processing of Cassava Leaves for Human Consumption. In International Workshop on Cassava Safety 375, pp. 203–8.

Bouis, H. E. and Welch, R. M. (2013). Biofortification – a sustainable agricultural strategy for reducing micronutrient malnutrition in the global south. *Crop Sci.*, 50, S21–32.

Burns, A. E., Gleadow, R. M., Zacarias, A. M., Cuambe, C. E., Miller, R. E. and Cavagnaro, T. R. (2012). Variations in the chemical composition of cassava (Manihot esculenta Crantz) leaves and roots as affected by genotypic and environmental variation. *J. Agric. Food Chem.*, 60(19), 4946–56.

Cassava Biz (2005). Information service provided by the Integrated Cassava Project (ICP) of the International Institute of Tropical Agriculture (IITA) to encourage, promote and expand agribusiness development in the cassava subsector in Nigeria. http://www.cassavabiz.org/index.asp.

Ceballos, H., Sánchez, T., Denyer, K., Tofiño, A. P., Rosero, E. A., Dufour, D., Smith, A., Morante, N., Perez, J. C. and Fahy, B. (2007). Induction and identification of a small-granule, high-amylose mutant in cassava (Manihot esculenta Crantz). *J. Agric. Food Chem.* 56(16), 7215–22.

Ceballos, H., Morante, N., Sánchez, T., Ortiz, D., Aragón, I., Chávez, A. L., Pizarro, M., Calle, F. and Dufour, D. (2013). Rapid cycling recurrent selection for increased carotenoids content in cassava roots. *Crop Sci.*, 53(6), 2342–51.

Charoenkul, N., Uttapap, D., Pathipanawat, W. and Takeda, Y. (2006). Molecular Structure of Starches from Cassava Varieties having Different Cooked Root Textures. *Starch*, 58(9), 443–52.

Chavez, A. L., Sanchez, T., Ceballos, H., Rodriguez-Amaya, D. B., Nestel, P., Tohme, J. and Ishitani, M. (2007) Retention of carotenoids in cassava roots submitted to different processing methods. *J. Sci. Food Agric.*, 87, 388–93.

Chávez, A. L., Bedoya, J. M., Sánchez, T., Iglesias, C., Ceballos, H. and Roca, W. (2000). Iron, carotene, and ascorbic acid in cassava roots and leaves. *Food Nutr. Bull. Suppl.*, 21(4), 410–3.

Chávez, A. L., Sánchez, T., Jaramillo, G., Bedoya, J., Echeverry, J., Bolanos, E. A., Ceballos, H. and Iglesias, C. A. (2005). Variation of quality traits in cassava roots evaluated in landraces and improved clones. *Euphytica*, 143(1–2), 125–33.

Djeni, N. T., N'Guessan, K. F., Toka, D. M., Kouame, K. A. and Dje, K. M. (2011). Quality of attiéké (a fermented cassava product) from the three main processing zones in Côte d'Ivoire. *Food Res. Int.*, 44(1), 410–6.

Dziedzoave, N. T. (1996). *Quality control studies on agbelima – development of quality specification and evaluation of cassava varieties for processing*. MSc thesis submitted to the Department of Biochemistry, University of Science and Technology, Kumasi, Ghana.

Dziedzoave, N. T., Ellis, W. O., Oldham, J. H. and Osei-Yaw, A. (1999). Subjective and objective assessment of agbelima (cassava dough) quality. *Food Control*, 10(2), 63–7.

Dziedzoave, N. T., Ellis, W. O., Oldham, J. H. and Oduro, I. (2000). Optimizing agbelima production: varietal and fermentation effect on product quality. *Food Res. Int.*, 33(10), 867–73.

Falade, K. O. and Akingbala, J. O. (2010). Utilization of cassava for food. *Food Rev. Int.*, 27(1), 51–83.

FAOSTAT (2013). Statistical Databases. Food and Agriculture Organization of the United Nations. *Cassava production and consumption in 2013*. http://faostat.fao.org/(accessed on 10 February 2015).

Hongbété, F., Mestres, C., Akissoë, N., Pons, B., Hounhouigan, D. J., Cornet, D. and Mathurin, N. C. (2011). Effects of cultivar and harvesting conditions (age, season) on the texture and taste of boiled cassava roots. *Food Chem.*, 126(1), 127–33.

Irtwange, S. V. and Achimba, O. (2009). Effect of the duration of fermentation on the quality of Gari. *Curr. Res. J. Biol. Sci.*, 1(3), 150–4.

La Frano, M. R., Woodhouse, L. R., Burnett, D. J., Burri, B. J. (2013). Biofortified cassava increases β-carotene and vitamin A concentrations in the TAG-rich plasma layer of American women. *Br. J. Nutr.*, 110, 310–20

Lancaster, P. A., and Brooks, J. E. (1983). Cassava leaves as human food. *Economic Botany*, 37(3), 331–48.

Latif, S. and Müller, J. (2015). Potential of cassava leaves in human nutrition: A review. *Trends in Food Sci. & Tech.*, 44(2), 147–58.

Mainasara, A. (2011). Scientists develop Vitamin A cassava in Nigeria, to tackle malnutrition in Africa. Africa Science News. http://www.africasciencenews.org/en/index.php/life-and-style/49-food/261-scientists-develop-vitamin-a-cassava-in-nigeria-to-tackle-malnutrition-in-africa-.

Maziya-Dixon, B., Dixon, A. G. O. and Ssemakula, G. (2009). Changes in total carotenoid content at different stages of traditional processing of yellow-fleshed cassava genotypes. *Int. J. Food Sci. Tech.*, 44, 2350–7.

Montagnac, J. A., Davis, C. R. and Tanumihardjo, S. A. (2009). Nutritional value of cassava for use as a staple food and recent advances for improvement. *Compr. Rev. Food Sci. Food Saf.*, 8(3), 181–94.

National Academy of Sciences/Institute of Medicine (2001). Vitamin A. In Dietary Reference Intakes for Vitamin A, Vitamin K, Arsenic, Boron, Chromium, Copper, Iodine, Iron, Manganese, Molybdenum, Nickel, Silicon, Vanadium, and Zinc. Washington, DC, USA: National Academy Press.

O'Brien, G. M., Mbome, L., Taylor, A. J. and Poulter, N. H. (1992). Variations in cyanogen content of cassava during village processing in Cameroon. *Food Chem.*, 44(2), 131–6.

Owuamanam, C. I., Iwouno, J. O., Ihediohanma, N. C. and Barber, L. I. (2010). Cyanide reduction, functional and sensory quality of gari as affected by pH, temperature and fermentation time. *Pak. J. Nutr.*, 9(10), 980–6.

Oyewole, O. B. and Odunfa, S. A. (1992). Extracellular enzyme activities during cassava fermentation for 'fufu' production. *World J. Microbiol. Biotechnol.*, 8(1), 71–2.

Oyewole, O. B. and Ogundele, S. L. (2004). Effect of length of fermentation on the functional characteristics of fermented cassava 'fufu'. *Journal of Food Technology in Africa*, 6(2), 38–40.

Padonou, W., Mestres, C. and Nago, M.C. (2005). The quality of boiled cassava roots: instrumental characterization and relationship with physicochemical properties and sensorial properties. *Food Chem.*, 89(2), 261–70.

Rodriguez-Amaya D.B. and Kimura M. (2004). *Harvestplus Handbook for Carotenoid Analysis*. Copyright HarvestPlus, Washington, DC, 63p.

Sánchez, T., Salcedo, E., Ceballos, H., Dufour, D., Mafla, G., Morante, N., Calle, F., Pérez, J. C., Debouck, D., Jaramillo, G. and Moreno, I. X. (2009). Screening of starch quality traits in cassava (Manihot esculenta Crantz). *Starch*, 61(1), 12–19.

Sánchez, T., Dufour, D., Moreno, I. X. and Ceballos, H. (2010). Comparison of pasting and gel stabilities of waxy and normal starches from potato, maize, and rice with those of a novel waxy cassava starch under thermal, chemical, and mechanical stress. *J. Agric. Food Chem.*, 58(8), 5093–9.

Sanni, L., Maziya-Dixon, B., Iwuoha, C., Ilona, P., Akoroda, M., Lemchi, J. and Ogbe, F. (2005). *Yield, proximate, and sensory properties of Kpokpogari from 43 Cassava Mosaic Disease varieties*. In African Crop Science Conference Proceedings, 7, 597–602.

Sayre, R., Beeching, J. R., Cahoon, E. B., Egesi, C. Fauquet, C., Fellman, J., Fregene, M., Gruissem,W., Mallowa, S. Manary, M., Maziya-Dixon, B., Mbanaso, A., Schachtman, D. P., Siritunga, D., Taylor, N., Vanderschuren, H. and Zhang, P. (2011). The BioCassava Plus Program: biofortification of cassava for sub-Saharan Africa. *Annu. Rev. Plant Biol.*, 62, 251–72.

Thakkar, S. K., Huo, T., Maziya-Dixon, B. and Failla, M. L. (2009). Impact of style of processing on retention and bioaccessibility of B-carotene in cassava (Manihot esculanta, Crantz). *J. Agric. Food Chem.*, 57, 1344–8.

Tomlins, K., Sanni, L., Oyewole, O., Dipeolu, A., Ayinde, I., Adebayo, K. and Westby, A. (2007). Consumer acceptability and sensory evaluation of a fermented cassava product (Nigerian fufu). *J. Sci. Food Agric.*, 87(10), 1949–56.

Treche, S., Avoumpo, E. and Adoua-Oyila, G. M. (1995). *Notoriété et acceptabilité de la chikwangue Agricongo à Brazzaville*, pp. 218–30. Eds., ORSTOM, Montpellier, France.

USDA (2015). *National Nutrient Database for Standard Reference* http://ndb.nal.usda.gov/ndb/foods/show/2943?fg=andman=andlfacet=andcount=andmax=25andsort=andqlookup=cassavaandoffset=andformat=Statsandnew=andmeasureby=. Accessed 10 February 2015.

Westby, A. (2002). Cassava utilization, storage and small-scale processing. In R. J. Hillocks, J. M. Thresh and A. C. Bellotti (eds), *Cassava Biology, Production and Utilization*, pp. 281–300, Oxon, UK: CABI Publishing.

Chapter 3

Cassava for industrial uses

Kuakoon Piyachomkwan and Sittichoke Wanlapatit, National Center for Genetic Engineering and Biotechnology, Thailand; and Klanarong Sriroth, Kasetsart University, Thailand

1 Introduction

2 Cassava root composition

3 Cassava chips

4 Cassava starch: production and properties

5 Cassava starch: products

6 Cassava starch: industrial applications

7 Summary

8 Where to look for further information

9 References

1 Introduction

Cassava is a shrubby perennial crop which produces underground edible roots rich in starch. The crop is extensively grown in tropical and subtropical areas including Latin America, Africa and Asia. Total root production was about 270 million tons in 2014 (www.faostat.fao.org). In the past, cassava was recognized as 'food for the poor' because the crop had been cultivated and utilized mainly as the staple food for human consumption in cassava-growing areas of many developing countries. In 2013, the Food and Agriculture Organization of the United Nations (FAO) announced that cassava had emerged as a multipurpose crop for the twenty-first century, responding to the priorities of developing countries, to trends in the global economy and to the challenge of climate change. This is because:

- Cassava has a good tolerance to poor growing environments, for example drought, infertile soil, pests and insects. The crop can still produce reasonable yields, using few if any inputs under poor growing conditions. Nevertheless, cassava also responds well to better cultivation practices including soil management, irrigation, fertilizer application and higher-yielding varieties. In some situations, cassava responds well to irrigation, but the crop is rarely irrigated, except extensively in southern India and in few other localized regions.

http://dx.doi.org/10.19103/AS.2016.0014.02

- Cassava roots are very rich in starch and can be employed as a staple food. It also serves as a raw material for the production of flour and starch. Cassava starch has a wide range of industrial applications, including food, paper, textile, plywood, pharmaceuticals and biofuel.

In this chapter, current uses of cassava in the industry as well as its potential uses in the future are discussed, in particular for the bio-based economy.

2 Cassava root composition

Fresh roots of cassava are harvested at 10–12 months after planting. The chemical composition of fresh roots of cassava versus other commercial crop sources is shown in Fig. 1a. Similar to potato tubers, fresh cassava roots contain very high moisture contents (>65%), whereas cereal grains – that is, maize, wheat and rice – contain very low moisture content (12–14%). Consequently, fresh cassava roots can readily deteriorate and have a very short shelf life. On a dry weight basis, starch is the most abundant component found in these starchy roots, tubers and grains (Fig. 1b). Relative to cereal grains, cassava roots contain fewer quantities of protein and lipid, but higher contents of fibre and ash.

Fresh cassava roots contain cyanogenic compounds, namely linamarin and lotaustralin. When plant tissues are damaged, these cyanogenic glucosides can be hydrolysed by an endogenous linamarase enzyme, to liberate glucose and acetone cyanohydrin compounds, which then further decompose to volatile hydrogen cyanide (Padmaja, 1995). The contents of cyanogenic compounds in cassava roots vary depending on variety, harvest time, environmental conditions of growth and cultural practices (Sriroth et al., 1999; Satisopasri et al., 2001). Based on the cyanogenic contents in edible roots, cassava can be classified into three broad groups: low toxicity (or sweet type), medium toxicity and high toxicity (or bitter type), with the cyanide content of <50, 50–100 and >100 mg HCN equivalent.kg^{-1} fresh weight, respectively (Jansz and Uluwaduge, 1997). Sweet cassava is generally used as a staple food for direct consumption, whereas the bitter types are mostly processed to chips and starch for applications in food and non-food industries.

The major deterioration of fresh cassava roots, known as post-harvest physiological deterioration (PPD), is rapid, beginning within 24–48 hours after harvest. The rate of PPD of fresh cassava roots depends on root quality and storage conditions, for example temperature and relative humidity. There are two stages in PPD: primary and secondary. Primary deterioration starts from the central vascular bundles of the root, spreads to the adjacent storage parenchyma and subsequently the stored starch undergoes structural changes. The visible signs of primary deterioration are vascular streaking with a blue or black discolouration that renders the inferior quality of roots. This initial, rapid deterioration is physiological and biochemical and has been strongly associated with mechanical damage, which occurs during harvesting and handling of fresh roots. Secondary deterioration is due to infection with microorganisms, leading to fermentation and softening of the root tissue. Consequently, cassava roots need to be consumed or processed soon after harvesting.

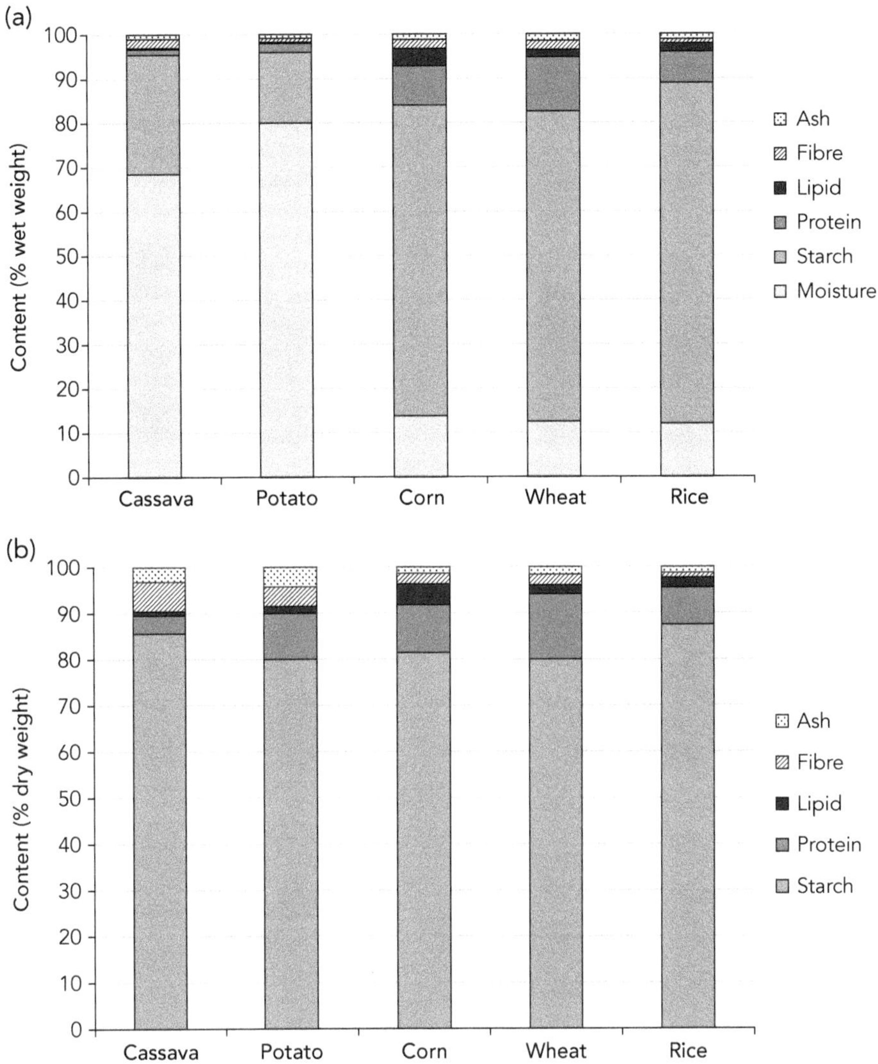

Figure 1 Chemical composition of important commercial crop sources: (a) on wet weight of fresh materials including cassava roots, potato tubers and grains of corn, wheat and rice and (b) on dry weight.

3 Cassava chips

Cassava chips are a dried form of cassava roots. Fresh cassava roots, after being harvested, get spoiled rapidly. The simplest means of extending the root's shelf life is by drying, which can lower the moisture content. The producers of cassava chips can be farmers or small businessmen, and are normally located in close proximity to the growing area.

3.1 Production

In a typical process as shown in Fig. 2, the roots are transferred from the growing area to the chip yards. Fresh roots are then loaded to the hopper of the chopping machine and are chopped into small pieces or chips. Optionally, in a large-scale processor where premium quality chips are produced, sand and soil removal is introduced prior to root chopping, by adapting machinery from cassava starch processing. After chopping, the

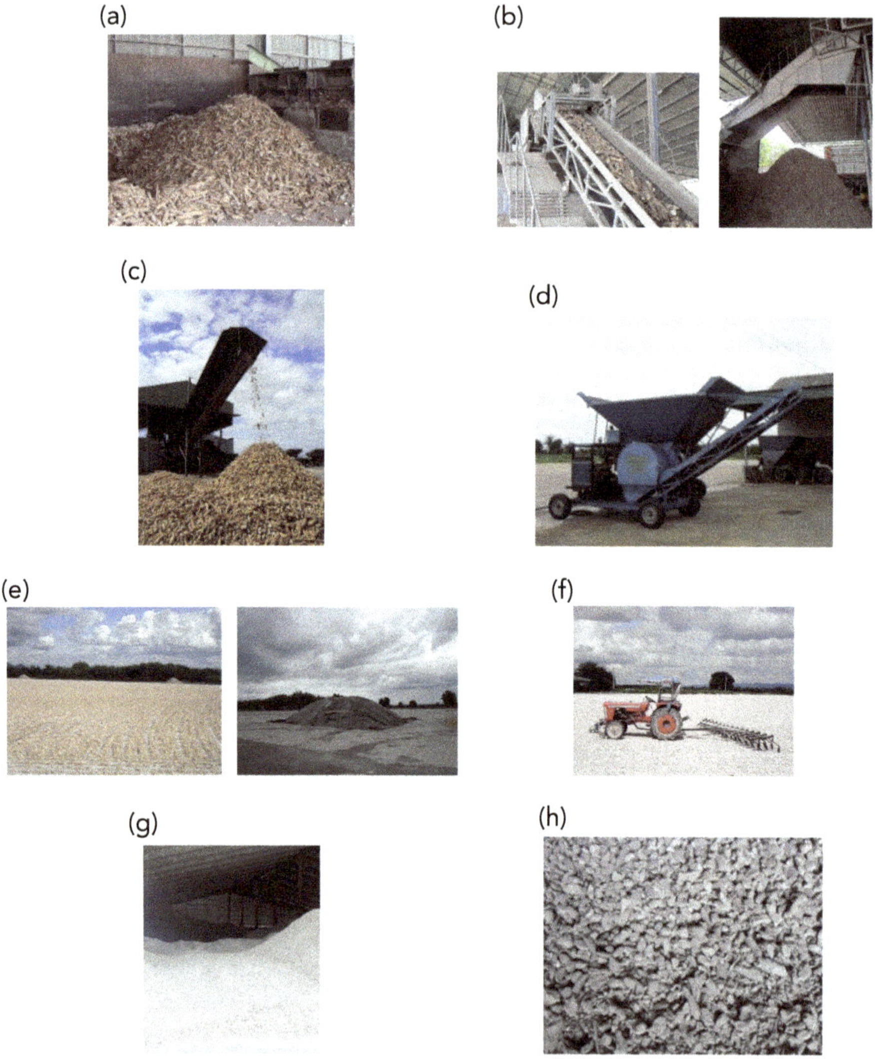

Figure 2 Process of making dried cassava chips: (a) roots are delivered to a factory; (b) soil and sand are removed; (c) clean roots; (d) root chopper; (e) chips are sun-dried and a stack of chips is covered with a canvas sheet to prevent from rain; (f) chips are turned over for proper drying; (g) chips are stored in a warehouse; (h) premium cassava chips. Source: http://www.fp7-gratitude.eu.

chips are sun-dried on a cement floor. By hands or a tractor with rakes, chips are spread (approximately 4–6 kg/m^2), and during drying, which typically requires 2–3 days, chips are turned over periodically, around 4–5 times a day, ensuring effective and uniform drying in order to obtain specified moisture contents of dried products. A very high moisture content of finished products is an undesired quality and it can create problems during chip storage in a warehouse. Moreover, it is too difficult to process dried chips during rainy periods. Before rain arrives, chips must be quickly piled up and must be covered with plastic sheets. This leads to longer drying time to properly dry the chips and accordingly results in lower chip quality. The final moisture content should be below 14%. It normally takes 2.40–2.50 kg of fresh roots (with 25% starch content) to produce 1 kg of chips (14% moisture content), depending on dry matter content of raw materials.

3.2 Quality

Chips should contain a moisture content less than 14% to prevent decomposition by heat, if stored with inefficient air ventilation. Sand and soil contamination is another critical quality of dried chips, since roots are in soil and are not washed when processed to chips, except for high-quality chips as described above. Physically, chips should not have unusual colour and odour, spoilage or moulds and other living organisms. According to the regular standard of cassava chips, the regulation of the Ministry of Industry of Thailand, the largest manufacturer and exporter of dried cassava chips is given in Table 1. Premium cassava chips contain less sand and fibre content than the standard products. Premium quality can be achieved by the additional step of sand and soil removal prior to root chopping (Fig. 2b). Premium cassava chips, although having a higher price, are preferred for some applications such as the bioethanol industry since sand can causemachine

Table 1 Thai standards for cassava chips

	Cassava chips*	Premium cassava chips**	Premium cassava chips for animal feed***
Starch (%w/w, minimum)			
- By polarimetric method	65	70	70
- By nitrogen free extract (NFE)	Not specified	75	75
Protein (%w/w, minimum)	Not specified	Not specified	2
Fibre (% w/w, maximum)	5	4	4
Moisture (% w/w, maximum)	14	13	13
Sand and soil (% w/w, maximum)	3	2	2
Dust (% w/w, maximum)	Not specified	Not specified	8
Unusual odour or colour	No	No	No
Spoilage or moulds	No	No	No
Living subjects or insects	No	No	No

*Source: Department of Foreign Trade, Ministry of Commerce, Thailand (2002)
**Source: Department of Foreign Trade, Ministry of Commerce, Thailand (2001)
***Source: Uthai Kanto and Sukanya Juttupornpong (2005) Cassava in animal nutrition

corrosion, shortening its shelf life. Additionally, when chips are used as animal feed, other qualities of chips, such as protein and cyanide contents, are recommended. In accordance with FAO/WHO, the cyanide content in cassava chips applied in animal feed should be not more than 100 mg/kg.

3.3 Uses

Animal feed

Cassava chips are a good source of energy due to their high content of digestible starch. In some countries, such as Thailand, where cassava is available at low cost relative to other energy sources, cassava chips are used to substitute maize and broken rice in animal feed. Furthermore, no or very low amounts of aflatoxin are found in cassava chips. Research has shown other benefits including healthier animals fed by cassava chips and less use of antibiotics in feed formulas (Kanto and Juttupornpong, 2002). However, as previously described, cassava contains less protein and lipid than corn and rice (Fig. 1). A supplement with alternative protein and lipid sources (e.g. *Leucaena* leaves, cassava leaves, soybean meal, fish meal, oil palm kernel and rice bran) is required to formulate complete nutritional formulas, which are varied for different types of animals. Feed formulas containing cassava chips for chickens (chick starter-layer, chick grower-layer, chick developer-layer, layer, breeder chicken, native crossed), duck (meat-type duck, duck layer, muscovy duck), swine, dairy cattle, beef cattle and goats have already been developed (Bureau of Animal Nutrition Development, Department of Livestock Development, Ministry of Agriculture and Cooperatives, Thailand). Some examples are presented in Table 2.

Cassava pellets

The development of this product was stimulated by a need to improve the uniformity in shape and size of cassava chips as required by the compound feed producers/users. Chips have very low bulk density and during transportation, loading and unloading of chips, dust generation causes serious air pollution, in particular when chips are handled at the ports. Therefore, the chips are pelletized to produce cassava pellets in order to improve the quality of export products. Production of pellets involves pressing ground chips and then extrusion through a large die. The heat and moisture in the chips helps in the formation of a pellet-like shaped product, known as a soft pellet. The process has then developed by grinding of chips to fine powder followed by steam extrusion; this produces strong pellets upon cooling, known as hard pellets (Fig. 3). Similar to cassava chips, the cassava pellets have the moisture and sand contents as the critical quality and these should not exceed 16% and 4%, respectively.

Bioethanol production

Ethanol or ethyl alcohol is an edible organic solvent that can be produced by chemical synthesis or by fermentation of fermentable sugars. Presently, around 95% of the world production (110 billion litres in 2015) is from bioprocess of sugar fermentation by yeast. This bioethanol can be used as potable alcohol, industrial solvent (e.g. for applications in pharmaceutical, cosmetics, paints and inks) and biofuel. The largest consumption is for biofuel, which accounts for more than 80% of total world production, as a result of the

Table 2 Feed formula for various animal types, using cassava chips

Raw material (kg)	Chick layer	Meat-type duck finisher (age 2–4 weeks)	Duck layer (older than 48 weeks)	Muscovy duck starter (age 0–3 weeks)	Muscovy duck layer (age 9 weeks-sell)	Fattening pig (60-100 kg)	Dairy cow (15-20 kg milk/day)
Milled cassava chip	42.4	46	39	43.2	50	65.6	42
Rice bran	7.45	13	24.5	10	20	-	12
Soybean meal (44% protein content)	26	25	17.45	34.8	15	27	16
Fish meal (55% protein content)	8	7.5	7.5	7	5	-	-
Rough rice bran/mung bean meal/fish meal (60% protein content)	-	-	-[1]	-[2]	5[2]	3[3]	-
Leucaena leaf meal	2.5	3	3	-	3	-	-
Oyster shell meal	8	0.6	1.1	0.4	0.5	-	28[4]
Dicalcium phosphate	0.8	0.8	6.5	0.7	0.7	2	-
Animal fats/vegetable oils	4	3	-	3	-	1.8	-
DL-Methionine	0.1	0.4	0.2	0.2	0.1	-	-
Salt	0.5	0.45	0.5	0.45	0.45	0.35	1
Mineral and multivitamin	0.25	0.25	0.25	0.25	0.25	0.25	1
Total	100	100	100	100	100	100	100

Source: Bureau of Animal Nutrition Development, Department of Livestock Development
[1] Rough rice bran; [2] Mung bean meal; [3] Fish meal (60% protein content); [4] Cottonseed meal

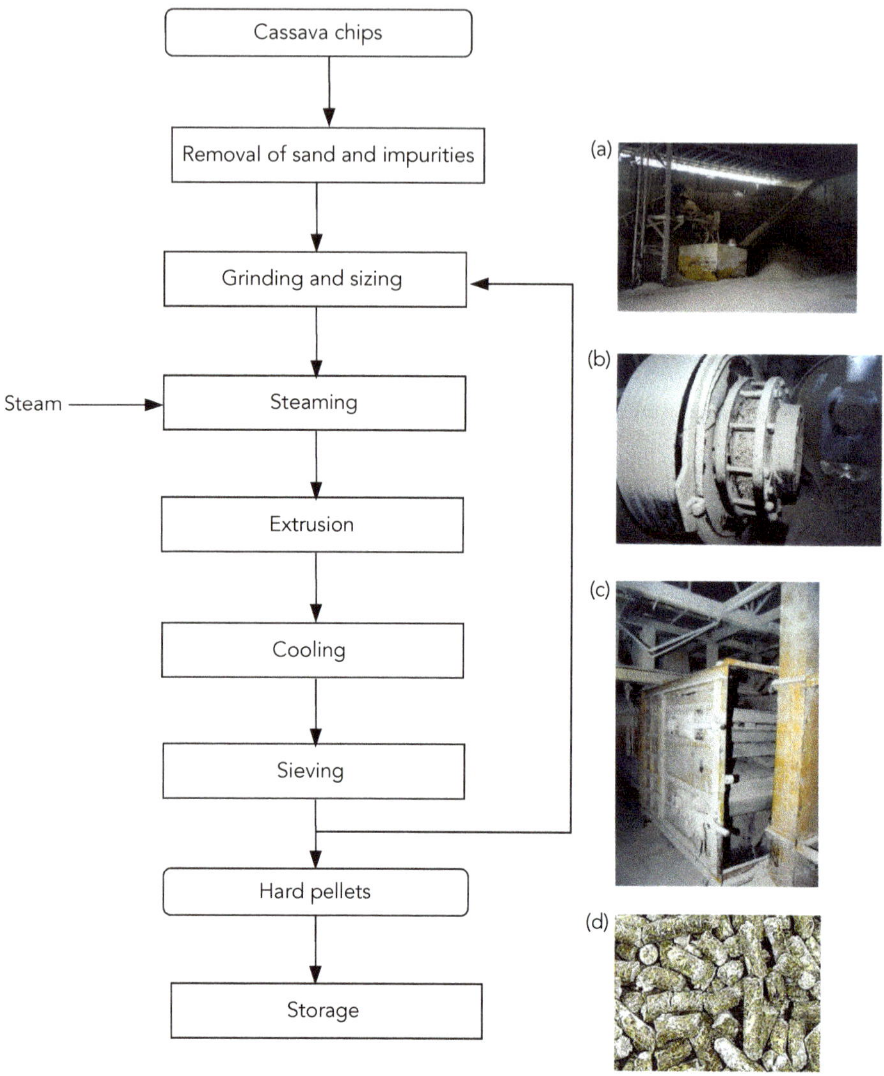

Figure 3 Process of making cassava pellets: (a) grinding of chips by hammer mill; (b) steam extrusion through a die; (c) cooling in a chamber; and (d) hard pellets. Source: http://www.fp7-gratitude.eu.

increasing demand for alternative, renewable and green energy, caused by the shortage and high fluctuating prices of gasoline and climate change concerns from use of non-renewable energy. As transport biofuel, two types of bioethanol can be used: hydrous ethanol (95%) and anhydrous ethanol (99.5%). The former one can be used for modified engine types called flexible fuel vehicles (or FFV) and the latter one can be used in a form of blends with gasoline (or petrol) at different proportions. This product is known as gasohol and is indicated by the 'E' numbers, that is, the percentage of anhydrous ethanol in the mixture by volume: E5, E10, E15, E20, E25 and E85 with 5%, 10%, 15%, 20%,

25% and 85% anhydrous ethanol in the blends, respectively. Gasohol with low ethanol concentrations (<25%) can be used in regular cars, but with high ethanol contents, it can be used only after engine modifications. Nowadays, more than 20 countries around the world, led by Brazil and the United States, have implemented renewable energy policy by using E10. The major feedstock for bioethanol is sugar cane in Brazil and maize in the United States. Other alternative feedstocks for bioethanol production are sugar beet, sweet sorghum and other starch crops including wheat, rice and cassava. In some countries where cassava roots are plentiful and reasonably priced, bioethanol is alternatively produced from cassava feedstock. Dried chips are considered to be the most promising feedstock for the process, with many advantages including extended shelf life, ease of storage and low transportation cost. Production technology of ethanol from maize can be effectively applied to dried cassava chips.

Starch, as a polymer of glucose, cannot be directly consumed by yeast. It has to be converted or depolymerized to glucose prior to yeast fermentation, which can be cost-effectively achieved by enzyme hydrolysis. Naturally occurring as semi-crystalline structures, starch granules, however, are insoluble in water and less susceptible to enzyme hydrolysis. Upon cooking in excess water, the granular structure of starch is disrupted, making glucose polymers become solubilized and more susceptible to enzyme attacks. At the same time, the starch slurry becomes more viscous. This process is known as gelatinization and the temperature at which the structural properties of starch are changed is the gelatinization temperature. Different starches have different gelatinization temperatures, implying the need of different cooking conditions. Cassava starch has a lower cooking temperature, relative to cereal starches; the pasting temperatures for cassava, maize, wheat and rice are 60–65, 75–80, 80–85 and 73–75°C (Swinkels, 1998; Thirathumthavorn and Charoenrein, 2005). The enzymatic hydrolysis of starch is a two-stage process involving liquefaction and saccharification. Liquefaction is a step in which starch is degraded by an endo-acting enzyme – namely α-amylase – which hydrolyses only the internal linkages (i.e. α-1,4) in the linear chain of starch, causing a dramatic drop in the viscosity of cooked starch. Typically, liquefying enzymes can have their activity at a high temperature (>85°C, depending on enzyme types) so that the enzyme can help reduce paste viscosity of starch during cooking. The dextrins – that is, products obtained after liquefaction – are further hydrolysed ultimately to glucose by glucoamylase enzyme (with the optimum temperature around 60°C, depending on the enzyme types), which can hydrolyse both glycosidic linkages in the linear (α-1,4) and branched (α-1,6) polymers.

A simple process of making bioethanol from cassava chips is presented in Fig. 4. Chips are initially ground to fine powder and mixed with water. Alternatively, ground fresh cassava roots are used during the harvest season due to cheap and plentiful supply. The slurry of ground cassava is cooked and the starch is initially converted to dextrin by the liquefying enzymes (α-amylase), at a temperature > 85°C, and then to glucose by saccharifying enzyme (glucoamylase), at 60°C. Glucose is subsequently converted to ethanol by yeast fermentation at 30–32°C; the most commonly used yeast is *Saccharomyces cerevisiae*. Most factories now use the process called simultaneous saccharification and fermentation (SSF), of which the second stage of hydrolysis by glucoamylase can be performed simultaneously with yeast fermentation at 30–32°C. SSF process can reduce the production time and energy consumption as well as improve enzyme and yeast performance by reducing the product inhibition and osmotic pressure from glucose. The conversion of ethanol from cassava chips is around 2.5 kg dried chip/litre of pure ethanol, depending on the amount of starch in raw materials and on the fermentation efficiency, as shown below:

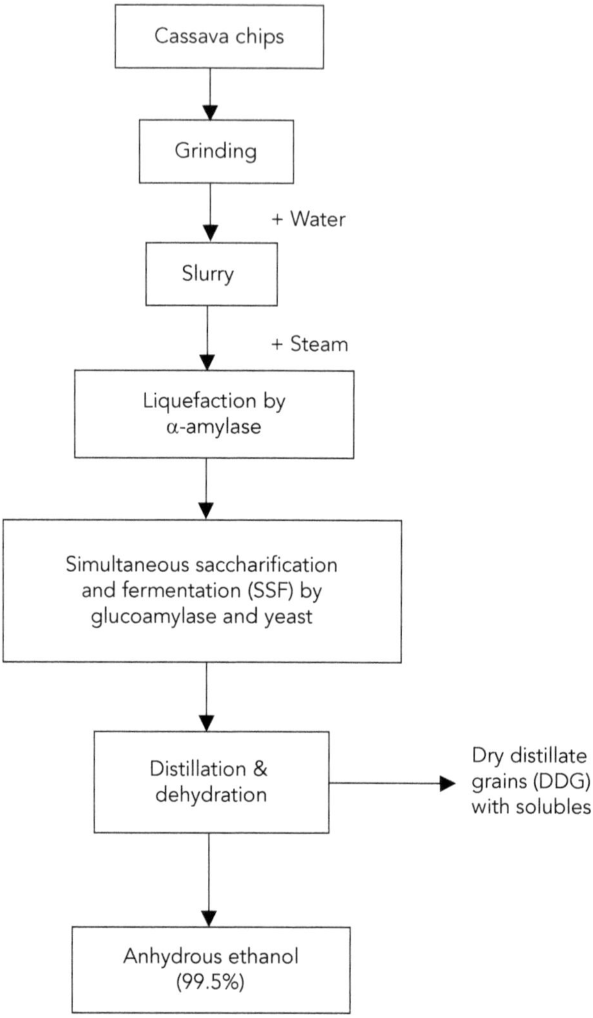

Figure 4 Bioethanol production process from dried cassava chips by the simultaneous saccharification and fermentation (SSF) process. Source: http://www.fp7-gratitude.eu.

1 Theoretically, 511 g ethanol is obtained from glucose 1 kg
2 Density of pure ethanol is 0.789 g/ml
therefore, 1 litre of ethanol (789.3 g) is obtained from glucose 1.5446 kg
3 Dried cassava chips of 100 kg contain starch 65 kg
or when hydrolysed to glucose (x 1.1) 71.5 kg
4 Glucose of 71.5 kg can be produced from chips 100 kg
therefore, glucose 1.54 kg is obtained from chips 2.2 kg
5 If the fermentation efficiency is 85%,
therefore, 1 litre of ethanol is produced from chips of 2.5 kg

During the process, a large amount of wastewater is produced (1400–1600 m^3 stillage/150 000 L anhydrous ethanol) and is treated by a biomethylation process to produce methane gas. In addition to liquid waste, a solid waste, called Dried Distillers Grains with Solubles (DDGS) is produced – approximately 150 tons/150 000 L anhydrous ethanol – which can be used as animal feed. Nevertheless, cassava DDG contains less protein and lipid contents than that obtained from maize. Maize DDGS contains around 28 and 10% protein and lipid, respectively, while cassava DDGS contains only 3.0 and 0.2% protein and lipid, respectively. (https://www.grains.org/sites/default/files/ddgs-handbook/Chapter-6.pdf)

Citric acid

Citric acid (2-hydroxy-propane-1,2,3-tricarboxylic acid) is a weak, edible, organic tricarboxylic acid, which can be present naturally in citrus fruits. Citric acid is labelled as 'generally recognized as safe' (GRAS), having pleasant taste, high water solubility and good chelating and buffering properties. It is used extensively as a flavouring, preservative, acidulant and emulsifying agent in food, confectionery and beverages, especially in carbonated beverages to provide taste and flavours and in jam and jellies for taste and pH adjustment. With an antioxidizing power, citric acid is used to preserve colour and to prevent flavour deterioration in processed fruits and oil-containing food products. In addition, it can be used as a chelating agent to deliver minerals, in the form of citrate salts, in dietary supplements, or as an acidifying agent to control pH in cleansing, personal care and pharmaceutical products. The annual world production of citric acid is approximately 2 million metric tons.

Citric acid can be extracted from unripe citrus fruits and juice from the waste of lemons, limes and pineapple. However, today over 99% of the world's production of citric acid is achieved by microbial fermentation processes. Similar to bioethanol production, citric acid can be produced from sugar or starch substrate. When starch is used, it has to be converted to glucose prior to fermentation by microorganisms. Industrially, citric acid is produced by the submerged or solid-state fermentation process with *Aspergillus niger*. Some yeasts of genus *Candida* can also produce citric acid. In the solid-state fermentation process by *A. niger*, cassava chips, moistened (65–70%) by soaking with water, are sterilized by steaming and then inoculating with *A. niger*. During fermentation at around 30°C, the fungus, with its own indigenous enzymes, can grow by utilizing glucose produced by starch hydrolysis. Optionally, fungus growth can be accelerated by adding extraneous enzymes. An effective alternative to solid-state fermentation processes is the submerged process. This process is similar to the production of ethanol from cassava chips. The slurry of ground chips is first hydrolysed by enzymes and the syrup of starch hydrolysate is then fermented by fungus inoculum. Although taking a longer fermentation time, the submerged fermentation process is more effective. After fermentation, the fungus is filtered out of the resulting solution and citric acid is recovered by a series of processing steps (Fig. 5). The normal process of citric acid recovery involves precipitating the acid with calcium hydroxide or lime to form the insoluble tricalcium citrate. The efficiency of the precipitation depends on citric acid concentration, temperature, pH and rate of lime addition. Calcium citrate is then filtered off and subsequently regenerated with concentrated sulphuric acid to obtain citric acid and the precipitate of calcium sulphate (gypsum). After filtering off the gypsum, a solution of citric acid (25–30%) is refined by a treatment of activated carbon and ion-exchange resins to remove impurities. The purified solution is then concentrated

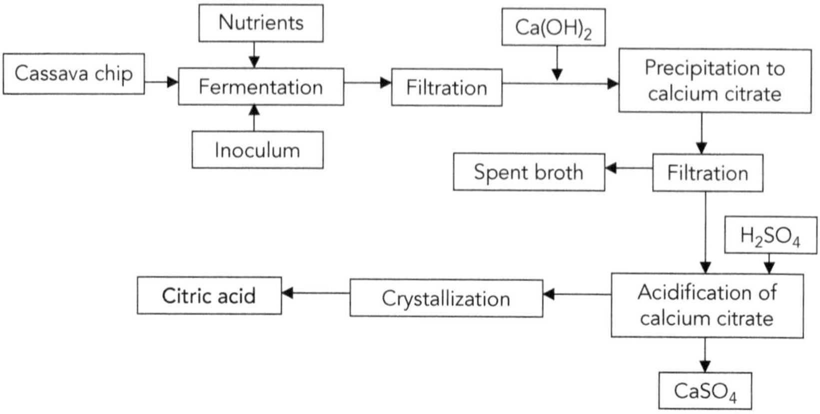

Figure 5 Simple scheme representing citric acid production from cassava chips. Source: http://www.fp7-gratitude.eu.

and crystallized. There are two forms of citric acid – that is, anhydrous (or water free) and monohydrate (containing one water molecule for every molecule of citric acid) – which are obtained by crystallization from hot water and cold water, respectively. The anhydrous form can also be produced by heating the monohydrate one. There are different grades of citric acid in the market, for example food, industrial and pharmaceutical.

4 Cassava starch: production and properties

Starch has long played an important role in human diet. Moreover, starch provides a lot of functionality that can be used as an additive in various applications including food, beverage, paper, adhesives, textiles, chemicals, cosmetics, pharmaceuticals, packaging and chemicals. Current global starch production is around 75 million tons in 2012 (Waterschoot et al., 2015) and continually increases. The most important commercial starches produced are from maize, cassava, potato, wheat and rice. Maize starch alone accounts for up to 80% of global production. Cassava starch is the second most-produced starch, major producers are mainly in South East Asia including Thailand, Vietnam, Indonesia and Cambodia.

4.1 Production

Unlike cereals (Fig. 1), cassava roots contain very small amounts of non-starch components, in particular, proteins and lipids, which must be removed effectively and economically by chemical processes. Consequently, cassava starch can be produced simply by a chemical-free process, using only water as the media of extraction. The process of cassava starch production is concisely illustrated in Fig. 6. After transportation to factories (mostly by farmers themselves), roots are sampled for determining starch content by a Reimann balance (Fig. 6a). These estimates are important in determining root prices. The estimation of starch content by a Reimann balance is based on the specific gravity, as determined by the weight of fresh roots in air and in water, which

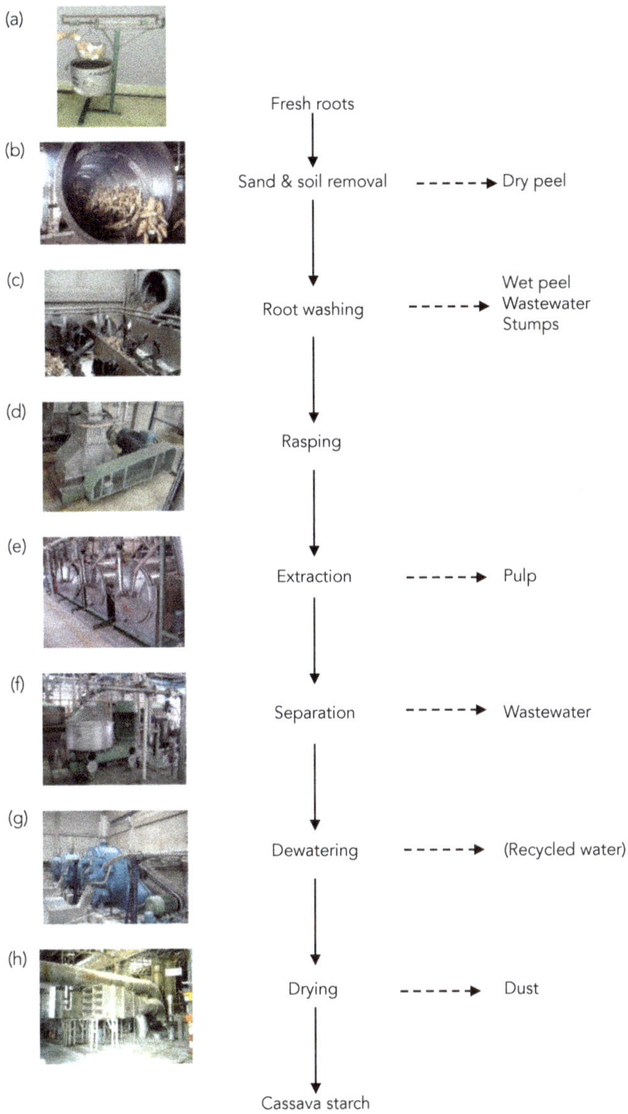

Figure 6 Typical cassava starch manufacturing process in Thailand: (a) measurement of starch content by a Reimann balance; (b) sand removal drum; (c) root washer; (d) rasper; (e) extractor; (f) separator; (g) dewatering centrifuge; and (h) flash dryer. Source: Breuninger et al. (2009)

relates to the dry matter as well as starch contents of roots. The correlation of dry matter and starch content of fresh roots to the specific gravity can vary, depending on varieties and environmental conditions of cassava planting, for example soil types, soil moisture, and age at harvest.

Once roots are delivered to the factory, most starch factories have applied a good manufacturing practice to process delivered roots within 1–2 days. Timing is critical since fresh roots quickly undergo deterioration by PPD and microbial attack. Processed roots are first passed through a drum (Fig. 6b) to get rid of soil, sand, pieces of peel and impurities; the waste generated in this stage is called *dry peels*. Roots are then passed through the root washer where recycled water is used (Fig. 6c) and some stones and wet peels are removed, while roots are moved along the washer by paddles. Washed roots are subsequently chopped into 1–2 cm chunks and ground with water by a rasper to a pulpy slurry (Fig. 6d). Optionally, sulphur water, containing a small amount of sulphur dioxide, is used as process water to prevent microbial growth, improve starch whiteness and increase extraction efficiency (Sriroth et al., 1998). Rasped root slurry from the rasper is passed through a series of coarse (equipped with the screen of 100 (150 microns) to 120 (125 microns) mesh) and fine extractors (equipped with the screen of 140–200 mesh) (Fig. 6e), either vertical or horizontal, where the pulp is eliminated from the starch slurry by the screen. During each extraction stage, starch slurry is passed through a sand-removing hydrocyclone to ensure complete removal of sand. Starch slurry received from the fine extraction is concentrated from 20 to 35% (or 10–17°Be' (Baume), a degree of starch concentration) to 40% (or 18–20 °Be') by a separator (Fig. 6f) or a series of hydrocyclones. Starch may be recovered from the final starch slurry and dried to produce native starch, or subjected to a chemical modification process. To produce native starch, water in the concentrated starch slurry is further reduced by a dewatering centrifuge (Fig. 6g), where the water content is reduced from 60% in slurry to 35–40% in cake. Alternatively, dewatering by a high-pressure filtration or filter press is applied in some native and modified starch factories since it reduces starch loss in the liquid stream and final moisture in the cake, which can in turn reduce the drying energy (Sriroth et al., 1999). Starch cake is then dried by a pneumatic conveying dryer, known as flash dryer (Fig. 6h) to yield the product with 12–14% moisture content. Normally it takes 4.40 kg of fresh roots (with 25% starch content) to produce 1 kg of starch. Actual amounts can vary depending on starch content in roots and efficiency of starch extraction.

During the starch manufacturing process, two primary wastes are produced, that is, liquid waste and solid waste.

Liquid waste: In cassava starch processing, the consumption of water and energy is very critical since it affects the production cost and starch quality. Although water consumption is optimized by water recycling at some stages, approximately 10–15 m^3 wastewater is generated per one ton of produced starch. The properties of wastewater from cassava starch manufacturing are given in Table 3. At present, most cassava starch factories have an effective means to manage this wastewater by biomethylation to produce biogas (methane). Biogas from wastewater is then employed as the energy source to replace imported, high-priced fuel oil, for drying starch and producing electricity used internally for the plant. After on-site treatment, final water with the quality that meets environmental requirements is then discharged to public channels or released to cassava-growing lands.

Solid waste: There are two basic types of solid waste, that is, peel and pulp. Peel can be obtained during removal of sands and soil (dry peel) and during washing (wet peel). The amount of peels removed mechanically is about 6–10 tons of sand and soil removal and 3–5 tons at root washing per 100 ton starch. Peels are currently used as fertilizer and as media for straw mushroom growing. The other solid waste – that is, cassava pulp or cassava bagasse – is generated at around 30–35% of the weight of rasped roots during starch extraction from the coarse extractor. Typically, the pulp from the coarse extractor

Table 3 Characteristics of wastewater from a cassava starch manufacturing process

Parameters	Values
Wastewater [m^3/ton starch]	10–20
pH	4.7–6.0
Total chemical oxygen demand [CODt, mg/l]	13,000–20,000
Soluble chemical oxygen demand [CODs, mg/l]	10,000–14,000
Total biochemical oxygen demand [BODt, mg/l]	10,000–13,000
Total solids [TS, mg/l]	12,000–20,000
Total suspended solids [TSS,[mg/l]	6,000–8,000

is repeatedly re-extracted until minimal starch content is achieved. The moisture content of the pulp is then reduced by a screw press before discharge, but still at high moisture content of around 70–80%. Therefore, it undergoes deterioration very rapidly, and can lead to environmental pollution if not appropriately managed. On a dry matter basis, starch is the most prominent component of pulp, at approximately 50–60%, depending on the quality of fresh roots and starch extraction efficiency of the rasper. Some factories have improved their processing efficiency which helps reduce the starch content in the pulp to 45% starch (dry weight basis). Currently, wet pulp is sold cheaply for animal feed uses. Although it is a promising carbohydrate source, the use of cassava pulp in animal feed is very limited due to the problem of storage and transportation. Alternatively, pulp is preserved as fermented pulp (silage) or sun-dried to extend the product's shelf life. The drying technology has also been developed in starch factories to produce dry cassava pulp for applications as animal feed and mushroom medium. This dried pulp has a maximum water content of 15%, minimum starch content of 55% and maximum fibre content of 20% (http://www.sonishstarch.com/th/products.asp). Cassava waste pulp has low economic value, and the energy cost for flash drying can be difficult to justify. Furthermore, dried powder of pulp has a low bulk density and is dusty, making transportation inefficient. Similar to conversion from chips to pellets, the pelletization of sun-dried pulp has been developed to increase the bulk density and the products are sold in a form of pellets for animal feed use. A commercial product, namely T-fibre® produced during the process of cassava starch production, has been introduced by a Thai company, which has employed advanced production technology from Sweden. T-fibres consist of non-soluble and soluble fibre at 41 and 8% by weight, respectively, and starch content of 37% by weight. With these components, T-fibre exhibits good water- and oil-holding capacity, which is highly favourable in food application. When applied in food products, it provides rich flavour, attractive appearance and great taste, with low production cost (Source: http://www.t-fibre.com/about_fiber.php).

The use of cassava pulp for producing alternative energy – that is, biomethane and bioethanol – has also been reported (http://www.thaigermanenergy.com/; http://www.asiabiogas.com; Carbon Energy Ltd., P & Paphob Renewable Ltd; Rattanachomsri et al., 2009; Poonsrisawat et al., 2014) at the pre-commercial scale. The value addition of cassava pulp to other products has been described, including extracted starch by enzymatic processes (Sriroth et al., 2000a; Wanlapatit et al., 2015), glucose and high fructose syrup (Johnson and Padmaja, 2011), microcrystalline cellulose (Menggred et al., 2013), and cellulose nanofibre for application as a reinforcing agent and filler in biodegradable materials, for example starch film or natural rubber (Teixeira et al., 2009; Pasquini et al., 2010; Wicaksono et al., 2013).

4.2 Properties

Since the roots of cassava contain very low amounts of other components, especially protein, extracted starch is very pure. Typically, the standards for cassava starch concern the contamination of ash, fibre and foreign matters. An example of cassava starch standards developed by Thailand, the largest exporter of cassava starch in the world, is summarized in Table 4 (Thai Industrial Standard, 1978).

In addition to application in non-food products, cassava starch is used as an ingredient in food products, where the standards are different compared to non-food uses. Some qualities including microbial content (e.g. total plate count, yeast and mould, *Staphylococcus aureus* and *Bacillus cereus*) and heavy metals (e.g. lead, arsenic and mercury) are now specified for trading purpose. Cassava starch has many unique properties that are appreciated in the food industry. Physically, cassava starch is very white; the degree of whiteness is >90. Commercial cassava starch contains no colour, no odour and few other impurities and its paste provides high viscosity with transparent appearance. The general specification of cassava starch for trade also includes other qualities, for example starch viscosity, whiteness, sulphur dioxide and cyanide content (Table 5).

Starch is a polysaccharide comprised solely of glucose monomers linked together by glycosidic bonds. It is composed of two types of glucans, namely amylose – a linear glucose polymer having only α-1,4 glycosidic linkage – and amylopectin, a branched glucose polymer containing mainly α-1,4 glycosidic linkages in the linear chain and a few α-1,6 glycosidic linkages at the branch structures (≈ 5%). The important properties of amylose and amylopectin are given in Table 6. Most starches contain approximately 20–30% amylose and the rest are amylopectin. Some starches such as waxy maize starch, waxy rice starch, amylose-free potato contain no amylose and some have very high amylose contents up to 50–70%, as in high-amylose maize starches (Table 7). Regular cassava starch contains approximately 17% amylose content. Recently, the development

Table 4 Thai standards for cassava starch

	Class		
Qualifications	1	2	3
Moisture content (% maximum)	13	14	14
Starch (% dry basis (db), minimum)	97.5	96	94
Ash (% db, maximum)	0.15	0.3	0.5
Acid-insoluble ash (% db, maximum)	0.05	0.10	0.15
Pulp (cm^3 in 50 g starch)	0.2	0.5	1.0
Residue on 150 μm sieve (% maximum)	1	3	5
pH	4.5 to 7	3.5 to 7	3.0 to 7
Protein (% db, maximum)	0.3	0.3	0.3

Source: Standard for tapioca flour/ starch (UDC 664.272), MOI (1978).

Table 5 General specifications for native cassava starch

Attribute	Specification
Moisture content (% maximum)	13
Starch content (% minimum)	85
Ash (% maximum)	0.2
pH	5.0 to 7.0
Whiteness (Kett scale, minimum)	90
Viscosity (Brabender Unit, BU, minimum using 6% dry weight)	600
Sulphur dioxide content (ppm, maximum)	30
Cyanide content (mg HCN equivalent/kg)	nil
Appearance	White, no dark specks, fresh odour

HCN = Hydrogen cyanide

of amylose-free cassava varieties by inhibition of granule-bound starch synthase I and later the discovery of a naturally occurring amylose-free cassava mutant have been reported by Raemakers et al. (2005) and by Ceballos et al. (2007). This natural mutant has been further applied successfully to develop high root yield varieties, which reinforces the commercial production of non-GM waxy cassava starch (Aiemnaka et al., 2012; Rojanaridpiched, 2013).

These two glucan polymers organize themselves into a semi-crystalline structure with alternating amorphous and ordered structure, induced by hydrogen bonding, and present in a form of minute granules. Starches from different botanical sources have different granule morphology. Under the microscopic view, cassava starch granules are smooth, spherical with 5–35 μm in diameter, and have one or more spherical truncations (Table 7). With the structural organization, native starch granules are insoluble in water. When heated in excess water, starch undergoes an irreversible structural transition, which includes the weakening of hydrogen bonds, and then the disruption of the ordered arrangement of glucan molecules. Consequently, heated starch granules can absorb water and become swollen, leading to the loss of granule integrity and leaching of glucose polymer. This concurrently introduces rheological property changes to cooked starch suspension, a phenomenon well known as starch gelatinization. As the heating continues, with shearing (mechanical forces), some starch granules absorb more water and become larger in size while some swollen granules are ruptured and dispersed into solution. Cooked starch paste is then a mixture of swollen granules, leached polymer and swollen granular fragments, all of which contribute to developed viscosity. Upon cooling, the polymer chains dispersed in starch paste can re-associate by hydrogen bonding. In this process the linear polymer (amylose) bonds more rapidly than the branched polymer (amylopectin). This phenomenon of chain reassociation (i.e. retrogradation) introduces starch gelation, a transformation of paste to gel, that is, a three-dimensional network (Fig. 7). Cassava starch has a low tendency to form gel because of low amylose content. A comparison in important structural and physico-chemical properties of normal and waxy cassava starch and other commercial starches is summarized in Table 7.

The physico-chemical properties of starches are affected by many factors. The gelatinization, pasting and gelation not only depend on botanical sources, but starches of the same origin also have various properties, partially affected by growth conditions. Many

Table 6 Some important properties of amylose and amylopectin

Property	Amylose	Amylopectin
Structure	Linear	Branch
Linkage (n = number of anhydroglucose unit, AGU)	α-1,4 glycosidic 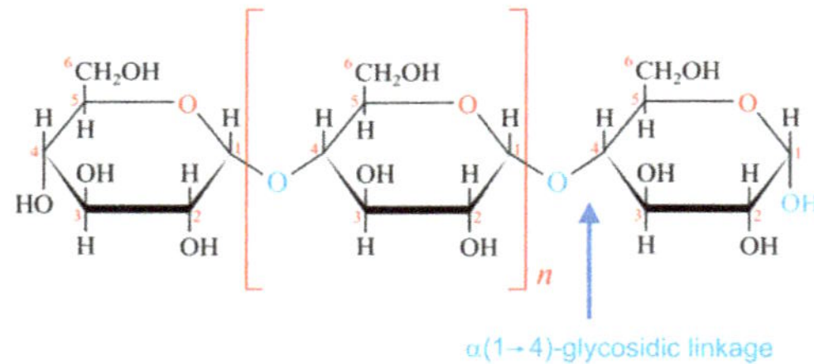 	α-1,4 (95%) and α-1,6 (5%) 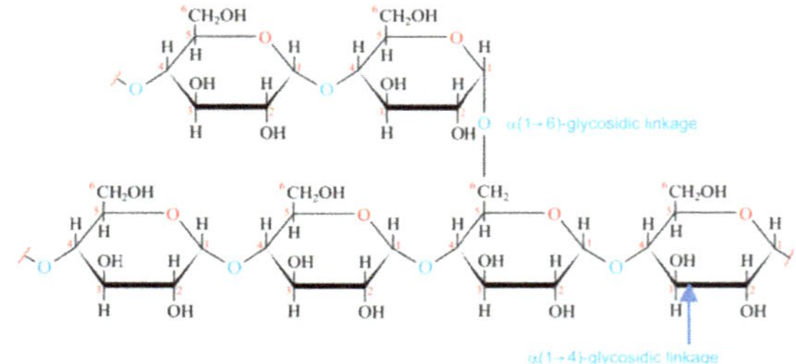
Complexing with iodine / colour	High affinity / Blue 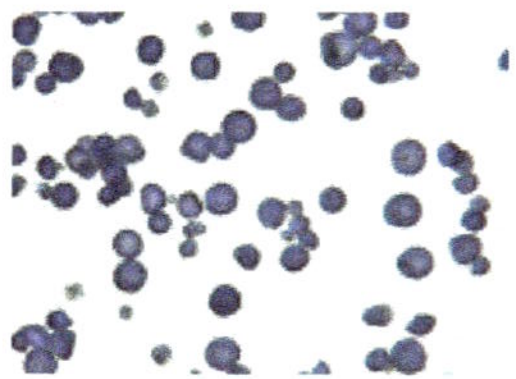	Low affinity / Reddish brown 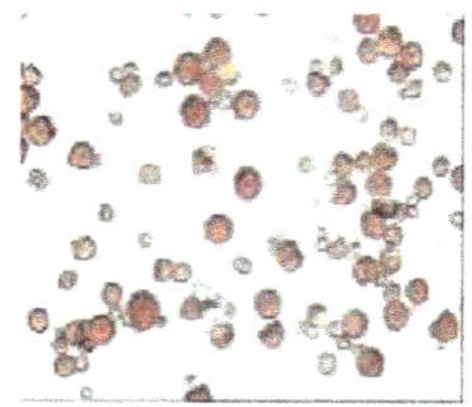
Water solubility	Low	High
Paste viscosity	Less viscous	High viscous
Gel formation	Firm	Non-gelling to soft
Retrogradation and syneresis	High tendency	Low tendency

Table 7 Some important structural and physico-chemical properties of major commercial starches

Property	Waxy maize	Regular maize	High-amylose maize	Potato	Wheat	Normal rice	Cassava	Waxy cassava
Source	Cereal grain	Cereal grain	Cereal grain	Tuber	Cereal grain	Cereal grain	Root	Root
Granule shape	Round, polygonal 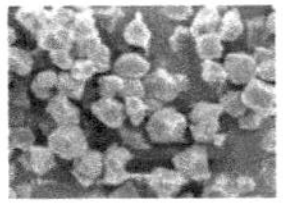	Round, polygonal 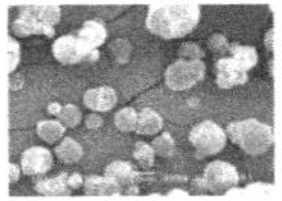	Long, irregular 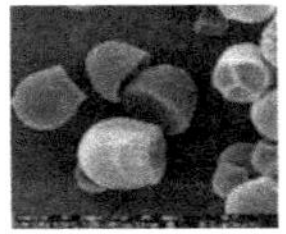	Oval, spherical 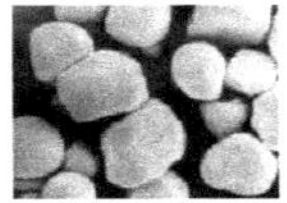	Round, lenticular 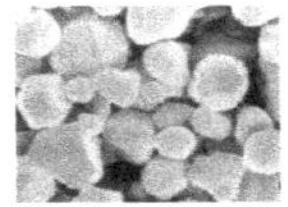	Polygonal, spherical compound granules 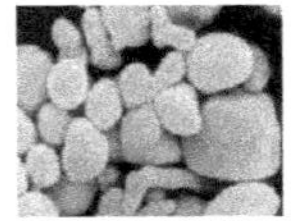	Oval, truncated 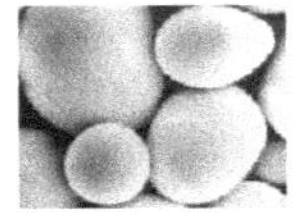	Oval, truncated 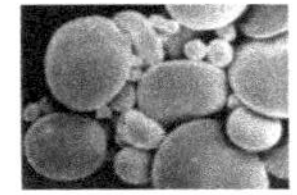
Granule size (microns)[1]	2–30	2–30	2–30	5–100	2–35	3–8	5–35	4–35
Amylose content (%db)[2]	<1	28	55-70	20	26	19	17	0
Protein content (%db)[3]	0.25	0.35	n.a.	0.1	0.4	0.5	0.1	0.1
Lipid content (%db)[3]	0.15	0.70	n.a.	0.1	0.80	n.a.	0.1	0.1
Phosphorus content (%db)[4]	0.002	0.02	0.02	0.09	0.05	0.07	0.01	n.a.
Gelatinization/ pasting temperature (°C)[5]	63–72	62–80	66–170	58–65	52–85	65–74[6]	52–65	63/68[7]
Paste viscosity	Moderate high	Moderate	Low	Very high	Moderate low	Moderate low	High	Very high
Paste texture	Long	Short	Short	Long	Short	Short	Long	Long
Paste clarity	Rather clear	Cloudy	Cloudy	Very clear	Cloudy	Cloudy	Clear	Very clear
Paste resistance to shear	Low	Moderate	High	Moderate	Moderate	Moderate	Low	Very low
Paste stability	High	Low	Very low	Moderate	Low	Low	High	Very high
Tendency of retrogradation	Low	High	Very high	Moderate	High	High	Low	Very low

db = dry weight basis; n.a. = not available.
[1] Tester et al., 2004; [2] Biliaderis, 2009; [3] Breuninger et al., 2009; [4] Lim et al., 1994; [5] BeMiller and Huber, 2007; [6] Mitchell, 2009; [7] Cellbalos et al., 2007.

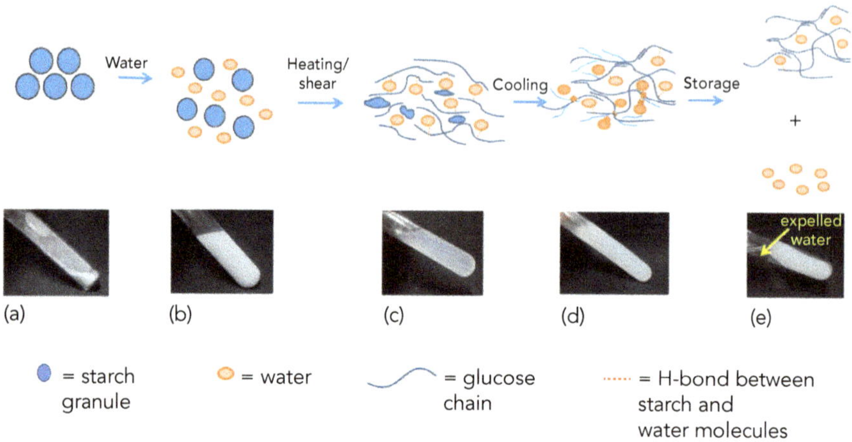

Figure 7 Phase transition of starch during cooking and cooling in water: (a) starch granules; (b) starch slurry; (c) cooked paste by gelatinization; (d) starch gel; and (e) retrograded starch with water syneresis.

studies have revealed that quantity and quality of cassava starch, including granule size, amylose content, gelatinization temperature and paste viscosity, depend on the variety, environmental conditions throughout the growth cycle, and root age (Asaoka et al., 1991; Moorthy and Ramanujam, 1986; Pardales and Esquibel, 1996; Defloor, et al., 1998; Sriroth et al., 1999; Santisopasri et al., 2000; Sriroth et al., 2000b). The application of sulphur dioxide as a processing aid has also been reported to affect some physico-chemical properties of cassava starch by increasing the gelatinization temperature and lowering the peak viscosity (Sriroth et al., 1998).

5 Cassava starch: products

Native cassava starch is directly used as food ingredients at both household and industrial levels for some applications such as noodles, bakery products and adhesives (e.g. for wallpaper and plywood). Furthermore, cassava starch is a starting material for making various intermediate chemicals/finished products which are then applied in many food and non-food industries (Fig. 8). Major commercial products from cassava starch are described in the following sections.

5.1 Cassava pearls

Pearls are a spherical form of starch with various diameters (2–8 mm), containing a portion of gelatinized starch (degree of starch gelatinization ranges from 30 to 80% as determined by an enzymatic assay and differential scanning calorimeter), caused by heat treatment of moistened starch. In the industrial process, the starch is wetted to reach equilibrium at 50% moisture content. Moistened starch is then pulverized, and a spherical shape is formed by horizontal shaking. Spherical, moistened pearls are roasted at 250–300°C for a certain period of time to partially gelatinize starch. After that, the pearls are further dried at a lower temperature (50–80°C) to reduce

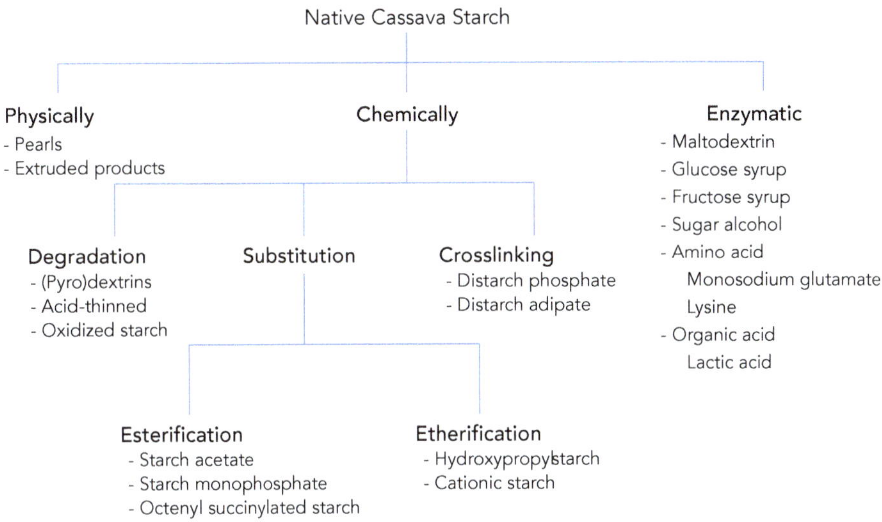

Figure 8 Important commercial cassava starch products.

the moisture content of finished products. When cooked, the pearls swell to about 1.5–2 times of their dried size (depending on the original sizes and cooking methods), and have unique characteristics, that is very transparent and chewy and are very popular in many dishes such as pudding.

5.2 Modified starches

Although native cassava starch has many distinct properties which can serve the specific needs for users, its more extensive application can be limited due to the lack of some functionalities, for example resistance to shear, heat and acid during processing; stability upon storage; and proper rheological properties of starch paste and gel. Starch is, therefore, mostly modified to improve its functional properties for specific uses. Furthermore, the modification can add some properties to starch such as ionic (cationic and anionic) groups, lipophilic groups, or non-enzyme hydrolysed bonds that broaden starch application. Starch modification can be classified into 3 groups: physical, chemical and enzyme processes (BeMiller, 1997).

Physical modification

Physical modification involves the treatment of cassava starch by physical means, for example, heat and shear. A combination of heat treatment and shear force has been widely used to produce many extruded snacks. The extrusion of cassava starch with the addition of plasticizers (e.g. glycerol) is employed to prepare thermoplastic starch (TPS), a material used in a bio-composite with other biodegradable materials such as polylactic acid (PLA). Another well-known physically modified product from cassava starch is alpha starch or pre-gelatinized starch. The unique property of pre-gelatinized starch is solubility in cold water without cooking (it is sometimes called instant or pre-cooked starch) and is often used as a thickening agent in products with minimal cooking process. Both native

Table 8 Specification of chemically modified starch for food application

Type of modified starches	INS-Number[1]	E-Number	Chemical reactants [4]	Standard/specification
Pregelatinized starch			-	-
Dextrin	1400[2]		Hydrochloric acid < 0.15% or orthophosphoric acid < 0.17%	pH 2.0 to 9.0
Thin-boiling starch	1401[2]		Hydrochloric acid or orthophosphoric acid < 7.0% or sulphuric acid < 2.0%	pH 3.0 to 7.0
Alkaline-treated starch	1402[2]		Sodium hydroxide or potassium hydroxide < 1.0%	pH 5.0 to 7.5
Bleached starch	1403[2]		Hydrogen peroxide and/ or peracetic acid with active oxygen < 0.45	Carboxyl groups increase not to exceed 0.1%
			Ammonium persulphate <0.075% and sulphur dioxide <0.05%	Sulphur dioxide < 50 mg/ka
			Sodium chlorite < 0.5%	No chlorite
			Potassium permanganate < 0.2%	Manganese < 50 mg/kg
Oxidized starch	1404	E1404[3]	Chlorine (as sodium hypochlorite) < 5.5%	Carboxyl groups < 1.1%
Distarch phosphate	1412	E1412[3]	Sodium trimetaphosphate	Phosphate (reported as phosphorus) < 0.04%
			Phosphorus oxychloride <0.1%	
Starch octenyl succinate	1450	E1450[3]	Octenyl succinic anhydride <3%	Octenyl succinic groups < 0.03%
Hydroxy-propyl starch	1440	E1440[3]	Propylene oxide < 10% or 25%[5]	Propylene chlorohydrin < 1 mg/kg Hydroxypropyl groups <7.0%
Acetylated starch	1420	E1420[3]	Acetic anhydride	Acetyl groups < 2.5%
			Vinyl acetate < 7.5%	
			Adipic anhydride and acetic anhydride <0.12%	

(Continued)

Type of modified starches	INS-Number[1]	E-Number	Chemical reactants [4]	Standard/specification
Monostarch phosphate	1410	E1410[3]	Orthophosphoric acid or Sodium or potassium orthophosphate or Sodium tripolyphosphate	Phosphate (reported as phosphorus) < 0.4%
Hydroxypropyl distarch phosphate	1442	E1442[3]	Sodium trimetaphosphate or phosphorus oxychloride and propylene oxide < 10%	Propylene chlorohydrin < 1 mg/kg or < 5 mg/kg[5] Hydroxypropyl groups < 4.0%
Acetylated distarch adipate	1422	E1422[3]	Acetic anhydride and adipic anhydride < 0.12%	Acetyl groups < 2.5% and Adipate groups < 0.135%
Acetylated distarch phosphate	1414	E1414[3]	Phosphorus oxychloride followed by acetic anhydride < 10% or 8%[5] or vinyl acetate < 7.5%[5]	Acetyl groups < 2.5% and Phosphate (reported as phosphorus) < 0.04%

[1] CCFAC International Numbering System (INS) 1989. ftp://ftp.fao.org/codex/standard/volume1a/en/CXxot04e.pdf
[2] Dextrin, bleached starch, starches modified by acid, alkali and enzyme or by physical treatment are not considered as food additives in the context of the EEC Directive 95/2/EC.
[3] Modified starches, Annex 1 of EEC Directive No. 95/2/EC (an ingredient category)
[4] According to Thai Industrial Standard Institute (TISI) unless specified by others (TISI, 1992).
[5] According to Food and Drug Administration (FDA) 21 CFR Ch. 1(FDA).

and chemically modified starch can be used to make pre-gelatinized starch and the most commonly used method is by drum drying. In this process, starch slurry (30–40% dry solid with or without some additives) is fed by the applicator roll to the heated drum where starch is gelatinized and dried. The dried film on the drum is scraped and pulverized to fine powder. Alternatively, pre-gelatinized starches can be produced by a spray dryer, where cooked starch paste is fed through the spray dryer. This process, however, is more costly.

Chemical modification

Cassava starch is easily modified by chemical processes already applied to other commercial starches (Fig. 8). Starch can be modified with chemicals in three different states: slurry (water > 40%), semi-dry (water < 40%) and paste (non-granular), which have different advantages and disadvantages. For economic reasons, the slurry process is usually applied to prepare most chemically modified starches, except for starch monophosphate, dextrins and sometimes for cationic starch. In the slurry process, starch slurry is mixed with required chemicals and the reaction occurs under the controlled conditions, that is, pH (usually at alkali), temperature (usually below the gelatinization temperature) and time. Salts such as sodium sulphate and sodium chloride are sometimes added to prevent granule swelling, induced by high pH of the medium. As soon as the reaction completes, it must be terminated, usually by neutralization with acid. The product of modified starches is then recovered, washed and dried by conventional equipment for native starch production. This process yields a product with good quality, but has environmental implications because it generates a lot of salt and wastewater.

The semi-dry or dry process has been introduced to reduce water consumption in the process of some chemically modified starches, for example cationic starch, mainly used in paper making. However, the semi-dry process results in products with inferior quality and more impurities. Table 8 provides the regulations, such as the amounts of chemicals used and allowed to be present in finished starch products for food applications. Improved properties of each type of modified starch are given in Table 9. It is a common practice to produce modified starches by more than one type of modification which are: dual modification (e.g. pre-gelatinized and chemically modified), two chemically modified (stabilized and cross-linked, oxidized and stabilized, acid-thinned and stabilized) or triple modification (e.g. pre-gelatinized, stabilized and cross-linked). Accordingly, by a combination of chemical types, degree of modification (named as degree of substitution (DS) or molar substitution (MS)) and integrated modification type (single, dual or triple), a broad spectrum of chemically modified starch products can be generated. Furthermore, custom-made products with specific and required functionalities for niche applications can be achieved by a design of integrated modifications.

Table 9 Properties and uses of some important modified starches in food products

Type of modified starches	Properties	Application
Pre-gelatinized starch	Soluble in water without cooking	Bakery, confectionery, dairy, batters and coatings, soups, sauces and gravies
Maltodextrin	- Soluble in water - Adjusted viscosity	Dairy, bakery, batters and coatings
Acid-thinned starch	- Low viscosity of cooked paste - High tendency to retrograde - Form hard gel	Confectionery, dairy
Oxidized starch	- Low viscosity of cooked paste - Stable paste - Low retrogradation - Good film properties	Confectionery, batter, coating
Cross-linked starch	- Adjusted viscosity - Good tolerance to processing conditions - Short paste texture - Hard gel	Canned food, sauces, soups, seasoning, dressing, pie filling, dairy, bakery, processed meat
Stabilized starch	- High viscosity - Low gelatinization temperature - Stable paste - Low retrogradation	Dairy, processed meat, bakery, soup, sauce, dressing
Pyrodextrins	- High viscosity - Soluble in water - Low retrogradation	Bakery, snack

5.3 Starch hydrolysates/sweeteners and derivatives

Starch, as a polymer of glucose, can be hydrolysed into smaller molecules. The degree of starch hydrolysis is usually described as Dextrose Equivalent (DE) – the amount of reducing sugars present in a product, relative to dextrose (glucose) – and is expressed as a percentage on a dry solid basis. Starch contains a negligible amount of reducing sugar and therefore its DE is zero, whereas glucose has DE of 100. With the range of DE from 0 (starch) to 100 (glucose), different products can be obtained by varying degrees of starch hydrolysis, including maltodextrin (DE<20), glucose syrup (DE 24, 38–42, >95) and dextrose sugar. Starch hydrolysates with different DE provide different attributes including osmolality, sweetness, hygroscopicity, freezing point depression, browning reaction and viscosity. The higher the DEs of maltodextrins, the greater those attributes are, except the viscosity. These diversified properties of starch hydrolysates can serve many applications in beverages, confectioneries, canned foods, seasonings, etc. In addition, there are different types of enzymes capable of hydrolysing starches with different specificity. Important commercial enzymes are α-amylase, β-amylase, glucoamylase, pullulanase and transglucosidase. Diversified products can be produced by varying enzyme types, degree of starch hydrolysis and derivatization. The actions and products of some important starch degrading enzymes are listed in Table 10.

A typical process of making starch hydrolysed products, as depicted in Fig. 9, can be classified into two major stages, that is starch hydrolysis and refinery. For starch hydrolysis, starch is cooked and gelatinized to facilitate enzyme hydrolysis. Types and dosages of enzyme used and the reaction time are varied greatly, depending on product specifications. Typically, starch is initially hydrolysed by a liquefying enzyme, that is α-amylase. This step is called liquefaction, which results in thinning of starch pastes. The products derived from liquefaction are called maltodextrins (DE<20). Further hydrolysis by α-amylase or by saccharifying enzymes such as glucoamylase and β-amylase, a process called saccharification, produces high-DE syrup. A complete hydrolysis by glucoamylase yields glucose while β-amylase gives maltose as the major end product (Fig. 10). Pullulanase enzyme, a debranching enzyme, is occasionally applied with saccharifying enzymes to increase the rate and yield of starch hydrolysis. Once the required products are obtained, the enzyme reaction is terminated, usually by heating and sometimes with pH adjustment. Subsequently, starch hydrolysis products are advanced to a refining

Table 10 Actions and products produced by important starch degrading enzymes

Name	Action of hydrolysis	Products
α-Amylase	Endo-acting enzyme, hydrolysing α-1,4 glycosidic linkage randomly	Maltodextrin (DE < 20), glucose syrup
β-Amylase	Exo-acting enzyme, hydrolysing α-1,4 glycosidic linkage from non-reducing ends	Maltose
Glucoamylase	Exo-acting enzyme, hydrolysing α-1,4 (and α-1,6 but at very slow rate) glycosidic linkage from non-reducing ends	Glucose
Pullulanase	Debranching enzyme, hydrolysing α-1,6 glycosidic linkage	Linear glucan
Glycosyltransferase	Intramolecular transglycosylation	Cyclodextrins

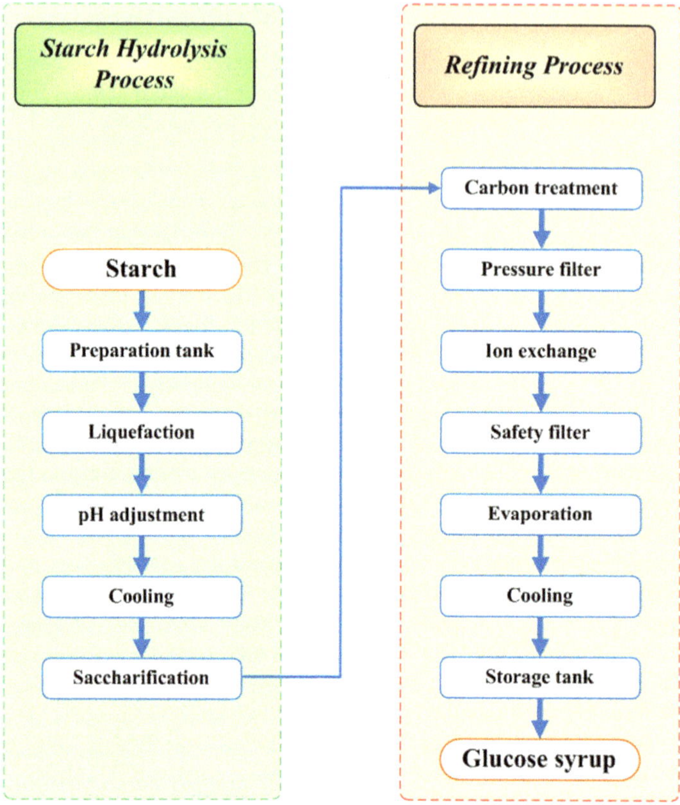

Figure 9 Typical process of glucose syrup production from starch by enzyme hydrolysis. Source: http://www.fp7-gratitude.eu.

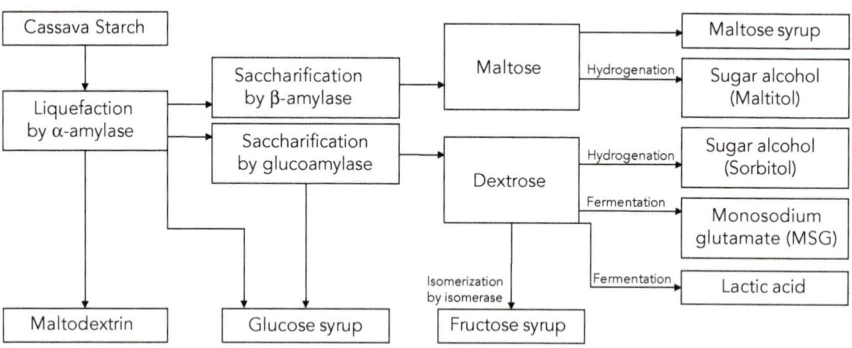

Figure 10 Simple scheme representing production of major sweetener products obtained from cassava starch hydrolysis. Source: http://www.fp7-gratitude.eu.

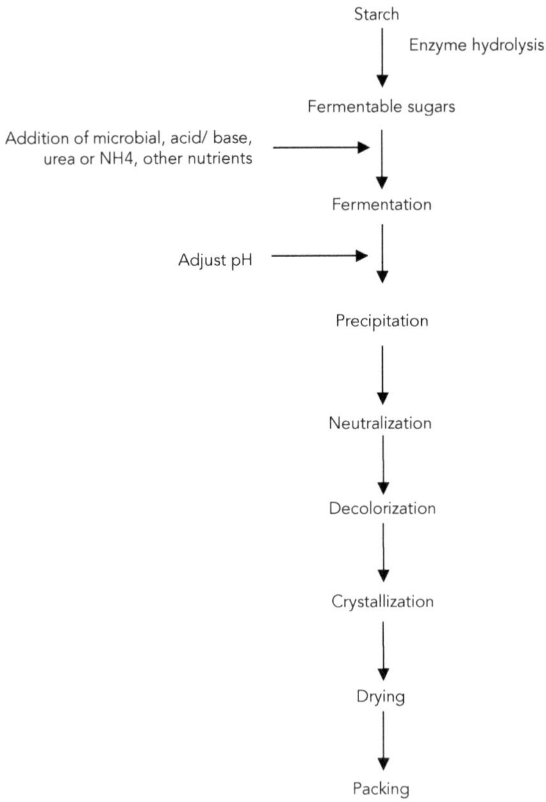

Figure 11 Simple scheme representing production of major amino and organic acids obtained from cassava starch hydrolysis.

step. The main objective of the refining process is to remove impurities such as metal ions, colour pigments, or enzyme residues that can have adverse effects on syrup qualities upon storage, such as colour and cloudy appearance. The refining process of syrup involves mainly carbon treatment and ion-exchange treatment by cation and anion exchange resin. The refined syrup is then dried (spray-dried for maltodextrins) or concentrated by evaporation. In some products, crystallization is applied and sugar crystals are produced.

More derivative products can be produced from hydrolysed sugar, either by chemical or by enzyme processes. One of the most-produced derivatives in the world is fructose. Fructose can be produced from glucose by the isomerase enzyme, in a process called isomerization, by which high fructose syrup (HFS, 42% fructose) is made. The Simulated Moving Bed (SMB) achieves chromatographic separation with ion-exchange resins, to produce very-enriched fructose syrup (VEFS, 90% fructose), made from HFS and then blended with HFS to prepare enriched fructose syrup (EFS, 55% fructose), the most common sweetener employed in the beverage industry.

The reduction of starch sugars by hydrogenation in the presence of metal catalysts, for example nickel, can produce sugar alcohols such as sorbitol from glucose, maltitol from maltose and hydrogenated starch hydrolysates (HSH), which are low-caloric, provide a cooling sensation and are not metabolized by oral bacteria, therefore preventing tooth decay. These sugar alcohols are favourably applied in many products such as chewing gum, confectionery, tablets and toothpaste. Another large promising industrial sector for cassava starch utilization is fermentation of glucose by microorganisms which can produce a wide range of chemicals, depending on the microorganism types. Important products are amino acids, alcohol and organic acids. In the fermentation process, glucose obtained by starch hydrolysis is fermented by microorganisms to produce various chemicals. Examples are ethanol by yeast (*Saccharomyces cerevisiae*); lactic acid by lactic acid bacteria; monosodium glutamate (MSG) by *Corynebacterium glutamicum*, *Brevibacterium flavum*, and *Arthrobacter paraffineus;* and L-lysine by *Brevibacterium lactofermentum* and *Brevibacterium flavum*. Once the fermentation process is completed, the next step involves a refining process including product recovery, purification using various technologies such as ion-exchange treatment, crystallization, distillation and filtration. Figure 11 shows the process of making MSG and lysine.

6 Cassava starch: industrial applications

Cassava starches, as native, modified and hydrolysed forms, have been employed in many industrial applications due to their versatile, advantageous physico-chemical properties and reasonable price. The important functionalities of starch are gelling, thickening, adhesion, binding, moisture retention, stabilizing, texturizing, anti-staling (staling is a change in both chemical and physical attributes caused by starch retrogradation that gives starch products a dry, leathery and hardened texture), film forming, emulsifying, flocculation and nutrition. Some major industrial applications of cassava starch are described.

6.1 Food products

Native and modified tapioca starches are widely applied as thickeners, binders, texturizers, stabilizers, fillers, sweeteners, flavour carriers and fat replacers in various food products such as canned food, frozen food, dry mixes, baked goods, snacks, dressings, soups, sauces, noodles, snacks, dairy products, meat and fish products, and infant food. Properties and applications of important modified starch are given in Table 9. Mostly, pre-gelatinized starch used in food products is prepared from chemically modified starch. Stabilized starches, mainly acetyl and hydroxypropyl, are typically employed as dual modified starches, that is, stabilized, cross-linked starches to serve the processor's need for both processing and storage of products. Uniquely, octenyl succinyl starch, with hydrophilic and hydrophobic properties, is used as an encapsulating and emulsifying agent.

6.2 Beverages

Modified cassava starch, as a colloid stabilizer and as a starch-based sweetener, is used in some beverages. Cassava-based syrups are also good sources of easily fermentable sugars for brewery applications.

6.3 Confectionery

Diverse types of modified cassava starches are used in confectionery applications for different purposes such as gelling, thickening, texture stabilizing, foam strengthening, crystallization inhibition, adhesion, film forming, glazing and casting moulds. Starch-based polyols are used in the manufacture of sugar-free chewing gum.

6.4 Adhesives and glue

Cassava starch and dextrins are excellent adhesives and used in many applications, including corrugated board, wallpaper, paper bags, laminated board, gummed paper, tapes, labels, stamps and envelopes. The consumption of cassava starch as an adhesive in making corrugated fibreboard is very large. Corrugated fibreboard consists of fluted and liner board that are stuck together by starch-based glue. The strength of corrugated board depends on flute types (height, number of flutes), board structure (single-faced, single wall, double wall and triple wall) and the quality of glue. Native cassava starch has demonstrated technical advantage over maize starch as the adhesive (ungelatinized) portion of the corrugating formulation by improving the tack and green bond strength (Breuninger et al., 2009).

6.5 Paper

Cassava starches, in both native and modified forms, are widely applied in papermaking industries due to their good film properties. In the papermaking process, stock of refined pulp is prepared with the addition of other compounds, for example filler pigments, dyes, sizing agents and strength additives. Cationic cassava starch carries positive charges, which can bind strongly with negatively charged cellulose. As a result it is commonly used as an internal sizing in papermaking to increase the paper strength. Recently, amphoteric cassava starch (starch containing both cationic and anionic charges) is utilized since most pulp used nowadays is recycled, not virgin pulp. In the part of the process where the paper sheet is formed on the wire screens, the sheet, with high water content (80%), is pressed to reduce water (to 50–60%) before drying. The use of ionic starches facilitates water removal and additive retaining. Afterward, the dried sheet continues to the finishing process to improve the paper surface. During this surface sizing, modified starch is applied to increase strength and printing quality. The commonly used external sizing agent is oxidized starch, which has good film properties. Some paper mills have developed on-site starch modification using an enzyme process.

6.6 Textile

Cassava starches are used in the textile industry as sizing agents to stiffen and protect the thread for improved weaving efficiency. They are also used as finishing agents to obtain smooth fabrics, and colour thickeners for printing to obtain sharp and durable printed fabrics.

6.7 Pharmacy and cosmetics

Native and modified cassava starches are used as binders, fillers and disintegrating agents for pharmaceutical tablet production. Specialty starches are also used as a carrier for skin moisturizers, encapsulating agents for vitamins, as sizing agents for hair mousse, and as thickeners in shampoo.

6.8 Chemicals

Cassava is a renewable feedstock that can convert solar energy to stored chemical energy in the form of organic compounds – that is, starch – via photosynthesis, using very inexpensive precursors, namely carbon dioxide and water. It is one of the crops that have gained a lot of interest as promising feedstock for facilities that integrate processes and equipment to produce fuels, power, chemicals and materials based on renewable biological resources, that is a bio-refinery. Cassava starch syrups are produced economically by acid and/or enzyme processes and used as feedstock to make various chemicals. Currently produced chemicals are amino acids (MSG, lysine), organic acids (citric, lactic and succinic acid) and alcohols (bioethanol, biobutanol). More value-added chemicals such as vitamins (e.g. ascorbic acid), antibiotics (e.g. penicillin, cephalosporin and griseofulvin), gum (e.g. xanthan gum by *Xanthomonas campestris*), protein (e.g. mycoprotein by *Fusarium graminearum*), growth promoter in feed can also be produced by various means, including chemical reaction, fermentation and other biotechnological processes.

7 Summary

Cassava has major advantages over other starch crops, in part because of its good tolerance and reasonable production under poor growing conditions, in particular drought and infertile soil. It is rich in starch and low in other components that are considered undesirable in starch. This makes cassava roots a good material for a cost-effective production of starch. Cassava starch has diversified functionalities, particularly when derivatized by physical, chemical and enzyme processes, which expand its applications in industries such as food, paper, textiles, adhesives and pharmaceuticals. Furthermore, dried chips are a very promising feedstock for bio-refineries, a process by which starch, a renewable energy source, is converted to many value-added, eco-friendly products, for example biofuel, biomaterials, and biochemicals by physical, chemical and biotechnological process. The advantages of using inexpensive cassava starch for syrup production in a bio-based economy are that the starch is renewable, very pure, easy for cooking due to a lower gelatinization temperature and has low tendency of retrogradation. Because of a combination of good attributes from the standpoint of cultivation, root traits, and uses for the starch, cassava is a versatile crop, not only for the twenty-first century, but beyond.

8 Where to look for further information

There are a number of research articles on cassava and cassava-related topics which can be accessed at some websites as follows:

- http://www.aseancassava.info/
- http://ciat.cgiar.org/what-we-do/breeding-better-crops/rooting-for-cassava/

General introduction to cassava plantation and utilization and updated statistics on cassava production and trading are available at following sources:

- http://www.fao.org
- http://www.gcp21.org
- http://www.tapiocathai.org
- http://www.fp7-gratitude.eu/images/ReportsPublications/Deliverable_reports/WP1_DRs/D1-1_revised.pdf

9 References

Aiemnaka P., Wongkaew A., Chanthaworn J., Nagashima S.K., Boonma S., Authapun J., Jenweerawat S., Kongsila P., Kittipadakul P., Nakasathien S., Sreewongchai T., Wannarat W., Vichukit V., Loppez-Lavalle L.A.B., Ceballos H., Rojanaridpiched C. and Phumchai, C. (2012), 'Molecular Characterization of a Spontaneous Waxy Starch Mutation in Cassava', *Crop Science*, 52 (5), 2121–30.

Asaoka M., Blanshard J.M.V. and Richard, J.E. (1991), 'Seasonal effects on the physico-chemical properties of starch from four cultivars of cassava', *Starch/Stärke*, 43, 455–9.

BeMiller J.N. (1997), 'Starch modification', *Starch/Stärke*, 49, 127–31.

BeMiller, J.N. and Huber, K.L. (2007) 'Carbohydrates', In Damodaran S., Parkin, K. L. and Fennema, O. R. (eds)., *Fennema's Food Chemistry*, 4th Editions, CRC Press, pp. 83-154.

Biliaderis, C.G. (2009), 'Structural transitions and related physical properties of starch, In BeMiller J. and Whistler, R. (eds)., *Starch Chemistry and Technology*, 3th editions, Elsevier., pp. 293-372.

Breuninger W.F., Piyachomkwan K. and Sriroth, K. (2009), 'Tapioca/Cassava starch: Production and use', In BeMiller J. and Whistler, R. (eds), *Starch Chemistry and Technology*, 3th editions, Elsevier, pp. 541–68.

Ceballos H., Sanchez T., Morante N., Fregene M., Dufour D., Smith A.M., Denyer K., Perez J.C., Calle F. and Mestres, C. (2007), 'Discovery of an amylose-free starch mutant in cassava (*Manihot esculenta* Crantz)', Journal of Agricultural and Food Chemistry, 55(18), 7469–76.

Defloor I., Swennen R., Bokanga M. and Delcour, J.A. (1998), 'Moisture stress during growth affects the breadmaking and gelatinisation properties of cassava (*Manihot esculenta* Crantz) flour', *Journal of Science, Food and Agriculture*, 76, 233–8.

FAO (2013), Chapter 1: *Cassava, a 21st century crop*, pp. 1–33. *In* Save and Grow: Cassava, A guide to sustainable production intensification. http://www.fao.org/3/a-i3278e/i3278e01.pdf.

FAOSTAT. (2016), 'World cassava production quantities in 2014', http://faostat.fao.org/ (accessed21 April 2016)

FDA, U.S. Code of Federal Regulations, Title 21, Chapter I, Part 172.892, Food Starch – Modified, U.S. Government Printing Office, Washington, DC.

Jansz E. R. and Uluwaduge, D. I. (1997), 'Biochemical aspects of cassava (*Manihot esculenta* Crantz) with special emphasis on cyanogenic glucosides: A review', *Journal of the National Science Council of Sri Lanka*, 25(1), 1–24.

Johnson R. and Padmaja, G. (2011), 'Utilization of cassava fibrous residue for the production of glucose and high fructose syrup', *Industrial Biotechnology*, DOI:10.1089/ind.2011.0015.

Kanto U. and Juttupornpong S. (2002), *Production of High Quality Cassava Chip For Animal Feed Uses*. TTDI 0101/45. Thai Tapioca Development Institute. Bangkok 20p.

Lim, S. T., Kasemsuwan, T. and Jane, J. L. (1994), 'Characterization of phosphorus in starch by 31P-nuclear magnetic resonance spectroscopy', *Cereal Chemistry*, 71, 488–493.

Menggred S., Chatakanonda P., Piyachomkwan K. and Sriroth, K. (2013), 'Preparation of microcrystalline cellulose from destarched cassava pulp', *Starch Update 2013 Proceedings*, Miracle Grand Convention Hotel, Bangkok, Thailand, pp. 291–7.

Mitchell, C.R. (2009), 'Rice starches: Production and properties', In BeMiller, J. and Whistler, R. (eds)., *Starch Chemistry and Technology*, 3th editions, Elsevier., pp. 569-78.

Moorthy S.N. and Ramanujam, T. (1986), 'Variation in properties of starch in cassava varieties in relation to age of the crop'. *Starch/Stärke*, 38: 58–61.

Padmaja G. (1995), 'Cyanide detoxification in cassava for food and feed uses', *Critical Reviews in Food Science and Nutrition*, 35(4), 299–339.

Pardales J. R. and Esquibel, C.B. (1996), 'Effect of drought during the establishment period on the root system development of cassava', *Journal of Crop Science*, 65, 93–7.

Pasquini D., Teixeira E.M., Curvelo A.A., Belgacem M.N. and Dufresne, A. (2010), 'Extraction of cellulose whiskers from cassava bagasse and their applications as reinforcing agent in natural rubber', *Industrial Crops and Products*, 32, 486–90.

Poonsrisawat A., Wanlapatit S., Paemanee A., Eurwilaichitr L., Piyachomkwan K. and Champreda, V. (2014), 'Viscosity reduction of cassava for very high gravity ethanol fermentation using cell wall degrading enzyme from *Aspergillus aculeatus*', *Process Biochemistry*, 49(11), 1950–7.

Rattanachomsri U., Tanapongpipat S., Eurwilaichitr L. and Champreda, V. (2009), 'Simultaneous non-thermal saccharification of cassava pulp by multi-enzyme activity and ethanol fermentation by *Candida tropicalis*', *Journal of Bioscience and Bioengineering*, 107(5), 488–93.

Raemakers K., Schreuder M., Suurs L., Furrer-Verhorst H., Vincken J., de Vetten N., Jacobsen E. and Visser, R.G.F. (2005), 'Improved cassava starch by antisense inhibition of granule-bound starch synthase I', *Molecular Breeding*, 16, 163–72.

Rojanaridpiched, C. (2013) 'Development of "Waxy Tapioca"', Paper presented at World Tapioca Exhibition on 2013, 19 June 2013, Ministry of Commerce, Nakhon Ratchasrima, Thailand.

Satisopasri V., Kurotjanawong K., Chotineeranat S., Piyachomkwan K., Sriroth K. and Oates, C.G. (2001), 'Impact of water stress on yield and quality of cassava starch', *Industrial Crops and Products*, 13(2), 115–29.

Sriroth, K., Chollakup R., Chotineeranat S., Piyachomkwan K. and Oates, C.G. (2000a), 'Processing of cassava waste for improved biomass utilization', *Bioresource Technology*, 71, 63–9.

Sriroth K., Piyachomkwan K., Wanlapatit S. and Oates, C.G. (2000b), 'Cassava Starch Technology: The Thai Experience', *Starch/Stärke*, 52(11–12), 439–49.

Sriroth K., Santisopasri V., Petchalanuwat C., Kurotjanawong K., Piyachomkwan K. and Oates, C.G. (1999), 'Cassava starch granule structure-function properties: influence of time and conditions at harvest on four cultivars of cassava starch', *Carbohydrate Polymers*, 38(2).

Sriroth, K., Wanlapatit S., Piyachomkwan K. and Oates, G.C. (1998) 'Improved cassava starch granule

Teixeira E. M., Pasquini D., Curvelo A.A.S., Corradini E., Belgacem M.N. and Dufresne, A. (2009), 'Cassava bagasse cellulose nanofibrils reinforced thermoplastic cassava starch', *Carbohydrate Polymers*, 78, 422–31.

Tester R. F., Karkalas J. and Qi, X. (2004), 'Starch-composition, fine structure and architecture', *Journal of Cereal Science*, 39, 151–165.

Thai Industrial Standard Institute (TISI). (1992), 'Standard for modified starch for food industry', UDC 547.458.61:591.133.1, Ministry of Industry, Bangkok.

Wanlapatit S., Wansuksri R., Chotineeranat S., Piyachomkwan K., Laoka B., Termvejsayanont N., Sriroth K. and Ruthaiyanont, A. (2015) 'Improving extraction yield in cassava starch factory by recovery of starch in pulp at high solid loading with enzyme cocktails as a continuous integrated process', Thai Patent No. TH1501002971.

Waterschoot J., Gomand S.V. and Fierens, E. (2015), 'Production, structure, physicochemical and functional properties of maize, cassava, wheat, potato and rice starches', *Starch/Stärke*, 67(1–2), 14–29.

Wicaksono R., Syamsu K., Yuliasih I. and Nasir, M. (2013) 'Cellulose Nanofibers from Cassava Bagasse: characterization and application on tapioca-film', *Chemistry and Materials Research*, 3(13), 79–87.

Chapter 4

New uses and processes for cassava

Keith Tomlins and Ben Bennett, Natural Resources Institute, University of Greenwich, UK

1 Introduction
2 Current and developing cassava production and use
3 New products from cassava: the case of High Quality Cassava Flour (HQCF) in Africa
4 Summary and future trends
5 Conclusion
6 Where to look for further information
7 Acknowledgements
8 References

1 Introduction

This chapter addresses the questions of what cassava is currently used for and what new processes might be introduced that promote greater utilisation, particularly products from small-scale farmers. Using the example of High Quality Cassava Flour (HQCF) in Africa, a framework of enabling factors that might help promoters of new technologies achieve sustainability is discussed. From this we highlight the key drivers of new uses and processes of cassava. Finally, we consider what new opportunities might arise from the research activity currently underway and highlight how actors might respond to these new approaches and norms.

2 Current and developing cassava production and use

Cassava is an important crop in sub-Saharan Africa and elsewhere in the world, being the third most important source of calories in the tropics and the sixth most important global food crop (after sugar cane, maize, rice, wheat and potato). Cassava's versatility, resilience to climate stress and huge potential for yield increase make it increasingly popular for both food security and industrial applications.

It has a diverse range of uses, mainly as food. More than half of the world's cassava production is for food use followed by a plethora of semi-industrial and industrial uses; for example, as an animal feeds ingredient, as a biofuel feedstock and as a source of

http://dx.doi.org/10.19103/AS.2016.0014.29

native starch for many other uses (Prakash 2008). In some countries (e.g. the Democratic Republic of Congo), cassava is grown for its leaves, an important source of protein and micronutrients.

The type of use varies by region of the world where it is produced and by the scale and efficiency of its production. For example, within Africa, cassava is produced on smallholder farms with family labour using hand tools and without use of external inputs (Lamboll et al. 2015), and yields are typically low (e.g. less than 10mt/ha). In Asia, Brazil and South East Asia production is more mechanised and cassava tends to be a cash monocrop (Naziri et al. 2014). Yield is higher in these areas (e.g., 20–30mt/ha). In Latin America, cassava production largely remains a marginal crop where most is for human consumption, but 10% is for animal feed (Henry and Hershey 2002). Southern Brazil is a notable exception, where expansive areas are highly mechanised. Cassava in the Caribbean is still primarily produced by smallholders for food use (Henry and Hershey 2002), although some is exported. Haiti and Cuba are the principal producers. It is of limited importance in the Pacific and most is grown for food use (Onwueme 2002). Cassava forms the basis of many upstream products, including foods (fermented and non-fermented), flour, animal feed, alcohol, starch, sweeteners, prepared foods and biodegradable products including paperboard, glue and building materials (Bechoff et al. 2016). These products are derived from a number of forms, ranging from fresh roots and leaves to modified cassava starch and different forms of glucose.

The production of cassava in the regions of the world is continually changing with the relative levels of production changing over time. To illustrate this, cassava production over a 20-year time period from 1994 to 2014 for different regions of the world is shown in Fig. 1 (Faostat 2016). Production has increased most rapidly in Africa, Caribbean and Asia

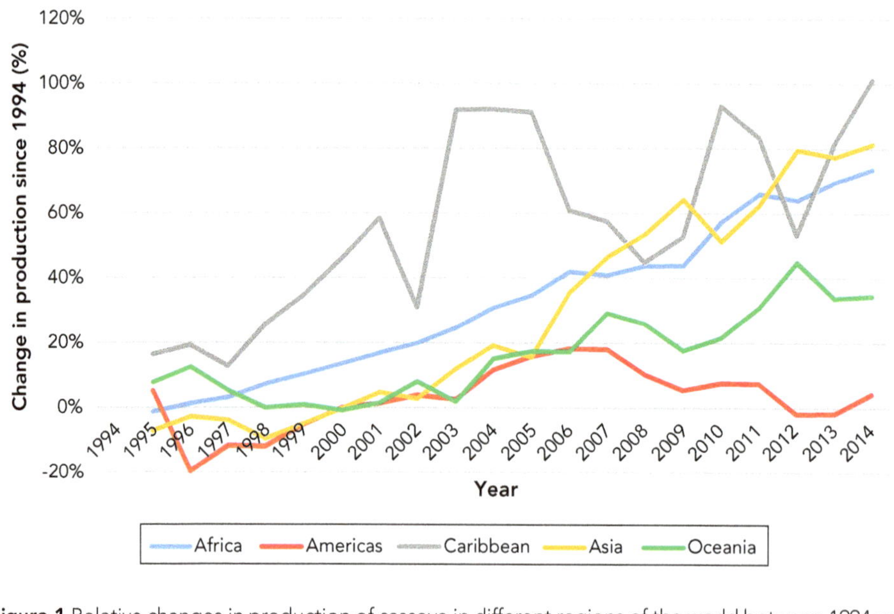

Figure 1 Relative changes in production of cassava in different regions of the world between 1994 and 2014 (authors' own calculations from FAO data).

where levels are currently up by between 70% and 100% compared to levels in 1994. Production has increased by less than 40% in Oceania and has stagnated in the Americas.

In addition to growing production and markets, the opportunities for new uses and processes are in part determined by trade across borders, particularly where domestic industries cannot benefit from economies of scale. The volumes of import and exports of cassava relative to production by country are illustrated in Fig. 2. Figure 2 suggested that only 9% of cassava produced is either imported or exported (as fresh root equivalent where the import and export figures for dry products were multiplied by 3). Almost all of the international trade in cassava is among only three countries – Thailand and Vietnam mainly export to China which has a growing demand for cassava for use in feeds and as biofuel. Africa currently does not trade much cassava internationally and much of its trade

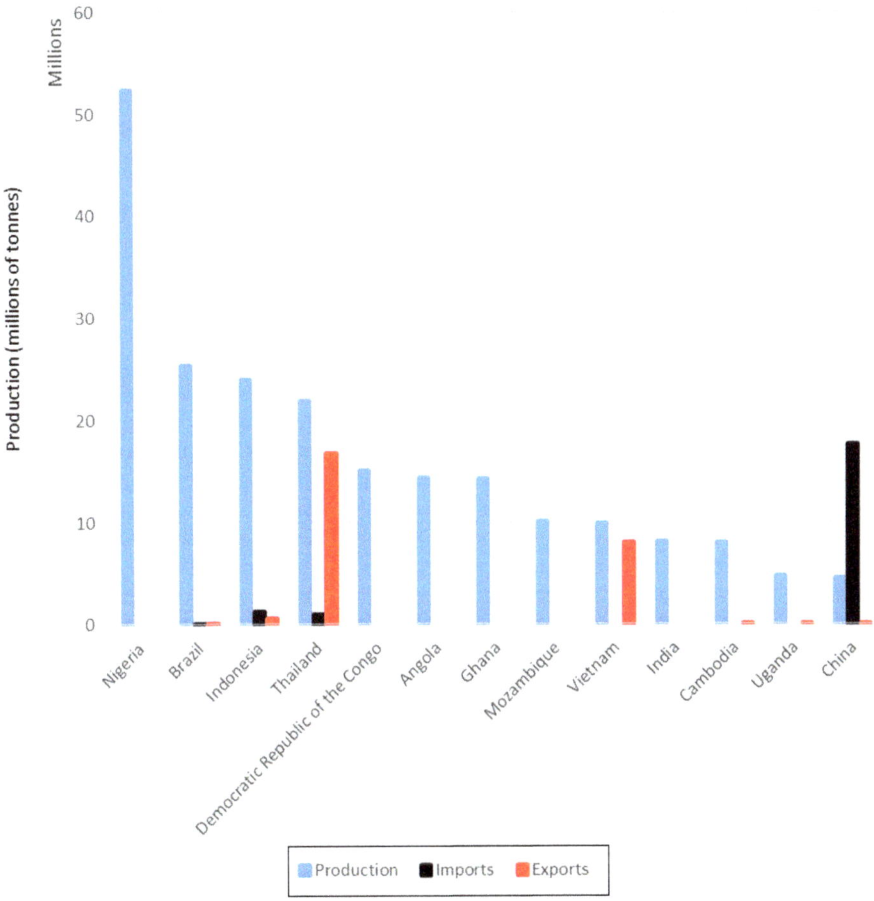

Figure 2 Production, imports and exports of cassava by country (authors' own calculations from FAO data).

Where volumes of imports and exports have been multiplied by a factor of three to convert to fresh root equivalents.

remains within the domestic informal sector (Bennett 2015), but precise figures are not available. The volume of international trade for cassava is lower compared to durables such as maize primarily because fresh cassava has a short shelf-life of less than three days (Bennett 2015) and the high moisture content adds to transport costs and moisture reduction costs. In addition, low yield and high factor costs (e.g. transport and processing) in key producing countries such as Nigeria make export uneconomic to cassava-deficient economies such as China in competition with high-yielding low-cost countries such as Thailand and Vietnam.

Since most of the cassava produced is consumed domestically/nationally, the expansion in production is mainly in response to increasing domestic demand (rapidly expanding population and increasing urbanisation) and supply factors (higher-yielding varieties, post-harvest technologies and switching to cassava in areas of high land pressure) (Lamboll et al. 2015). In domestic cassava economies, the issues of perishability and high cyanide content need to be addressed (Westby 2002). In traditional cassava-producing and -consuming economies, methods of processing have evolved to address the perishability/cyanide challenge. These include different forms of fermentation (e.g. baton de manioc in Cameroon and Chiquangue in the Democratic Republic of Congo), manufacture of fermented flours (e.g. fufu in Ghana) and production of fermented, dried cassava pellets (e.g. garri from Nigeria).

Processing of cassava has the added advantage of reducing bulkiness of fresh cassava roots by removing water, resulting in improved storability and lower transport costs to urban market centres. Examples of product types, intermediate processes and market sectors for cassava are summarised in Fig. 3.

The high potential that cassava has for increased productivity allies with a downstream set of products for which a strong market pull exists. For example, for traditional African products, this has led to the expansion of urban and peri-urban value chains and growth in the number of actors engaged in selling fresh cassava (e.g. 'kokonte' in Ghana), converting cassava into traditional foods for sale to households and the growing urban restaurant sector (e.g. 'garri' in Nigeria). However, huge scope exists for increasing cassava consumption as an intermediate good in a wide range of industries at many different scales (Bennett 2015). The hypothesis is that, by increasing yields and individual farm productivity for cassava, farm and factory gate prices can be reduced and new products developed using cassava that can compete against more traditional food/starch sources such as maize and wheat. In the next section we consider one example of a new product driving growth in cassava productivity and the challenges that this process has involved.

3 New products from cassava: the case of High Quality Cassava Flour (HQCF) in Africa

As we have seen, the drivers of new product development for cassava are perishability, food safety (the existence of cyanide in the final product), quality and the price of substitutes such as wheat and maize. Until recently, limited options were available to address these issues. Most of the options involved extracting starch from raw cassava, an industrial process with huge economies of scale. With a few exceptions, the large-scale processing of cassava for starch has not been successful in Africa. Reasons cited by industry include:

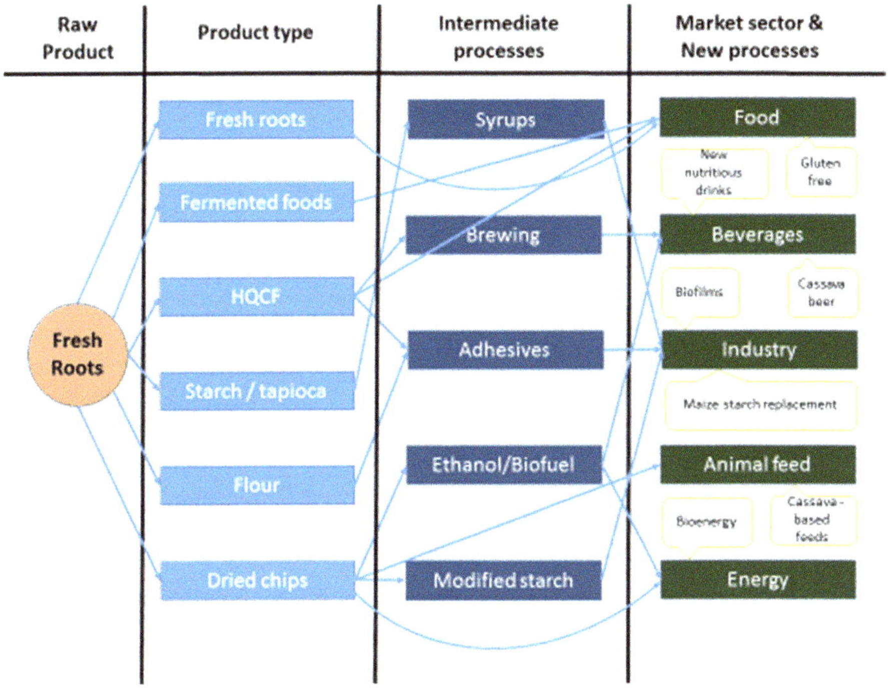

Figure 3 Examples of product types and markets for cassava.

high transaction costs, intermittent energy supply, competition with traditional uses for raw materials and much more efficient global competitors such as India and Brazil for key markets (Bennett 2015).

Westby and Choo (1994) confirmed that an ancient solution to the cyanide issue is to grate fresh cassava and promote sufficient tissue disruption to allow the endogenous linamarase present in cassava cells to react with linamarin and create the conditions for the cyanide to volatize spontaneously (Westby 2002). Once dried and sieved, this free-flowing flour is nearly cyanide free. This insight led to the development of a new intermediate product called HQCF with potential for household, small industry and large industry application.

Subsequently, there have been significant efforts, supported by donors and the private sector, to establish industries on the basis of HQCF, the most notable of which is the Cassava: Adding Value for Africa (C:AVA) project (C:AVA; http://cava.nri.org and http://cava2.unaab.edu.ng/) funded by the Bill and Melinda Gates Foundation in Nigeria, Ghana, Tanzania, Uganda and Malawi and the EU funded Cassava Growth Markets project. C:AVA aims to increase incomes and livelihoods of small-scale cassava growers and thus drive demand for cassava productivity tools such as better planting material, improved farm management and mechanisation. To do this, the project sought to increase the supply of HQCF at different scales, while also promoting demand for HQCF as a safe and cost-effective raw material for a range of industries, largely focusing on replacing more expensive ingredients such as wheat flour. New information about how women can engage with new cassava commercial opportunities has been gained (Forsythe et al. 2015, 2016). Women can gainfully participate in new commercial cassava opportunities while maintaining, if not increasing, food security. However, this is highly dependent on gender norms and household relations.

Table 1 Factors and characteristics enabling HQCF upscaling

Enabling factor	Characteristic	Description
Leadership	Flexibility	The ability to develop a range of models and approaches best suited to the market conditions
	Diversity of approach	Not all production or scaling models should be the same
	Led by 'champions'	Pioneers and risk takers are needed to get a new process working at scale
	Diversity of chain drivers	Lead actors are needed across the range of drivers: producers, buyers, facilitators, integrators, processors and policy makers
	Longevity	It can take many years before scale is achieved
Incentives	Competitive	Having better qualities and/or price than alternative products
	Acceptable to consumer	Having the same or better appeal to consumers
Institutional capacity	Collective production	Groups of farmers forming an HQCF processing business
	Clusters of producers	Groups of farmer-led HQCF processing businesses supplying end users
	Sector-specific intermediaries	Bulking agents for fresh roots and distributors of HQCF
	HQCF end users ready to buy	Industries committed to investing in HQCF supply
	Service providers able to support	Bodies supporting farmers, processors and end users
Infrastructure	Roads to production areas	Easy access to areas of raw root production
	Transport available from production areas to processors	Farm-to-road and road-to-processing site
	Services available and affordable	Sufficient and reliable power and water supply

Enabling factor	Characteristic	Description
	Communications between actors possible	Mobile phones to manage markets and supply
Technology	Production approach optimal: village vs SME vs industrial scale	HQCF bulked from many locations or from one location
	Drying approach optimal: sun drying vs mechanical drying	Simple technology with seasonal constraints vs expensive and complex technology with consistent supply
	Utilisation of waste and by-product	Contribution of cassava processing by-products and waste to income optimised
Finance	Capital costs appropriate to scale and chain	Greater scale requires more capital investment at start-up All approaches need some capital
	Working capital sufficient for production	Money to buy roots and pay for recurrent costs
Policy	Quality assurance, regulation and standard setting appropriate	Setting and regulating product standards to prevent free-loading
	Institutional leverage maximised	Using policy to promote demand: e.g. tariffs on substitutes and minimum inclusion rates for government procurement
Social & cultural	Consumers' acceptance	Acceptance or preference for products by end users and consumers
	Gendered	New processes take into account the role and impact upon women and girls

Source: Adapted from (Lamboll et al. 2015).

A number of lessons can be learnt from the efforts to upscale HQCF (Lamboll et al. 2015). These 'enabling/constraining' contexts are described in Table 1.

Some additional important lessons have been learnt about new cassava product development from the C:AVA HQCF experience as follows:

- **Market research**. Fully understand the potential market and market segments before developing supply patterns, to ensure that the scale needs of end users are planned ahead of production.
- **Scale correctly**. Ensure that the supply model can meet the demand identified by the market research.
- **Technology support**. Ensure that local firms have the ability to supply and support any equipment needed for processing. An important aspect of success for C:AVA

was its encouragement of the development of small-scale flash dryer fabrication in Nigeria (Marchant et al. 2015).

One aspect of HQCF development that may have been overlooked is the promotion of the new product as an ingredient. This arose because the availability of HQCF generated several new value chain opportunities that, to date, have not been fully exploited. These include:

- Using HQCF as a binding agent in the paperboard industry, including HQCF as a replacement for starch and fibre from other sources in processed food and beverages (although some early success has been achieved in the beer and processed meat sectors)
- Utilising HQCF as specialised aquaculture feed ingredient to promote pellet floating
- As an ingredient in the emerging gluten-free market in Asia
- At the right price against alternative ingredients, there are clearly huge opportunities to include HQCF as a wheat replacement in a range of different bakery products. At low inclusion levels (e.g. less than 20%) consumers are unable to see the difference in many breads. Higher rates of inclusion are possible with specialist enzyme additives.

In conclusion, HQCF has started to achieve its aim of driving demand for cassava production and promoting efficiency gains. However, there is still some road to travel. Other new processes and approaches that can drive demand for cassava production are emerging.

4 Summary and future trends

What HQCF has shown in recent years is that cassava processed on a sub-industrial scale, can, against a background of many challenges, supply industries at significant scale. There are two big drivers of process development in the cassava sector. The first is cost. Higher energy costs open up opportunities for cassava if it can be produced and processed efficiently. The government of China has supported the emergence of a bioethanol industry. In Africa, there are, to date, only a few examples of successful bioethanol production. With the right incentives and supply models, this can change. As world food prices rise in real terms and transport costs also go up, domestic or regionally produced cassava can become the optimal ingredient for industrial production of a wide range of products. The second driver is scale. Traditional starch extraction-based industries seem to be just too large to work effectively in the African context. New, cheaper and more robust technologies that allow efficiency at a smaller scale to meet the demands of medium-sized upstream industries that supply small, but rapidly growing, new market segments will be the new norm (e.g. bioethanol, aquaculture feed and bakery products). In addition, a possible third area for new product development that is emerging now is new areas of consumer preference. We can see, for example, that gluten-free products are increasingly in demand in higher-income countries, and cassava is an ideal ingredient for these. Scope exists in emerging markets for new food products based on locally sourced ingredients, for example, in the rapidly expanding snack sector. Opportunities to combine the advantages from cassava with other potential locally sourced ingredients have hardly been explored. Composite flours with improved nutritional benefits are a good example of this approach.

Finally, one way of predicting where cassava processing will take us in the near future is to consider activity in patenting new processes (see https://patentscope.wipo.int/search/en/result.jsf). What this tells us is that patents for new uses and processes cover a range of areas that include agronomy, ethanol production, amino acids (potentially food supplements or medical applications), biodegradable films, plastics and packaging, plant breeding, cosmetics, environment improvement, feeds, new food products, reducing postharvest deterioration and processing methods. A trend was that patents applied for in China tend to focus on agricultural production, whereas those published in the United States tended to focus on higher-value end-user products.

5 Conclusion

This chapter has considered the current uses of cassava and its processed products and explored this in the context of the development of new processes from the point of view of introducing HQCF in Africa. A framework for understanding the various enabling factors to developing such a sub-sector was provided in the ambit of this case study. Looking at the research activity through the lens of recent patents shows the likely frontier of cassava-processing technologies in the near future.

Some conclusions can be drawn from this investigation. Firstly, as has been shown by Bechoff et al. (2016), there is a critical link between efficient processing and development of appropriate cassava varieties. While traits such as productivity and resistance to disease will always be important, considerations of value to processors such as starch content and quality should not be overlooked.

We also conclude that individual technical advances can be successful only in the context of an efficient cassava supply system (Lamboll et al. 2015) and the encouragement of the development of locally produced and owned processing equipment fabrication (Marchant et al. 2015). No amount of technology will overcome competitive disadvantage and this needs to be rigorously assessed on an ongoing and case-by-case basis.

Finally, considering the range of sectors and subsectors where cassava can be utilised, it is felt that much more could be done to systematically address the enabling factors that make a subsector successful. Cassava platforms in Thailand and China demonstrate the power of this approach at national and subregional levels.

6 Where to look for further information

There are a number of sources for further information and these are suggested as follows:

- Attending forums specifically for cassava and other root crops such as the International Society for Tropical Root Crops (ISTRC) (www.istrc.org and Facebook group) and the Global Cassava Partnership for the 21st Century (GCP21) (www.gcp21.org/)
- Attending conferences related to cassava and new products such as the triennial ISTRC and ISTRC-Africa Branch Symposia which have been running since 1967. These conferences and others often include speakers who will present new information about new and emerging cassava products and market trends.

- Attending Trade Fairs where cassava products are exhibited
- Reading scientific literature
- Reading industry magazines
- Contacting projects on cassava and enterprises directly

7 Acknowledgements

The information in this chapter is the result of many years of international collaboration between natural and social science researchers at the Natural Resources Institute, University of Greenwich, UK, and our valued partners in many countries, notably Ghana, Malawi, Nigeria, Tanzania and Uganda.

8 References

Bechoff, A., K. Tomlins, G. Fliedel, L. A. Becerra Lopez-Lavalle, A. Westby, C. Hershey and D. Dufour (2016). Cassava traits and end-user preference: relating traits to consumer liking, sensory perception, and genetics. *Critical Reviews in Food Science and Nutrition.*

Bennett, B. (2015). Guest editorial: Smallholder cassava production and the cassava processing sector in Africa. *Food Chain*, 5, 1–3.

Faostat (2016). Agriculture Organization of the United Nations Statistics Division (2014). *Production Available in: http://faostat3.fao.org/browse/Q/QC/S [Review date: April 2015].*

Forsythe, L., A. Martin and H. Posthumus (2015). Cassava market development: a path to women's empowerment or business as usual? *Food Chain*, 5, 11–27.

Forsythe, L., H. Posthumus and A. Martin (2016). A crop of one's own? Women's experiences of cassava commercialization in Nigeria and Malawi. *Journal of Gender, Agriculture and Food Security*, 1, 110–28.

Henry, G., and C. Hershey (2002). Cassava in South America and the Caribbean. *Cassava: Biology, Production and Utilization*, 17–40.

Lamboll, R., V. Nelson, H. Posthumus, A. Martin, K. Adebayo, F. Alacho, N. Dziedzoave, G. Mahende, V. Sandifolo and L. Sanni (2015). Practical lessons on scaling up smallholder-inclusive and sustainable cassava value chains in Africa. *Food Chain*, 5, 28–52.

Marchant, A., A. Graffham, L. Sanni and I. Adeoya (2015). Dramatic efficiency gains through improved heat exchangers: the case of cassava flash drying in Nigeria. *Food Chain*, 5, 91–104.

Naziri, D., W. Quaye, B. Siwoku, S. Wanlapatit, T. Viet Phu and B. Bennett (2014). The diversity of postharvest losses in cassava value chains in selected developing countries. *Journal of Agriculture and Rural Development in the Tropics and Subtropics*, 115, 111–23.

Onwueme, I. (2002). *Cassava in Asia and the Pacific.* CABI Publishing: Oxon, UK, pp. 55–65.

Prakash, A. (2008). Cassava: international market profile. *Background paper for the competitive commercial agriculture in Sub-Saharan Africa (CCAA) study. Trade and Markets Division: Food and Agriculture Organisation (FAO), Rome.*

Westby, A. (2002). Cassava utilization, storage and small-scale processing. *Cassava: Biology, production and utilization*, pp. 281–300.

Westby, A., and B. Choo. (1994). Cyanogen reduction during lactic fermentation of cassava. In *International Workshop on Cassava Safety*, 375, 209–16.

Part 2

Current cassava cultivation and opportunities for improvement

Chapter 5

Cassava cultivation in Asia

Tin Maung Aye, CIAT, Vietnam

1 Introduction
2 Cassava production in Asia: an overview
3 Varietal improvement
4 Other key research areas for developing cassava production in Asia
5 Case study: developing sustainable cassava production in Cambodia
6 Summary
7 Where to look for further information
8 References

1 Introduction

Cassava (*Manihot esculenta* Crantz) was traditionally grown as a subsistence crop in Asia, but has become one of the most popular crops, especially in upland environments. Cassava can be used as a primary source of food and animal feed, and its production provides income for many rural communities, especially those living in remote rural areas of tropical and sub-tropical regions. Cassava often represents the cheapest source of carbohydrate (energy) and supports many smallholder farming (crop–livestock) systems in the tropics. This starchy root crop is frequently substituted as the staple food for rural people during periods of rice shortage.

In fact, cassava was initially grown in Asia as a food crop for local consumption. This crop was introduced to many parts of Asia by European explorers in the late eighteenth and the early nineteenth centuries (Taco Bottema and Henry, 1992; Onwueme, 2002). There is little evidence that the crop made its way across the entirety of Asia, but the crop did arrive early to India, Java of Indonesia, and the Philippines. However, cassava production systems in Asia have been developing at a rapid rate due to a range of drivers including increasing regional market integration, changing labour availability for agriculture, and government policies that encourage commercial agriculture alternatives to subsistence agriculture. Cambodia, Indonesia, Thailand and Vietnam are among both the top cassava producer countries and exporters in the world.

http://dx.doi.org/10.19103/AS.2016.0014.05

Many farmers grow cassava under a wide range of conditions, even in areas that have often suffered from drought and infertile, erosion-prone soils, characteristic of many upland soils in Asia. In addition to their low fertility, the uplands are also characterized by diverse land use systems (Dierolf et al., 2001). When smallholders grow cassava on their marginal lands, there are often minimal to no inputs with the exception of their family labour (Howeler, 1988, 2014; Howeler and Aye, 2014). Fertilizers and manure are not commonly applied in smallholder cassava systems, although in recent years, fertilizer use is increasing. Moreover, soil degradation by erosion can be a serious concern since cassava is often cultivated on sloping lands with light-textured soils with low organic matter and nutrient content. Soil erosion carries away not only plant nutrients in the soil and fertilizer but also a disproportionate amount of clay particles, organic matter and beneficial soil microorganisms.

In many cassava-growing areas in Asia, smallholder farmers still use traditional or low-yielding varieties. Recently, many smallholder farmers have begun to grow newly introduced varieties from different parts of Asia especially from China, Thailand and Vietnam. However, when these varieties are grown outside their adapted areas, they often fail to provide optimum yield. The diverse agro-ecological conditions of the uplands represent a challenge for developing varieties with sufficiently broad adaptation, or developing specific varieties for each agro-ecological sub-region. Research on cassava varietal improvement has successfully focused mainly on yield and starch content. In addition, research to improve crop and soil management practices have resulted in a better understanding of sustainable cassava production and suitable agronomic practices.

2 Cassava production in Asia: an overview

2.1 Cassava production systems in Asia

While cassava is one of the most abundantly produced cash crops in the tropical and sub-tropical uplands in Cambodia, southern China, Thailand and Vietnam where most of the continent's cassava is produced, it is more thought of as a subsistence food crop for smallholder farmers in East Timor, various parts of Indonesia, southern India and northern Laos. Cassava is also produced in rice-based systems and in lowland environments before or after rice cultivation. Although in many parts of Asia cassava is a rain-fed crop, it is irrigated in certain parts of India, Thailand and southern Vietnam. In the hilly and mountainous regions of Asia, especially in remote ethnic areas, the crop is mostly grown in mixed cropping systems with other food crops. More than nine million farmers grow cassava in Asia, including more than four million farmers in the Greater Mekong sub-region. Another 1.5 million households in southern China and three million in Indonesia grow cassava. Since many smallholder cassava farmers are women, the crop has significant implications for gender equity. Farm size in Asia can vary but tends to be very small, for example 0.2–0.8 ha/family in China, Kerala State of India, Java island in Indonesia, and Vietnam and 2–3 ha/family in Cambodia, Laos, Myanmar and Thailand.

While cassava varieties vary from country to country and farmer to farmer, a few varieties from the Thai and Vietnamese breeding programmes have been broadly successful across the region, especially KU50 from Kasetsart University in Thailand, perhaps the world's most widely grown variety. Cassava is often grown in association with other upland crops such as rice, maize and legume crops such as cowpea, mung bean, peanut and soybean. However, it is primarily cultivated as a sole crop in major cassava-growing regions in Asia. Farmers commonly plant cassava at the beginning of the rainy season, which in most Southeast Asian countries is in April or May, but in Indonesia, this season can range from October to December. Some farmers may also plant cassava towards the end of the rainy season, which in most countries is in September and October, or even in November. In some regions of Cambodia, Laos, Thailand and southern Vietnam, farmers grow cassava year-round.

Farmers prepare the land by hand (by hoe), cattle and buffalo or by two- or four-wheeled tractors. Most farmers in India, Indonesia and Thailand plant cassava stakes vertically, while most farmers in China, Vietnam and Cambodia plant stakes horizontally or in an inclined position. Nowadays, fertilizers and manure are applied to cassava in commercial growing areas, but not necessarily in balanced ratios or in sufficient quantity to maintain an optimum yield in the long term. In addition, herbicides are commonly used for weed control in the region, but many smallholders still manually control weeds with a hoe 2–4 times during the first 3–4 months. The average yield in the Southeast Asian region is highest in India; however, the production cost per ton of roots is also fairly high because of the intensive use of labour and other inputs such as fertilizer and supplementary irrigation. It is estimated that cassava farmers in Asia received more than US$1.5–2.0 billion per year of additional gross income due to higher yields obtained in 2012 compared to 1994 (Howeler, 2014).

2.2 Climate and cassava-growing soils in Asia

Cassava can generally grow only between 30° S and 30° N of the equator, in the regions from sea level to about 2300 m altitude. This range allows the production of cassava in many parts of tropical and subtropical Asia, particularly in Southeast Asia. Cassava productivity is driven by three climatic variables – temperature, light and precipitation – and by soil properties. In tropical Asia temperature and light generally do not limit cassava production. However, effective soil moisture, especially in regions where potential evaporation exceeds precipitation many days of the year, is the dominant factor determining the cassava growth rate and starch accumulation in the roots. The crop grows best in areas where the temperature is between 25 and 29°C. Its growth stops when temperatures drop below 15°C, but cassava will tolerate temperatures up to 38°C (Alves, 2002). Cassava is highly drought tolerant but still grows best when soils have sufficient moisture. Approximately 295 million ha of the uplands in Southeast Asia are made up of low pH soils with low phosphorus (P) status and potential for aluminium (Al) toxicity. The main soil types are Ultisols, and to a lesser extent Oxisols, Entisols, Inceptisols, Histosols and Andisols (Dierolf et al., 2001). Ultisols usually contain low levels of OM, N, P, K and Mg. Al-toxicity and P-fixation are also common in these soils (Dierolf et al., 2001). Moreover, Ultisols are fine-textured with a red or yellow subsoil and associated with undulating or hilly landscapes, which make them susceptible to erosion. Cassava is predominantly grown on Utilisols because of its ability to grow reasonably well in very acidic and low-fertility soils compared to grain crops (Anue and Lar, 1997; Howeler, 2014). The negative effects of growing cassava on sloping lands and light-textured soils without soil conservation measures include soil erosion, plant nutrient depletion and loss of organic matter and beneficial soil microorganisms. Erosion results in the deterioration of the soil's

Table 1 Cassava production, area and yield in various countries in Asia in 2003 and 2013

	Area harvested ('000 ha)		Production ('000 tons)		Fresh root yield (t/ha)	
Country	2003	2013	2003	2013	2003	2013
Total South America	2,282	2,145	30,405	28,776	13.3	13.4
Total Africa	11,947	14,177	104,664	157,987	8.8	11.1
Total Asia	3,384	4,182	55,452	88,221	16.4	21.1
Cambodia	25	350	331	8,000	13.2	22.9
China	250	286	4,015	4,599	16.0	16.1
India	207	207	5,426	7,237	26.2	35.0
Indonesia	1,245	1,066	18,524	23,937	14.9	22.5
Lao PDR	1	45	4	1,120	5.8	25.2
Malaysia	5	3	93	45	20.2	14.5
Myanmar	12	49	138	630	11.2	12.9
Philippines	209	217	1,622	2,361	7.7	10.9
Sri Lanka	26	24	229	293	8.7	12.2
Thailand	1,022	1,385	19,718	30,228	19.3	21.8
East Timor	10	7	41	27	4.1	4.1
Vietnam	372	544	5,309	9,743	14.3	17.9

Source: FAOSTAT, 2015.

physical and chemical properties, which in turn affects the soil's productivity and cassava yields, with shallow soils or those having unfavourable subsoil being most affected.

2.3 Cassava production, area and yield in Asia

Asia contributes about one-third of the world's total cassava supply with Southeast Asia taking the lead both in terms of yield and harvested area (Table 1). In Southeast Asia, the total harvested area of cassava is currently about 4.1 million ha with the largest areas in Thailand (34%), Indonesia (26%), Vietnam (13%) and Cambodia (8%) in 2013 (FAOSTAT 2015). The average cassava yield in Asia increased from about 12.7 t/ha in 1984 to 21.1 t/ha in 2013 (FAOSTAT 2015). However low yield still constitutes one of the main challenges of cassava production in Asia. Cassava production data of individual countries in Asia are presented in Table 1.

3 Varietal improvement

Most traditional cassava varieties are the outcome of selection by generations of farmers (Howeler and Aye, 2014). Since the mid-1970s, cassava breeders in national programmes

Table 2 Cassava varieties suitable for human consumption that have been grown or released in Asia, and their most important characteristics

Country	Variety name	Year of release	Main characteristics
Cambodia	Damlong Kor	–	Good eating quality
	Damlong Mi	–	Good eating quality
China	Bread cassava	–	Low yield, high starch content, eating quality
	SC 102	–	Low yield, high starch content, eating quality
	SC 6068	1980	Low yield, high starch content, eating quality
	SC 9 = yolk cassava	2005	High yield, high starch content, high β-carotene*, eating quality
East Timor	Mantega	–	Good eating quality; high β-carotene
	Lesu	–	Good eating quality
	Ai Luka 2	2007	High yield, good eating quality
	Ai Luka 4 = Gading	2007	High yield, good eating quality
India	M4	–	Good eating quality, relatively low yield
	Sree Visakham	1977	High β-carotene, suitable under coconut
	Kalpaka	1996	Early bulking, good cooking quality
	Sree Jaya	1998	Early maturing, good cooking quality, suitable for rice-based rotation
	Sree Vijaya	1998	Early bulking, β-carotene, for rice-based rotation
	Sree Rekha	2000	High yield, β-carotene, for lowland and upland
	Sree Prabha	2000	High yield, β-carotene, for lowland and upland
	Vellayani Hraswa	2002	Early bulking, good cooking quality, for rice-based rotation
Indonesia	Adira 1	1978	High yield, good eating quality
	Malang 2	1992	High yield, sweet
	Darul Hidayah	1998	High yield, sweet, specific adaptation
	Litbang UK2	2012	High yield, high ethanol yield, early bulking, sweet
Malaysia	Medan	–	Popular old eating variety, good when steamed
	Sri Pontian	2003	Edible, for snack food
Philippines	Golden Yellow	–	Good eating quality, high β-carotene
	VC-2	1988	High yield, edible

(Continued)

Table 2 *(Continued)*

Country	Variety name	Year of release	Main characteristics
	VC-3	1990	Dual purpose
	VS-4	1990	High yield, dual purpose
	PSB Cv-11	1995	Dual purpose
	PSB Cv-12	1995	Dual purpose
	PSB Cv-15	1999	Dual purpose
	NSIC Cv-48= R 72	2013	High yield, drought tolerant, dual purpose
Thailand	Hanatee	–	Relatively low yield, good eating quality
	Rayong 2	1984	Good eating quality, for snack food
Vietnam	Vinh Phu	–	Good eating quality
	Gon	–	Good eating quality
	KM 95	1995	High yield, dual purpose
	KM 95-3	1998	High yield, dual purpose
	KM 98-7	1998	High yield, dual purpose
	KM 98-1	2005	High yield, dual purpose
	KM 140	2007	High yield, dual purpose, early bulking
	KM 98-5	2008	High yield, dual purpose, early bulking

*β-carotene: Beta-carotene converts into vitamin A in the human diet.

across Asia have developed improved varieties with specific characteristics. Asia has lesser genetic diversity than Latin America and the introduction of germplasm from the centre of origin of cassava into Asia has allowed for more rapid improvement in many desired characteristics. After the establishment of the CIAT cassava programme regional office in Bangkok, Thailand in 1983, cooperation greatly expanded between the CIAT cassava programme and national programmes across Asia. Five common objectives of cassava varietal improvement in Asian programmes have been: 1) higher yielding capacity under low- and medium-input conditions; 2) higher root dry matter content; 3) early harvestability; 4) resilience though incorporating disease and pest resistance and tolerance to adverse soil and climatic conditions; and 5) lower root cyanide content, especially in areas where cassava is grown for food. The number of cassava varieties available has continually increased, with more than 50 improved varieties officially released in major cassava-growing countries in Asia. New varieties typically have a higher harvest index (more efficient at allocation of photosynthates to the roots), higher yields, and higher root dry matter (DM) content, along with early root bulking. Better taste and low HCN content are important in areas where roots are consumed directly or with little processing. Some of the main eating varieties and industrial varieties in Asia are shown in Tables 2 and 3. Continued adoption of new, higher yielding varieties is likely to further increase cassava yields in many countries in Asia.

Table 3 Cassava varieties released in Asia suitable for processing, and their most important characteristics

Country	Variety name	Year of release	Main characteristics
Cambodia	Damlong narrow leaf	--	High yield
	Malaysia = KU50	1	High yield, high starch content
China	SC 201	--	High yield, suitable for poor soils
	SC 205	--	High yield, suitable for high-fertility soils
	Nanzhi 188	1987	High yield, poor cold tolerance
	Nanzhi 199	1987	High yield, high starch content
	SC 124	1988	High yield, cold tolerant
	SC 8002	1994	High yield
	SC 8013	1994	High yield, typhoon resistant
	GR 891	1998	High yield, high starch content
	GR 911	1998	High yield
	SC 5	2002	High yield, typhoon resistant
	SC 6	2002	High starch content, typhoon resistant
	SC 7	2005	High yield
	SC 8	2005	High yield
	SC 10	2006	High yield, cold tolerant
	Gui Re 3	2006	High yield, high starch content
	Gui Re 4	2008	High yield
	SC 11 = MBra 900	2009	High yield
India	H-97	1971	High yield, high starch content
	H-165	1971	Early maturing, erect, suitable under coconut
	H-226	1971	High yield, high starch, suitable for processing
	Co-1 = ME-7	1976	High starch, CMD[2] tolerant, suitable for processing
	Sree Sahya	1977	High yield, drought tolerant
	Co-2	1985	Drought, CMD and root rot tolerant
	Sree Prakash	1987	Early maturing, suitable for rice-based rotation
	Co-3	1992	CMD tolerant, good cooking quality, suitable for processing
	Nidhi	1993	Early maturing, for sandy soils along coast
	H-119	1995	Early maturing

(Continued)

Table 3 *(Continued)*

Country	Variety name	Year of release	Main characteristics
	Sree Harsha	1996	High yield, high starch, drought tolerant, suitable for processing
	Co-4	2002	CMD tolerant, high starch content, erect
	Sree Padmanabha	2006	First CMD-resistant variety, drought tolerant
	Sree Athulya	2006	High yield, high starch, triploid, for processing
	Sree Apoorva	2006	High yield, high starch, triploid, for processing
Indonesia	Adira 2	1978	High yield, bitter
	Adira 4	1987	High yield, bitter
	Malang 1	1992	High yield, bitter
	UJ-3 = Thai = Rayong 60	2000	High yield, early bulking, very bitter
	UJ-5 = Kasetsart = KU 50	2000	High yield, high starch content
	Malang 4	2001	High yield, bitter
	Malang 6	2001	High yield, bitter
Malaysia	Black Twig	–	Commercial starch variety, highly adaptable
	Sri Kanji 1	2003	High yield, relatively high starch content
	Sri Kanji 2	2003	High yield, relatively high starch content
Philippines	Lakan	–	High yield
	VC-1	1986	High yield
	VC-3	1990	Dual purpose
	VC-4	1990	High yield, dual purpose
	VC-5	1990	High yield, bitter
	PSB Cv-11	1995	Dual purpose
	PSB Cv 12	1995	Dual purpose
	PSB Cv-15	1999	Dual purpose
	PSB Cv-19	2000	Mite resistant
	NSIC Cv-22=KU 50	2008	High yield, high starch content
	NSIC Cv-48=R72	2013	High yield, drought tolerant
Thailand	Rayong 1	–	High yield, rather low starch content
	Rayong 3	1983	High starch content, very branching

Country	Variety name	Year of release	Main characteristics
	Rayong 60	1987	High yield, early bulking, very bitter
	Sriracha 1	1991	High starch content
	Rayong 90	1991	High starch content, relatively high yield
	Kasetsart 50 (KU 50)	1992	High yield, high starch content
	Rayong 5	1994	Relatively high yield, high starch content
	Rayong 72	1999	High yield, drought tolerant
	Huay Bong 60	2003	High yield, high starch content
	Rayong 7	2005	High yield, high starch content
	Rayong 9	2005	High yield, high starch, high ethanol yield
	Huay Bong 80	2008	High yield, high starch content
	Rayong 11	2011	High starch content, high yield
	Rayong 86-13	2013	High starch content, high yield
Vietnam	La Tre = SC 205	–	High yield, mainly suitable for fertile soils
	KM 60=Rayong 60	1993	High yield, early bulking, very bitter
	KM 94=KU 50	1995	High yield, high starch content
	SM 937-26	1995	High yield, high starch content
	KM 95	1995	High yield, dual purpose
	KM 95-3	1998	High yield, dual purpose
	KM 98-7	1998	High yield, dual purpose
	KM 98-1	2005	High yield, dual purpose
	KM 140	2007	High yield, dual purpose, early bulking
	KM 98-5[3]	2008	High yield, dual purpose, dual purpose
	Sa21-12	2012	High yield, high starch content
	Sa06	2012	High yield, high starch content
	KM419	2013	High yield, high starch content
	HL-S10	2013	High yield, early bulking
	HL-S11	2013	High yield, high starch content, early bulking
	KM 101	2013	High yield, early bulking

[1] Introduced from Vietnam in the eastern provinces and from Thailand in the west, but never officially released.
[2] CMD = Cassava Mosaic Disease.
[3] KM 98-5 released in Tay Ninh and Dong Nai provinces.

4 Other key research areas for developing cassava production in Asia

4.1 Improvement in crop management

The rapid adoption of high-yielding cassava varieties and good agronomic practices in Asia has led to significant yield increases in the favourable rain-fed cassava-growing areas (Lefroy et al., 2015). Varietal improvement alone can substantially increase cassava productivity in some cases. However, in most situations an integrated approach is required to simultaneously improve agronomy and soil management, to take advantage of the potential of new varieties (Kawano, 1995). The overall cassava production increase from 1995 to 2005 in Southern China, Thailand and Vietnam and from 2005 to 2015 in Cambodia, Laos and Myanmar was mainly due to adoption of improved cassava varieties (such as KU 50, Rayong 72, KM95, KM98-1, SC124, Nanzhi 199, GR891), combined with better management.

Many commercial farmers use good-quality planting materials to grow healthy plants and obtain high yields. Planting materials derived from the mother plants that are between 8 and 18 months old, and have about 1–2 cm diameter stems result in better germination and produce better yields. In monoculture cassava systems, the distances between rows vary from 80 to 120 cm, and the distances between plants in the row from 60 to 120 cm. The appropriate distances between rows and between plants will vary according to the soil's fertility status, the branching habit of the variety, climatic conditions, and whether the crop is grown in monoculture or intercropping systems. Plant density of about 12 000 plants per ha (90 x 90 cm) of a semi-branched variety provided the highest yield with well-fertilized soil and favourable climatic conditions, but plant population of 18 000 plants per ha (75 x 75 cm) of the non-branching variety provided highest yields with well-fertilized soil in India. Three different locations in Thailand reported that the highest yields were obtained with spacing of about 15 000 plants per ha (100 x 66 cm) with two different cassava varieties (Tongglum, et al., 1987).

4.2 Time of planting and harvest

In Asia, planting of cassava is traditionally done at the beginning of the monsoon season (early May until June in most Southeast Asia countries but October and November in Indonesia). To obtain good growth and yield, cassava should be planted as early as possible, just before the rains or after the rains begin. Delayed planting leads to reduced cassava root yield. Although cassava can remain in the field for up to a year or more, the key part of the growing season is when local weather conditions (i.e. rainfall and temperature) allow normal plant growth. However, some farmers are now planting cassava throughout the year especially in Thailand and southern Vietnam, where the areas have sufficient soil moisture or access supplementary irrigation. Because the cassava industry has continued to grow and evolve over a number of years in Thailand and Vietnam, there is increasing need to extend availability of sufficient amounts of raw materials by extending the production period. However, in such a region it would be impossible to establish cassava during the dry season without enough soil moisture and/or rainfall.

Many farmers harvest cassava at 10–11 months after planting. The harvesting period coincides with the dry season (from January to April). Harvesting is easier when the soil is moist, and late harvesting of cassava will result in the roots becoming woody or more fibrous. Cassava planted on ridges and in sandy soils is also easier to harvest than cassava planted on flat ground and in heavy soils. However, the right time for harvesting cassava varies according to varieties, planting time, climate and soil conditions as well as market price of cassava.

4.3 Soil fertility maintenance in Asia

In Asia, cassava is frequently planted on soils with extremely low fertility that are often light textured with low soil organic matter. The physical and chemical properties of the soils in any type of climate will determine the relative yield of cassava and its quality. Results of experimental trials have been variable in the region, but deficiency of particular minerals (i.e. N, P, K) is known to reduce both quantity and quality of cassava roots. Most cassava-growing soils in Asia are inherently low in available N and P, and moderately low in available K. In general, N is the most limiting nutrient in the first few years of cassava production but K becomes the most limiting nutrient in continuous growing of cassava in many soils. Nevertheless, over-fertilization should be avoided, especially of N, which will result in excessive leaf formation at the expense of root growth (Howeler, 2014).

In spite of tolerance to low-fertility conditions compared to many crops, cassava is also highly responsive to N, P and K nutrient application in Asia. But which of these nutrients is the most limiting depends on the local soil characteristics. High-yielding cassava varieties are typically more responsive to mineral fertilizers than traditional ones. To produce optimum yields for maximum profits while conserving soil fertility, cassava requires the careful application of plant nutrients at the right rate and time, and in the right place. In the long term, the available nutrients are maintained at optimum levels in most soils when the amount of nutrient applied equals the amount lost. In poor soils, there may need to be a period in which nutrients applied surpass nutrients extracted, in order to first build soil nutrients to optimum levels. In practice, inorganic fertilizers and organic fertilizers can be used as available nutrient sources for cassava plants to achieve maximum profits and environmental sustainability. Depending on the age of the plant at harvest, a considerable amount of nutrients (N, P, K and Mg) remain in the leaves and stems. A combination of these cassava residues and mineral fertilizers should be applied to soils to supply sufficient nutrients to the cassava plants and improve the overall quality of the soils. The nutrient content of the leaves and stems should be considered when calculating fertilizer recommendation rates (Table 4).

In general a combination of 100 kg N, 20 kg P and 80 kg K per hectare, plus returning plant tops to the soil at harvest, can achieve up to 40–50 t/ha of cassava root yields in many parts of Southeast Asia. The optimum fertilization rates for specific production conditions

Table 4 Average nutrient content of cassava roots, stems and leaves, total plant (kg/t)

Material	N	P	K	Mg	Ca
Roots	2.33	0.52	4.11	0.61	0.34
Stems + leaves	2.58	0.56	1.72	1.22	0.45
Total plant	4.91	1.08	5.83	1.83	0.79

Sources: Howeler, 1981.

depend on individual variety, soil characteristics, climatic conditions and crop management practices. A systematic evaluation of current farmers' crop management practices and the development of more site-specific fertilizer recommendations are needed to assist farmers in increasing yields and profitability in a sustainable manner.

4.4 Pest, disease and weed incidence in Asia

The most important pests found in Asia include mealybugs, red mites, whiteflies, scale insects, white grubs, termites and several pests that attack dried cassava during storage. Various species of mealybugs especially cassava mealybug (*Phenacoccus manihoti*) are considered the most devastating cassava pests in Cambodia, southern Laos and Thailand. They can cause 20–40% losses during cassava cultivation. The red mites (*Tetranychus* sp.) are the most prevalent dry season pests; however, there have also been reports of green mites. Depending on the age of the cassava plant and the duration of attack, yield losses can vary from 20 to 50%.

Cassava diseases include cassava bacterial blight (CBB), root rots, brown leaf spot, cassava witches' broom (CWB). CBB is widespread and can be serious during the rainy season. It is caused by the bacteria *Xanthomonas axonopodis* pv. *manihotis*. Cassava root rots can be caused by a wide range of fungal or bacterial pathogens. Root rots are commonly found in heavy or compacted soil and in areas with substantial rainfall. The most common root rots are caused by a variety of species within the genus *Phytophthora*, especially *Phytophthora drechsleri*. The disease can attack both young and mature cassava plants, causing sudden wilting, severe defoliation and soft root rot. The infected roots exude a pungent, watery liquid, and they decompose completely. Southeast Asia has had serious CWB disease problems since 2010. CWB causes excessive sprouting, and small leaves with short petioles. In general, when infection takes place early in the season, plants are dwarfed and show an exaggerated proliferation of buds, as well as shoots and small branches growing from a single stake. Sprouts have short internodes and many small leaves. The roots of affected plants are thinner and smaller with drastically reduced starch content. If cassava plants are infected early in the growth cycle, they remain small.

In late 2015 there was an outbreak of the Sri Lankan cassava mosaic virus (SLCMV) reported in the Rattanak Kiri province of eastern Cambodia. SLMV is a distinct begomovirus, closely related to the virus that causes cassava mosaic disease (CMD) in Africa. The symptoms of these two diseases include chlorotic mottling of green leaves with leaf deformation, which may lead to leaf fall and severe stunting of plants. The disease is spread mainly through the use of infected planting material, as well as by the whitefly *Bemisia tabaci*. SLCMV has symptoms and aetiology similar to the devastating cassava mosaic virus (CMV) of Africa and represents a potentially devastating disease for the region.

Weeds are one of the major production constraints for cassava in the region. The major weed species in Asia are *Imperata cylindrica*, *Cyperus rotundus*, *Cynodon dactylon*, *Chromolaena odorata*, *Amaranthus viridis*, *Cleome viscosa* and *Portulaca oleracea*. Weed-related yield losses ranging from 40% to 80% have been recorded. Weeding consumes about 150–200 labour days per hectare in India (Nedunchezhiyan et al., 2015), whereas 25–30 labour days in China and Vietnam (Howeler, 2014). In general, the crop should be weed-free during the early stages of plant growth (i.e. the first three months or until canopy closure). Weeding is most often done by hoe, by animal-drawn cultivator or hand tractor, and/or with herbicides. In smallholder cassava farms, hand-weeding is mostly done by women and children. As labour is scarce and expensive in many parts of Asia, many farmers are now using pre-emergence and post-emergence herbicides, especially in Cambodia, China, Thailand and Vietnam.

Control from the threats of insects, diseases and weeds can be achieved through several methods. However, since cassava is a long-season crop, and is exposed to these problems over a long period, it is hardly ever economic, nor effective, to control these problems through frequent applications of pesticides and herbicides. Where the use of chemicals seems to be the only way of solving an immediate problem they should be used within the context of integrated pest, disease and weed control to increase their efficiency and minimize unwanted side effects such as damage to natural enemy communities and non-target species, as well as residue problems and hazards to humans and the environment. To minimize yield loss due to pests and diseases, it is important to incorporate pest and disease resistance and tolerance characteristic as a high priority objective in the national breeding and evaluation programmes.

5 Case study: developing sustainable cassava production in Cambodia

In Cambodia, agriculture is the primary economic activity, accounting for 36% of the GDP and employing 71% of the population. Until recently, the dominant agricultural crops were rice, maize and cassava. Cassava is an important crop that carries significant risks and rewards for smallholder farmers in Cambodia. A recent needs-assessment of cassava production in Cambodia has highlighted that the cassava sector plays a significant role in economic growth, creating the largest share of employment, promoting local production growth and thus contributing to poverty alleviation in Cambodia (Ou Wenjun et al., 2016). However, additional research is still needed to identify gaps encountered by smallholder farmers in the cycle of cassava production, and to improve cassava production technologies in sustainable smallholder farming systems for the future development of the country (Lefroy et al., 2015). This case study will highlight the knowledge and tools required to address challenges to sustainable cassava production in Cambodia and similar cassava production areas in Asia.

5.1 Current cassava production systems in Cambodia

While Cambodia is now one of the largest cassava producers in the region, its history of cassava cultivation is relatively brief. The crop was first grown in Cambodia primarily for human consumption, but has become increasingly popular in the past 10–15 years, especially for resource-poor farmers, due to expanding market opportunities, especially for export to China, Thailand and Vietnam. In Cambodia, this crop is now the second most important crop after rice (MAFF 2015, 2016).

Research conducted in the last decade (2005–15) to better understand the principles of cassava cultivation has led to more productive systems in Cambodia (Huang Jie et al., 2007; Sarom et al., 2015; Sopheap et al., 2011b). The findings have been most relevant for cassava cultivated in high-input production systems especially for high-yielding industrial varieties planted in less fertile soils. The research however has been less applicable for the majority of Cambodia's cassava producers – smallholder resource-poor farmers in the marginal uplands. Nevertheless, in Cambodia, cassava is still viewed by farmers as a crop with significant development potential since it is well adapted to poor soil. Formulating a sustainable cassava production strategy is indispensable due to the continued scarcity of good agricultural land, the need to reduce deforestation and the support needs created by rapid population growth.

5.2 Geography, climate and cassava-growing soils in Cambodia

Cambodia is located in Southeast Asia in the southwestern corner of Indochina, between latitudes 10° and 15° N and longitudes 102° and 108° E. The country covers an area of 181 035 square km and shares borders with Lao PDR, Thailand and Vietnam. Cambodia has a tropical monsoon climate with distinct dry and wet seasons. The dry season is from November to April and the wet season is from May to October. Annual rainfall ranges from 1250 to 4000 mm. The mean temperature ranges from 21 to 35°C, with the hottest month being April and the coolest month being December. There are six different soil types used for growing cassava, namely sandy, loamy, clayey, sandy clay, rocky and sediment (El, 2010). Many of the light-textured soils (i.e. sandy and loamy) are highly vulnerable to erosion. For good growth and yields, cassava requires friable, light-textured and well-drained soils (Ekanayake et al., 1997). It is clear that cassava is well suited to the biophysical conditions across much of the country.

5.3 Cassava production, area and yield in Cambodia

Cassava is cultivated, at both small and large scales, in almost all provinces in the country, with major production areas found in the northeast (Kampong Cham, Kratie), northwest (Battambang, Banteay Meanchey, Pailin), and central regions (Kampong Thom). Currently, cassava uses the most agricultural land compared to other major upland crops (maize, soybean and mungbean) (MAFF, 2016). Table 5 shows production and yield in each province of Cambodia in 2015. The total area of cassava has expanded dramatically from 15 380 ha in 2000 to 626 030 ha in 2015, while annual production has markedly increased from 147 763 t in 2002 to 14.4 million t in 2015. The average yield has also increased significantly from 9.6 t ha-1 in 2000 to 22.9 t/ha in 2015, a result of higher yielding varieties and adoption of more effective agronomic practices by commercial better-off farmers. However, much of this improvement reflects the expansion into new production areas where soils are relatively fertile. The rapid increase of cassava production in Cambodia is mainly attributable to the adoption of high-yielding industrial varieties and the expansion of planted areas (Sopheap et. al., 2012). In farmer surveys, cassava farmers did not identify soil degradation (erosion and nutrient depletion) as a major concern (Sopheap et. al., 2011a). However, research shows that without careful soil fertility management, cassava yields cannot be improved and the same levels of cassava production sustained.

Several factors have contributed to the conversion of cassava from a subsistence to a commercial crop, and to its status as the second most important crop in the country (MAFF, 2015). A rapid increase in demand for cassava chips by Thailand and Vietnam was most likely the lead contributor. Moreover, further growth in cassava production in Cambodia is expected due to stronger demand from China and newly evolved domestic processing industries, as well as continued demand from neighbouring countries. Cassava production has several distinct advantages for Cambodian farmers compared to other food and cash crops, including cheaper production costs, more consistent yields and

Table 5 Cassava cultivated area, production and yield in the provinces of Cambodia in 2015

Province	Cultivated area (ha)	Production (t)	Yield (t/ha)
Bunteay Meanchey	76,411	1,608,372	21.0
Battambang	155,424	4,879,854	31.4
Kampong Cham	17,062	312,906	18.3
Kampong Chhnang	478	62,677	5.6
Kampong Speu	124	3,780	30.5
Kampong Thom	43,254	703,646	16.3
Kampot	2,166	8,063	3.7
Kandal	54	445	8.2
Koh Kong	263	4,327	16.5
Kratie	62,474	1,438,734	23.0
Mondulkiri	10, 727	195,305	18.2
Phnom Penh City	12	NA	NA
Preah Vihear	18,045	265,898	14.7
Prey Veng	519	9,290	17.9
Pursat	21,703	651,090	30.0
Ratanakiri	16,088	249,415	15.5
Siem Reap	20,050	320,439	16.0
Preah Sihanouk	205	2,665	13.0
Steung Treng	12,117	48,400	4
Svay Rieng	8,477	154,734	18.3
Takeo	1,355	12,249	9.0
Oddar Meanchey	35,643	754,440	21.2
Kep	98	1,354	13.8
Pailin	70,861	1,670,025	23.6
Tbaung Khmum	52,420	1,058,126	20.2
Total	626,030	14,356,232	22.9

Source: MAFF, 2016.

lower general risk. In Cambodia, traditional cassava cultivation practices are no longer capable of meeting production requirements, and consequently, widespread application of science-based sustainable production methods is essential.

5.4 Varietal improvement in Cambodia

In the past, farmers grew mainly the eating varieties *Damlong Kor* and *Damlong Mi*. At present, eating varieties are planted in small patches, and industrial varieties (bitter varieties) are processed into dried chips or sold to starch factories. Most popular varieties come from Thailand and Vietnam, and adoption varies across and within regions. Farmers are usually very interested in trying new varieties that may provide better yields. Farmers use closer spacing for less-branching varieties and relatively poor soils compared to highly branching varieties and fertile soils. Generally, plant density of about 15,000 per ha is sufficient to provide a high yield of cassava.

The CIAT cassava programme and national partners in Cambodia conducted strategic research at the Chamkar Leu Seed Farm in Kampong Cham Province, and at some of the satellite stations under CARDI from 2008 to 2012. The results of varietal testing clearly indicated that the two introduced varieties from Thailand (Huay Bong 60 and KU50) gave significantly higher fresh root yields (49 and 48 t/ha, respectively) with high starch content (29.5%), whereas the local Damlong Kor produced the lowest yield (6 t/ha) with the lowest starch content (24%). Because the genetic potential of cassava is very high, yield differences among varieties can be high even under the same environmental conditions and with the same crop management practices. The top six varieties were Huay Bong 60, KU 50, Dam Lektouch, Rayong 1 from Thailand, and SC 201 and SC 8 from China (Fig. 1). This experiment suggested that most of the high-yielding varieties from Thailand and China are suitable for production conditions in Cambodia. However, the varietal trials must be conducted on-farm at specific locations in Cambodia to optimize variety selection. Released varieties from China, Thailand and Vietnam could successfully be evaluated for yield and dry matter content in Cambodia, with a potential to select varieties best suited for local smallholder farmer conditions. Greater genetic diversity with respect to pest and disease resistance is needed for the long-term stability of cassava production. Moreover, the demand for improved low-HCN varieties for

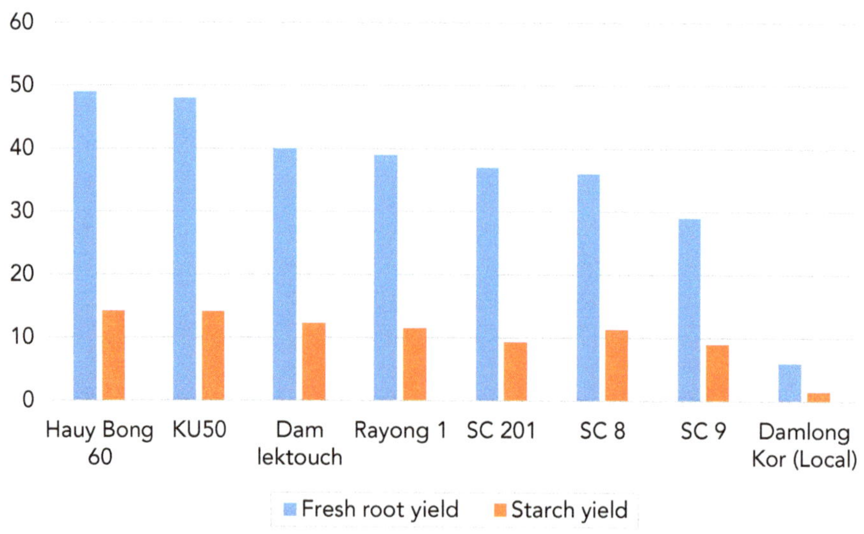

Figure 1 Yield results (t/ha) of the cassava variety trial conducted at Chamkar Leu Seed Farm in Kampong Cham, Cambodia (2010/11).

Table 6 Selected cassava varieties suitable for human consumption and for processing that have been grown in Cambodia, and their most important characteristics

Variety name	Year of release	Country of origin	Main characteristics
Damlong Kor	–	Cambodia	Good eating quality
Damlong Mi	–	Cambodia	Good eating quality
Damlong narrow leaf	–	Cambodia	High yield
Malaysia = KU50	–	Thailand	High yield, high starch content
Rayong 1	–	Thailand	High yield, rather low starch content
Rayong 60	1987	Thailand	High yield, early bulking, very bitter
Rayong 5	1994	Thailand	Relatively high yield, high starch content
Rayong 72	1999	Thailand	High yield, drought tolerant
Huay Bong 60	2003	Thailand	High yield, high starch content
SC 205	–	China	High yield, mainly suitable for fertile soils
KM 95–3	1998	Vietnam	High yield, dual purpose
KM 98-7	1998	Vietnam	High yield, dual purpose

human consumption continues to be significant in Cambodia as a means of providing food security.

5.5 Improved soil fertility management in Cambodia

Sustainable soil conservation and fertility management is crucial for long-term economically viable cassava production. In the past, many farmers in Cambodia did not apply any fertilizer to cassava, and fertilizer use is still limited in many cassava-growing areas. Only some farmers in the long-standing cassava-growing provinces (e.g. Kampong Cham, Battambang) use mineral fertilizers and manures. Nevertheless, the amount applied is often minimal and not well-balanced. Farmers are becoming better informed about the economic returns to fertilizer application, and use is increasing. In Cambodia, as in other Southeast Asian countries, technical support such as soil and plant analysis is low or even absent on farm, and still largely inadequate even in many research stations in the provinces. In low-fertility soils in Kampong Cham province, a balanced fertilizer application to Thai variety KU50 increased its fresh root yield from 9.3 t/ha (without fertilizer) to 33 t/ha (with 180 N: 90P2O5; 180K2O), and also increased its starch content from 21.3% (without fertilizer) to 32.2% (with 180 N: 90P2O5; 180K2O) (Sophearith, 2014). The farmer participatory research (FPR) fertilizer demonstration trials in Kampong Cham clearly indicated not only increased cassava fresh root yield and starch content but also increased net income from fertilizer application (Table 7). Methodologies that would enable development by local researchers and extension workers, in collaboration with farmers, of fertilizer recommendations along with necessary soil conservation measures

Table 7 Results of the FPR fertilizer demonstration trial conducted at Cham Karkor village, Chork commune, Orang Ov district, Kampong Cham in 2010/11

Treatment	Fresh root yield (t/ha)	Starch content (%)	Production cost (US$)	Gross income (US$)	Net income (US$)
[a]Farmers' practice	13.9	22.9	602	1,487	885
[b]Improved practice (with fertilizer)	17.4	24.5	666	1,861	1,194

[a] 100 kg of Urea per ha
[b] 100 kg of 15N:15P:15K + 50 kg of KCl per ha

that are specific to particular environments and soil types can contribute significantly to maximization of returns from cassava production and maintenance of cassava yield growth over time.

6 Summary

CIAT scientists and Asian national partners have developed widely adopted cassava varieties for Asia's various agro-ecological zones. The average yield of cassava in Asia is highest in the world, at 21.1 t/ha. This includes a wide range of varieties (industrial and traditional) and growing conditions, including areas with infertile acid uplands or extensive farming. The high-yielding varieties were developed with yields up to triple those of traditional varieties in many cassava-growing areas in Asia. In recent years, the majority of farmers in favourable areas have switched to industrial varieties for their superior starch yield and profits. However in comparison with experimental yield potential, there is a large scope for further increasing cassava production by closing this yield gap.

Knowledge regarding yield potential, exploitable yield gaps and constraints to improving productivity at the field level is still not fully exploited at the farm level. The main limitation to achieving higher yields and associated higher profitability per unit of arable land is often the ineffective use of inputs (particularly nutrients, quality planting materials and herbicides). Better nutrient, weed, pest and disease, and water management practices, along with the use of germplasm with a higher yield potential, are required to ensure profitable cassava production while meeting growing demand. To maintain the attractiveness of cassava as a feedstock for a range of processing industries, cassava production must be highly cost effective and more environmentally friendly so as to allay concerns about land degradation.

Sustainable cassava production in Asia can be approached at both the farm level and the national level. At the farm level, a number of important measures are necessary: execution of appropriate land preparation, timely planting and harvesting, planting of the best-adapted varieties, balanced fertilization, efficient weed control, effective control of pests and diseases, and proper use of family labour. At the national level, several interventions are essential: enhancing the cassava productivity through improving the existing cassava varieties, developing techniques to improve rainwater capture and utilization, developing effective residue management practices, boosting of agricultural research and extension services, increasing agricultural credit facilities to enable farmers to purchase the necessary inputs, and putting in place cassava price policies that encourage farmers to increase cassava production in a sustainable manner.

Ultimately CIAT's cassava programme has contributed significantly to the capacity building of the national cassava research systems in Asia. Major effort must be made to transfer sustainable technology through education, training and raising the levels of knowledge of the smallholder farmers. More farmers should be involved in the development of new research packages since they usually understand better the possible impacts of new cassava varieties and better agronomic practices on their farming systems.

7 Where to look for further information

If you are looking for more information about cassava cultivation in Asia, the best first contacts are listed below. Although these addresses will change with time, they will guide you in the right direction to find the information you want.

For general enquiries visit the CIAT website: http://www.ciat.cgiar.org or contact Regional Office of CIAT for Asia:

c/o Agricultural Genetic Institute, Pham Van Dong Street, Bac Tu Liem, Hanoi, Vietnam
Tel: +84 4 37576969 Fax: +84 4 37570999

Guangxi Subtropical Crops Research Institute (GSCRI)
No. 22 Yongwu Road, Nanning
Guangxi province, P. R. China
http://www.gxrzs.com

Chinese Academy of Tropical and Agricultural Sciences (CATAS)
No. 4 Xueyuan Road, Haikou
Hainan, P. R. China
http://www.at0086.com/CATAS

Faculty of Agriculture
Brawijaya University
Malang 565567
East Java, Indonesia
http://www.ub.ac.id

Research Institute for Legumes and Tuber Crops (RILET)
Jl. Raya Kendalpayak KM 8, Malang,
East Java, Indonesia
http://en.litbang.pertanian.go.id
Dept. of Agronomy
Kasetsart University
Chatuchak, Bangkok 10900
Thailand
http://www.ku.ac.th

Rayong Field Crops Research Center
320 Sukhumvit Road

Huai-Pong Amphur Muang
Rayong 21150, Thailand
http://www.doa.go.th

Hung Loc Agricultural Research Center
Hung Thinh, Thong Nhat
Dongnai province, Vietnam
http://iasvn.org/en

Root Crop Research and Development Center
Van Dien, Thanh Tri
Hanoi, Vietnam
http://fcri.com.vn/

8 References

Alves, A. C. (2002). Cassava botany and physiology. In Hillocks, R. J., Thresh, J. M. and Bellotti, A. C. (eds), *Cassava: Biology, Production and Utilization*. CABI publishing, chapter 5, pp. 67–89.

Aune, J. B. and Lal, R. (1997). Agricultural productivity in the tropics and critical limits of properties of Oxisols, Ultisols, and Alfisols. *Tropical Agriculture* 74: 96–103.

Dierolf, T., Fairhurst, T. and Mutert, E. (2001). Soil fertility kit: A toolkit for acid, upland soil fertility management in Southeast Asia. Potash & Phosphate Institute (PPI), Singapore. P149 Direrolf et al., 2001.

Ekanayake, I. J., Osiru, D. S. O. and Porto, M. C. M. (1997). Agronomy of cassava, IITA Research Guide 60. Training Program, International Institute of Tropical Agriculture (IITA), Ibadan, Nigeria, p. 30.

El, S. (2010). Result of Cassava Survey in Cambodia. In: Howeler, R. H (ed.), *A New Future for Cassava in Asia: Its Use as Food, Feed and Fuel to benefit the Poor*. Proceeding of the 8th Regional Workshop held in Vientiane, Lao PDR, 20–24 October 2008, pp. 470–89.

Howeler, R. H. (1981). Mineral nutrition and fertilization of cassava (*Manihot esculenta* Crantz). CIAT, Cali, Colombia, p. 52.

Howeler, R. H. (1988). Agronomic practices for cassava production in Asia. In: Howeler, R. H and Kawano, K. (eds), *Cassava Breeding and Agronomy Research in Asia*. Proceeding of the 2nd Regional Workshop held in Thailand, 26–28 October 1987, pp. 313–40.

Howeler, R. H. (1992). Agronomy research in Asian cassava network – An Overview 1987–1990. In Reinhardt, R. H (ed.), *Cassava Breeding, Agronomy and Utilization Research in Asia*. Proceeding of the 3rd Regional Workshop held in Malang, Indonesia, 22–27 October 1990, pp. 260–85.

Howeler, R. H. (2014). Sustainable Soil and Crop Management of Cassava in Asia – A Reference Manual. Centro Internacional de Agricultura Tropical (CIAT), Cali, Colombia, p. 280.

Howeler, R. H. and Aye, T. M. (2014). Sustainable Management of Cassava in Asia: From Research to Practice. CIAT, The Nippon Foundation. Cali, Colombia, p. 148.

Huang, J., Seng, S., Kang, A., Chiv, P., Suon, S. and Tian, Y. (2007). Cassava production and utilization in Cambodia. In Howeler, R. H. (ed.), *Cassava Research and Development in Asia: Exploring New Opportunities for an Ancient Crop*. Proceeding of the 7th Regional Workshop held in Bangkok, Thailand, 28 October–1 November 2002, pp. 39–47.

Kawano, K. (1988). CIAT Cassava germplasm development in Asia: Cooperation with national programs. In Howeler, R. H and Kawano, K. (eds), *Cassava Breeding and Agronomy Research in Asia*. Proceeding of the 2nd Regional Workshop held in Rayong, Thailand 26–28 October 1987, pp. 117–26.

Kawano (1995). *Green Revolution And Cassava Breeding*, 4th Regional Workshop, pp. 355–67. Kerala, India, 2–6, November 1993.

Lefroy, R. D. B., Fahrney, K. S. and Aye, T. M. (2015). Developing an R&D strategy for cassava in Southeast Asia: From a CIAT perspective. In Howeler, R. H (ed.), *Sustainable Cassava Production in Asia for Multiple Uses and for Multiple Markets*. Proceeding of the 9th Regional Workshop held in Nanning, Guangxi, China, 27 November to 3 December 2011, pp. 322–36.

Ministry of Agriculture, Forestry and Fisheries (MAFF) (2015). Annual Reports. Phnom Penh, Cambodia.

Ministry of Agriculture, Forestry and Fisheries (MAFF) (2016). Annual Reports. Phnom Penh, Cambodia.

Nedunchezhiyan, M., Misra, R. S., and Naskar, S. K. (2015). Weed management in cassava. In Howeler, R. H. (ed.), *Sustainable Cassava Production in Asia for Multiple Uses and for Multiple Markets*. Proceeding of the 9th Regional Workshop held in Nanning, Guangxi, China, 27 November to 3 December 2011, p. 274.

Onwueme (2002). Cassava in Asia and the Pacific. In Hillocks, R. J., Thresh, J. M. and Bellotti, A. C. (eds), *Cassava: Biology, Production and Utilization*. CABI Publishing, chapter 4, pp. 55–65.

Ou Wenjun, L. M., Aye, T. M. and Srey, S. (2016). Current situation of cassava production, constraints and opportunities in Cambodia. *Journal of Agriculture, Forestry and Fisheries* 5 (3): 64–70.

Sarom, M., Chheng, N., Leang, S., Nou, P., Sok, S., Kong, T., Chhay, K., Aye, T. M. and Fahrney, K. S. (2015). Recent development in cassava production and utilization Cambodia. In Howeler, R. H. (ed.), *Sustainable Cassava Production in Asia for Multiple Uses and for Multiple Markets*. Proceedings of the 9th Regional Workshop held in Nanning, Guangxi, China, 27 November to 3 December 2011.

Sopheap, U., Patanothai, A. and Aye, T. M. (2011a). Farmers' perception on cassava production in Cambodia. *Khon Kaen Agriculture Journal* 39: 279–94.

Sopheap, U., Patanothai, A. and Aye, T. M. (2011b). Nutrient balances for cassava cultivation in Kampong Cham province in Northeast Cambodia. *International Journal of Plant Production* 6(1): 37–58.

Sopheap, U., Patanothai, A. and Aye, T. M. (2012). Unveiling constraints to cassava production in Cambodia: An analysis from farmers' yield variations. *International Journal of Plant Production* 6(4): 409–28.

Sophearith, S. (2014). Performance of cassava (KU50) on different chemical fertilizer rates and plant density. MSc Thesis. Royal University of Agriculture, Phnom Penh, Cambodia.

Taco Bottema, J. W. and Henry, G. (1992). History, Current status and potential of cassava in Asia. In Howeler, R. H. (ed.), *Cassava Breeding, Agronomy and Utilization Research in Asia*. Proceeding of the 3rd Regional Workshop held in Malang, Indonesia, 22–27 October 1990, pp. 3–26.

Tongglum, A., Tiraporn, C. and Sinthuprama, S. (1987). Cassava cultural practices research in Thailand. In Howeler, R. H. and Kawano, K. (eds), *Cassava Breeding and Agronomy Research in Asia*. Proceedings of 2nd Regional Workshop, held in Rayong, Thailand. 26–28 October 1987, pp. 131–44.

Chapter 6

Cassava cultivation in sub-Saharan Africa

Dunstan S. C. Spencer, Enterprise Development Services Ltd, Sierra Leone[1]*; and Chuma Ezedinma, UNIDO, Nigeria*[2]

1 Introduction
2 Cassava's contribution to incomes in SSA
3 Cassava's contribution to food and nutrition security
4 Drivers of change in cassava production in SSA
5 Future trends and conclusion
6 References

1 Introduction

Cassava (*Manihot esculenta*) was introduced to sub-Saharan Africa (SSA) by Portuguese traders from Brazil in the sixteenth century. The crop is now produced in 40 of the 53 countries of SSA, accounting for 61% of global production (Fig. 1).

In SSA, cassava is cultivated under a wide range of ecological and agronomic conditions. Cassava's adaptability to relatively marginal soils and erratic rainfall conditions, its high productivity per unit of land and labour, the certainty of obtaining some yield even under the most adverse conditions and the possibility of maintaining continuity of supply throughout the year make the crop very adaptable to rain-fed agriculture conditions.

1.1 Production to consumption system

Figure 2 presents a schematic representation of the cassava *production to consumption* system in SSA. Over 90% of production takes place in small farms. Inputs cover the whole spectrum for agricultural production in SSA. Imported inputs consist of small hand tools (used by small-scale farmers), tractors, machinery, fertilizer and other agricultural chemicals (used mainly by medium- and large-scale farmers).

In cassava-growing areas of SSA, multiple crops are typically produced in multiple fields by each farmer. Results from the Collaborative Study of Cassava in SSA (COSCA), the most recent comprehensive study of the economics of the crop in the region (Nweke

1. Senior Partner, Enterprise Development Services Ltd, Freetown, Sierra Leone; formerly Director, Resource and Crop management Division, International Institute of Tropical Africa (IITA), Ibadan, Nigeria.
2. Officer in Charge, UNIDO, Nigeria; formerly Impact, Monitoring, and Evaluation Officer and scientist with the Integrated Cassava Project of the International Institute of Tropical Agriculture (IITA), Ibadan, Nigeria.

http://dx.doi.org/10.19103/AS.2016.0014.06

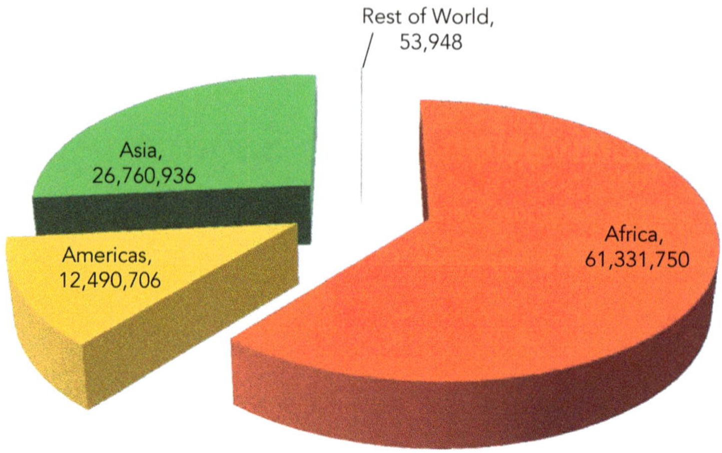

Figure 1 Distribution of global cassava production in 2013, mt. Source: FAOSTAT, 2/12/2015.

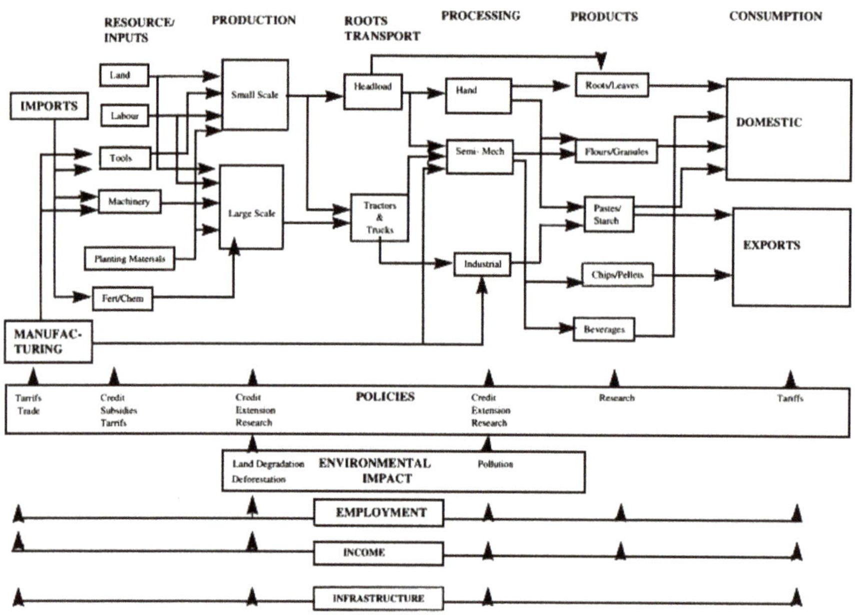

Figure 2 The cassava production to consumption system in sub-Saharan Africa. Source: Spencer (2005).

et al., 2001), show that farmers grow an average of 6–7 crops, with a range of 1–15 crops. Only about 25% of the fields were planted with a single crop. Rice, yam and cassava were the crops grown most often as sole crops. It is clear that for systems that include cassava, mixed cropping is the usual practice.

Large-scale agro-industries, which produce fertilizer, agricultural chemicals and farm machinery (tractors, ploughs, etc.), are not widespread in SSA. Small-scale and artisanal agro-industries that produce hand tools and processing machinery are more important in the cassava system and provide a link to the industrial sector. Roots are transported by head load, as well as by trailers and small trucks, where the road network allows them to be used. Cassava roots are processed into a wide variety of products for human and industrial consumption, ranging from simple boiling to fermented products and beverages. Most of the products are consumed domestically within the countries in which they are produced. However, there is a small but growing export trade in dried cassava chips and other industrial products.

Agricultural policies affect the cassava system at different stages and in different ways. The system contributes significantly to employment creation and income generation. It has an impact on the environment and its development will be affected by infrastructure constraints, especially the state of the rural road network.

1.2 Trends in production

Globally, the five largest cassava producing nations are Nigeria, Thailand, Indonesia, Brazil and Democratic Republic of Congo (DRC). In SSA the major cassava producing countries include Nigeria (53 million mt in 2013), DRC (16 million mt), Angola (16.4 million mt), Ghana (15.9 million mt) and Mozambique (10 million mt) (Fig. 3).

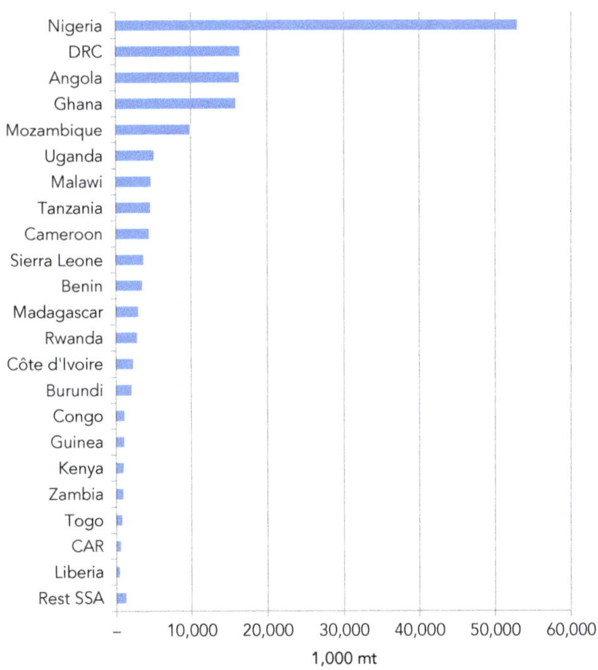

Figure 3 Cassava production by country in sub-Saharan Africa, 2013. Source: FAOSTAT, 2/12/2015.

Cassava production in SSA has increased nearly five-fold since 1961 from 31.5 million mt to 157.8 million mt in 2013 (Fig. 4). This is due to almost a tripling in area cultivated (from 5.5 million ha to 13.75 million), and a much smaller increase in average yield (from 6.6 mt/ha in 1961 to 10.4mt/ha in 2013). However, there has actually been a decline in area cultivated in the semi-arid Sahelian countries of Burkina Faso, Mali, Niger, Senegal and Sudan, and the island countries of Réunion and Seychelles. But some traditional maize-producing countries such as Malawi have seen a more than tenfold increase in cultivated area (from 21 000 ha in 1960 to 211 089 in 2013).

Table 1 shows that as cassava production is expanding in SSA, the crop is to a large extent replacing fallow, confirming that most of the production increase has been due to

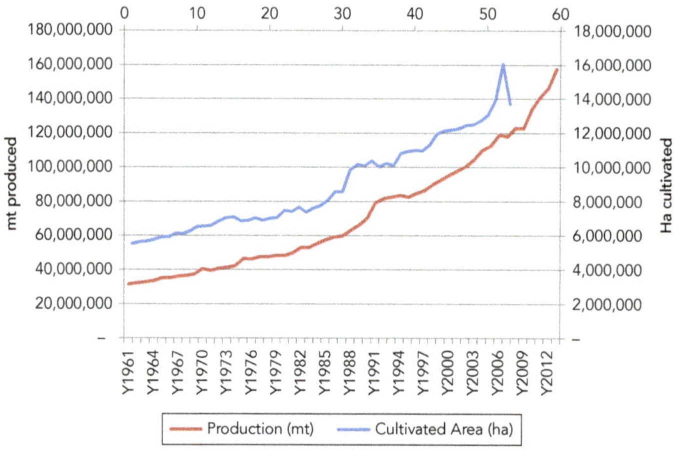

Figure 4 Trends in production of cassava in sub-Saharan Africa. Source: FAOSTAT, 2/12/2015.

Table 1 Percentage of farmers reporting that cassava is replacing alternative crops by agro-ecological zone in sub-Saharan Africa

Crop replaced	Humid	Sub-humid	Non-humid	Highland humid
Yams	16	16	6	0
Cocoyams	11	4	0	0
Potato	4	2	6	0
Plantain	5	0	2	0
Banana	2	0	0	0
Maize	1	9	21	0
Rice	2	4	0	0
Other food crops	8	20	21	12
Cash crops	11	6	3	2
Fallow	40	39	42	43

Source: Spencer (2005) Table 3.

increase in crop area. The fact that COSCA data also show that cassava is often planted just before land is allowed to go into fallow indicates that the crop is also being used to increase land use intensity. Cassava is also replacing other root crops, especially yam in the humid zone, maize in the non-humid zone and other food crops in the sub-humid zone at important rates.

1.3 Cassava yield

Figure 6 shows the trend in cassava yields. In 1954, the average cassava yield in SSA was between 5 and 10 tonnes per ha (Jones, 1959). The COSCA study found that the average yield was between 10 and 15 tonnes per ha in the six COSCA countries in the early 1990s. Currently, as shown in Fig. 6, a few of the countries are achieving over 15 mt per hectare. It is evident that cassava yields have increased significantly since the early 1990s. This is due to the planting of high-yielding varieties and the adoption of better agronomic practices, as discussed later.

Figure 6 shows the yields of cassava roots in the different SSA countries grouped by agro-ecological zones[3] and by region, showing that the crop is able to achieve similar yields in all agro-ecological zones, demonstrating its adaptability.

1.4 Pest and disease constraints

Historically, cassava had few serious pests and diseases in SSA. However, the situation changed as cassava cultivation intensified and exotic pests were introduced. The major cassava pests in SSA include relatively few phytophagous arthropods, plant pathogens and weeds compared to the pest complex found in the neotropics. Together, these species could reduce cassava production by as much as 50%. The most severe pests are the exotic species accidentally introduced into areas where the local germplasm is susceptible to attack, where effective natural enemies/antagonists are absent and where

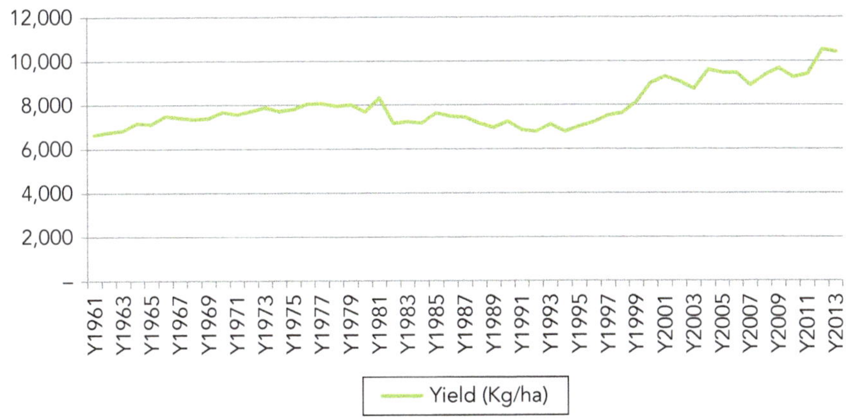

Figure 5 Trend in cassava yield in sub-Saharan Africa. Source: FAOSTAT, 2/12/2015.

3. Countries grouped into an agro-ecological zone which represents around 50% or more of its cropland.

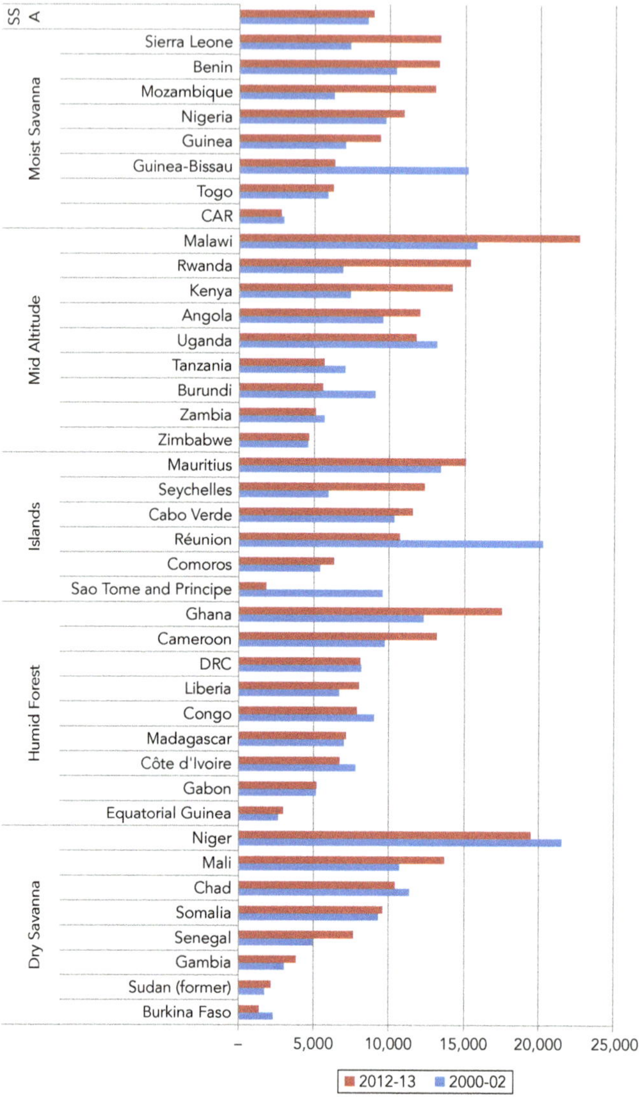

Figure 6 Cassava yield in sub-Saharan Africa countries (fresh roots), grouped by agro-ecological zones. Source: FAOSTAT, 2/12/2015.

a tradition of practices to cope with the introduced pests had not had time to evolve. In addition, pest problems are being created where intensification of cassava production erodes the environmental stability inherent in balanced agro-ecosystems. The major pest concerns are cassava green mites, variegated grasshopper during outbreak episodes, root mealybug in the rain forest ecozones, cassava mosaic disease (CMD), cassava bacterial blight, cassava anthracnose disease and root rots in the humid lowlands. The role of termites, nematodes and certain weed species particular to specific ecozones has been reported as constraints but have not received adequate attention.

The longer shelf life of fresh cassava and extended storage of dried cassava (chips/flour) will aggravate the problems caused by post-harvest pests to which the products are exposed for longer periods. These include the devastating larger grain borer and a number of root rots. Larger grain borer can consume as much as 74% of cassava chips in only four months after harvest. Fungi are also known to infest cassava chips during processing and handling, in the field or during storage; they may lead to the formation of mycotoxins, making the chips unable to meet trade and health standards.

1.5 Gender in cassava production

Cassava has been described as a women's crop by some scholars (Ikpi, 1989 and 1989a; Okorji, 1983). For example, Okorji (1983) found in the Abakaliki area of southeast Nigeria that women owned more cassava fields than men. However, the more recent COSCA studies (Nweke et al., 2001) show that the proportion of the household cassava field area (hereafter field) owned by women ranged from 4% in the DRC to 24% in Côte d'Ivoire. By contrast, the proportion of cassava fields owned by men ranged from 15% in Côte d'Ivoire to 72% in Uganda and 81% in Nigeria. Joint ownership by both men and women account for the balance of the percentages in each country.

There is an important exception in tree crop-dominated rural economies. Among the six COSCA countries, women in Côte d'Ivoire owned a higher proportion of the cassava fields than men. In Côte d'Ivoire and in most other tree crop dominant-farming systems, the men concentrate on producing tree crops such as cocoa and coffee. Hence, although there are specific locations in certain countries where women owned more cassava fields than men (Okorji, 1983; Ezumah and Domenico, 1995), these locations are exceptions and not the norm.

The COSCA studies found that both men and women make significant contributions of their labour to the cassava industry. However, men and women were found to specialize in different tasks. Men worked predominantly on land clearing, ploughing and planting, while women specialized in weeding, harvesting, transporting and processing (Fig. 7). This was later confirmed by Ezedinma et al. (2006). Categorization of cassava as a women's crop is therefore a misleading half-truth that is based on anecdotal evidence and isolated village studies.

The COSCA studies also found that the introduction of labour-saving technologies in cassava production and processing has led to a redefinition of gender roles in the cassava food system. In Nigeria, on farms where land clearing or ploughing was mechanized, men increased their labour inputs in planting, weeding and harvesting. As the processing tasks became mechanized, the contribution of male labour to cassava processing increased because men operated all of the processing machines. When processing became mechanized, women shifted their labour to production tasks such as weeding and men managed the mechanized processing tasks.

Furthermore, the COSCA studies found that women who wanted to plant cassava were not constrained by the lack of access to new cassava production technologies or to essential production inputs such as farmland, cuttings of improved varieties or hired labour. Both men and women were able to plant cassava on land that they had inherited, purchased or rented. Finally, although most of the cassava processing machines were owned by men in most of the COSCA countries, both men and women had access to the

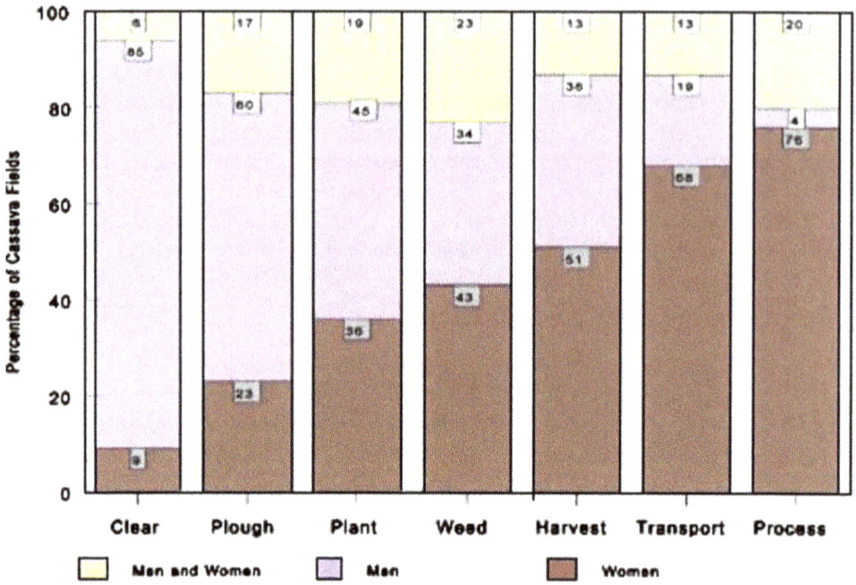

Figure 7 Proportion of labour for different tasks supplied by gender in cassava fields in sub-Saharan Africa. Source: Nweke (2005), Figure 3.

machines by paying a fee to the machine owners. More studies are needed to corroborate these findings.

2 Cassava's contribution to incomes in SSA

2.1 Cassava as a source of household cash income

Cassava is an important source of farm income in many SSA countries. The COSCA study showed that in both Ghana and Nigeria, cassava cash incomes were the highest in farms with access to mechanized cassava processing equipment for the preparation of *gari*. The contribution of cassava to farm incomes was found to be low in Tanzania and Uganda because the majority of farmers in both countries lacked access to improved cassava processing equipment.

In cassava producing areas, food crops contribute about 40% of household cash income, industrial crops and non-farm activities about 25% each, while livestock contributes about 10% (Table 2). About 26% of cash income from all food crops in cassava-growing households (91% of all household) was derived from sale of cassava. This proportion was exceeded only in rice growing households (29% of total households), where 29% of food cash income was derived from rice sales. It is clear that in cassava-growing areas, where the most important source of cash income is food crops, cassava is the most important food crop generating cash income.

Earlier studies by Fresco (1982) estimated that women farmers in southern Zaire sold 20–40% of their cassava, while Tollens (1992) and Gossens (1996) estimated that the

Table 2 Percentage distribution of cash income of cassava producing households by source in major cassava producing countries in sub-Saharan Africa

Source	Côte d'Ivoire	Ghana	Nigeria	Tanzania	Uganda
Cash income per capita (US$)	114	108	177	51	48
			Percentage		
Food crops	15	55	55	16	28
Cassava	5	13	12	4	6
Maize	2	13	8	3	8
Rice	5	3	6	2	2
Yam	1	5	8	0	0
Cooking banana	0	0	0	3	2
Plantain	2	4	1	0	1
Sweet potato	0	0	1	0	2
Other food crops	0	17	19	4	7
Other crops*	76	21	20	25	11
Livestock	3	3	7	4	7
Non-farm	6	21	18	55	54
Total	100	100	100	100	100

* Other crops means industrial crops such as cotton and so on and tree crops such as cocoa, coffee, oil palm and so on. Source: Nweke (2005), Table 7.

proportions of total cassava production marketed were 42–46% in Bandundu and 55% in the Bas-Zaire regions of the DRC. Spiro (1980) estimated that an average of 72% of cassava was planted for the market in south-western Nigeria, compared to 68% for maize.

The potential for future growth of cassava consumption in SSA is shown by the fact that the income elasticity of demand estimates for cassava products[4] have been found to be positive at all income levels (Table 3). The COSCA studies showed that for some cassava products, the income elasticity of demand was above one, and in some cases, for example, for *gari* the main cassava product in Nigeria, significantly higher than that of maize at all income levels. In Ghana, the income elasticity of demand for cassava (all products combined) was higher among urban than rural households. These estimates provide convincing evidence that cassava has a strong market demand in the COSCA study countries. However, since the income elasticities are generally higher for the cereals and bananas in all the countries, except as indicated above for Nigeria, it is expected that in SSA in general, over time, the demand for cassava products would grow less than that for the other food products.

4. Income elasticity of demand measures the responsiveness of the quantity demanded for a good or service to a change in the income of the people demanding the good. It is calculated as the ratio of the percentage change in quantity demanded to the percentage change in income.

Table 3 Income elasticity of demand for cassava and other food staples for rural households in Nigeria, Tanzania and Uganda; rural and urban households in Ghana

	Nigeria'			Tanzania			Uganda			Ghana	
Staple	All	Low Income	High Income	All	Low Income	High Income	All	Low Income	High Income	Rural	Urban
All cassava	0.78	0.84	0.76	0.77	0.80	0.66	1.00	1.00	1.00	0.73	1.46
Fresh roots	1.24	1.28	1.21	0.79	0.79	0.67	0.95	0.96	0.94	–	–
Gari	0.85	0.85	0.77	–	–	–		–	–	–	
Dried roots	0.55	0.57	0.53	0.75	0.80	0.66	1.17	1.15	1.13	–	
Maize	0.71	0.74	0.65	0.98	0.98	0.97	0.91	0.85	0.91	0.84	0.83
Rice	1.12	1.13	1.13	1.14	1.25	1.11	1.36	1.59	1.25	1.00	1.50
Pulses	1.02	1.01	1.02	1.06	1.05	1.07	0.78	0.84	0.69	–	–
Banana	2.06	1.97	1.69	0.94	0.93	0.95	1.22	1.38	1.16	0.13	1.10
Yam	0.91	0.90	0.92		–		–		–	–	
Sweet potato	–	–	–	0.97	0.97	0.74	0.79	0.64	–	–	–

Source: Nweke (2005).

2.2 Cassava as a source of foreign exchange

Virtually all of cassava production in SSA is used domestically, so cassava has not played any role as a foreign exchange earner, or in import substitution in the past.[5] However, there now appears to be a window of opportunity opening up for export of cassava products, as the traditional Asian exporters appear to be having difficulties in satisfying demand particularly in the European Union market due to changes in the relative costs of production and comparative advantage. Some SSA countries are already taking advantage of this trend. For example, exports of cassava chips from Ghana which commenced with 500 tons in 1993, reached 29 000 tons in 1996, and averaged 6300 mt per year in 2010–13 in Uganda.

AQ: Note that both 'tons' and 'mt' are used in the sentence 'F example, exports of cassava...'. Please advis if 'mt' shoul read 'tons'.

There are also some import substitution possibilities for cassava flour and industrial starch. For example, Djoussou and Bokanga (1995) reported that cassava flour produced by a new method is in high demand by large biscuit manufacturers around the city of Ibadan in Nigeria. They estimated that with a 15% substitution rate of wheat flour with cassava, Nigeria could save up to $14.8 million in foreign exchange annually, with $12.7 million going to cassava processors and $4.2 million to cassava farmers. The efforts by Nigeria to realize the import substitution potential of cassava flour are discussed later in Section 4.3.4.

3 Cassava's contribution to food and nutrition security

3.1 Cassava as a robust crop for the future

The resilience of cassava noted earlier makes it a robust crop under the predicted climate change conditions of the future. Nweke (1996) noted that famine rarely occurs in areas where cassava is widely grown, since it provides a stable food production base. This is illustrated in Fig. 8.

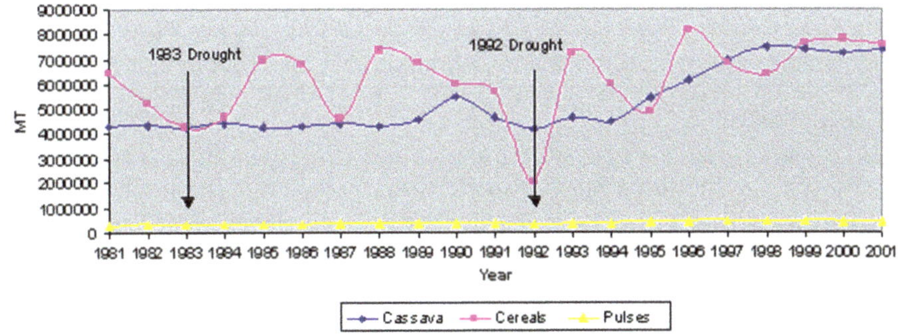

Figure 8 Production of cassava, cereals and pulses in five drought affected countries in southern sub-Saharan Africa. Source: Adebayo Abass et al. (2016).

5. There is of course some inter-regional trade in cassava products, especially the flours, but this trade is unrecorded and difficult to quantify.

3.2 Cassava-derived foods[6]

Cassava is processed by a variety of methods depending on locally available processing resources, local customs and preferences. In SSA, there are five common groups of cassava products: fresh root, dried roots, pasty products, granulated products and cassava leaves (Nweke, 2005).

3.2.1 Fresh root

The roots of sweet cassava varieties are eaten raw, roasted in an open fire, or boiled in water or oil. The cyanogens in the roots are destroyed by slowly cooking the roots. Starting with cold water, gradual heating promotes the hydrolysis of the cyanogen (Grace, 1977).

3.2.2 Dried roots

Dried cassava roots are stored or marketed as chips, balls and flour. Chips/flours are called kokonte in Ghana; lafun in Nigeria; cossette or unga in Congo; chinyanga, ugali, makopa, abutega, omundyenge, udaga, chikandole, bada, dimwehe, uhu, udunule, mbute, makonga, obusanobwe or kivunde in Tanzania; obuite, udilimbe, habalogala, ajobo or malwa in Uganda. Chips and balls are milled into flour at home by pounding with a pestle and mortar in preparation for a meal. There are two broad types of dried cassava roots: fermented and unfermented. Preparing unfermented dried cassava roots by sun- or smoke-drying is the simplest method of cassava preparation. Since this method is inefficient in the elimination of cyanogens, it is used mostly for preparing sweet cassava varieties, which have low cyanogen content.

In the case of fermented dried cassava roots, the fermentation is accomplished in one of two ways: stacking in heaps or soaking in water for a number of days. The fermentation process, whether in water or in heaps, influences the taste of the final product. The longer the fermentation period, the stronger is the sour taste. Taste is an important attribute, especially for consumers who eat fermented cassava products and who desire the strong sour taste.

The recent introduction of a grater has eliminated stacking and fermentation and therefore saves time. The roots are simply peeled, washed and grated. The pulp is placed in a perforated container, covered and a weight put on it for about three hours and the cyanogens are squeezed out along with effluent. The half-dried pulp is then dried in the sun (Alyanak, 1997).

In SSA, cassava is sun-dried on virtually any surface in the open air such as a large flat rock in the field, on the shoulders of a paved road, on flat rooftops, in a flat basket or even on bare ground. Bright colour is an important attribute desired by consumers[12]. The colour of dried roots depends on the colour of the raw roots (white, cream or yellow), the method and duration of the fermentation, the method of drying, the efficiency of the drying energy and the cleanliness of the drying environment.

3.2.3 Pasty products

Two forms of pasty cassava products are common in SSA: unsteamed and steamed pastes. The most popular is the unsteamed paste because it is stored or marketed without

6. This section is mainly from Nweke (2005).

cooking. Unsteamed wet paste is also known as *akpu* or *fufu* in Nigeria, *agbelima* in Ghana and *water fufu* in Cameroon. To prepare the uncooked paste, the roots are soaked in water for three to five days, during which time the roots soften and ferment. The soaked roots are manually crushed and sieved by shaking in a basket which is in a sack under water, thereby separating the pulp into the sack while collecting the fibre in the basket.

Steamed paste known as *chikwangue* in DRC, *placali* in Côte d'Ivoire and *myondo* or *babolo* in Cameroon is made from fine cassava mash. To prepare the paste, fibre is removed by hand from roots fermented by soaking in water. The roots are then stacked in a heap to further ferment. The pulp is ground with a stone or pounded in a mortar. The resulting fine pulp is firmly wrapped in leaves and steamed. The double fermentation as well as the steaming improves the shelf life of steamed paste. The sour flavour achieved through extended fermentation is a characteristic that is cherished by regular customers but is a turn-off for new consumers.

Steamed paste is stored or marketed in a ready-to-eat form. Preparing steamed paste is expensive because many steps are involved and each one requires additional inputs. For example, grinding and sieving are labour-intensive. The soaking step is water use intensive and steaming is fuel wood use intensive. In Congo, steamed paste – *chickwangue* – is prepared entirely by women.

3.2.4 Granulated products

In SSA, there are three common types of granulated cassava products: *gari*, *attieke* and *tapioca*. The methods for making granulated cassava products originated in Brazil. To prepare *gari*, fresh cassava roots are peeled, washed and grated. The resulting pulp is put in a porous sack and weighted down with a heavy object for three to four days to express effluent from the pulp while it is fermenting. The de-watered and fermented lump of pulp is pulverized and sieved and the resulting semi-dry fine pulp is toasted in a pan. The grating, effluent expressing, pulverization, toasting and the addition of palm oil are adequate to reduce cyanogens to a safe level (Hahn, 1989). Fermentation gives a sour taste to *gari*. The duration of fermentation varies depending on consumer preference for sour taste. The COSCA study found that commercial *gari* processors in Nigeria ferment cassava for different lengths of time depending on the market. Toasting extends the shelf life so that *gari* can be easily transported to urban markets. If kept in a dry environment, *gari* will store better than grain because *gari* is not known to be attacked by weevils (Okigbo, 1984).

The second type of granulated cassava products is *attieke*, a type of steamed cassava that was originally found only in Côte d'Ivoire. *Attieke* is made in much the same way as *gari* with more or less the same inputs. However, instead of toasting, *attieke* is steamed. *Attieke* is available in a wet form and it has a shorter shelf life than *gari*.

The third type of granulated cassava product is *tapioca*, which is primarily consumed in Benin and Togo. To prepare *tapioca*, cassava is grated and then put in water, pressed and kneaded to release the starch. The starch is permitted to settle at the bottom of the container and the water is drained off. The operation is repeated several times to prepare a high-quality product. The damp starch is spread on a pan and toasted in the same way as *gari*, to form a coarse granular product.

3.2.5 Cassava leaves

Cassava leaves are edible and a more convenient food product than the fresh roots. Cassava leaves are storable in dry form and since they have lower water content, they are less expensive to dry than the roots. If leaf harvesting is properly scheduled, it does not have an adverse effect on cassava root yield (Dahniya, 1983; Lutaladio and Ezumah, undated). Cassava leaves have a nutritive value similar to other dark green leaves and are an extremely valuable source of vitamins A (carotene) and C, iron, calcium and protein (Latham, 1979). The consumption of cassava leaves helps people in many SSA countries to compensate for the lack of protein and some vitamins and minerals in the roots. Cassava leaves are prepared by leaching them in hot water, pounding them into pulp with a pestle and mortar before boiling in water along with groundnuts, fish and oil. This process eliminates cyanogens from the leaves, making them safe for human consumption.

Cassava leaves are an important vegetable in Congo and Tanzania. In countries where cassava leaves are consumed (DRC, Congo, Tanzania, Guinea, Sierra Leone) farmers often select cassava varieties with a large canopy of leaves, and in Tanzania, farmers plant tree cassava – *mpiru* – for the production of leaves. In such countries, producers earn additional income by selling cassava leaves in addition to the roots. Truckloads of cassava leaves, locally called *pondu* in the Congo, are a common sight on the roads from the provinces to urban markets in Kinshasa.

3.3 Cassava's contribution to SSA diets

The main staple foods in the average SSA diet are (in terms of energy) cereals (46%), roots and tubers (20%), and animal products (7%). By comparison, in Western Europe the main staple foods in the average diet are (in terms of energy) animal products (33%), cereals (26%), and roots and tubers (4%).

The data presented in Table 4 shows the nutrient content of major staple foods that are consumed in SSA. Because of its high water content, raw roots contain less energy and

Table 4 Nutrient content of major staple foods

	Water (g)	Energy (kJ)	Protein (g)
STAPLE	Component (per 100 g portion)		
RDA	3000		50
Maize/Corn	10	1528	9.4
Rice (white)	12	1528	7.1
Rice (brown)	10	**1549**	7.9
Wheat	13	1369	12.6
Potato	**79**	322	2
Cassava	60	670	1.4
Soybean (green)	68	615	**13**
Sweet potato	77	360	1.6
Sorghum	9	141	11.3
Yam	70	494	1.5
Plantain	65	511	1.3

Source: USDA, 'Nutrient data laboratory', 2015.

Table 5 Nutritional composition of different kinds of foods (100 g) for comparison with cassava root

Food	Water (g)	Energy (kcal)	Protein (g)	Fat (g)	Ash (g)	Carbohydrate (g)	Dietary fibre (g)
Cassava, raw root	59.68	160	1.36	0.28	0.62	38.06	1.8
Cassava leaves (raw)**	70.0	91	7.0				4.0
Cassava leaves (dried)**	27.0	194	32.5				–
Cassava leaves		91	1.0 to 10.0	0.2 to 2.9	0.7 to 4.5	7 to 18.3	0.5 to 10.0
Cassava bâton or chikwangue*		86	0.18	0.02	0.21	21.2	0.4
Cassava gari*		119	0.37	0.2	0.34	28.8	0.6
Cassava flour (retting, no peel)*		85 to 102	0.16 to 0.22	0.04 to 0.06	0.16 to 0.19	20.9 to 25.1	0.4
Cassava flour (dried)**	12.0	342	1.5				1.4
Potato, raw	79.34	77	2.02	0.09	1.08	17.47	2.2
Wheat flour, unenriched	11.92	364	10.33	0.98	0.47	76.31	2.7
Rice, white, unenriched	12.89	360	6.61	0.58	0.58	79.34	–
Maize, yellow	10.37	365	9.42	4.74	1.2	74.26	7.3
Sorghum	9.2	339	11.3	3.3	1.57	74.63	6.3
Raw egg (white)	87.57	52	10.9	0.17	0.63	0.73	0
Raw fish (trout)	71.42	148	20.77	6.61	1.17	0	0

* Nutritional value after processing 100 g of cassava root.
** Agricultural Research Council/LNR, Cassava, http://www.arc.agric.za/arc-iic/Pages/Cassava.aspx.
Source: Montagnac et al. (2009).

protein than the cereals. However, when the comparison is made between the processed cassava food products in which a lot of moisture has been removed, the gap is much less (Table 5).

Detailed country data in SSA by agro-ecological zones (Fig. 9) shows that cassava currently contributes 10% to 30% of calorie supply in the diet of countries outside the dry savanna zone in SSA. Because of its relatively low protein content the contribution to dietary protein is much lower, usually 2% to 10% (Fig. 10) .

Using FAO data, Philipp Aerni (2005) showed that whereas production and consumption of cassava per capita in SSA as a whole slightly increased or at least remained stable, Central SSA experienced a steep decline in consumption and production of cassava for the last 40 years. In Western SSA, production and consumption/capita experienced a strong upward trend at the beginning of the 1990s but then remained stagnant, whereas in Eastern SSA, a slight decline in production and consumption/capita can be observed from 1991 to 1995 which subsequently reached the previous levels. The strong decline in Central SSA might be related to plant diseases and pests as well as a breakdown of local cassava trade due to the civil wars, especially as almost all major food crops production are in decline in Central

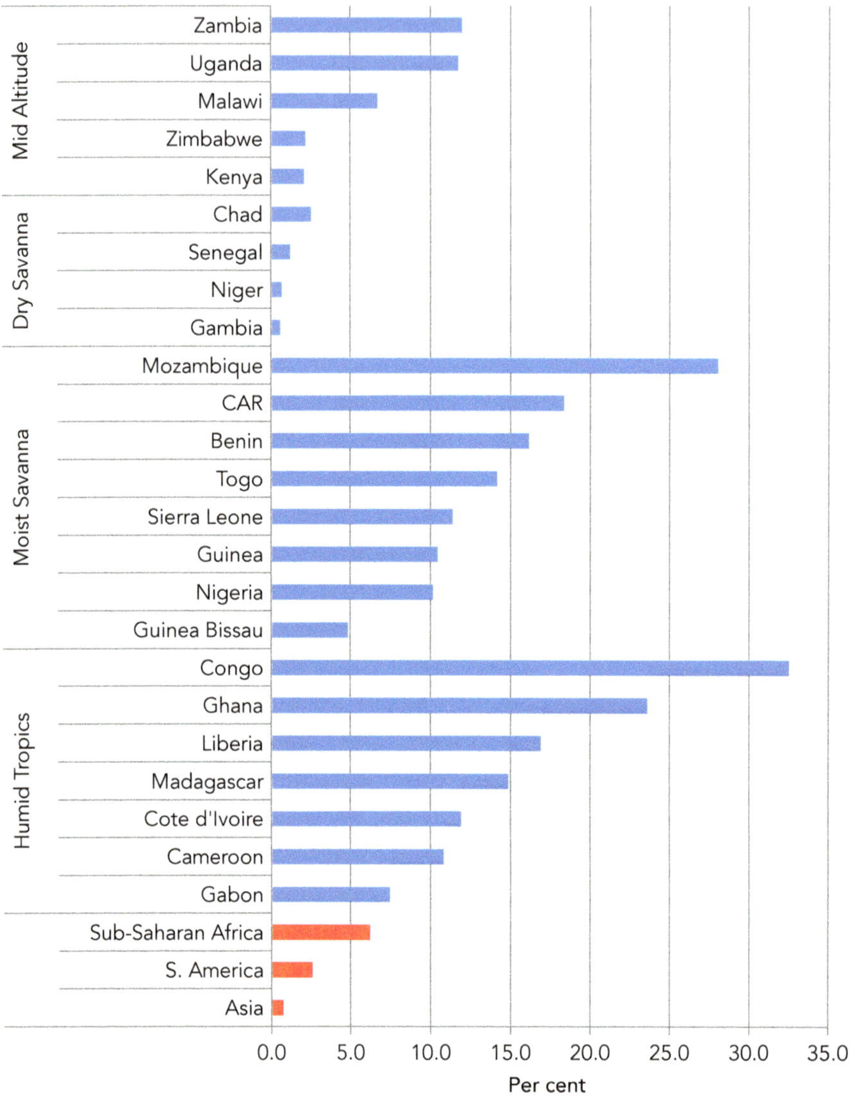

Figure 9 Contribution of cassava and products to supply of total Kcal per day in sub-Saharan Africa in 2011, by agro-ecological zones. Source: FAOSTAT, Food Balance Sheets, 25/03/2016.

SSA (with the exception of rice, which still provides a very low share of the daily intake of calories).

The data in Fig. 11 confirms the decline in contribution of cassava to energy supply in the SSA diet over the last 30 years, despite the increases in production of the crop, indicating that consumption has not responded to income increases as much as the cereals and animal food sources. By contrast the contribution of rice has increased while that of other staples has remained more or less the same.

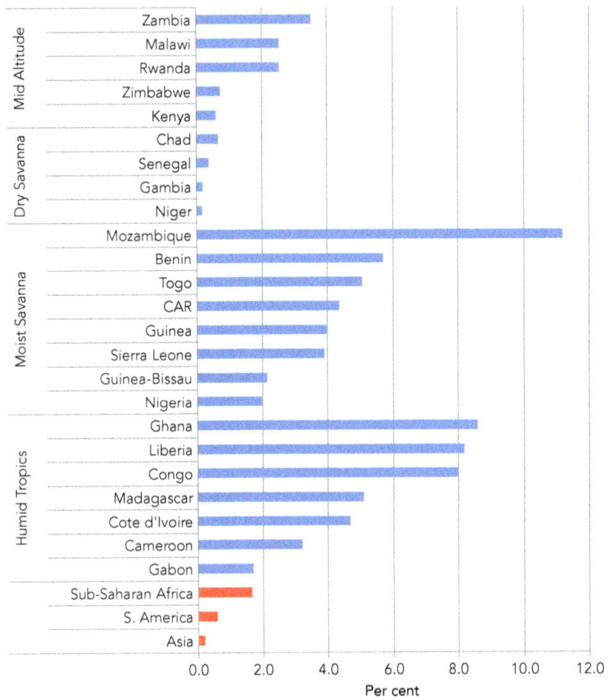

Figure 10 Contribution of cassava and products to total protein supply per day in sub-Saharan Africa in 2011, by agro-ecological zones. Source: FAOSTATS, Food Balance Sheets, 25/03/2016.

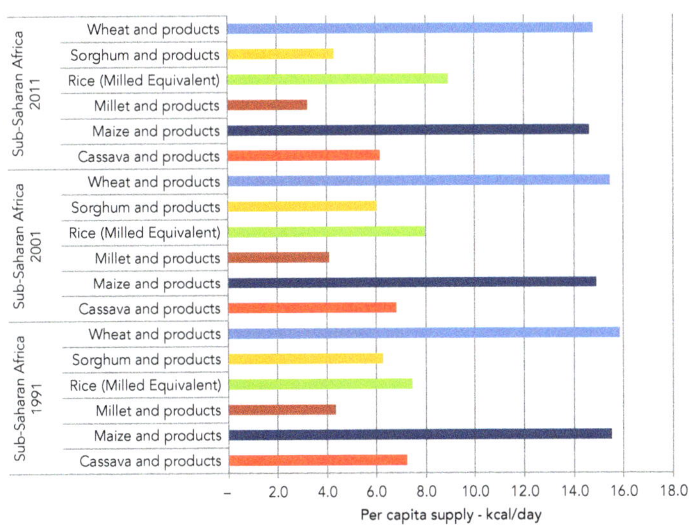

Figure 11 Trends in energy supply from main staples in the sub-Saharan Africa Diet. Source: FAOSTAT, Food Balance Sheets, 25/03/2016.

4 Drivers of change in cassava production in SSA

4.1 Adoption of improved cassava varieties

The International Institute of Tropical Agriculture (IITA) located in Ibadan, Nigeria, has played a leading role in developing improved cassava varieties for SSA that are disease- and pest-resistant, low in cyanide content, drought-resistant, early maturing and high yielding (Dixon et al., 2011). Disease-resistant varieties give sustainable yields of about 50% more than local varieties. Distribution of CMD-resistant varieties in response to the CMD outbreak in East and Central SSA resulted in production levels recovering to pre-epidemic levels in less than five years. Improved cassava varieties are now used in most cassava-growing countries in SSA. IITA's biological control programme resulted in a 95% reduction in cassava mealybug damage and a 50% reduction in damage caused by the cassava green mite.

Until recently, supply side factors such as yield, disease/pest resistance, age at maturity and physiological factors such as weed control were considered as critical for release of elite cassava genotypes by research institutions. Recently, other demand side factors based on industrial utility and nutritional functionality guide the release of elite cassava genotypes. These factors include starch and vitamin A content.

Many factors affect the rate of adoption of improved cassava varieties by farmers in SSA. Under normal circumstances, the willingness of farmers to test and continue to grow new varieties increases with real yield differences between cassava varieties when they have planted in their own fields with their own level of crop management; the availability of local markets that can absorb all or most of farm output at fair prices; reduced or no increase in demand for family or hired labour for production, processing, transportation and marketing of produce; and little or no need for new inputs or difficult operations within prevailing cropping systems. One important point is that farmers in many localities differ in their principal motive for cultivating cassava. There are also several microenvironmental variables that combine to influence adoption of new varieties. These variables specifically include (a) road access for output from cassava farms, (b) transport system, (c) cost of product relative to that of transportation, (d) the absorptive capacity of local markets to which farmers sell their cassava roots, (e) quantity of usable stems that reach a cassava farmer at planting time and (f) quantities of planting stems from the field of a farmer that are replanted after each harvest of the season's crop.

4.2 Expanded use of processing machinery

Cassava roots typically have a shelf life of 24–48 hours after harvest so fresh roots must be processed within 2–3 days after harvest. This transformation requires equipment for one or more of the following processes: peeling, grating, boiling, fermenting, drying, frying and milling.

A wide range of cassava processing techniques exists along with relevant equipment for processing the roots into various products. The quality, standards and specialization of equipment differ by cassava product, fabricator and country. Equipment may be designed to produce multiple products, while others are designed for specific cassava products.

The key cassava processing machines include cassava peelers, washers, graters, chippers, pressers, extruders, milling machines, dryers, pelletizers and packaging machines. Many of these have been developed for microlevel use either manually or motorized. At the industrial scale these machines are combined together and automated for specific uses in starch, cassava flour and animal feed processing.

4.3 Industrial uses of cassava

The future of cassava in SSA is linked to its diversification into industrial uses including cassava starch, animal feed, ethanol and cassava flour. The key challenge to setting up industrial cassava processing plants in SSA is the availability of energy and a consistent supply of raw materials at low costs from the farm level. In the latter case, the development of nucleus farms, out-grower schemes and buffer farms in places where there are private sector agribusiness investments becomes very important. This is currently the highest development challenge for the cassava subsector, as it offers the best opportunity for moving cassava up the value chain.

4.3.1 Starch

Among all the industrial commodities that can be made from cassava roots, production of starch has the largest employment generation capacity due to the diverse number of industries that utilize starch as a raw material (Fig. 12). Globally, cassava is the fourth main source for starch production, after maize, wheat and potato. Four to five tons of roots are normally required to produce one ton of cassava starch, but the ratio may be as high as ten to one, depending on the quality of the root and the efficiency of the processing. Starch is normally further transformed into value-added products and used for the preparation of glucose syrup, glues, alcohol and other industrial uses.

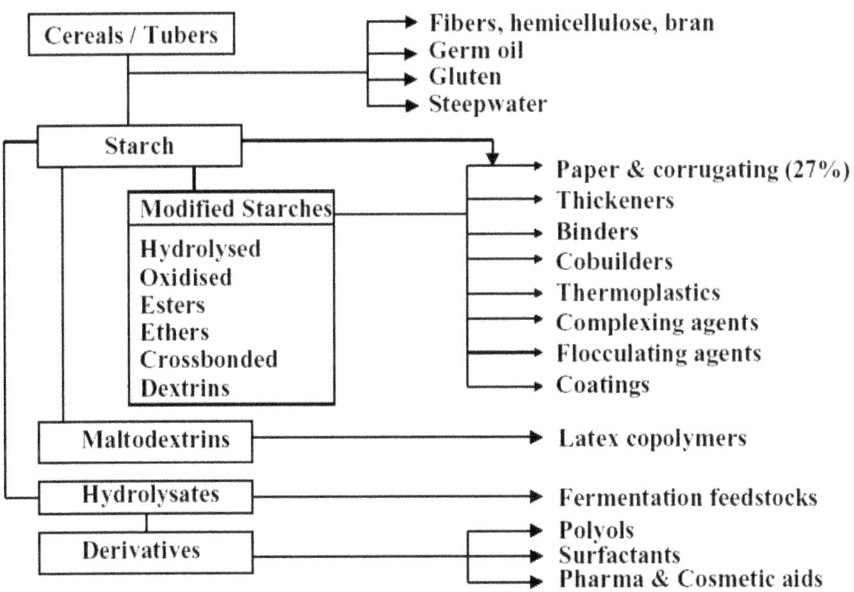

Figure 12 The various processes and industrial uses of cassava starch.

4.3.2 Animal feed

There is rapidly growing demand for cassava pellets for use as animal feed, primarily as a source of calories in balanced rations. Pellets are easier to transport and pack and easier for animals to consume than whole cassava. A critical factor in the use of cassava as livestock feed is its cross-price elasticity effect with maize (Knipscheer et al., 2004). While maize is a preferred feed stock it becomes unattractive when there is a rise in its price, giving rise to an increased use of cassava chips or pellets. Livestock feed being promoted by the International Livestock Research Institute has also been formulated with shredded, dried and pelleted cassava peels.

4.3.3 Ethanol

Ethanol is produced by fermenting and distilling cassava. Ethanol has various industrial uses: it can be mixed with petrol or used on its own as a transport fuel. It can also be used as a base for alcoholic beverages. Finally, ethanol can be utilized as industrial alcohol which is important in the pharmaceutical and cosmetic industry. The competitive position of cassava as a raw material to produce ethanol is highly dependent on the price of maize, and in the case of fuel ethanol, on the price of crude oil. Nonetheless, many countries are adopting policies requiring certain percentages of renewable energy for fuel, and this can give advantage to cassava even in the face of lower oil prices.

4.3.4 Flour

High-quality cassava flour (HQCF) offers several benefits. It is completely gluten-free and can be used as a partial substitute for wheat flour by up to 20% in bread. There has been considerable effort by the government of Nigeria to introduce composite flour (a mixture of 80% wheat flour and 20% HQCF) for baking purposes. This has been pursued as a policy since 2002 and later reinforced by an Act of Parliament in 2014. The objective was to reduce the cost of wheat imports into Nigeria (which as of 2002 was Nigeria's largest food import followed by rice), reduce pressure on foreign exchange, create jobs and improve local incomes. Leading master bakers and industrial corporate bakers have entrenched the policy, mainly because it is cheaper to use composite flour at 20% HQCF with no change in taste and a low glycaemic index (a positive health aspect for such bread to consumers). Some successes have been recorded as a result of this policy.

One is the adoption and adaptation of the drying equipment to suit the local environment in terms of fuel and scale efficiency. The fabrication of the flash dryer has been domesticated in Nigeria and several local manufacturers have the capacity to produce the equipment, although the cost of fabrication is determined by the international price of stainless steel which is imported.

A second emerging success story from this policy is the introduction of enzymes to compensate for the outstanding gluten deficiency in cassava flour in such a way that they build the same consistency and volume as with 100% pure wheat. Companies in Netherlands and South Africa (Olivier van Lieshout Bakery Initiatives, pers. comm., January 2017) are involved in the development of these enzymes but details are currently kept as a company secret. The use of enzymes may improve the quality of baked products from cassava flour and will allow the use of a higher percentage of cassava flour beyond the recommended 20% in the composite flour. However, it is the additional input cost of utilizing the enzymes (especially if it has to be imported) that will decide the adoption rate and utility for cassava flour in the food industry.

4.4 Commercial production of cassava

Commercial farms are beginning to emerge in SSA to support agro-industrial concerns using cassava fresh roots as raw material for starch and starch-derived products such as glucose, cassava flour and animal feed. Under such conditions, cassava is often produced in large fields under sole cropping conditions and with improved inputs. To achieve efficiency in production under commercial production requires a high degree of mechanization. Mechanized cassava planters and harvesters exist in many commercial farms in SSA. However, such farms still face significant challenges in the availability of energy, skills and maintenance of machinery and farm infrastructure.

However, large plantations will not be the only model for agro-industrial cassava production. There are many examples from Asia and some from Latin America, demonstrating efficient systems of aggregating the production from many small farms into large centralized processing plants. In the case of SSA, one of the main impediments to such systems is poor transportation infrastructure in rural areas. In addition, in value chains where dry cassava chips are required as an initial step in processing, the cost of artificial drying is often prohibitive. In Asia, most drying is done under natural sun drying due to the favourable weather conditions in many cassava-growing areas. In the humid tropics, natural drying is much more challenging, but there is ongoing research to develop effective and economic mixed natural/artificial drying.

4.5 Emerging trade in cassava food products

Cassava and cassava-based food products are traded across national borders, especially for villages in close proximity to each other. Most of this trade is informal and unrecorded. The Kano-Katsina-Maradi axis in Nigeria is a major cassava cross-border trade route in West SSA. Gari and chips later converted to flour are the most important products traded across borders. The rural supply markets are located in south and central Nigeria. Kano, especially the Dawanau market, acts as an assembly market in the movement of cassava chips and gari across the borders to the Sahelian countries of Burkina Faso, Chad, Mali, Niger and northern Cameroon (Ezedinma et al., 2006; Ospina-Patiño and Ezedinma, 2015). A major research and development implication of this observation is that efforts at the dissemination of high-yielding varieties with improved disease resistance in the central and southern parts of Nigeria can also have positive outcomes and impact on the lives of the poor in the Sahelian regions of SSA. Improving local cassava production in the south and central regions of Nigeria will therefore help create wealth, especially for the key actors in the commodity chain for gari and cassava chips and also help to reduce risks arising from cyclical drought in the Sahelian countries.

4.6 Effect of population dynamics

Africa is the second largest and second most populous continent with an estimated population in 2013 of 1.033 billion people. By 2050, population in Africa is projected to increase to 2.4 billion, making it the continent with the largest population increase. Africa currently has a very low population density of about 65 people per square mile, which puts it behind Asia, Europe and South America. The high fertility rate, rapid population growth and young populations of which three-fifths are under the age of 25, will put pressure on the demand for food in SSA.

In addition, urbanization is growing in both developed and developing countries. At the continental level, the share of the African urban population in 2010 was about 36%. It is projected to increase to 50% and 60% by 2030 and 2050, respectively. It is projected that between 2010 and 2025, some African cities will account for up to 85% of national populations. The challenge of the projected change in the continent's demographic landscape is how to find innovative and sustainable ways to feed the bulging population, especially the urban population in SSA. In this respect, cassava will continue to play a major role, especially with the transformation into value chains more oriented towards supplying industrial and urban food needs.

4.7 Effect of climate change

SSA is one of the most vulnerable continents to climate change. This situation is worsened by the poor state of economic development and low adaptive capacity. Extreme poverty, frequent natural disasters such as droughts and floods, and heavy dependence on rain-fed agriculture further increases the continent's vulnerability.

Agricultural production in many SSA countries and regions will be severely affected by climate change. Agricultural losses are estimated to be possibly severe for several areas (e.g. the Sahel, East SSA and Southern SSA) accompanied by changes in the length of growing periods, which will especially impact mixed rain-fed, arid and semi-arid systems under certain climate projections. In some countries, yields from rain-fed agriculture could be reduced by up to 50% by 2030. This could lead to loss of livelihood, and possibly even social anarchy amongst the people.

While cassava is considered a drought-tolerant crop, and there is some evidence that its cultivation expanded in the last two decades, we are beginning to witness a decline in production area, especially in the Sahelian countries. This is an indication of a major challenge not just for cassava production but for the national security of the countries in the Sahel.

4.8 Science, technology and innovation

Science and technology play a huge role in today's modern agriculture. Mechanization has enabled farmers to work on larger fields with less labour in shorter amounts of time. Other advances include more precise and more powerful genetic manipulations which enable breeders to develop plants that will grow better in tougher conditions, be resistant to diseases and produce more product. While there are some concerns about the safety of genetically engineered crops, they have helped to reduce some of the risks inherent in agriculture as well as helped to improve nutritional requirements which otherwise would have been unable to reach poorer households in remote rural areas.

In the case of cassava, there are currently no commercial areas planted to genetically engineered varieties. The newly emerging capacity for more precise gene manipulation through gene editing should allow rapid progress in the incorporation of game-changing traits increasing such as disease resistance or starch functional properties, without increasing the concerns among the public.

Many of the issues associated with meeting the demand for food in the face of climate change, potentially on less land, with less water, and with lower environmental impacts require research and innovation. There will not be any single or easy solution to tackle food security fully. In the past years, the focus has been on increasing food production to meet growing demand but more recently it has also been emphasized that up to a third of the world's food production is lost or wasted. Moreover, increasingly food consumption patterns in combination with sedentary lifestyles turn into significant and rising burdens on public health. Thus, it is necessary to take a 'food systems' view giving equal attention to improving agricultural productivity, reducing the negative environmental impacts of production, reducing waste at all stages in the food chain and helping citizens of all countries eat more healthy diets. This is possible with cassava given the multiple uses in food, feed and industry.

4.9 Biofortification of cassava

Cassava is a target for biofortification because of its importance as a staple crop. There is a global effort to develop cassava germplasm enriched with bio-available nutrients since 2005.[7] Major objectives include an increase in the minerals zinc and iron, protein, and vitamins A and E. HarvestPlus is the principal driver for use of biofortified crops worldwide. Biofortification is used as a mechanism for reduction of malnutrition, especially in Africa and South Asia, and leading to improvements in the livelihoods of millions of people around the world. Biofortification is the process of increasing the density of vitamins and minerals in a crop, through plant breeding or agronomic practices, so that when consumed regularly will generate measurable improvement in vitamin and mineral nutritional status.[8] Biofortified crops offer a production-based intervention that, by design, initially reaches remote populations, which comprise a majority of the under-nourished in many countries, and then penetrates to urban populations as production surpluses are marketed (Ezedinma and Nkang, 2008). In this way, biofortification complements fortification and supplementation programmes, which work best in centralized urban areas and then reach into rural areas with good infrastructure.

5 Future trends and conclusion

Future needs and directions for cassava in Africa have been discussed throughout the body of this chapter. In this section some of the important points are highlighted.

There is a need for improved statistics on cassava in SSA. The accuracy of published statistics on agriculture in Africa in general and root crops in particular leaves much to be desired. Time series data from FAO, the most readily available source, often do not correlate with national statistics and other sources. The COSCA study has provided a wealth of first hand information on cassava production, processing and marketing, but now needs to be updated and expanded.

Preparing to meet increased demand: The data on the likely outcome of population and income growth on cassava in SSA show that demand is likely to increase under all scenarios. This provides justification for substantially increased investment in the crop. In individual countries cassava market development should be centred around the creation

7. 'BioCassava Plus' supported by The Bill and Melinda Gates Foundation [http://biocassavaplus. org/]).
8. http://www.harvestplus.org/

of an enabling environment through, for instance, improved access of farmers to both input and output markets, as well as introduction and adoption of improved cassava varieties that are resilient to climate change and have the characteristics required in local communities and other improved and sustainable practices. Adaptation and adoption of improved post-harvest practices, including biofortification of cassava products, are also likely to play an increasing role in the response of the crop to the changing industrial and consumer demand.

6 References

Abass, A., J. Legg, I. Okike, P. Ntawuruhunga, S. Hauser, P. Pypers, N. Mahungu, R. Okechukwu and V. Okoruwa. 2016. Intensification of Cassava: Making Cassava an Agro-industrial Crop in SSA, Presentation, Multinational – Technologies for Sub-Saharan Sub-Saharan Africa Agricultural Transformation (TAAT), Project Identification/Preparation Workshop, 12–14 April 2016, IITA, Ibadan, Nigeria.

Aerni, P. 2005. Mobilizing Science and Technology for Development: The Case of the Cassava Biotechnology Network (CBN) 9th ICABR International Conference on Agricultural Biotechnology: Ten years later Ravello (Italy), 6 July to 10 July 2005.

Alyanak, L. 1997. A modest grater means safer and quicker cassava in Uganda. News Highlights. News/1997/970508-e.rtf. FAO, Rome.

Bouis, H. E., C. Hotz, B. McClafferty, J. V. Meenakshi and W. H. Pfeiffer. 2011. Biofortification: A new tool to reduce micronutrient malnutrition. *Food and Nutrition Bulletin*32 (1) (supplement), The United Nations University.

Dahniya, M. T. 1983. Evaluation of cassava leaf and root production in Sierra Leone. Proceedings of the International Society for Tropical Root Crops, Lima.

Dixon, A. G. O., M. O. Akoroda, R. U. Okechukwu, F. Ogbe, P. Ilona, L. O. Sanni, C. Ezedinma, J. Lemchi, G. Ssemakula, M. O. Yomeni, E. Okoro and G. Tarawali. 2007. Fast track selection approach to release of elite genotypes of cassava for various uses in Nigeria's cassava economy. *Euphytica*. DOI 10.1007/s10681-007-9487-2

Dixon A. G. O, R. U. Okechukwu, M. O. Akoroda, P. Ilona, F. Ogbe, C. N Egesi, P. Kulakow, G. Ssamakula, B. Maziya-Dixon, Iluebbey, M. O Yomeni C. Geteloma, B. James, O. N. Eke-Okoro, L. Sanni, Ntawuruhunga, G. Tarawali, N. Mahungu, J. Lemchi, C. Ezedinma, E. Okoro, E. Kanju, A. A. Adeniji and K. Nwosu. 2011. *Improved Cassava Variety Handbook*, International Institute of Tropical Agriculture, Ibadan, Nigeria, 129pp.

Ezedinma, C. I., and N. M. Nkang. 2008. The effect of quality on gari prices in Nigeria: a Hedonic Analysis, *Journal of Food, Agriculture and Environment* 6 (1): 18–23.

Ezedinma, C. I., I. Ojiako, J. Lemchi, R. Okechukwu, L. Sanni, M. Akoroda, F. Ogbe, E. Okoro and A. Dixon. 2007. *The Cassava Food Commodity Markets and Trade Network in Nigeria*, International Institute of Tropical Agriculture, Ibadan, Nigeria, 296pp.

Ezedinma, C. I., C. Okarter, G. Asumugha, and F. Nweke. 2006. Trends in Farm Labour Productivity and implications for Cassava Industrialisation in Nigeria, Proceedings of the 40th Conference of the Agricultural Society of Nigeria, held at National Roots Crops Research Institute, Umudike, Abia State, 2004.

Ezumah, N. N. and Domenico, C. M. D. 1995. Enhancing the role of women in crop production: A case study of Igbo women in Nigeria. *World Development* 23 (10): 1731–44.

FAO and IFAD – Food and Agriculture Organization of The United Nations and International Fund For Agricultural Development. 2005. A Review of Cassava in sub-Saharan Africa with Country Case Studies on Nigeria, Ghana, The United Republic of Tanzania, Uganda and Benin. *Proceedings of The Validation Forum on The Global Cassava Development Strategy*, Volume 2.

Fresco, L. 1986. *Cassava in Shifting Cultivation: A Systems Approach to Agricultural Technology Development in Sub-Saharan Africa*, Royal Tropical Institute, Amsterdam, the Netherlands.

Gossens, F. 1996. *Cassava Production and Marketing in Zaire – The Market of Kinshasa*. Leuven University Press, Leuven.
Grace, M. R. 1977. *Cassava Processing*. FAO Plant Production and Protection Series No. 3. Food and Agriculture Organization of the United Nations, Rome.
Ikpi, Anthony E. 1989. Cassava development in Nigeria. In *Cassava: Lifeline for the Rural Households*. Edited by Anthony E. Ikpi and Nathalie D. Hahn. UNICEF, Lagos, pp. 1–15.
Jones, William O. 1959. *Manioc in Sub-Saharan Africa*, Stanford University Press, Stanford, 315pp.
Knipscheer, H., C. Ezedinma, P. Kormawa, G. Asumugha, K. Makinde, R. Okechukwu and A. Dixon. 2007. *Opportunities in the Industrial Cassava Market in Nigeria, A Sub Sector Study*, International Institute of Tropical Agriculture, Ibadan, Nigeria, 42pp.
Latham, M. C. 1979. *Human Nutrition in Tropical Sub-Saharan Africa*, FAO, Rome.
Lutaladio, N. B., and H. C. Ezumah. undated. Cassava leaf harvesting in Zaire (Congo). M'vuazi.
Montagnac, Julie A., Christopher R. Davis and Sherry A. Tanumihardjo. 2009. Nutritional value of cassava for use as a staple food and recent advances for improvement, *Comprehensive Reviews in Food Science and Food Safety* 8 (3): 181–94.
Nweke, F. I. 1996. Processing Potential for Cassava Production Growth in Sub-Saharan Africa COSCA Working Paper no 11, Collaborative Study of Cassava in Sub-Saharan Africa, IITA Ibadan, 47p.
Nweke, F. I. 2005. The Cassava Transformation In Sub-Saharan Africa. Chapter 1 in FAO & IFAD, 2005.
Nweke, Felix, I., Dunstan, S. C. Spencer and John K. Lynam. 2001. *The Cassava Transformation: Sub-Saharan Africa's Best Kept Secret*. Michigan State University Press, East Lansing.
Nweke, Felix I. 2004. New challenges in the cassava transformation in Nigeria and Ghana. EPTD discussion papers 118, International Food Policy Research Institute (IFPRI), Washington DC.
Nweke, Felix I. 2010. Controlling cassava mosaic virus and cassava mealybug in sub-Saharan Sub-Saharan Africa. In *Proven Successes in Agricultural Development: A technical compendium to millions fed*. Edited by David J. Spielman and Ranjul Pandya-Lorch. International Food Policy Research Institute, pp. 99–121.
Nweke, Felix I. and Steven Haggblade. 2010. The cassava transformation in West and Southern Sub-Saharan Africa. In *Successes in Sub-Saharan Sub-Saharan Africa Agriculture: Lessons for the Future*. Edited by Steven Haggblade and Peter B. R. Hazell Editors. Johns Hopkins, Baltimore, pp. 29–70.
Okigbo, Bede N. 1980. Nutritional implications of projects giving high priority to the production of staples of low nutritive quality: The Case for Cassava (*Manihot esculenta, Crantz*) in the Humid Tropics of West Sub-Saharan Africa, International Institute of Tropical Agriculture, Ibadan, Nigeria. The United Nations University Press, Food and Nutrition Bulletin, Volume 2, Number 4, October 1980.
Okigbo, Bede N. 1984. Improved permanent production systems as alternative to shifting intermittent cultivation. In *Improved Production Systems as an Alternative to Shifting Cultivation*. FAO Soils Bulletin 53. Soils Resources Management and Conservation Service, Land and Water Development Division, Food and Agriculture Organization of the United Nations, Rome, Italy, pp. 1–100.
Okorji, E. C. 1983. Consequences for agricultural productivity of crop stereotyping along sex lines: a case study of four villages in Abakaliki area of Anambra state. M.Sc. thesis, Department of Agricultural Economics, University of Nigeria, Nsukka, Nigeria.
Ospina-Patino, M. T. and C. Ezedinma. 2015. Understanding the linkage of urban and rural markets of cassava products in Nigeria, *Sub-Saharan Sub-Saharan Africa Journal Of Agricultural Research*, 3805–13.
Ouraga-Djoussou, L. H., and M. Bokanga. 1995. Cassava Flour Production – Background Information to the Site Visits.
Paiva, S. A. R., and Russell, R. M. 1999. Beta-carotene and other carotenoids as antioxidants. *Journal of the American College of Nutrition* 18: 426–33.
Shrimpton, R. 1989. Vitamin A deficiency in Brazil, perspectives for food production oriented strategies. *Ecology of Food and Nutrition* 23: 261–71.
Spencer, D. 2005. Cassava In Sub-Saharan Africa: Past, Present and Future, Chapter 2 in FAO & IFAD, 2005. http://www.fao.org/docrep/009/a0154e/A0154E04.HTM

Tanumihardjo, S. A. 2008. Food-based approaches for ensuring adequate vitamin A nutrition. *Comprehensive Reviews in Food Science* and *Food Safety* 7: 373–81.

Tanumihardjo, S. A., Bouis, H., Hotz, C., Meenakshi, J. V. and McClafferty, B. 2008. Biofortification of staple crops: an emerging strategy to combat hidden hunger. *Comprehensive Reviews in Food Science* and *Food Safety* 7: 329–34.

Tollens, E. 1992. *Cassava Marketing in Zaire-an Analysis of its Structure, Conduct and Performance*. K. U. Leuven, Belgium (Mimeograph).

United States Department of Agriculture 2015. Agricultural Research Service, National Nutrient Database for Standard Reference, Release 28 November 2015.

WHO/FAO. 2003. *Diet, Nutrition and the Prevention of Chronic Diseases*. Report of the joint WHO/FAO expert consultation; WHO technical report series, Geneva, Switzerland.

Chapter 7

Cassava cultivation in Latin America

Olivier F. Vilpoux, Denilson de Oliveira Guilherme and Marney Pascoli Cereda, Catholic University of Campo Grande, Brazil

1 Introduction
2 Global and Latin American approaches to cassava production
3 Cassava competitiveness in Latin America
4 Boosting cassava production in Latin America
5 Improving the Latin American starch industry
6 Brazilian cassava flour (farinha)
7 Consumption of boiled and fried cassava
8 Future trends: cassava for animal feed
9 Where to look for further information
10 References

1 Introduction

Cassava was domesticated from its wild species ancestors in the New World, probably in South America. Several theories of origin have been posed in the past, including southern Amazon basin, or from the region between the Brazilian and the Venezuelan Amazon rainforest. There may have also been a parallel meso-American origin. Hence, it was natural for the indigenous people throughout the New World to appropriate this plant for food use. The domestication of this species by the indigenous people provided plants with modifications in their roots to store starch. While many of the some 100 wild *Manihot* species produce starchy storage roots, cassava does so in a highly efficient manner, and is the only cultivated species in the genus. The propagation of these domesticated plants is always executed through the use of stakes (stem pieces), by vegetative propagation, which makes the cultivated plants clones of a single species, *Manihot esculenta* Crantz. This origin, clearly connected to indigenous cultures, has been presented as an advantage as well as a disadvantage. The clear association to social factors has maintained the production of cassava over the years, but has also been the cause of the difficulty to modernise this crop, out of traditional ties in which it is immersed. A well-known contrast is Thailand, a country with little or no tradition with food use of cassava, but which managed to establish more commercial and industrial avenues of production and use. Other Asian

http://dx.doi.org/10.19103/AS.2016.0014.07

countries, such as Indonesia and Vietnam, are also increasingly following the industrial-based development of the crop.

Some of the challenges that cassava systems need to address to move beyond traditional history, for greater impact in Latin America are: (1) break through the stagnated level of agricultural productivity, (2) develop new varieties, including replacement of earlier ones because of the accumulation of viruses, (3) clonal *cleaning*, especially of viruses, as is done in other vegetative crops, (4) use of rapid multiplication techniques in addition to the use of clonal cleaning to obtain large quantities of more productive propagation material, (5) selection of architecture-appropriate varieties for mechanical harvesting, (6) enhanced markets for cassava for human consumption, with different processes and products (7) elucidation of cooking mechanisms for fresh cassava roots, (7) development of new products more adapted to the needs of modern lifestyles, and (8) use of the entire cassava plant in animal feeding.

Implementation of these actions can help achieve a higher level of productivity and societal benefits of cassava to Latin America, similar to advances in Asia.

2 Global and Latin American approaches to cassava production

To understand the potential of cassava production in Latin America it is necessary to quantify the international production. In 2014, Africa dominated the production of cassava, with more than half of the world production, compared to Asia, with a third. Latin America, as the place of origin of this crop, ranked third with only 12.2% (Fig. 1).

Cassava is the fifth most important in terms of global production of the main starchy crops between 2000 and 2014, far behind maize. However, from a production of 176 million tonnes in 2000 it increased to 270 million tonnes in 2014 (Fig. 2).

Thus, when considering the evolution of production, cassava is the second among major crops in the past 15 years, only behind maize, whose production was driven by, among others, the production of ethanol in the United States and by demand for animal feed in Asia. The annual growth rate of cassava production in the world since the beginning of the century was 3.1% and a total growth of 53.5% (Fig. 3).

Despite the high growth of world cassava production since the early twenty-first century, the analysis by continent indicates a very different reality (Fig. 4). While Asia experienced growth well above the world average, Africa was in the middle and the Americas remained stable, with a total growth of 5.7% and an annual growth of just 0.4%. Thus, this continent has seen its share in world production falling from 17.6 to 12.2%. Meanwhile, Asia has gone from 28.1 to 33.4%.

Although cassava is grown in nearly every country of South and Meso-America and the Caribbean, only seven countries in Latin America have more than 1% of the hemisphere's production. In total, these seven countries represent 96% of the production in the Americas (Table 1). The three largest producers alone – Brazil, Paraguay and Colombia – account for 88.1% of the region's production, and Brazil alone, 70.8%.

Detailed analysis indicates a slight decrease of production in Brazil and a small growth in Paraguay (with an annual rate of 0.8%), in the first 15 years of this century, well below the global annual growth (Table 2). Among the top three producers, only Colombia experienced substantial growth, close to the world average.

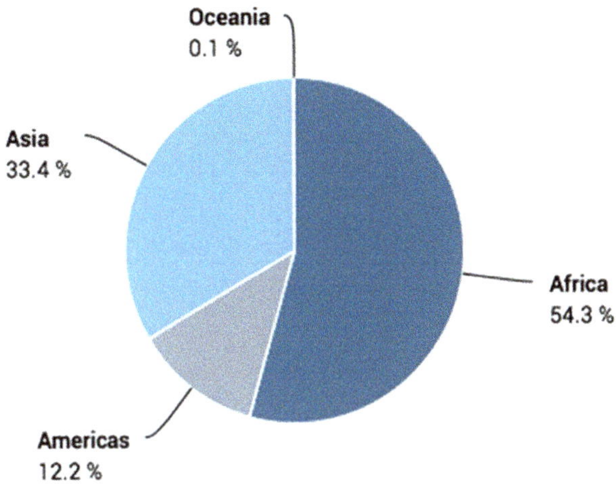

Figure 1 Participation of each continent in the world cassava production, in 2014. Source: FAOSTAT (2016)

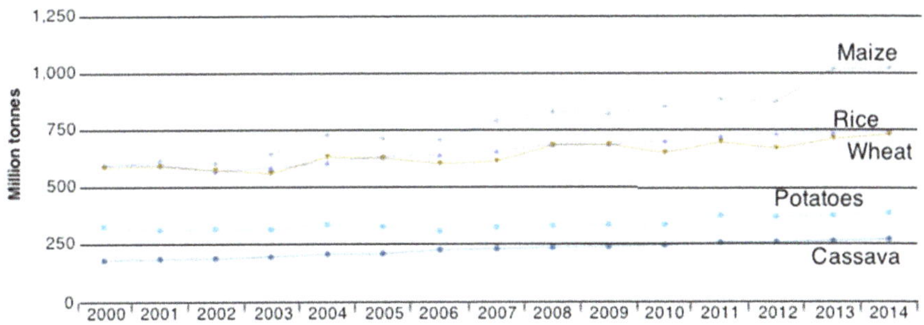

Figure 2 World growth production of the main starch crops between 2000 and 2014. Source: FAOSTAT (2016)

Haiti is the only country that had an increase in production above the world average, but this country represents less than 2% of production in the continent. Growth in Haiti is driven by cassava's adaptation to increasingly degraded soils and population growth. In parallel, Venezuela experienced a sharp decrease of more than 37%, as the country has largely failed to invest in agricultural development and relied more on imports.

In addition to production, it is important to analyse the evolution of productivity. For the main producing countries of the Americas, it is possible to differentiate the countries of Central America, with a yield of less than 8 tons per hectare (Fig. 5), and of South America, with a yield above 10 tonnes per hectare.

Further analysis indicates a high stability from 2000 to 2104, with similar production values across the period, for most of the assessed countries. Cuba has achieved a 30% productivity increase in this period, but the yield in this country are very low. Paraguay also managed an increase of 26%, but Fig. 5 indicates a great instability in the productivity of that country. In parallel, productivity in Haiti fell 23.7%, probably as a consequence of

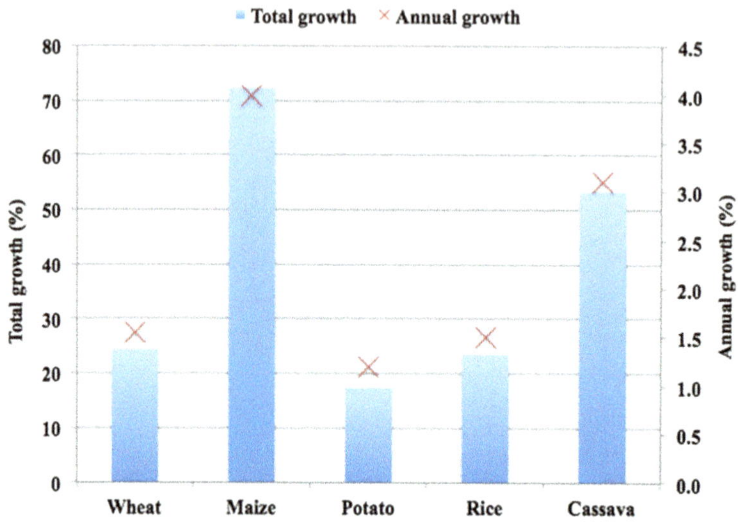

Figure 3 Total growth of world production and the average annual growth rate of the main starch crops, between 2000 and 2014. Source: FAOSTAT (2016)

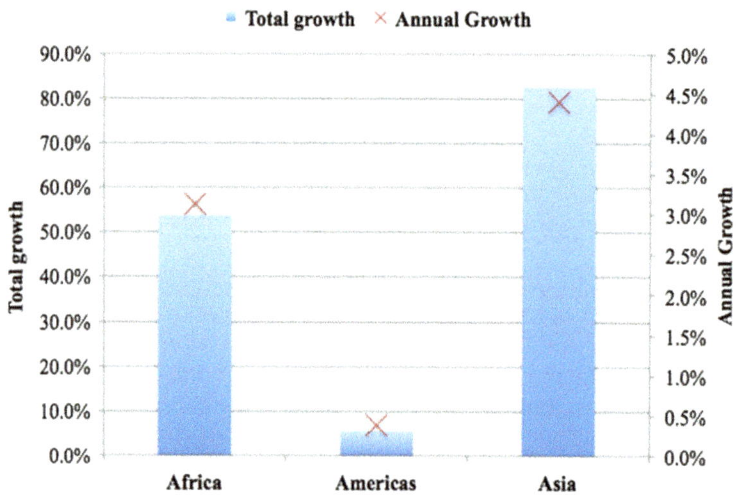

Figure 4 Total growth of cassava production and average annual growth rate in the main producing continents, between 2000 and 2014. Source: FAOSTAT (2016)

the difficulties after the 2010 earthquake, and possibly resulting from the movement of cassava to ever-more degraded soils.

The analysis should also compare the evolution of productivity with the major cassava producing countries in the world, for example, Nigeria, with 54.8 million tonnes in 2014,

Table 1 Cassava production in major countries of the Americas and percentage of production in relation to total production of the region, in 2014

	Tonnes (000)	Production %
Brazil	23,242	70.8
Paraguay	3,060	9.3
Colombia	2,635	8.0
Sub total	**28,937**	**88.1**
Peru	1,188	3.6
Haiti	615	1.9
Cuba	436	1.3
Venezuela	358	1.1
Sub total	**2,598**	**7.9**
Total	**31.54**	**96.0**

Source: FAOSTAT (2016)

Table 2 Total growth of cassava production and average annual growth rate in the major producing countries of the American continent, between 2000 and 2014

	Total growth (%)	Annual growth (%)
Brazil	−0.4	0.0
Paraguay	12.5	0.8
Colombia	47.0	2.8
Peru	34.6	2.1
Haiti	82.1	4.4
Cuba	2.1	0.1
Venezuela	−37.3	−3.3

Source: FAOSTAT (2016)

Thailand, with 30 million, Indonesia with 23.4 million (slightly ahead of Brazil), and finally, the Democratic Republic of Congo, with 16.6 million tonnes.

The trends in productivity differentiate three groups of countries (Fig. 6): the first one with a positive growth and a productivity in 2014 above 22 tonnes per hectare (23.4 for Thailand); the second one, with Brazil, presenting a stable productivity around 15 tonnes per hectare; and the last group, showing a productivity under 10 tonnes per hectare, with stable or declining yield, as in the case of Nigeria.

These results clearly indicate a loss of competitiveness of Brazil, which was surpassed by Indonesia and Thailand in recent years.

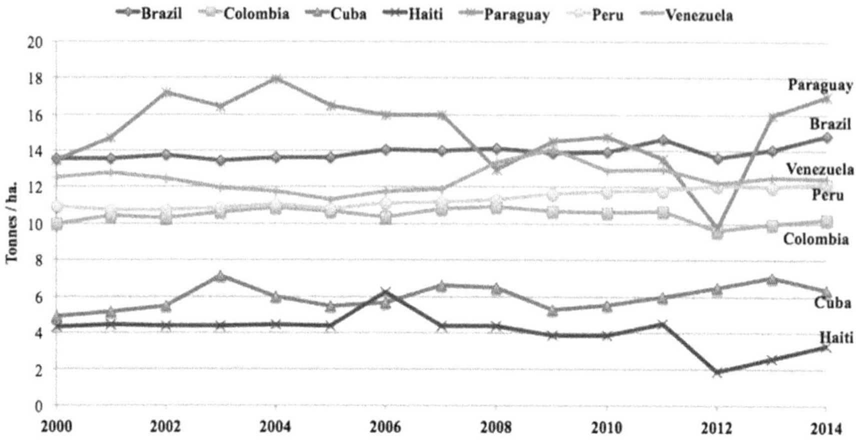

Figure 5 Cassava productivity growth in major producing countries in Latin America, between 2000 and 2014. Source: FAOSTAT (2016)

Table 3 further supports the grouping of countries based on the evolution of cassava productivity, with a strong distinction among Group 1 (Thailand and Indonesia), Group 2 (Brazil) and Group 3 (Nigeria and Democratic Republic of Congo).

We hypothesize that the main markets of cassava explain some of the differences of development in production and productivity among the different groups of countries:

- Group 1 (Thailand and Indonesia - Asia) starch, animal feed and human food in Indonesia;
- Group 2 (Brazil - South America): starch, flour for human consumption and fresh cassava. Cassava is also used for animal feed, but this use is concentrated among small producers and no statistics or official data are available;
- Group 3 (Nigeria and the Democratic Republic of the Congo - Africa): flour and fermented products for human consumption.

Based on this classification it is useful to assess the evolution of the main markets of the products consumed in Latin America.

3 Cassava competitiveness in Latin America

The main products to be evaluated are starch, cassava flour for human consumption and fresh cassava. Cassava is also used for animal feed in Latin America, but mainly on small farms, without the existence of a developed market. Dried chips for animal achieved modest success in Colombia, but the competitiveness of cassava chips with maize has suffered since the implementation of the free trade agreement between Colombia and the US in 2012. The dried chips industry has failed in Brazil, even in the northeast where the climatic conditions are very favourable for natural drying.

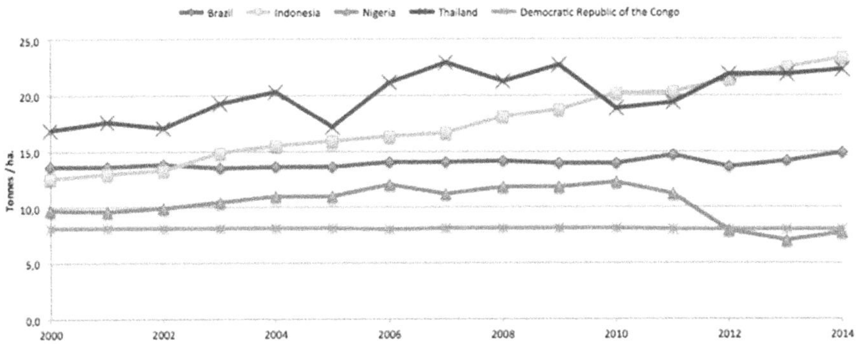

Figure 6 Cassava productivity growth in major producing countries, between 2000 and 2014. Source: FAOSTAT (2016)

Table 3 Total growth of cassava production and average annual growth rate in the major world producing countries, between 2000 and 2014

	Total growth (%)	Annual growth (%)
Indonesia	86.4%	4.5%
Thailand	32.0%	2.0%
Brazil	9.4%	0.6%
Democratic Republic of the Congo	−0.5%	0.0%
Nigeria	−20.4%	−1.6%

Source: FAOSTAT (2016)

In Brazil the main food product from cassava is a coarse flour called farinha. A similar product is called *gari* in West Africa and both are very different from cassava chips meal.

3.1 Starch

Although cassava starch is produced throughout South America, generally on a small scale, only Brazil and Paraguay produce it with larger scale technology. However, sour starch (fermented) has reached medium and large scale production (CEREDA & BRITO, 2016), a product described in more detail later.

Based on Vilpoux (2002), the starch produced in Brazil at the beginning of the twenty-first century was derived from two sources – maize and cassava – with 70 and 30% of production, respectively. The distribution between the two raw materials depends mainly on their costs. A comparison of productivity trends gives a good indication of the evolution of production costs in recent years (Fig. 7). While maize yield increased 89% between 2000 and 2014, cassava grew only 9%, which indicates a loss of competitiveness of this raw material in the starch industry.

Figure 8 shows the results of the decline in competitiveness of the cassava starch market in Brazil, with a steady fall in the Brazilian production between 2005 and 2013, far from the

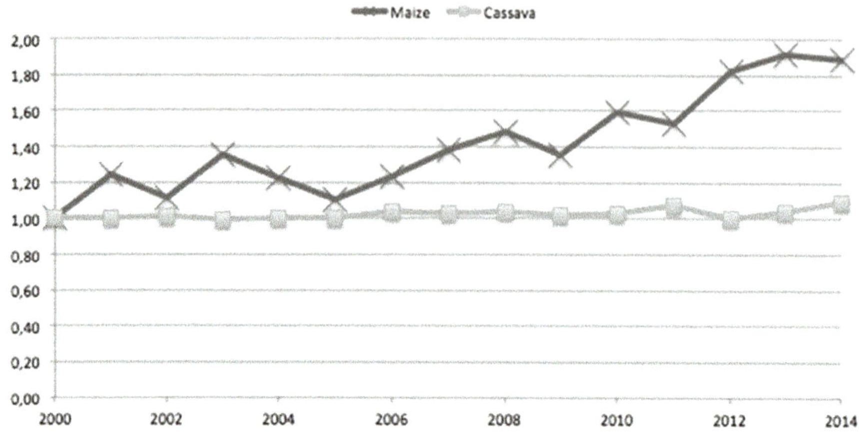

Figure 7 Maize and cassava productivity growth in Brazil, between 2000 and 2014. Source: FAOSTAT (2016)

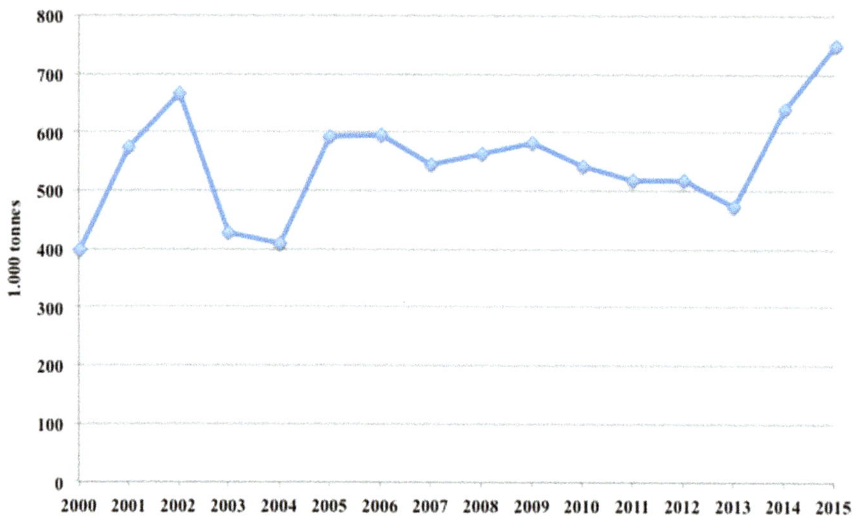

Figure 8 Evolution of cassava starch production in Brazil, between 2000 and 2015. Source: ABAM (2016) and CEPEA (2016b)

peak in 2002 of nearly 700,000 tonnes. However, the production has grown considerably between 2014 and 2015, which corresponds to a massive drop in cassava prices (Fig. 9). Thus, the recovery is not due to an improvement in the industry structure or raw-material productivity, but to the falling prices, a similar situation to the one that took place from 2001 to 2002.

An important factor influencing the cassava competitiveness is the wide variation in prices (Fig. 9). Between January 2006 and December 2015, prices of cassava roots in Paraná (the main Brazilian state producer of starch), ranged from 135 to 725 R$ per

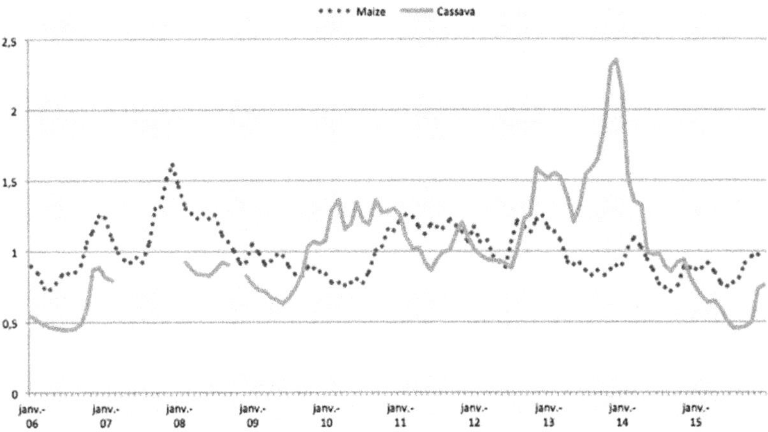

Figure 9 Variation in prices* of maize (São Paulo Brazilian state) and cassava (Paraná Brazilian state), between January 2016 and December 2015, based on the average price from January 2006 to December 2015. *Prices standardized based on INPC. Value of Brazilian Real of January 2016. Source: based on CEPEA (2016a,b)

ton (standardized values based on currency value on January 2016). Compared with maize, the cassava price variations, based on the average price from January 2006 to December 2015, are much higher, with a peak at the end of 2013 and high falls in 2006 and 2015.

3.2 Cassava flour for human consumption and fresh cassava

Despite the lack of recent data it is possible to identify, from the Brazilian Institute of Geography and Statistics (IBGE) Survey on Family Budget (POF) in 2002 and 2008, a fall in consumption of cassava flour and fresh cassava in Brazil, the main consumer of these products in South America (Fig. 10).

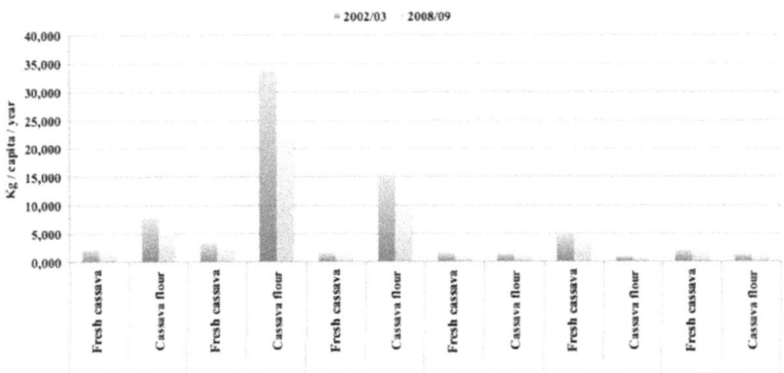

Figure 10 Evolution of the annual consumption *per capita* of cassava flour and fresh cassava, from 2002/2003 to 2008/2009. Source: IBGE (2003) and IBGE (2009).

This drop occurs in all the surveyed areas and confirmed a trend already identified by Vilpoux (2002) since the 1980s.

The Family Budget Survey of 2002/2003 shows that Brazilians are diversifying their food, reducing direct consumption of traditional products such as rice, beans, potatoes, cassava flour and fresh cassava. In parallel, the consumption of prepared foods is growing (IBGE, 2004).

4 Boosting cassava production in Latin America

Improving cassava production, and the resulting benefits to producers and consumers, entails research and development in different steps in the cassava value chains.

The production of cassava in the Americas comprises a wide range from traditional to modern operations for its cultivation, such as soil preparation, acidity correction, selection, planting of cuttings, cultural operations and root harvest. A range of approaches are being employed to increase productivity, which has lagged behind other major crops in many countries, including Brazil, by far the region's largest producer.

4.1 New varieties

Development of new varieties is a time-consuming process. The breeder should know clearly the goals for these new varieties, including knowledge of the traits that need to be improved, as well those traits in the current varieties that need to be preserved varieties. The genebanks (referred to *BAG*s in Brazil), and especially the diverse landrace varieties within those genebanks, are an important source of genetic diversity. In general the development of new varieties takes into account productivity, resistance to pests and diseases, use (flour, starch, human consumption or animal feed), resistance to drought or excessive rainfall, and soil fertility. Little or no effort has been spent in selecting material to facilitate mechanized processing, e.g. with more rounded and straighter root shape.

The use of new varieties is a common practice to increase productivity, especially when the focus is the production of raw material for the starch industry. New candidate varieties are typically tested throughout the target region before being released and disseminated, in order to exploit the high potential for improved productivity. Productivity can vary considerably depending on the cropping area (Rós et al., 2011; Tironi et al., 2015).

New varieties may produce well initially, but then decline after a few years as a function of pest and pathogen accumulation, especially viruses, due to vegetative multiplication. One of the solutions is to replace these *degenerated* varieties with new *clean* ones. This ongoing necessity for renovation may partly explain why, despite the regular release of new varieties, the Brazilian productivity does not increase.

One solution could be the *cleaning*, or the removal of systemic pathogens such as viruses, of the varieties in use, in order to restore productivity. Nevertheless, in most production systems there is not good information on the impact of pathogens accumulated in planting material, on yield over time. This information is fundamental to designing a strategy to recover productivity by cleaning of planting material.

4.2 Traditional cassava propagation

The propagation of cassava worldwide is based on stem cuttings. Well-lignified stems (typically from the middle part of the plant) are cut from plants in the previous production cycle and cultivated in the next one. Regularly, stems cut from one cassava plant allow the production of about 10 new plants.

This propagation system has the advantage that it allows precise duplication of the inherited (genetically controlled) traits of the parent plant. However, there are also some drawbacks such as the low multiplication rate and high risk of pests and diseases being transmitted from generation to generation, ultimately reflected in crop yield decline. This system also facilitates the movement of pests and diseases to different regions (Silva et al., 2002; Fukuda and Carvalho, 2007; Santos et al., 2009). While the degeneration of planting material may be a universal phenomenon in cassava, there is a high potential for widespread effects in the Americas. Because cassava originated in the region, the pests and diseases affecting the crop also evolved and diversified here. The Americas have a far wider range of pests and diseases affecting the cassava crop, in comparison with either Africa or Asia, even though in Africa there are a few very devastating pests and diseases. For this reason, there is a special need and a special responsibility for researchers in the Americas to provide leadership in finding solutions to the problems of seed degeneration.

Besides the direct effect of accumulated pests and diseases, stem pieces may become physiologically weakened from nutrient deficiency, weed competition and/or drought, resulting in lower establishment rates and less vigorous plants. The use of strong and healthy selected cassava stems is a key factor for increasing productivity. Many producers are unaware of the strong connection between healthy planting material and productivity, or do not know how to take proper measures to improve quality of planting material.

4.3 Tissue culture

Techniques such as tissue culture can offer solutions to the problem of planting material degeneration, and are already a reality for other crops such as banana, pineapple and eucalyptus.

Cassava tissue culture is being employed for the cleaning of propagative materials and to increase the availability of high quality seedlings (Escobar et al., 2012). Nonetheless, the use is still very limited. In most situations, it has been difficult to develop systems that are commercially viable, i.e. where farmers are willing to pay for clean propagation material at a level that will sustain a profitable business for entrepreneurs. Based on global experiences, in most cases cleaning through tissue culture techniques needs to be applied mainly at the early stages of propagation, and more traditional stem-based propagation, using good quality control techniques, can be used to produce the commercial-level seed.

In vitro techniques, when properly managed, can achieve high plantlet acclimatization and establishment rates after their initial hardening and adaptation, reaching 100% of vigorous plantlets ready to be transplanted to the production field (Marín et al., 2008). More importantly however is the reports that it is possible to get more than 50% increase in productivity of the varieties that were propagated many generations in the field.

Although tissue culture techniques allow overcoming the problems of contamination caused by systemic diseases (Silva et al., 2002), their cost is still high. But if used in conjunction with more conventional rapid multiplication, they can allow the installation of healthy seed nurseries. To improve the dissemination of plantlets produced through the use of virus cleaning techniques, the technology of tissue culture could be used in addition to other rapid multiplication propagation methods. This system was used in Costa Rica with good results. This technique was first proposed in 1982 (CIAT, 1982) and again promoted in the 2000s, resulting in a book chapter (Silva et al., 2002).

4.4 Rapid multiplication

There are many systems to increase the multiplication rate of cassava over the conventional farmer system. One of the simplest and most effective ones is the use of *mini-stems* or *mini-setts*. This system begins with cutting and planting small pieces (typically 2–5cm) of cassava stems with only two lateral buds. Each bud will produce about four sprouts (total of eight sprouts per mini-stem) each of which can be cut and rooted in water, often supplemented with a rooting hormone treatment. A mature cassava plant produces on average more than 20 mini-stems, which multiplied by eight will originate a total of 160 adult plants. Considering that each plant originated through this rapid multiplication will be vigorous enough to produce 10 cassava stems (of 20 centimetres each) per plant, after 12 months it is possible to obtain, from only one plant, enough propagative material to establish 1600 plants (Silva, Cereda & Fiorini, 2002; Fukuda & Carvalho, 2007).

Rapid multiplication includes a set of techniques that can be used to expand the availability of propagative materials well beyond the quantity and quality of the cassava stems otherwise available in the market (Silva et al., 2002, Fukuda and Carvalho, 2007, Rodrigues et al., 2008). In addition rapid multiplication can also easily incorporate methods to eliminate contamination caused by systemic pathogens, in order to increase productivity and to continue using the same varieties. The technique is also important when trying to introduce new varieties, which are not yet available in large plantations.

It should be noted that, in spite of the apparent simplicity and effectiveness of rapid propagation systems in cassava, their use outside of experimental applications has been limited. This low level of adoption and application calls for additional research and promotion to improve the knowledge about losses resulting from poor quality or pathogen-infected planting material, and methods for improving even further the efficiency and cost-effectiveness of rapid propagation systems for cassava.

4.5 Mechanization of planting and harvesting

In Latin America cassava is still typically a traditional crop cultivated in small areas, often less than a hectare, for fresh cassava, but sometimes more than 50 hectares for industrial cassava, for example produced in the Brazilian South, for starch production.

In the case of fresh cassava, which essentially reaches the local market, the small scale does not affect the competitiveness of the production system, being basically sold in local markets or to small processing companies. For industrial cassava destined for the flour (Brazilian farinha) markets in the Brazilian North and Northeast, the small scale is also not typically a constraint, mainly because the products are essentially destined to the local trade, where alternative products do not strongly compete. In these cases, traditional planting methods and the resulting yields are sufficient.

For industrial cassava designed for the national or international markets, such as starch and some types of Brazilian flour produced in the South of Brazil, the traditional small-scale, labour-intensive agriculture production practices negatively affect the competitiveness.

Figure 11 Large-scale cultivation of industrial cassava in the Brazilian Northeast region.

As already discussed, the main cassava competitor in the starch industry is maize, which benefits from important economies of scale.

Starch factories also benefit from economies connected to large-scale production. In Brazil, coordinating the supply from many small suppliers complicates the organization of production, typically increasing transaction costs and introducing greater price variation.

The large-scale production of cassava is already attempted in various regions of Brazil, such as in the South, Centre-West and Northeast. These examples demonstrate the possibility of large-scale cultivation (Fig. 11).

Currently, the great majority of industrial cassava produced in the Brazilian South is partly mechanized (land preparation, planting, (Fig. 11) weeding (mechanical or chemical), and part of harvesting), which allows the effective reduction of labour costs.

Nonetheless, full mechanization of cassava harvest to increase the scale of production remains a challenge, . Industrial systems use equipment similar to a subsoiler to facilitate roots removal, but that does not fully solve the high labour input in harvest, such as removing roots from stems, cleaning soil and debris from the roots, and loading onto trucks or wagons for transport to the factory.

The development of a fully mechanized harvester has been attempted many times. However, typically the projects were isolated, discontinuous, and with low communication between entrepreneurs, farmers and researchers.

4.6 Opportunities for improvement

The increase in cassava productivity and profitability in Latin America will rely on systems improvement and not just changes in one or a few specific techniques. Further, the entire value chain of production, processing and marketing need to be simultaneously considered in order to achieve success.

Clonal cleaning is fundamentally important to achieve the benefits of healthy plants, standardized form of roots and with high vegetative vigour. However, this practice needs to have the capacity to produce plants with affordable commercial prices, and to achieve this in combination with rapid multiplication techniques.

The cassava root morphology is another aspect with great relevance for the development of the industrial sector and must be taken into consideration in order to facilitate mechanization. Cassava roots of some varieties are elongated and irregular, which promotes their breakage in the soil during harvest, especially in heavy or compacted soils and in the dry season. Furthermore, the root generally has an irregular surface, with constrictions and unevenness that complicate peeling. Plant breeders should identify and select varieties with more appropriate shape for processing and mechanization, such as shorter, more rounded and smooth-surfaced roots. Another point to be considered in root selection is the ease of removal of the peel from the flesh (parenchyma), a practice required for the Brazilian flour and other industries for human consumption.

Equally important, it will be necessary to facilitate the stabilization of farmgate prices by having continuous access to information about planting and production across seasons. Vasconcelos, Vilpoux and Paranhos Filho (2013) conducted a survey of cassava production, in the Northwest region of the Brazilian state of Paraná, through the use of images captured by a Landstat Sensor TM5. This technology could be more widespread in starch production regions, especially once it has a level of precision that can determine stage of growth and make qualitative evaluations of plant health (pests and diseases) and predict yield. Such technologies would certainly first be applied to large plantations, but, based on use of drone and satellite imagery in other crops, could eventually be used at small farm level as well.

5 Improving the Latin American starch industry

The Latin American starch industry is relatively underdeveloped. By far the main starch sources are maize and cassava, and only a few countries such as Peru, Argentina, and South of Brazil, can extract potato starch, but only occasionally and in small quantities. For this analysis there is the necessity of considering both native and modified starches. In industrial markets, native cassava starch can only compete with maize starch through a lower price or by its value-added functional properties. In a scenario of low productivity and variable prices, the only element that discriminates this starch is its facility for the elaboration of modified starches. For this reason, when the price is favourable, native cassava starch is purchased even by multinational companies who are traditional producers of maize starch. This price advantage explains the increase of cassava starch production in 2014 and 2015, years of cassava price decline (Fig. 8 and 9).

The private sector has traditionally invested little in research. Companies in Brazil, Paraguay and, more recently, Bolivia continued with a portfolio of 5 or 6 modified starches, basically with processes and products reproduced from multinational companies. On the other hand, most of these products were developed in international platforms having maize starch as the basis. Since there is limited technical knowledge of starch properties, cassava starch is often submitted to processes that are suitable for maize, but when applied to cassava starch greatly reduce its viscosity.

Another important secondary product of cassava is fermented cassava starch, a starch modified by lactic fermentation and sun dried (Cereda and Brito, 2016). Although cassava fermentation is an ancient process to enhance shelf life and flavour of certain products, it has gained additional market position since 2006, as predicted by Cereda and Vilpoux (2003). Nonetheless, it has also lost much of its exclusivity in the products that valued it, e.g. the cheese bread and the sour starch biscuits.

Fermented cassava starch is no longer specific to the South American continent, having been widely introduced, modified and disseminated as well in Africa. Besides Brazil, there is also literature about this modified starch manufacturing and its expansion properties in countries such as Paraguay, Bolivia and Colombia. Fermented, or *sour* starch, is used to produce different types of cheese bread and sour starch biscuits under the names of *pan de bono, besitos, buñuelos, cuñapé,* and others (Cereda and Brito, 2016). The major obstacle to increase the use of fermented cassava starch in food products is the lack of standardization of its main feature, which is its expansion during baking without having to use baking powder or biological yeast. Another constraint is that the artisanal manufacturing process requires sun exposure in order to achieve expansion properties during baking. The use of more controlled and more sanitary artificial drying changes the properties of the starch such that it loses the expansion properties conferred by sun drying.

Because of difficulties in producing fermented cassava starch under improved quality standards, yet having the desired traditional characteristics, the starch industry developed chemically modified starches (oxidized and acidified) in an attempt to replicate the key traits of fermented starch. These processes are mainly held as trade secrets but, according to clients, the products do not fully meet the consumer expectations, in part because the expansion capacity is not preserved for more than 2 or 3 months. In addition, there is increasing consumer interest in natural products, and chemically modified starches as a substitute for fermented native cassava starch may become less appealing to an important segment of consumers.

5.1 Training in technologies and types of modified starches

Although the literature provides information confirming the ability to modify cassava starch more easily than maize starch, Brazilian starch business managers tend not to be experts on starch chemistry. They do not master starch analyses and the differences between cassava and maize starch, especially traits that could be valued in the market. Examples are greater transparency and gelatinization at a lower temperature, allowing for energy economy during starch modification. We have provided courses adjusted to the companies' needs for more than 10 years, but in practice many starch companies started using maize as their raw material as it gained competitive advantage over cassava starch, mainly because of price stability and better raw material availability.

5.2 Modernising production of fermented cassava starch and modified products with higher added value

To overcome the difficulties mentioned earlier, there is now available a process for production of fermented cassava starch in line with the technology of a modern starch extraction industry. The system and the starch treatment were both registered under patent application BR10201402768. A commercial continuous flow process is proposed and evaluated, involving use of a modular ultraviolet type C (UV-C) lamp to substitute for the sun exposure, which imparts expansion capacity to traditional fermented starch dried in the sun. The doses may vary from 1,000 to 10,000 Joules per litre ($J\ L^{-1}$) applied to a starch/lactic acid suspension in water. The modified cassava starch produced by this technology showed an 11 $cm^3\ g^{-1}$ expansion and allows standardization of the starch expansion, a trait of interest for companies and consumers. In addition, the process results in a fermented cassava starch free of impurities and with less microbial contamination compared to traditional processes (Brito et al., 2015a).

For other modified cassava starches there is a continuing need for new research. At the core of the research is to understand the needs of the starch customers. Greater involvement of companies can drive the process of defining product traits and the full steps to bring such products to fruition within a value chain approach.

6 Brazilian cassava flour (farinha)

Farinha is the traditional Brazilian cassava flour, used throughout the country and especially in Amazonia and the Northeast. Its quality is difficult to improve because it varies considerably and corresponds more to personal and regional preferences that have developed over many years.

Farinha is not a common product throughout Latin America, unlike fermented cassava starch. Farinha is culturally intimately connected to Brazil, from where it was introduced to Africa, became known as gari, and is now widespread in West Africa. Although flour types are well characterized in terms of technical standards, these standards only include the most common types found. In Brazil there are criteria for at least 6 major flour groups – types of farinha recognized in the Brazilian legislation: thick, thin, medium, raw, *farinha d'água* and *beijusada* (flocked or *Paulista*). To make the flocked farinha, the grated and pressed mass with around 30% water is automatically distributed in a thin layer on a rotating circular iron plate. This plate is heated by direct fire with wood or gas. When the mass falls on the plate the temperature may reach 280°C. The starch located in the mass layer in contact with the heated plate gelatinizes to form a 'flock', or aggregated particles. When the plate ends one rotation, the drying is complete and the farinha is removed. This flocked farinha is called beijú throughout all Brazil. By breaking during handling, gelatinized particles become a flour.

Farinha is known as a high calorie food. This is indeed true, but it is now known that the physical characteristics of carbohydrates, mainly the starch, may vary considerably according to type of flour. Brito et al. (2015b) simulated the human digestion in *in vitro* conditions to evaluate the digestibility of the different types of flour recognized in the Brazilian legislation. This test clarified that flours are very different in glucose quantity, and also present very different behaviour in time of glucose release.

In the case of *farinha d'água*, the release of a glucose peak occurs 15 minutes after its consumption, while raw flour presents various small peaks during one hour after its ingestion. The latter represents a more *resistant* starch, which does not give a sharp glucose peak and will therefore be better for people with compromised capacity to regulate blood sugar, such as diabetics.

Although farinha is undoubtedly a high-calorie food, the evidence of the presence of a resistant starch in some types of flour could change consumer perceptions as well as have significant impact on health.

Cassava fibre also has good nutritional properties. Leonel et al. (1999) found that this fibre has a nutritional quality comparable to wheat bran and could be used as health food.

6.1 Opportunities for improvement

To overcome the challenges of how farinha is perceived and how it is consumed, there is the necessity to educate nutritionists, policy-makers and the general public about these research results.

Although farinha is a Brazilian staple, there are also other cassava products that can become more important in modern diets and can be more widespread in use. One example, again from Brazil, is *farofa*, a seasoned type of cassava flour. This is the same as farinha, but into which seasonings are added. *Farofas* include many types, such as sweet *farofas* or seasoned *farofas*, based on different cultural preferences. The most common is the *paulista* type, which uses flocked Brazilian cassava flour (previously cited), typical of the

Figure 12 *Bammy* production in a small industry in Jamaica and final product at the right.

Brazilian state of São Paulo. To its base are added fried chopped bacon, pepper, garlic, onion, and other seasonings. In addition to the taste and attractive appearance, *farofas* can be packaged in order to facilitate their transportation and use. Our experience has shown that it is a product of easy preparation and can be adapted to the local cuisine. However, because they have high oil content, *farofas* can become rancid during storage.

Another product developed to adapt to changes in consumption habits is a cereal bar substitute (Silva et al., 2013). Well-known by consumers, they can be presented in sweet or salty forms, and are well-accepted even under principles of good nutrition. Typically they are considered high-fibre-content food, which can present balanced nutritional composition by adding dried fruits, nuts, honey, etc.

Both *farofas* and food bars based on cassava flour can attract some attention for new products. These products typically have a rustic appearance, and therefore can be ideal for incorporating an irregular-coloured cassava flour, a typical characteristic in many processing systems.

Another product that could be adapted to boost the flour market is *bammy*, a typical food of Jamaica. *Bammy* is made with shredded and pressed cassava flour mass, then cooked, cut into small sticks that are fried and consumed similarly to French fries (Fig. 12).

7 Consumption of boiled and fried cassava

Unlike flour, fresh cassava is consumed throughout Latin America, with the exception of Argentina and Chile. The fresh root is consumed boiled, or fried after boiling, as it is preferred in the state of São Paulo (Brazil) and in Cuba.

Traditionally in Brazil, fresh cassava for cooking is harvested early, after 6 to 12 months, when roots are not too fibrous. In the state of São Paulo, in Southeast Brazil, cassava for cooking purposes is yellow, which consumers identify as high quality. In other parts of Brazil and most other countries, the preference is for white-fleshed roots. In Brazil, maybe because of the São Paulo state influence, yellow cassava is becoming more common over time for fresh consumption. Yellow cassava has the advantage of having varying levels of provitamin A (precursor to vitamin A) described below.

7.1 Some important research

The literature (Siritunga and Sayre, 2003) reports research results from the evaluation of the cassava genebanks for cyanogenic potential, amylose/amylopectin content, post-harvest deterioration resistance, higher nutritional value (especially for varieties with high provitamin A content), and other traits.

A long-time concern that has been extensively studied as a limiting factor for the direct consumption of cassava is the presence of cyanogenic compounds (linamarin and lotaustralin) in the entire cassava plant.

Several research and plant breeding attempts have been made to improve the cassava nutritional quality, reducing its cyanide content and retain its high calorie value. There have been attempts to obtain cyanide-free cassava roots, especially through transgenic approaches. Transgenic plants were obtained in which the expression of the cytochrome P450 genes (CYP79D1 and CYP79D2), that catalyse the first dedicated step in linamarin synthesis, was inhibited. The results showed that the linamarin content of roots was reduced by 99% in transgenic plants, with a 60 to 94% reduction in leaf linamarin content (Siritunga and Sayre, 2003). These transformed plants have not been approved for commercial use, and there is currently little ongoing research to eliminate cyanogenic compounds. The rising availability of gene editing techniques, especially if they have fewer regulatory barriers, could reinvigorate research into acyanogenic cassava.

Although it may be possible in the future to obtain plants free from cyanide, the cyanide content is typically not a real problem. There is international literature confirming that the syndromes sometimes related to cyanide poisoning, especially in African countries, are due to malnutrition and stress. Toxic effects are dependent on poor nutritional status of consumers, particularly those with low consumption of animal proteins. Detoxification mechanisms of cassava cyanide are enzymatic and if well managed, leave no residues, as cassava cyanide is not cumulative.

Another problem that generally reduces the value of raw cassava roots is the enzymatic browning (typically referred to as physiological post-harvest deterioration, or PPD), initially manifested as blue dots and streaks that can compromise their appearance, and then progressing toward fully darkened flesh that no longer has value in commercial markets.

Uarrota et al. (2015) investigated the emergence of symptoms in cassava roots from 3 to 5 days of storage and showed guaiacol peroxidase and total protein amounts to be correlated with PPD in cassava roots. Since it is an enzymatic oxidation reaction, the development of blue-colour dots occurs only when in contact with oxygen. Preserving the roots protected from the presence of oxygen overcomes this problem. For example simple storage under water prevents the development of the blue colour. Boiling the roots stops the PPD process. Even in those roots that already present the initiation of PPD, boiling may cause slight symptoms to disappear. Symptoms from more advanced deterioration will not be eliminated.

Vitamin A deficiency is regionally or locally endemic in most Latin American countries, though the problem is far less prevalent than in Africa. The presence of carotene (provitamin A) in yellow roots was evaluated and extensive breeding is being done to raise levels in several research centres, including EMBRAPA in Brazil, CIAT in Colombia, and IITA in Nigeria.

Ceballos et al. (2012) reported on the importance of carotenoids retention after processing. This study was conducted to quantify the dry matter content and carotenoids found in different sections of cassava roots from six clones and to assess the true retention of carotenoids after 30 min of boiling. The average true retention (dry matter basis) was 86.6% and ranged from 76.0 and 96.7% (averages per clone and section of the root, respectively). The study showed that during boiling, the weight of samples changed from slight losses to gains of up to 40% (depending on original DMC of the uncooked root), resulting in an 'apparent dilution' of the carotenoids.

Carotenoids are easily lost by oxidation. Flour such as *farinha* is especially vulnerable to loss because it is less dense and air can penetrate to its interior.

7.2 Cassava cooking

Fresh cassava roots are consumed in different ways throughout Latin America, sometimes with simple boiling, and sometimes subsequently fried, for example. The consumption of cooked roots is probably the most common way to use cassava throughout Paraguay and part of the Western region of Brazil. In the fresh roots market, the cooking characteristics of cassava are the main aspects to be considered.

Konan et al. (2003) reported that boiling is one of the most common forms of consumption of yam and cassava in Côte d'Ivoire and that cooking properties are one of the main attributes required by consumers. They also reported that certain cassava varieties showed preferred agronomic traits, but poor cooking quality remains a limitation to their consumption. Kouadio et al. (2011) noted that cooking quality is usually related with cooking time. In Brazil, a cassava variety is considered of good cooking quality when it cooks and obtains a final friable texture in less than 20 min (Pereira et al., 1985). In Africa the main quality attributes of boiled cassava are whiteness (in the case of white roots), sweetness and friability (also locally referred to as 'mealiness'. all of which should be high (Padonou et al., 2005). In Brazil yellow cassava roots are appreciated because consumers think that the yellow colour is related to cooking properties.

There is so far no standardized and internationally accepted methodology to establish the quality of cooking for cassava roots. This parameter is reported as an important point for the consumer and may hinder fresh roots consumption. The best-known methods are traditional cooking tests where the texture is evaluated with a fork. Also the apparatus of evaluation of the cooking for beans (Mattson) and texturometer was tested. Nevertheless, the method that establishes the amount of water absorbed during cooking seems to be more appropriate (Kouadio et al., 2011). These authors report the comparison between the cooking quality of yam tubers and cassava roots, using the method of water absorption. Results showed that cassava and yam have different cooking quality. Two quality types were described: hard cooking and mealy cooking for both cassava and yams. They found, for eight cassava varieties, that the dry matter does not influence the cooking quality. Water absorption by mealy cooking cassava (27.6%) was significantly higher compared to water absorbed by hard cooking cassava (12.4%).

Acceptability of the cassava food product depends on the final texture after processing and it varies based on factors such as variety, age of the roots at harvest, soil type and other environmental factors (Lorenzi, 1994; Ngeve, 2003). Cassava roots have short cooking periods, mainly in the dormant phase, for example, resulting from low winter temperatures or drought. During the poor cooking period, cassava is characterized by a waxy texture with sweet flavour and chestnut aroma. Similar problems are also reported for other tubers and rhizomes (Kouadio et al., 2011).

Safo-Kantanka and Owusu-Nipah (1992) studied cooking quality of 13 cassava varieties from three successive annual harvests and determined that the *mealiness*, a desirable quality factor of the cooked root and frequently cited as a quality factor in Africa, was associated with high dry matter content, larger starch granules and less cell cohesiveness after cooking than non-mealy varieties. Eggleston and Asiedu (1994) found a relationship between compressive strength of boiled cassava roots and intercellular adhesion, but concluded that many other factors were involved in cooked root quality. Neither of the authors presented data on starch gelatinization or described the kinetics of texture degradation during processing of the samples, which were cooked for a fixed period of time (20 min).

Our experience with companies that produce cooked cassava suggests that cassava cooking at the industrial level might be classified in 3 types:

Type A: normal cooking, up to 30 minutes in boiling water, at room pressure;
Type B: with cooking conditions at normal pressure, cassava consistency is vitreous or waxy; the root opens into layers, and it is translucent, without typical cassava flavour or aroma. Under pressure, the cooking has normal or close to normal development.
Type C: appropriate texture is not achieved even with cooking under high pressure.

Although cassava roots typically present water content of 60–70% this internal water is not enough to obtain the desirable cooking quality. This is not the case of potato tubers that can cook well in microwave without addition of water. The people of South American know and solve this problem by cooking cassava roots in a proportion of at least 50% of its weight in water.

Cooking problems can in some cases be explained by the lack of additional water to penetrate the cells and to gelatinize the starch. To test this hypothesis we made transverse thin slices of cassava roots harvested at 12 months. Microscope cuts of raw roots thin slices, coloured with Lugol, show individuals starch granules as small and dyed dark blue (Fig. 13a). Cuts presenting cooked slices show that the starch granules are swollen, occupying a larger space (Fig. 13b). The experiment was made with a type B root, of intermediate cooking. It is possible to identify regions where starch granules are blown out and vegetal cells walls are destroyed (a). In these regions it is not possible to visualize starch granules, which are all perfectly gelatinized. It is also possible to observe poorly cooked regions (b), where the starch granules are swollen but remained in the vegetal cells, which are deformed but perfectly visible. In these cells, starch granules have not reached the total gelatinization level and didn't break the cell walls. Finally, part of the root did not cook (c), where cells and starch granules are slightly deformed.

Table 4 shows the average area of cassava root cells, calculated with the Zeiss microscope software. The volume of cassava root cells increased more than 10-fold with cooking in

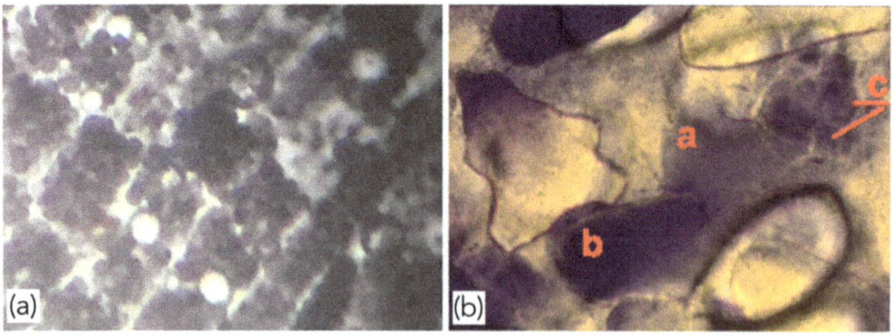

Figure 13 Microscope cuts of raw root coloured with Lugol (a) and cooked root coloured with Lugol (b).

Table 4 Area and pH of vegetal cells of cassava under different cooking regimes (average of three replicates, after 20 minutes of cooking at 97°C)

Material	Area (μm²)	pH
Raw cassava	246	4.8
Cooked only in water	3,815	5.9
Cooked in water + 2% sodium bicarbonate	5,922	6.4
Cooked in water + 5% sodium bicarbonate	5,236	9.5

Note: Root slides examined and measured with an optical microscope Axio Scope.A1 Zeiss

water, as a function of the swelling of the starch granules that can be clearly observed inside the raw cassava cells (Fig. 13). The cooking in bicarbonate solution caused an almost 25-fold volume increase in the volume of raw cassava cells and 1.5 times compared to cooking in water.

These results indicate that the swelling during boiling, with a distension of the cell wall which facilitates cell separation, as mentioned by McComber et al. (1988) in potatoes, can also occur in cassava.

Table 4 indicates that the cooking effect in water slightly increased the natural pH of the cassava root from 4.8 to close to 6.0. However, the bicarbonate solution increased much more the pH, which reached 9.5 with 5% bicarbonate. The effect of gelatinization by addition of alkaline substances is well known in starch chemistry and allows explaining the effect on cooking. The sodium bicarbonate was able to cross the cell wall, helping the heat to gelatinize the starch granules. The two factors are necessary because without heat the bicarbonate has no effect.

Table 4 shows a larger area increase of cells with 2% bicarbonate. However, the small reduction of cellular area observed with 5% bicarbonate may be caused by rupture of cell walls with partial leakage of the gelatinized starch mass.

None of the assumptions currently found in the literature explain why in certain periods of the year cassava does not properly cook. The popular knowledge associates the period of poor quality cooking to a certain time of the year, and not to age of the plant per se. This period corresponds to the Brazilian summer, characterized by high temperatures and heavy rains, which for cassava is the period of greatest vegetative activity. According to farmer tradition, this corresponds to the months, which present the letter 'R' (September, October, November, December, January, February, March and April). The period in which good cooking quality occurs would be the remaining months, from May to August, the period corresponding to the Brazilian winter, when the temperatures are lower, with less rain and reduced photoperiod. Based on this hypothesis the best cooking period occurs when the cassava plant is in vegetative dormancy.

7.3 Opportunities for improvement

The main criteria used by researchers who evaluate the quality and the characteristics of fresh cassava for food use are: total content of the cyanogenic glucosides linamarin and lotaustralin (or of cyanide released from them), flavour, texture, and cooking quality. Among these characteristics, the cooking quality is a limiting factor for the development of commercialization of cassava for human consumption, not only in national but also in international markets. In parallel, the cyanogenic glycoside content might be controlled during the processing.

The biggest problem for processing of fresh cassava for food is the irregular quality after cooking and the lack of control over this characteristic. This variability complicates access to high quality product all year. Companies are required to stock the roots under freezing during the seasons in which cassava cooked well, in order to attend the market's demand during the rest of the year. Some companies manage fresh cassava all year long, but in that case the product quality is unstable, which harms cassava's image for consumers.

The fact that cassava does not cook well throughout the year, in many growing environments, has been known for a long time, but the specific causes and the solutions haven't been properly studied yet. Storage also influences cooking quality. A frozen cassava industry, processing one tonne of fresh cassava per day, has noted a cooking difference between roots harvested at the same place and time, but processed in different periods of the day. In that situation, the roots processed in the morning cooked quickly, while those processed in the afternoon needed more time, with some roots not achieving the level of cooking required for consumption. This seems to indicate that the cooking quality is also related to postharvest physiology and not only with the physiology of the plant.

All over the world it is possible to observe that cassava often cooks better after it goes through a period of stress and enters into a reduced growth rate. These stress periods might be caused either by dry weather, such as occurs in many parts of the tropics, or cold weather, as in South of Brazil and other subtropical regions. These periods have different durations depending on the region, but they were noticed in all Brazilian regions as well as in other major producing countries (e.g. Costa Rica, Venezuela, Colombia, China, and Thailand). At the end of the stress period, at the start of a new period of growth, cassava cooking quality will typically go through its worst phase, as carbohydrate reserves in the roots are mobilized for new foliage development.

Among the numerous factors impacting cooking quality, the most explored is the fact that it is influenced by some sorts of stress, such as drought, temperature stress (increase or reduction), or mechanical action stress (damage, cultural operations and pruning). The literature also indicates situations where cassava roots show either improvement or lower quality of the cooking characteristics, but without indicating the growing conditions.

Based on current knowledge, it is possible to summarize the following observations:

- In many growing environments there is a period when cassava cannot be suitably cooked;
- Even roots that might be well cooked (type A or B (see Section 3.4)) can decline in quality to the immediately lower stage when stored – e.g. from A to B, or from B to C;
- Slow freezing improves the cooking of roots that are in type B (intermediate cooking), but it has no effect on those in type C (poor cooking);
- Storage before cooking might modify the cooking quality of fresh roots;
- The pH of roots stored at room temperature tends to increase (Sarmento, 1994);
- Experiences in Brazil indicate that it is possible to improve the cooking of roots by adding calcium or sodium bicarbonate to the cooking water. This process works essentially for intermediate cooking roots (type B), but it is more difficult to improve cooking when cassava cooks very poorly (type C). A possible hypothesis is that in type C cassava, water as well as bicarbonate, does not penetrate in the cells, preventing

starch gelatinization. Sodium bicarbonate should not exceed 5% as an additive to improve the cooking; despite the roots becoming softer they also develop a strong aroma and a dark brown colour, which are negative consumer traits;
- The period of lower quality cooking often coincides with the beginning of sprouting after the end of a period of stress;
- The roots at B and C cooking types do not develop the characteristic aroma and flavour of cooked cassava, even if they become soft after pressure cooking;
- Low temperatures, such as occur in subtropical winters, do not cause poor cooking quality. On the contrary, the cold period is exactly when the roots are better cooked. The period of poor cooking occurs when cold weather stops and rains begin, coinciding with plant sprouting.
- There is evidence that cultural treatments and field conditions might influence the root cooking phenomena, and may either improve or worsen those characteristics.

Based on this information, it is possible to establish two hypotheses to explain difficulties in achieving good cooking quality, at the cellular level. The first one is the non-penetration of the water into the root cells, preventing total starch gelatinization. That could be explained by *waterproofing* of the cellular wall, in a process fast enough to generate differences in cooking quality between cassava roots of the same field, harvested together and processed within a matter of hours of each other. Another hypothesis proposes alterations in the cell wall, causing it to becomes rigid and inflexible, and thus not allowing the starch granules to swell and, consequently, to have a proper gelatinization for good cooking. Testing these hypotheses is an important next step to solve the problem of improving cooking quality of cassava.

8 Future trends: cassava for animal feed

The world average *per capita* income is expected to grow 84% between 2010 and 2050. One of the expected consequences is that *per capita* meat consumption will increase by about 61%. If population growth rate is considered in this period, the world meat demand should increase 120% (FAO, 2015). This data indicates the strong potential for animal feed production, whose demand is expected to grow in parallel with meat consumption.

Early studies have demonstrated that the cassava plant can be fully used in animal feeding for ruminants and non-ruminants. Even in poultry production, where fibrous feeds are not recommended, good results were obtained for chickens exclusively fed with dried and ground cassava. The only change observed was in relation to the egg yolks, which became lighter (Yoshihara et al., 2008).

Vilpoux et al. (2013) compared the feeding of sheep with maize-based and with cassava-based diets. Animals fed with cassava ate three times a day, while the animals fed with maize could eat only twice a day and remained ruminating in the third period, which negatively affected the animals' weight gain when compared to the animals on cassava-based diets. This fact was explained by the researchers as probably due to the higher oil content in the maize-based feed, which leads to the impermeability of starch granules, resulting in a lower efficiency of the enzymes present in the rumen.

With all the demonstrated potential for use of cassava in animal feed, it is imperative that research be further directed toward sustainable and competitive systems that will bring benefits to all players in the animal feed value chains, from producers to feed manufacturers to livestock growers.

9 Where to look for further information

For more information on production, processing and use of cassava in Latin America it is possible to consult the series 'Cultivation of Latin American starchy tubers' published by the Cargill Foundation, São Paulo, Brazil.

This series is composed of 4 books, published in Portuguese, but with a digital version available in English for book 2 (*Agriculture: Latin American Starchy Tubers*) and book 3 (*Processing, Uses and Potential of Latin American Starchy Tubers*):

- Book 1: Cereda, M. P. (Ed.), *Propriedades gerais do amido (General Properties of Starch)*. São Paulo: Cargill Foundation, 2001. 220p.
- Book 2: Cereda, M. P. (Ed.), *Agricultura: Tuberosas Amiláceas Latino Americanas (Agriculture: Latin American starchy tubers)*. São Paulo: Cargill Foundation, 2002. 540p.
- Book 3: Cereda, M. P.; Vilpoux, O. F. (Eds), *Tecnologia, usos e potencialidades de tuberosas amiláceas latino americanas (Processing, Uses and Potential of Latin American Starchy Tubers)*. São Paulo: Cargill Foundation, 2003. 711p.
- Book 4: Cereda, M. P. (Ed.), *Manejo, uso e tratamento de subprodutos da industrialização da mandioca (Handling, Use and Treatment of Byproducts Derived from Cassava Industrial Processing)*. São Paulo: Cargill Foundation, 2001. 320p.

10 References

Abam – Associação Brasileira De Amido De Mandioca. (2016). *Estatísticas de produção*. Paranavaí, Brasil, 2016. Available in: http://www.abam.com.br/estatisticas-producao.php. Consulted in May of 2016.

Brito, V. H. dos S., Demiate, I. M. and Cereda, M. P. (2015a). Continuous flow process to make fermented cassava starch substitute and its physical, chemical and functional properties. *Starch* (Submitted).

Brito, V. H. dos S., Silva, E. C. and Cereda, M. P. (2015b). Digestibilidade do amido in vitro e valor calórico dos grupos de farinhas de mandioca brasileiras. *Braz. J. Food Technol., Campinas*, 18(3), pp. 185–91. dx.doi.org/10.1590/1981-6723.2714.

Ceballos, H., Luna, J. L., Escobar, A. F., Ortiz, D., Pérez, J. C., Sánchez, T., Pachón, H. and Dufour D. (2012). Spatial distribution of dry matter in yellow fleshed cassava roots and its influence on carotenoid retention upon boiling. *Food Research International*, 45, pp. 52–9.

Centro Internacional de Agricultura Tropical (CIAT) (1982). Multiplicación aceleradade material genético promisorio de yuca. *Guia de estudio*. Cali, p. 28.

CEPEA. (2016a). Centro de Estudos Avançados em Economia Aplicada. Departamento de Economia, Administração e Sociologia – DEAS, Escola Superior de Agricultura Luiz de Queiroz – ESALQ, Universidade de São Paulo, Brazil. Available in http://www.cepea.esalq.usp.br/. Consulted in May of 2016.

CEPEA. (2016b). Mandioca/CEPEA: Produção de fécula ultrapassa 750 mil t em 2015. Departamento de Economia, Administração e Sociologia – DEAS, Escola Superior de Agricultura Luiz de Queiroz – ESALQ, Universidade de São Paulo, Brazil. 2 p. Available in http://www.cepea.esalq.usp.br/comunicacao/Cepea_Mandioca_Maio16.doc. Consulted in May of 2016.

Cereda, M. P. and Brito, V. H. S. (2016). Fermented Foods and Beverages from Cassava (Manihot esculenta Crantz): South America and Brazil. In: Ana Lucia Barretto Penna; Luis A. Nero; Svetoslav D. Todorov. (Org.). Fermented Foods of Latin America: From Traditional Knowledge to Innovative Applications. 1ed.Boca Raton: CRC Press – Taylor & Francis Group, vol. 1, pp. 192–213.

Cereda, M. P. and Vilpoux, O. F. (2003) Polvilho azedo, critérios de qualidade para uso em produtos alimentares. In Cereda, M. P. and Vilpoux, O.F. (Org.). *Tecnologia, usos e potencialidades de tuberosas amiláceas Latino Americanas*. 1ed.São Paulo: Fundação Cargill, vol. 3, cap. 13, pp. 333–354.

Cereda, M. P. (Org.). (2003). *Tecnologia, usos e potencialidades de tuberosas amiláceas Latino Americanas*. 1[ed].São Paulo: Fundação Cargill, vol. 2, p. 540.

Eggleston, G. and Asiedu, R. (1994). Effects of boiling on the texture of cassava clones: a comparison of compressive strength, intercellular adhesion and physicochemical composition of the tuberous roots. *Tropical Science*, 34, 259–73.

Escobar, R. H., Caicedo, E., Muñoz, L., Ríos, A., Azcárate, A., Dorado, C. and Tohme, J. (2012). *El cultivo in vitro*: otra manera de propagar la yucca. Centro Internacional de Agricultura Tropical (CIAT). *Área de Investigación en Agrobiodiversidad*, p. 56.

Food and Agriculture Organization of the United Nations – FAO. (2016). FAOSTAT data.

Food and Agriculture Organization of the United Nations – FAO. IFAD and WFP (2015). *The state of food insecurity in the world 2015*. Meeting the 2015 international hunger targets: taking stock of uneven progress. Rome, FAO.

Fukuda, W. M. G. and Carvalho, H. W. L. (2007). Propagação rápida de mandioca no nordeste brasileiro. *Embrapa Tabuleiros Costeiros*, (circular Técnica).

IBGE – Instituto Brasileiro DE Geografia E Estatistica. (2003). *Pesquisa de Orçamentos Familiares–POF*. Available: http://www.ibge.gov.br/home/estatistica/populacao/condicaodevida/pof/2002aquisicao/default.shtm. Consulted in May of 2016.

IBGE – Instituto Brasileiro DE Geografia E Estatistica. (2004). *Pesquisa de Orçamentos Familiares–POF*. Comunicação Social, 19 de maio de 2004. Available in: http://www.ibge.gov.br/home/presidencia/noticias/19052004pof2002html.shtm. Consulted in May of 2016.

IBGE – Instituto Brasileiro DE Geografia E Estatistica. (2009). *Pesquisa de Orçamentos Familiares –POF*. Available: http://www.ibge.gov.br/home/estatistica/populacao/condicaodevida/pof/2008_2009_aquisicao/default_zip.shtm. Consulted in May of 2016.

Konan, A.G., Nindjin, C., Agbo, N.G. and Otokore, O.D. (2003). Paramétres de qaulité des produits de consomation de l'igname en milieu urbain de la Cóte d'Ivoire. Atelier sur l'igname. *Agronomie Africaine*, 4, pp.15–19.

Kouadio, O. K., Nindjin, C., Bonfoh, C. B., N'drib, D. and Amanib, G.N. (2011). Water absorption as an evaluation method of cooking quality for yam (*Dioscorea alata*) and cassava (*Manihot esculenta*, Crantz). *Procedia Food Science*, 1, p.153–9.

Leonel, M., Cereda, M.P. and Roau, X. (1999). Aproveitamento do resíduo da produção de etanol a partir de farelo de mandioca como fonte de fibras dietéticas. *Ciênc. Tecnol. Aliment.*, 19(2), pp. 241–5.

Lorenzi, J. O. (1994). Variação na qualidade culinária das raízes de mandioca. *Bragantia*, 53(2), pp. 237–45.

Marín, A., Perdomo, D., Albarrán, J.G., Fuenmayor, F. and Zambrano, C. (2008). Evaluación agronómica, morfológica y bioquímica de clones élites de yuca a partir de vitroplantas. *Interciencia*, 33(4), pp. 365–71.

McComber, D.R., Osman, E.M. and Lohnes, R.A. (1988). Factors related to potato mealiness. *Journal of Food Science* 53, pp. 1423–6.

Ngeve, M. J. (2003). Cassava root yields and culinary qualities as affected by harvest age and test environment. *Journal of the Science of Food and Agriculture*, 83, 249–57.

Padonou, W., Mestres, C. and Nago, M. C. (2005). The quality of boiled cassava roots: instrumental characterization and relationship with physicochemical properties and sensorial properties. *Food Chemistry*, 89, pp. 261–70.

Pereira, A. S., Lorenzi, J. O. and Valle, T. L. (1985). Avaliação do tempo de cozimento e padrão de massa cozida em mandiocas de mesa. *Revista Brasileira da Mandioca*, 1(4), pp. 27–32.

Rodrigues, A. R., Alves, J. M. A., Uchôa, S. C. P., Albuquerque, J. A. A, Rodrigues, G. S. and Barros, M. M. (2008). Avaliação da Capacidade de Enraizamento, em Água, de Brotações, Ponteiros e Estacas Herbáceas de Clones de Mandioca de Mesa. *Agro@mbiente On-line*, 2(1).

Rós, A. B., Hirata, A. C. S., Araújo, H. S. and Narita, N. (2011). Crescimento, fenologia e produtividade de cultivares de mandioca. *Pesq. Agropec. Trop. Goiânia*, 41(4), pp. 552–8.

Safo-Kantanka, O. and Owusu-Nipah, J. (1992). Cassava varietal screening for cooking quality: relationship between dry matter, starch content, mealiness and certain microscopic observations of the raw and cooked tuber. *Journal of the Science of Food and Agriculture*, 60, pp. 99–104.

Santos, V. S., Souza, A.S., Viana, A. E. S., Ferreira Filho, J. R., Souza, K. A. and Menezes, M. C. (2009). *Multiplicação rápida, método simples e de baixo custo na produção de material propagativo de mandioca. Cruz das Almas*: Embrapa Mandioca e Fruticultura Tropical, p. 23, Boletim de Pesquisa e Desenvolvimento.

Silva, E. C. da, Sobrinho, V. dos S. and Cereda, M.P. (2013). Stability of cassava flour-based food bars. *Food Sci. Technol*, Campinas, 33(1), pp. 192–8.

Silva, M. N. da, Cereda, M. P. and Fiorini, R. A. (2002). Cereda. M. P. Multiplicação rápida de mandioca In: Cereda, M.P. Coord. *Agricultura: tuberosas amiláceas latino Americanas*. Fundação Cargil, São Paulo, livro 2, cap. 9, p. 187 a 197.

Sarmento, S. B. (1994). Alterações que ocorrem nas raizes da mandioca durante o armazenamento, em condições ambientes e condições de refrigeração. 1994. 120p. Dissertação (Mestrado) em Energia na Agricultura, Universidade Estadual de São Paulo.

Siritunga, D. and Sayre, R. T. (2003) Generation of cyanogen-free transgenic cassava. *Planta*, 217(3), pp. 367–73.

Tironi; L. F., Uhlmann, L. O., Streck, N. A., Samboranha, F. K., Freitas, C. P. O. and Silva, M. R. (2015). Desempenho de cultivares de mandioca em ambiente subtropical. *Bragantia*, 74(1), pp. 58–66.

Uarrota, V. G., Moresco, R., Schmidt, E. C., Bouzon, Z. L., da Costa Nunes, E., Nunes, E. C., de Oliveira Neubert, E., Peruch, L. A. M., Rocha, M. and Maraschin, M. (2015). The role of ascorbate peroxidase, guaiacol peroxidase, and polysaccharides in cassava (*Manihot esculenta* Crantz) roots under postharvest physiological deterioration. *Food Chemistry*, 197, pp. 737–46.

Vasconcelos, B. R. de, Vilpoux, O. F. and Paranhos Filho, A. C. (2013). Estimativa da área de mandioca industrial na região de Paranavaí, Estado do Paraná por meio do sensor Landsat TM5. *Bol. Goia. Geogr*: Goiânia, 33(2), pp. 259–77

Vilpoux, O. F. Produção e uso de amido. IN: Cereda, M.P. (Org.). (2002). *Tuberosas amiláceas Latino Americanas*. 1ed. São Paulo: Fundação Cargill, 1(2), pp. 21–56.

Vilpoux, O. F., Yoshihara, P. H. F., Pistori, H., Ítavo, L. C. V., Cereda, M. P. and Ítavo, L. C. V. (2013). Criação de ovinos com ração a base de mandioca integral com tecnologia apropriada para agricultura familiar. *Revista Brasileira de Gestão e Desenvolvimento Regional*, 1, pp. 211–35.

Yoshihara, P. H. F., Abreu, A. P., Viilpoux, O. and Cereda, M. P. (2008). Frango caipira curado alimentado com mandioca integral como inovação visando a sustentabilidade para agricultura familiar. In: Forum Ambiental da Alta Paulista, 4, 2008, Tupã. *Fórum Ambiental da Alta Paulista*. Tupã: ANAP - Associação Amigos da Natureza da Alta Paulista, v. 4. p. 1634–40.

Chapter 8

Drivers of change for cassava's multiple roles in sustainable development

Clair H. Hershey, formerly CIAT, Colombia

1 Introduction

2 The development of cassava as a crop

3 The role cassava can play in development

4 Constraints on cassava cultivation and production

5 Addressing constraints: genetic diversity and breeding

6 Addressing constraints: crop and soil management

7 Addressing constraints: processing, market and gender issues

8 Addressing constraints: policy issues

9 A new research paradigm for cassava

10 Cassava and climate change

11 Conclusions

12 Where to look for further information

13 References

1 Introduction

Humans have long held concerns about the sustainability of the earth's capacity to produce enough food. But, on the whole, the dire forecasts of our inability to keep pace with basic calorie needs, going back to Malthus and continuing to the present, have been consistently wrong. They have been wrong because of human creativity directed toward more productive agriculture – a combination of expanding land under production and increasing yields on existing lands. But demographic projections clearly show the urgent need to accelerate these improvements, especially in developing countries. To 2050, global population is projected to increase at 0.7% per year, which in itself seems a significant challenge. But this is not globally uniform. Africa will continue with the highest rate of increase – West and East Africa at 2.21 and 2.15 %, respectively, with some countries, for example Tanzania, reaching nearly 3% per year growth (Fig. 1).

http://dx.doi.org/10.19103/AS.2016.0014.08

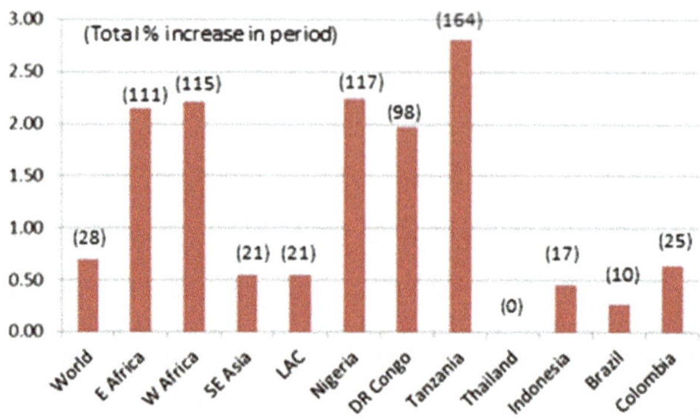

Figure 1 Average annual per cent projected population growth in selected regions and countries, 2015–2050. Source: http://esa.un.org/wpp/index.htm.

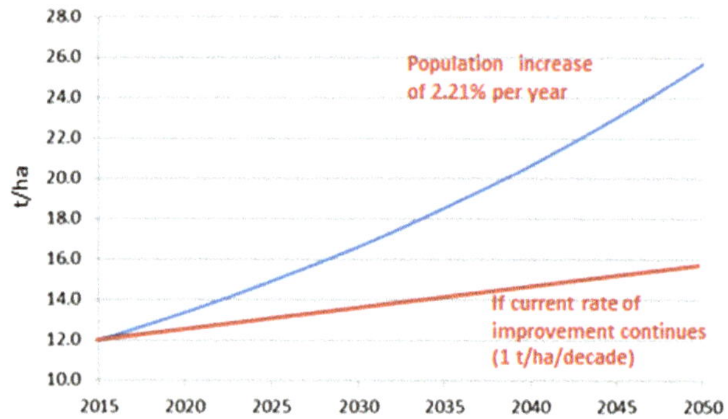

Figure 2 Cassava yields required to keep up with population increase in West Africa (2015–2050).

In addition to population growth, there are other key drivers of food production and demand: the increasing cost, and sometimes scarcity, of the productive resources for agriculture (land, water, labour, energy, fertilizers, seeds); changing food demands created by urbanization, and by overall greater wealth (e.g. more meat consumption and added-value products); and the multiple effects of climate change. Broadly across the tropics, crop production has kept pace with demand by a combination of about 1% per year productivity growth and increased land under production. Going forward, the pressure will be on increased productivity, as there is ever less new land that can be brought under production without high monetary cost or environmental risks. Figure 2 illustrates the special challenge faced by cassava production in keeping pace with expected population growth in West Africa. If current trends in yield improvement continue at only about 1 t/ha/decade, the gap between production and demand from population growth at 2.21% per year, if only the current land is cultivated, would leave a huge deficit.

This chapter reviews the options and strategies to sustainably improve cassava yields to keep pace with demand. It also looks more broadly at improving the role cassava can play in improving farmers' and consumers' lives – through income generation, improved food security, better nutrition, and a healthier environment.

2 The development of cassava as a crop

Improving cassava productivity requires an understanding of its development as a global crop. Because of its long growing season – usually nine months or more – cassava cultivation is restricted to the tropics and subtropics, where near year-round growth is possible if sufficient water is available. A range of evolutionary, agronomic and commercial factors have defined where and how cassava is grown, how it is used, and the challenges growers face. The crop originated in the Americas and was widely distributed throughout the tropics and subtropics of the western hemisphere before the arrival of Europeans in the fifteenth century (Allem, 1990, 2002; Allem et al., 2001; Olsen and Schaal, 2001; Nassar and Ortiz, 2008). Traders carried it to Africa relatively quickly after Columbus. While the introduction to Asia is not well-documented, it appears that Spanish traders introduced the species from Mexico to the Philippines in the nineteenth century, and independently from Africa to India.

While about 100 countries grow cassava (FAOSTAT, 2016), production is skewed toward a relatively few major ones. Four countries produce almost half of global output of fresh roots (Nigeria, Thailand, Brazil and Indonesia), and three-quarters of production comes from just ten countries. Over half the production area is in Africa, but only one of the top four producers is located there. The remainder of production consists of about 30% from Asia and 16% from the Americas (Fig. 3).

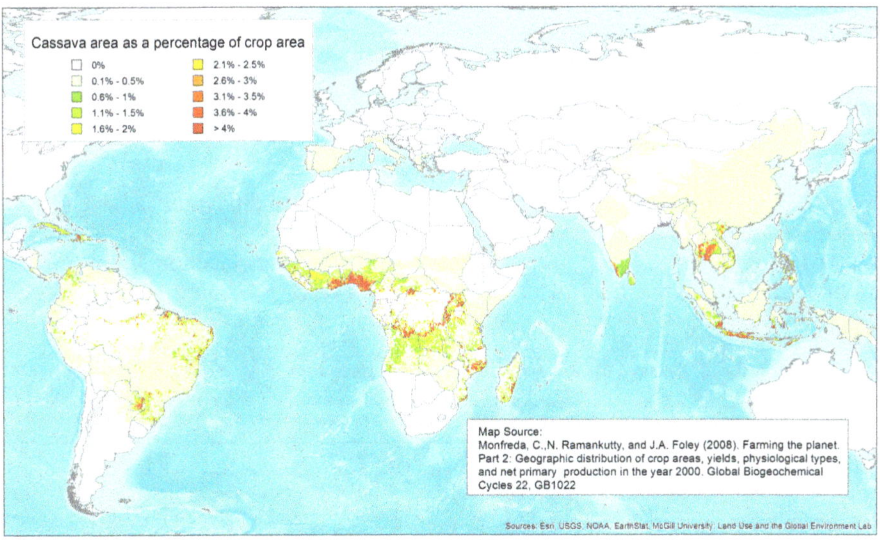

Figure 3 Distribution of global cassava planted area.

While every cassava-producing continent encompasses a wide array of production systems and uses for cassava, some broad generalizations apply (see Table 1). In Africa, most of the production is on small farms (often less than one hectare) for processing roots for food. Processes include grating, de-watering, fermenting, toasting, pounding and boiling. Areas where fresh consumption is common include Ghana and Uganda. The leaves are also an important source of protein, vitamins and several minor nutrients, most notably in the Democratic Republic of Congo. In recent years there has been a growing commercial interest in cassava production on a larger scale, especially for high-quality flour and for beer making. For example, in early 2017, Tanzania announced that China was investing US$ one billion in cassava production, in part to supply Chinese markets and in part to meet local needs.

Production in Asia is also mainly by smallholders, with a few exceptions such as some large plantations for starch production in Indonesia, Laos, Cambodia and others. Uses are highly diversified within and across countries. Indonesia, India and the Philippines produce mainly food products. Thailand, Vietnam and China produce mainly animal feed and industrial starch, and China is also moving aggressively into biofuels from cassava. In the Americas, Brazil is by far the largest producer. Production systems range from the large plantations (up to a few thousand hectares) in the South, to the small landholdings for local markets in most of the rest of the country. In most other countries of the Americas, production is predominantly on small farms.

In all continents, and especially Asia and Africa, the vibrant market situation of recent years is attracting new, large investors. There are increasingly examples of interest by companies buying or contracting large land areas for industrial use of cassava, for example, in Colombia, Guyana, Indonesia, Laos, Cambodia and Nigeria. Organizations focused on development-oriented support to cassava research will need to closely monitor the impact of such trends, especially on farmer welfare and the environment. There is often inadequate planning for the management implications of scaling up quickly in large plantations, and many of them have experienced early difficulties in production.

Typically large farms have advantages over small farms in marketing their products. This is especially a challenge for cassava, since small farms remain the norm around the world even as industrialization of the harvested crop progresses. However, the move to more intensive, industry-oriented processing has not necessarily meant a move toward large farms in the case of cassava. Southern Brazil and NE Thailand present two contrasting cases in this regard.

Southern Brazil produces cassava mainly for the starch market, based on large farms, often over several hundreds or even thousands of hectares. It is an environment where large farms have been the norm for many years, and cassava production and processing have been adapted to this land tenure system. In Thailand, on the other hand, cassava

Table 1 Consumption of cassava as percentage of total for each region

Region	Food	Feed	Export	Other
Africa	91	8	0	1
Asia	50	8	32	10
S. America	43	51	1	5

Source: Lynam, 2008.

farms remain small, usually a few hectares or less. Large centralized processing plants need to coordinate and aggregate the production from many farmers. Cassava processing plants typically locate near the production areas because of the bulkiness and perishability of the roots. This is an important way for processing operations to contribute to rural development in cassava production areas.

3 The role cassava can play in development

Although little-known in much of the developed world, cassava has been among the world's fastest-growing food crops in the past 25 years. It currently ranks as fourth most important food crop after rice, wheat and maize, and first among starchy staples (Table 2).

One of the key reasons for cassava's success as a crop is its resilience. Cassava produces better than many crops on acid and low fertility soils, and under periodic or even extended droughts. Because it has no specific maturity period, there is no specific period of growth during which it is especially vulnerable to environmental stresses. Cassava is thus more resilient than most crops in the face of multiple abiotic and biotic constraints. This resilience allows farmers to be moderately productive, at least in the short term, with low inputs, or even with relatively poor crop and soil management. These crop characteristics mean that the people who rely on cassava to provide a significant part of their income or nutritional needs are typically among the world's poorest. They are often the urban poor who subsist on the lowest-cost sources of calories, or farmers who earn their living cultivating less-favoured agricultural areas, for example, drought-prone and/or marginal lands with poor native fertility, high acidity, or on slopes vulnerable to erosion.

In many areas cassava tends to be a *second choice* for farmers. If growers have better land, or have access to inputs that will improve growing conditions, they will often plant higher value crops. However, expanding new markets for cassava products – especially in Asia, but increasingly in Africa and the Americas – are providing unprecedented opportunities for farmers to improve their income and well-being. These changes are creating both opportunities and challenges across an array of cassava value chains. Not

Table 2 Change in production, area and yield for cassava in developing countries and comparison with other starchy staples (roots, tubers and bananas)

	Production (million t)			Area (million ha)			Yield (t/ha)		
Crops	1961–63	1986–88	2011–13	1961–63	1986–88	2011–13	1961–63	1986–88	2011–13
Cassava	74.6	139.0	266.1	9.9	14.3	21.3	7.5	9.8	12.5
Banana	21.1	42.8	105.0	2.0	3.2	5.0	10.5	13.3	21.0
Plantain	13.4	23.9	37.5	2.5	4.1	5.4	5.3	5.8	6.9
Potato	28.8	72.0	224.5	3.5	6.1	12.6	8.1	11.9	17.8
Sweet potato	93.0	123.4	101.2	12.6	9.0	8.0	7.4	13.7	12.6
Yam	8.6	14.0	56.5	1.2	1.8	6.2	7.3	7.6	9.1

Source: FAOSTAT, 2015. FAOSTAT reports banana and plantain separately; however, no systematic criteria are used to make this separation.

all cassava growers are linked to markets; some are subsistence farmers, who grow only for family use. For these farmers, food security is often the first concern. However, most cassava farmers grow at least part of their crop for sale. Even in West Africa, where cassava is key for food security, it is estimated that 95% of farmers sell some of their crop as a source of income.

Access to markets drives farmer investments to increase productivity and reduce unit production costs, and underpins improved livelihoods for the majority of growers. Entry into the marketplace generates income to improve the family's ability to obtain a diversified and healthy diet, as well as broadly improve livelihoods through more options for education and other life choices. There is now a widespread interest, even for those countries where cassava's role is mainly for food security, to gradually transform it into a cash crop. Value addition of cassava, to bring benefits to growers, is currently a key objective in many countries in Africa, including Nigeria, Ghana, Tanzania, Uganda and others. This transformation should result in poverty alleviation, rural development and strengthened links between the producers and their markets (Nweke et al., 2002).

Although cassava is mostly cultivated under low input and sub-optimal soil and environmental conditions, the crop, in fact, has a very high production potential when provided optimum conditions. Both hypothetical models and field data show that cassava has a yield potential on the order of 80–90 tons per hectare per year (El-Sharkawy, 2012; El-Sharkawy et al., 1990). With a global average yield of about 12 t/ha, it is easy to see that there is a large yield gap that needs to be addressed to bring cassava's potential benefits to both producers and consumers. The challenge for those researching cassava is how to improve yields in a way that is sustainable, can benefit smallholders and allow them to exploit the new market opportunities that are emerging.

4 Constraints on cassava cultivation and production

4.1 Abiotic and biotic constraints on cassava cultivation

Cassava production predominates on acid and less fertile soils (Howeler, 2011a). It has the well-known ability to tolerate high soil Al concentrations and soil acidity without lime amendments. This adaptation to soil acidity favours cassava production across large areas of the cassava belt of Africa, in the southern cone of South America, the savannas of Colombia and Venezuela, in Meso-America and the Caribbean, and in southern China. Cassava is especially adapted to soils with low P availability. The association with root mycorrhizae allows the plant to very effectively extract P from soils with very low levels. However, without these mycorrhizal associations, cassava grows poorly even where P levels are moderately high (Howeler, 2002). Moreover, improving yields requires a proactive approach to optimally supply crop nutritional requirements rather than the more traditional reliance on cassava's ability to tolerate a range of nutritional deficiencies. Another important factor is the production of cassava on marginal, sloping land unsuitable for other crops which creates potential problems of soil erosion. Improving cassava cultivation suggests the need for more emphasis on maintaining and improving overall soil health.

Cassava is adapted in the tropics and subtropics from some of the driest (e.g. 400 mm annual rainfall) in the *Sertao* of NE Brazil, to some of the wettest agricultural environments (e.g. 4000 mm annual rainfall) in the Pacific coast region of Colombia. Cassava uses several complementary mechanisms to tolerate long dry periods, including deep rooting to access sub-soil moisture reserves; stomata that respond quickly to low ambient humidity, thereby reducing transpiration when water is limiting; and the ability to draw on carbohydrate and water reserves in the starchy roots (El-Sharkawy, 2012). Some varieties also appear to tolerate drought by an excessive leaf area index under favourable conditions, which is reduced to ideal levels (about 4) under drought stress, thereby maximizing yield. Some of these mechanisms come at the expense of optimum yields, but they do allow the plant to survive where other crops may fail completely.

Cassava is intolerant of flooding. Relatively short periods of submersion, of only a few days or less, can destroy a plantation. In heavy soils and in poorly drained soils, cassava often suffers from root rots and generally performs poorly. Irrigation is infrequently applied to cassava (southern India is a major exception) due to its high drought tolerance and long growing season (requiring many irrigation events). If irrigation is available, farmers usually prefer to grow higher value crops as compared to cassava.

More than 200 arthropod pests and pathogens (mites, insects, bacteria, fungi, phytoplasms and viruses) affect cassava (Bellotti, 2002; Bellotti et al., 1999). Most do not reach economic threshold levels of damage though a few are highly destructive in particular areas. The traditional belief that cassava was generally not vulnerable to pests and diseases came from a period when most of the crop was cultivated on small and isolated plantations, often in intercropping systems, and before there was extensive international travel that readily carried pests and pathogens among regions. Most of the destructive pests and pathogens in Africa and Asia, for example, were introduced from the Americas via unauthorized movement of planting material. In addition climate change is opening new possibilities for distribution and adaptation in new areas where these organisms may not have existed, or to allow pests to increase in numbers due to more favourable conditions for their aetiology (Ceballos et al., 2011; Herrera et al., 2011).

Because of its long crop cycle, cassava may be exposed to pests and diseases over many months, which means that their chemical control needs to be repeated many times, and is therefore also non-economic. This places greater emphasis on integrated pest management (IPM) systems. Effective pest and pathogen monitoring and diagnosis systems are essential to early detection and a precondition for effective management. The *PlantWise* system of CABI, for example, may be a good model to incorporate cassava data and take advantage of a system that is applicable for a broad range of crops (https://www.plantwise.org/).

IPM systems for cassava focus on three main components: host plant resistance, crop management, and biological control. The combination of these approaches can be effective for most pests and pathogens of economic importance in cassava. Use of chemical control has a low priority for research, with the exception of highly targeted applications such as for planting material (stake) treatment or infestation focal points.

4.2 Processing and market constraints on cassava production

An important constraint on processing of cassava is rapid deterioration of the roots once harvested. Cassava roots can be stored in the ground for many months as part of an intact growing plant. There is no well-defined maturity period, although root quality may vary over time due to plant age and environmental factors. However after harvest, roots begin to deteriorate quickly, often within a day to a few days (Beeching et al., 1993; Reilly et al., 2007). While in some subsistence production systems, farmers may harvest only a few roots each day as needed, most farmers harvest roots on a larger scale, for processing and/or marketing.

Another key factor driving processing is a poisonous component that occurs at lower or higher levels in all cassava varieties – cyanogenic glucosides that enzymatically break down to release HCN when cell structure is compromised (Du et al., 1995; McMahon et al., 1995; Wheatley and Chuzel, 1995; Andersen et al., 2000; Mkumbira et al., 2003). Roots that are boiled and eaten without additional processing need to be low cyanogenic potential (CnP) types. Processing steps which break cells and cause release of HCN are required to make high CnP varieties suitable for human consumption. The main primary processes involve one or more of the following: grating or grinding and drying to produce flour; soaking and fermentation; slicing or chipping and drying; and starch extraction. Variations include fermentation before or after grinding; forms of compressing to remove water; sun or artificial drying; and toasting or baking. Secondary processes include the production of a wide array of pellet-, flour- and starch-based products for food, feed and industry (Cock, 1985). Primary processes must therefore convert not only a potentially toxic product into a safe one, but also a highly perishable product into a more easily stored and transportable raw material for further processing.

Cassava producers have several challenges with regard to their capacity to respond quickly to market fluctuations, with important implications for their crop management options and decisions. After a year of low prices, farmers tend to reduce planting area. However, this also reduces the supply of planting material, and constrains capacity to respond quickly to plant larger areas if prices increase. Stakes can only be stored for very limited periods, typically from several weeks to a few months, depending on the storage system. Lack of close synchronization between production and market demand, especially in emerging markets, often creates wide fluctuations in farm-gate prices.

A potential solution is greater market diversity, which means a greater chance of stable demand and consequently more stable prices. Whilst farmers will almost always argue that prices are not high enough and buyers will argue that they are too high, both producers and buyers agree that widely fluctuating prices prevent proper planning and financial management. The mature markets of Asia include animal feed, starch for food and industry, biofuels and processed products for human food. While any one of these markets may experience considerable fluctuation in demand and in prices, together they provide some stability to the prices that farmers receive. Stable markets encourage farmers to adopt new technologies (varieties, use of fertilizers and soil conservation measures), which result in enhanced productivity and ultimately in more competitive prices, which in turn consolidate the competitiveness of cassava in these markets. A key challenge is supporting the market to achieve this greater price stability. As an example, the boom and bust nature of cassava markets is a key reason that sustainable formal seed systems have been so difficult to implement. In turn, the lack of agile seed systems contributes significantly to the slow response time of the whole production system to market opportunities.

5 Addressing constraints: genetic diversity and breeding

In developing cassava as a crop, some of the greatest assets and opportunities are the germplasm collections around the world. CIAT holds the largest of these as an in vitro collection at headquarters in Cali, Colombia. The CIAT genebank holds about 5500 landrace accessions, along with another approximately 600 advanced varieties and breeding lines. The collection is available to all interested parties, under the conditions of exchange and use of the International Treaty for Plant Genetic Resources for Food and Agriculture (ITPGRFA) (http://www.fao.org/plant-treaty/en/).

The global ex situ genebanks, in total, probably represent most of the genetic diversity that exists in cassava, although this assumption needs to be confirmed through further investigation (Hershey, 2008). It is highly likely that some rare genes are not captured in the current collections. As an example CIAT's collection has limited representation from Central America or Bolivia, and no accessions from Suriname or French Guyana. With the decreasing costs of sequencing and molecular marker development, the time is right to use the genome-wide characterization of cassava genetic diversity and to fill gaps in the collection. The coordinated phenotyping and genotyping of the cassava genetic resources held in genebanks will also help to accelerate improvements in breeding.

In addition to cultivated cassava, there are some 100 wild *Manihot* species that are poorly collected and poorly evaluated. Their potential contributions to cassava improvement are nearly completely untapped. Many populations are at risk in their native habitats due to urbanization, expansion of agriculture and climate change. It is imperative to extend the collection of these species for their future potential contributions to sustainable production solutions.

Africa has had limited exchange of germplasm with the Americas or with Asia due to the presence of some viruses in Africa that do not exist in other continents, and several viruses in the Americas that do not exist in Africa or Asia. Modern molecular-based diagnostic methods such as real time PCR and deep sequencing now allow a very high level of security for the detection and subsequent cleaning of viruses, but it is still very difficult to exchange vegetative material between Africa and the Americas because of strict quarantine regulations.

These high precision detection tools have also been a double-edged sword. The converse of the safety benefits is that the techniques have allowed the detection of many organisms whose role in disease is unknown and which may present no quarantine threat. However, their presence can create long delays and high costs for the important exchange of material across continents. Quarantine authorities in some countries have allowed huge volumes of planting material to move unregulated across national boundaries, while in other countries they are so restrictive as to inhibit fully safe movement from qualified labs.

Genetic diversity is only useful if it is well characterized. CIAT has carried out multi-year germplasm screening for a broad range of traits, including yield, nutrient use efficiency, drought tolerance, pest and disease resistance, plant architecture, and root quality traits.

CIAT has routinely selected under low nutrient levels, to allow the more efficient types to express their favourable traits. However, establishing selection nurseries that specifically allow measurement of nutrient use efficiency, with treatments at both low and high levels, is expensive and complex for large-scale germplasm screening (El-Sharkawy, 2012). This is an area of research with potential to benefit from development of molecular markers and the use of marker-assisted selection or genomic selection.

Cassava is in the field for long periods, and it has no post-establishment critical period of drought vulnerability. This means that drought tolerance becomes very difficult to define and measure in germplasm evaluation. Drought can be comprised of a wide range of variables, for example, total rainfall during the growing season; length of period(s) with low or zero rainfall; and the growing phase during which drought stress occurs (e.g.

early , mid , late season). While there would be clear advantages to better understanding of the mechanisms involved and the genetic control of tolerance to water deficits, this understanding will require much more controlled environments than is possible under natural and variable conditions.

One of the key research approaches has been to compare varietal responses under irrigated and non-irrigated conditions in dry environments. There appears to be wide genetic diversity among germplasm accessions (reviewed by El-Sharkawy, 2012).

Breeders have capitalized on this genetic variation through various strategies, but mainly by planting breeding nurseries under drought stress conditions. This strategy has both advantages and disadvantages. The advantages include simplicity of management, and the possibility to simultaneously select different mechanisms through exposure to conditions that are representative of where new varieties will actually be grown. Disadvantages include the fact that the specific conditions of drought tend to be highly variable from year to year. This means that in any given year, it may not be possible to target the specific desired varietal traits. CIAT, for example, has had a few experiences of 'drought' trials in environments with historical severe drought stress, where the trials have been destroyed by flooding (L.A. Becerra, pers. comm.).

Cassava evolved under pressure from many pests and diseases, and as a consequence genetic resistance co-evolved, a process reinforced through selection by farmers. In many of the major crops, plant breeders protect nurseries with pesticides, generation after generation, such that many resistance genes were probably lost due to genetic drift. In the case of cassava, this has rarely happened. First, cassava breeding has been practised on a more limited scale and for a shorter period than many other crops. Secondly, at least in recent years, most cassava breeders allow natural infestations of pests and pathogens as a way of selecting for resistance. These strategies have allowed a remarkable opportunity for capitalizing on host plant resistance in cassava, without breeders having to use exotic material or wild species in lengthy pre-breeding programmes. Host plant resistance is a clear and successful example of the development of sustainable practices. Nonetheless, as new pest challenges arise, especially as a result of climate change, there is greater likelihood of the need to delve further into germplasm collections and engage in pre-breeding to extract new resistance genes from more diverse genetic resources.

Breeders have made excellent gains in developing resistance to several key pests and pathogens, including cassava bacterial blight, cassava mosaic disease, super-elongation disease, *Phoma* leaf spot, thrips, cassava green mite and whiteflies (Jennings and Iglesias, 2002). In recent years, molecular tools have shown promise to aid in selection, specifically with cassava mosaic disease in Africa. A molecular marker for a single gene resistance not only allows speeding up the breeding process, but also has allowed the selection for resistance in Colombia, where the disease does not exist. Breeders now have a greater ability to combine desired traits from the Americas with the virus resistance needed for adaptation in Africa (Okogbenin et al., 2011). While molecular-assisted selection is so far very limited for cassava, this should change quickly in the next few years as the costs of various –*omics* technologies decline rapidly.

Another area where breeding has improved cassava through exploiting wide genetic diversity is in its nutritional quality. Cassava is known primarily as a low-cost source of calories. To achieve a balanced diet, consumers need to also include sources of protein, vitamins and minerals. However, one of the opportunities that was relatively unappreciated until the turn of the twenty-first century is the capacity to increase carotenoid content, through breeding, to levels that would significantly address vitamin A deficiency in key populations, especially West Africa. In normal diets in Nigeria and DR Congo, high carotene cassava can provide 40% or more of daily needs. Much of the effort to achieve this goal has been funded by the biofortification programme of Harvest Plus, a multidonor organization targeting key crops and key nutrients to improve human nutrition in developing countries (http://www.harvestplus.org).

High carotenoid cassava is moderately to deeply yellow. These types of roots have been used for millennia in some parts of the world, and especially by the indigenous people of Amazonia. This region has been the main source of germplasm for breeding programmes aiming to further increase carotenoid contents. In CIAT's breeding programme, initial sources contained on the order of 7–8 ppm carotenoids, and through a concerted and steady selection process, these levels are currently over 20 ppm in best materials. Harvest Plus nutritionists identified a target of 15 ppm of carotenoids, so current efforts are shifting toward combining better agronomic traits with high carotenoids (Sánchez et al., 2014). At IITA in Nigeria, breeding for high carotenoid content is now mainstreamed in the breeding programme, while most programmes in Asia and the Americas give considerably lower priority due to the generally lower importance of cassava as a food, and the lower incidence of vitamin A deficiency in the cassava-consuming regions of those continents.

Successful efforts have also been made through transgenic methods to improve protein content of cassava (up to 3.2-fold increase), but to date this has not proceeded beyond the laboratory (Stupak et al., 2008). Certainly there are other macro- and micronutrients that can be addressed through conventional or molecular breeding, but currently the main interest in the marketplace is for the high-energy product of cassava – its starch. Leaves are also a highly nutritious source of protein, vitamins and minerals, but there is limited work on the improvement of any nutritional aspects of leaves.

6 Addressing constraints: crop and soil management

Since cassava is more tolerant of drought and poor nutrient conditions than many other crops, the high inputs of fertilizers and pesticides and improvements in irrigation that helped boost yields of rice and wheat during the Green Revolution were not generally applied to cassava cultivation (Kawano and Cock, 2005). There have been a number of obstacles. Because of the crop's drought tolerance, it is often not cost-effective to invest in irrigation systems. Many cassava farmers have limited access to inputs such as fertilizers or pesticides and the credit with which to purchase them. They also often lack opportunities for mechanization of processes such as land preparation, planting or harvesting. When market prices are sufficiently high and stable, farmers will be more easily convinced to invest in inputs that increase their productivity and profitability.

However, cassava does respond to soil fertility improvement, as shown in many experiments, and widely on farms in India, for example, where farmers have achieved high yields with moderate fertilizer inputs and irrigation. Soil fertility maintenance is a fundamental component of successful cassava cultivation. Crops extract nutrients from the soil, and without their replenishment, yields in most soils will decline over time. Low soil fertility may be the single most pervasive constraint to high and sustainable cassava production worldwide. But it is highly amenable to improvements through management. Results from many cassava soil fertility trials have demonstrated, first, that yields steadily decline without soil amendments, and second, that yields can be stable when appropriate amendments are made. Substantial improvements to crop productivity usually include the application of exogenous nutrients in organic or inorganic form (Howeler, 2011a).

In addition to practices that may be more broadly applicable to many crops, there are several innate characteristics of cassava that allow us to design sustainable agronomic management approaches. The root association with mycorrhizae allows a very efficient extraction and uptake of soil phosphorous. The fungus exists naturally in virtually all cassava-growing areas, and usually no special management is required to achieve good root infection for efficient P absorption. In some situations, where cassava is newly introduced into an area where it has not previously been planted, there may be an economic advantage to inoculation (Howeler et al., 1987).

Because of cassava's relatively slow early growth, canopy closure can take up to three months or more, leaving the crop particularly vulnerable to weed infestation. Weeds can be a serious constraint to crop growth and yield, and their economic control a major challenge. Typically, manual weed control requires about 40% of labour inputs to produce a cassava crop. Manual weeding is often done by women, especially in Africa and Asia. Research on sustainable weed management has received relatively little emphasis to date, but new projects in West Africa are attacking the problem. More farmers are looking for ways to reduce the high labour inputs and cost of growing cassava.

Chemical weed control is possible, and herbicide use is rising, but mainly in Asia and in larger plantation systems elsewhere. Mechanized weeding is somewhat difficult in cassava except during the earlier stages of crop growth. Researchers face multiple challenges to integrate effective and economical weed control, with sustainability principles, and gender-sensitive approaches. It is a research area that will become increasingly important and will require greater research emphasis. Herbicide resistant cassava could be a popular option for farmers, as it has been for crops like maize, soybeans and canola. Technically it will probably not be very difficult to incorporate resistance (e.g. to glyphosate) through transformation protocols. But the licensing, regulatory, and the socio-economic issues (e.g. gender implications; consumer acceptance) will likely mean that any such technology is many years from applied, widespread use.

The long growth cycle of cassava is conducive to the build-up of many types of pests and pathogens. This creates challenges, but also opens up many opportunities during the crop's long period in the field, to introduce variable management packages for suppressing pest and pathogen damage. Some of the common practices that can contribute to pest suppression include adjusting planting date, plant spacing, and intercropping. As an example, early trials in the eastern plains of Colombia showed that planting near the end of the rainy season was a viable strategy for reducing losses from bacterial blight and super-elongation disease (CIAT, 1980). One of the challenges of using management practices to control pests and pathogens is to assure that the changed management does not reduce yields even more than would the pest.

Biological control is one of the most effective practices for cassava pest management. The development time can be relatively rapid (in contrast to the long lead time for developing resistant varieties, for example); there is virtually no trade-off in yield or quality with the application of bio-control methods; and in many cases, the control can be long-lasting without the continued need to reintroduce the bio-control organisms. In the Americas, bio-control agents (parasites and predators) evolved along with the crop during many millennia. However, when traders introduced cassava to Africa and Asia, most of these beneficial organisms were left behind. When new pests were introduced, they were often able to spread uninhibited by the natural enemies they had faced in their evolutionary homeland. There have been several examples of the introduction of natural enemies to successfully control mites and insects. In Africa, the cassava mealybug (*Phenacoccus manihoti*) caused devastating losses until an effective natural enemy, *Anagyrus lopezi (a parasitic wasp)* was introduced in the 1980s, saving billions of dollars in potential crop losses (Zeddies et al., 2001). Biological control never results in *complete* control, which leaves open the potential for fluctuations in levels of pest populations (similar to most types of host plant resistance as well). In some years and in some locations, economic damage levels may be significant. Like other types of pest management, biological control must be accompanied by constant monitoring, preparation for additional releases, and preparation for supplemental management within an IPM system.

Because cassava is among the most tolerant of crops in marginal conditions, it often occupies lands that are prone to erosion. This is true worldwide, but is particularly an issue in the Andean zone of South America and in SE Asia. Slow early growth and relatively wide spacing among plants mean that canopy closure can take 2–3 months – a period when the soil remains exposed to the heavy rains which typically occur near planting time. This situation can lead to severe soil erosion with devastating environmental and social consequences. Soil erosion in cassava systems is one of the most urgent problems for the long-term sustainability of cassava-based farming systems to support smallholder farmers.

Erosion control can be accomplished through soil preparation practices (e.g. ridge planting; conservation tillage, which leaves soil-protecting residue on the surface and soil-holding roots below the surface); strip cropping; intercropping; terracing; live barriers; practices that allow good ground cover (mulching; use of herbicides instead of hoeing); and practices that promote rapid canopy closure to protect exposed soil from direct rainfall impact (e.g. high early-vigour varieties; fertilization to promote rapid early growth). One of the most successful technologies is planting of vetiver grass barriers (Howeler, 2011b). Very little research has been done on conservation tillage systems for cassava. Clearly there are challenges – with the need to plant a large stem piece instead of a small seed, the inevitable soil disturbance that takes place at harvest, and the scarcity of good weed management systems without soil disturbance. Nonetheless, the potential payoff in lowering costs of production, in soil conservation, and in energy conservation makes this a critical area of research.

However, it is important to note that farmers are often reluctant to invest in soil management practices that do not provide short-term payback, especially if land is rented, or otherwise not securely available for the long term. The impact of erosion control is often not immediately evident to farmers, nor easily quantified. Their concept of long-term income loss may not be based on real, field-level data over time. This is also a management area which will depend almost wholly on the public sector initiatives; there are, in a broad sense, few options that can be offered that will be brought about through a relatively short-term profit motive of the private sector. This gives the public sector a heavy responsibility to thoroughly research practical erosion control methods, to educate growers, and to educate policy-makers on the need for policy support.

7 Addressing constraints: processing, market and gender issues

Cassava conversion to marketable products can involve a wide range of processing techniques and some of them produce large amounts of waste that can contribute significantly to environmental pollution and depletion of water resources (FAO, 2001). Key environmental issues are:

- Use of large quantities of water for starch extraction
- Environmental risks of waste water disposal, especially when discharged into streams or bodies of water
- Potential pollution from residues of processing (including peels and starch wastes)
- High-energy use for artificial drying of chips for animal feed, starch, flour or other end products (cost and generation of greenhouse gases)
- High-energy use for ethanol distillation (and generation of greenhouse gases)

With the rise of the starch industry throughout SE Asia, and the ethanol industry in China, the management of wastes and by-products is a growing concern, and many creative new technologies and systems are being developed to minimize environmental impact and increase profitability. The treatment of effluent waters is a major issue in the process of starch extraction. It results in major economic costs (if the effluents are not properly recycled or otherwise managed) or environmental costs (if effluents are dumped into the surrounding environment). Capture of methane as fuel energy from sealed waste storage tanks or ponds is becoming more common in large modern processing facilities. Technology also exists to efficiently produce high-quality flour that can substitute for starch for many uses, but whose processing has far less impact on the environment. Water is used only in the whole root washing, while the flour is extracted simply by grinding dried roots, with separation through screens and air flow. In West Africa, for example, most cassava is peeled before processing and the peels contribute to environmental pollution. Recent research on use of peels for animal feed both solves the issues of negative environmental impact and adds value along the chain from processors, feed manufacturer and feed purchasers/users (Iheanacho et al., 2015).

As cassava markets continue to develop and diversify, breeding can be a powerful tool to target opportunities for new product properties. One example of this is the identification of a natural mutation of amylose-free cassava (Ceballos et al., 2007), which will allow industry to develop certain starch-based products without the chemical modification that is currently required, with potential benefits to processing efficiency, the environment and human health. In another case, a different starch mutation (Ceballos et al., 2008) was generated through mutagenesis, resulting in the production of small starch granules (about one-third of the normal size). This mutation is useful for the bioethanol industry as the starch is more easily degraded into simple sugars, a necessary step before fermentation can be initiated, resulting in potentially lower energy use in the conversion process.

As has been noted, although still widely produced in very traditional ways, cassava is moving quickly from being mainly a food security crop to being a source of both food security and income. Traditional cassava cultivation, especially in Africa, typically involves women. In many resource-limited farming households it is the women who stay on the farm attending to the different chores while the men go to the villages in search of other income-generating activities. However, mechanization and intensification often change the gender balance with men becoming more involved. Where cassava becomes a cash crop, and becomes more intensively cultivated, it is likely that the roles women and men play will change.

Cassava systems for the future need to take account of the crop's potential to benefit women, men and families. Researchers need to actively search for potential areas where gender plays an important role. The study of gender implications of specific technologies need to be part of research design from the outset, rather than an afterthought after a technology is already developed and disseminated. As an example, increasing use of herbicides, and especially the development of herbicide tolerant cassava, will have highly gender-differentiated impact. In principle, for example, herbicide resistance should benefit women as they could re-direct their labour away from manual weeding to other more productive tasks. Whenever the trait is identified or induced, however, careful analysis of its expected advantages will be tested through participatory approaches to make sure that the technology benefits women in practice.

Another activity typically linked to women and children is the peeling of cassava, for example in the production of farinha in NE Brazil or gari and fufu in West Africa. Mechanization of peeling is notoriously difficult to achieve well, but progress is slowly being made. The success of mechanized peeling, or the growth of processing methods which do not require peeling, will have immense gender implications in Africa. Plant breeders can intervene to facilitate manual peeling by selecting for varieties with even root surfaces, and with peel that is easily removed. A given trait may be either complementary for both manual and mechanical peeling, or may be conflicting. In addition, a thin peel is desirable for those industries where the entire root is processed, since it maximizes the proportion of valuable tissue. However, a thick peel facilitates the operation of manual peeling, reducing the overall cost of such operations and therefore maintaining its competitiveness.

8 Addressing constraints: policy issues

Policies aimed specifically at the sustainability effects from the outputs of crop research are emerging more prominently in recent years, even in developing countries. The scientific community has a major challenge to educate, inform and advocate for such policies. A few examples follow, although this is not by any means a comprehensive list.

8.1 Policy on food security and equity

Developing countries that support technological and economic progress as a means of addressing food security and equity will find that cassava, where it is adapted, is often an effective vehicle toward significant improvements in food security and equity.

8.2 Policy on market development

Policies that favour development of new industries can open opportunities for cassava markets. The broad range of products that can be derived from cassava provides an ideal vehicle for new industry development, to the advantage of cassava growers. Multiple market opportunities for farmers mean that there are likely to be better prices and smaller variations in market prices. A key example of this kind of intervention is the policy in Nigeria to mix 10% cassava flour with wheat flour for the baking industry.

8.3 Trade policy

Open versus protectionist trade policies will impact the kinds of markets where cassava can be competitive. The main competing commodity to cassava in global markets is maize, especially as a source for starch and animal feed. Policies that favour or penalize either maize or cassava trade through tariffs or other added costs will affect the overall health of the cassava economy and those who depend on it. Because maize typically has a much higher political profile than cassava, it is more common that policies favourable to maize are adopted, causing comparatively more challenging conditions for those in the cassava value chain.

8.4 Policy on biofuels

China is leading the way in biofuels from cassava, as a result of a dual policy that aims, on the one hand, to reduce reliance on fossil fuels, and on the other hand aims to keep basic foods from competing in the biofuels market. Thus, cassava, as an efficient energy producer, and with a very minor role as a food in China, is an ideal option. In addition, there are several producers in the region anxious for export markets (e.g. Cambodia, Laos, Vietnam, Thailand, Indonesia). In Africa the situation is more complex, where cassava for biofuels is likely to compete directly with food markets, and where most of those who rely on cassava for food are not able to absorb cost increases even of small levels without suffering serious consequences. Also, production costs in Africa tend to be too high currently to compete in the biofuels market.

8.5 Policy on agriculture in fragile ecosystems

Thailand attempted for many years to support crop diversification in the Northeast of the country, to prevent the continued spread of cassava into fragile soils. The policy had limited success because cassava is so much better adapted than most other crop options for providing a profit to farmers. Likewise, Vietnam has policy restrictions on the cultivation of cassava in the uplands where there is high erosion risk. These types of policies are however rather rare on a global basis. In order to succeed, they need to either strictly prevent the growing of cassava in inappropriate environments with adequate enforcement, or provide equal or better alternatives through technology support and/or subsidies that give farmers attractive alternatives. In fact, effective policies that address the use of fragile landscapes are sorely needed in many countries, but along with that, enforcement capacity, education of growers, and the offer of sustainable technologies for positive impact. Enforcement alone, where farmers are given no alternative means of livelihood, is generally ineffective.

8.6 Policy on models for research and extension support

Until recent times, there was nearly no private sector support for cassava research. This is changing, but slowly. In Thailand, for example, the private sector provides modest support for development of specialty starch varieties, provides extension services in the form of advice on management practices, and provides growers with biological control organisms for the cassava mealybug, a newly emerging pest problem. There are examples elsewhere as well of important but quite limited industry support to technology development. This means that public support is the main determinant of the success of cassava research in most countries. The international agricultural research centres (CIAT and IITA) have strong multidisciplinary research programmes, but they rely completely on the capacity of national or private sector partners to jointly develop that technology and deliver it to farmers or to industry. Policy that supports a sustainable research and extension system is essential to the ability of cassava to play its full role as a vehicle for sustainable development.

Table 3 Estimated research intensity by crop in sub-Saharan Africa in 2010

Crop	FTE scientists per MT of production	Crop	FTE scientists per MT of production
Chickpea	83.6	Rice	10.1
Pigeon pea	68.9	Potato	8.7
Lentil	59.0	Maize-W&C Africa	8.1
Soybean	45.6	Groundnut	7.2
Dry beans	43.3	Banana	4.2
Field pea	31.4	Sweet potato	3.6
Wheat	20.5	Sorghum	3.3
Faba bean	20.5	Pearl Millet	1.6
Barley	15.1	Yams	1.0
Maize-E&S Africa	12.3	Cassava	0.9
Cowpea	11.3		

Source: Gollin, et al, 2014.

The public sector for cassava research is seriously underfunded in most countries. Table 3 illustrates the investment in research on a wide range of crops, relative to their production in sub-Saharan Africa. In this analysis, cassava is right near the bottom in terms of research investment relative to production. There needs to be much more support from studies on the impact of investment in research by national and local governments.

8.7 Policy on credit and crop insurance

The long cycle of cassava from planting to harvest often implies a heavy burden for farmers because of the corresponding long time required to recover his/her investments. It is becoming a common practice for governments, through different banking systems, to provide soft credits to farmers, particularly in cases where they have some sort of agreement with the processing sector and after it has been demonstrated that proper inputs and management practices will be used in growing the crop. This practice offers several advantages as it promotes linkages between the production and processing sector and encourages the adoption of technologies for the sustainable and competitive production of cassava. Within the same policies farmers can also have access to crop insurance. For insurance to have more widespread impact, however, more data on production risks are necessary. To date, crop insurance applied to cassava is relatively rare.

9 A new research paradigm for cassava

Cassava research was very limited prior to the development of the Consultative Group on International Agricultural Research System (CGIAR), which included two centres working on cassava – the International Centre for Tropical Agriculture (CIAT) in Cali, Colombia and the International Institute of Tropical Agriculture in Ibadan, Nigeria in the late 1960s. In 2011, CGIAR introduced a major shift in its approach to research by organizing into CGIAR

Research Programmes (CRPs), a system designed to optimize inter-centre collaboration and synergy. Cassava became part of the Root, Tuber and Banana CRP (CRP-RTB), including four CGIAR centres and five major crops (cassava, potato, sweet potato, yam and banana/plantain). Later, other research centres such as CIRAD (France) and Wageningen Agricultural University (Netherlands) became key global partners of the programme, along with partnerships with many regional and national R&D programmes, universities, NGOs and private sector companies. This new approach recognized the synergies that could be achieved by coordination of research among organizations working on key starchy staples, with common problems in breeding (vegetative propagation, high heterozygosity, polyploidy (except cassava)); common issues in seed systems (vegetative propagation, low participation of the formal seed sector); a common feature as crops of the poor, with high participation of women in production and processing, and other commonalities. CRP-RTB, having begun its second 5-year phase of research in 2017, is now the main driving force for cassava research at the global level, and coordinates with a wide range of other national and international programmes and initiatives (http://www.rtb.cgiar.org/).

10 Cassava and climate change

It is widely accepted that responding to climate change represents one of the great challenges of agricultural research. Climate change may have direct effects on crop growth and development (temperature, rainfall, CO_2 levels) or indirect effects (soil organic matter, soil erosion, pest and disease patterns). There is an emerging consensus that cassava is one of the most promising options among tropical crops in dealing with changes such as rising temperatures, rising CO_2 and increasingly uncertain rainfall patterns. This can be seen in climate maps combining temperature and rainfall parameters specific to cassava's growth responses (Fig. 4) which indicate that cassava will probably continue to be grown in nearly all areas where it is currently adapted. This is largely because of its tolerance to

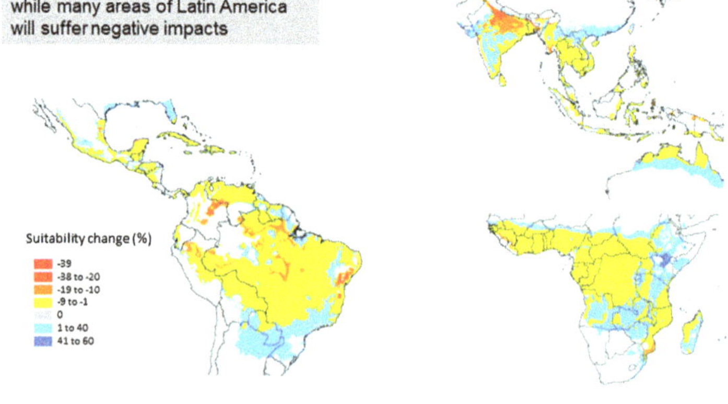

Figure 4 Impact of climate change on areas of cassava cultivation. Source: Decision and Policy Analysis Unit, CIAT.

both high temperature and drought, even in some areas where these changes will create severe stress for other crops.

Areas that become too hot or too dry for other crops could become new growing areas for cassava. Some climate models show that India could be especially affected by rising temperatures, with broad shifts away from grains and pulses in some areas (Ceballos et al., 2011). Possibly the most significant effect of temperature rise on cassava's adaptation will be to allow it to move into higher altitude and higher/lower latitude regions. Currently, cassava's limit at the Equator is at about 2000 masl, and this just for a narrow range of germplasm accessions from the Andean zone of the Americas, especially Colombia. These new highland areas for cassava are likely to be most important in both East Africa and the Andes. Currently cassava can be grown in latitudes to near the Tropics of Cancer and Capricorn. Global warming may extend this range, as winters become more mild. This is of particular interest in China, which is looking for options to expand planted area but has a relatively limited region within the subtropics that is suitable for cassava. While global warming is almost certainly having overall negative effects on agriculture, there will be opportunities for farmers to adapt by introducing new varieties, new crops and new practices if science has the resources to develop the appropriate technological support.

Atmospheric CO_2 is one of the major causes of climate change and has increased by 40% from a pre-industrial revolution baseline. It has been suggested that increases in atmospheric CO_2 concentration could result in a reduction in root production. Free-air-CO_2-Enrichment (FACE) methods allow field evaluation of crops under elevated CO_2 concentrations which simulate the predicted levels for the decades to come (El-Sharkawy, 2009). These studies suggest that photosynthetic efficiency would increase more in C_3 (like potatoes and cassava) than in C_4 crops (like maize and rice) (Long et al. 2004, 2006). Modelling and FACE results could guide the molecular optimization of the photosynthetic apparatus to maximize carbon gains without increasing crop inputs (Rosenthal and Ort, 2012).

Climate change models and crop models also suggest that constraints other than the direct effects of drought, brought about by climate change, and especially the effects of pests and diseases, are likely to be more severe. Climate change modelling, layered with pest adaptation maps, illustrates the potential spread to new areas in the context of climate change. This allows the application of resources in highest risk areas for monitoring, diagnosis and management. It is expected that pests affecting cassava will evolve in a dynamic pattern, particularly as a result of increased temperatures that reduce the relevance of diapause and/or shortens their life cycle (Ceballos et al. 2011). Figures 5 and 6 illustrate areas where cassava green mite and whitefly, respectively, are likely to increase or decrease in severity due to climate change by 2020. For both species, there will be widespread effects in the Americas and Africa, but less so in Asia.

Some of the new soil-related challenges likely to be exacerbated by climate change are: more rapid loss of organic matter due to higher soil temperatures; planting in areas more vulnerable to erosion (e.g. further up hillsides as temperatures rise); and greater nutrient leaching in areas where rainfall has increased. For cassava in an era of climate change, one of the greatest challenges for sustainable soil management is in areas where the crop expands to replace species that are less adapted to drier conditions. Unless this expansion into new areas is accompanied by appropriate management and technologies, there is a risk that growers without the experience of growing the crop will use practices that exacerbate erosion and nutrient depletion. There should also be attempts to introduce diversification programmes, such as the planting of perennial

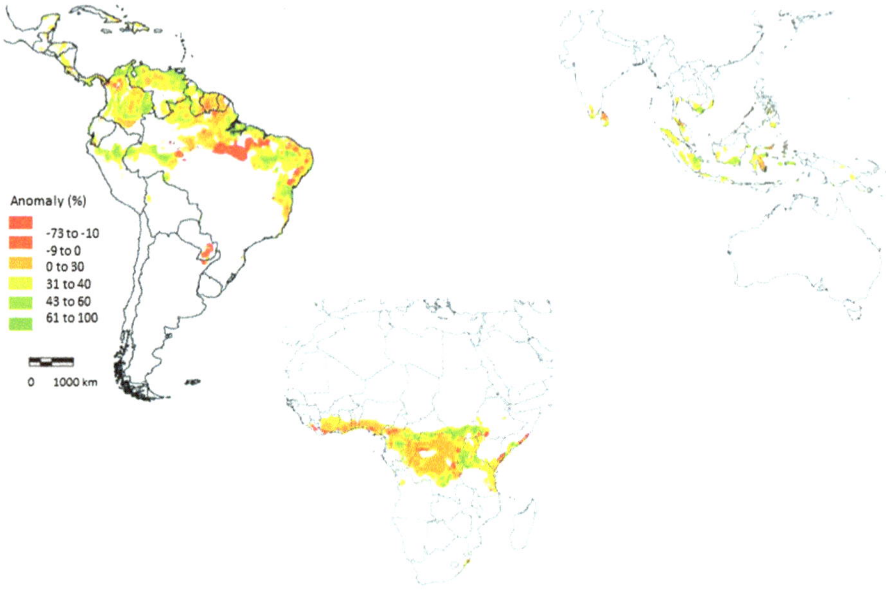

Figure 5 Climate change impacts on cassava green mite to 2020. Source: Decision and Policy Analysis Unit, CIAT.

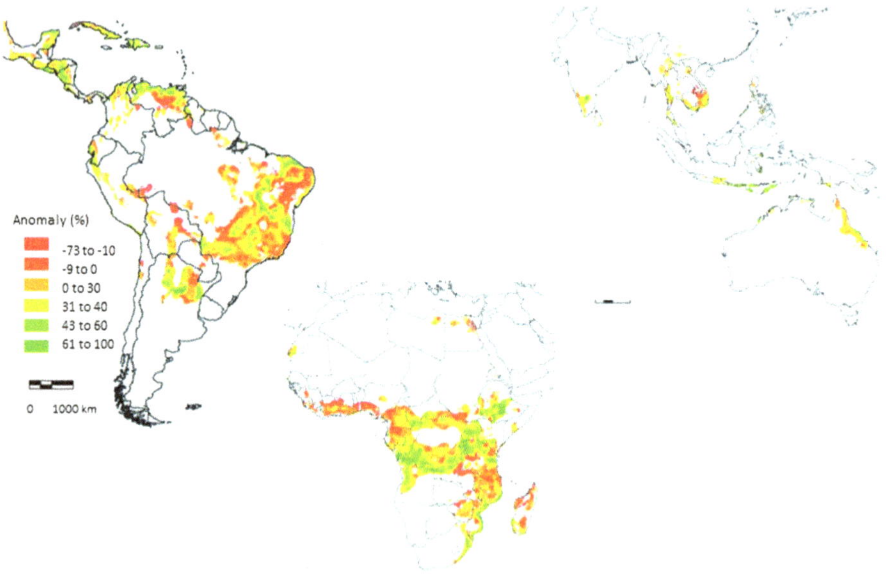

Figure 6 Climate change impacts on cassava whitefly to 2020. Source: Decision and Policy Analysis Unit, CIAT.

crops/pastures/trees in the most vulnerable areas. As noted, conservation tillage or no-tillage systems have had relatively little application in cassava. Alternatives to conventional tillage will be important both in areas of reduced and increased rainfall. In reduced rainfall, conservation tillage conserves soil water. Under heavy rainfall, it reduces erosion and improves soil structure for better drainage. These advantages need to be weighed against the possibility of sacrificing yield or income as a result of adoption of these practices.

11 Conclusions

This chapter has identified a number of priorities for cassava research. These include making the crop:

- **More productive** – higher yields, and especially higher yields per unit of time and unit of input (e.g. labour, fertilizer, water).
- **More profitable and competitive** – creating new opportunities for growers to increase their incomes.
- **More resilient** – reversing land degradation and adapting to the new challenges caused by climate change.
- **More equitable –** providing new opportunities for the rural poor, especially women, through rural enterprise development, and access to inexpensive and nutritious food.

Addressing these challenges will help to make cassava a key crop in addressing the challenges of the twenty-first century.

12 Where to look for further information

Several national and international organizations maintain websites that provide up-to-date information on many cassava research and development topics:

- International Centre for Tropical Agriculture (www.ciat.cgiar.org)
- International Institute for Tropical Agriculture (www.iita.org)
- Root, Tuber and Banana Program of the CGIAR (www.rtb.cgiar.org)
- Global Cassava Partnership for the 21st Century (www.gcp21.org)
- International Society for Tropical Root Crops (www.istrc.org)
- Integrated Cassava Project (www.cassavabiz.org)
- Thai Tapioca Development Institute (www.tapiocathai.org/English/Mainpage_e.html)

A few reviews provide a broad perspective on recent research and development issues, and recommendations on production and processing applications:

- FAO Global Cassava Development Strategy (www.fao.org/ag/agp/agpc/gcds)
- The Cassava Handbook (https://cgspace.cgiar.org/handle/10568/54299)
- Save and Grow Cassava: A guide to sustainable production intensification (www.fao.org/3/a-i3278e.pdf)

- Sustainable Soil and Crop Management of Cassava in Asia (https://cgspace.cgiar.org/handle/10568/51590)

13 References

Allem, A. C. 1990. The closest wild relatives of cassava (*Manihot esculenta* Crantz). *Euphytica* 107:123–33.

Allem, A. C. 2002. The origins and taxonomy of cassava, pp. 1–16. In R. J. Hillocks, J. M. Tresh and A. C. Bellotti (Eds), *Cassava: Biology, Production and Utilization*. CABI Publication: Wallingford, UK and New York, USA.

Allem, A. C., R. A. Mendes, A. N. Salomão and M. L. Burle. 2001. The primary gene pool of cassava (*Manihot esculenta* Crantz subspecies *escultenta*, Euphorbiaceae). *Euphytica* 120:127–32.

Andersen, M. D., P. K. Busk, I. Svendsen and B. L. Møller. 2000. Cytochromes P-450 from cassava (*Manihot esculenta* Crantz) catalyzing the first steps in the biosynthesis of the cyanogenic glucosides linamarin and lotaustralin. *J. Biol. Chem.* 275:1966–75.

Beeching, J. R., P. Marmey, M. C. Gavalda, M. Noirot, H. R. Haysom, M. A. Hughes and A. Charrier. 1993. An assessment of genetic diversity within a collection of cassava (*Manihot esculenta* Cranz) germplasm using molecular markers. *Ann. Bot.* 72:515–20.

Bellotti, A. C. 2002. Arthopod pests, pp. 209–36. In R. J. Hillocks, J. M. Tresh and A. C. Bellotti (Eds), *Cassava: Biology, Production and Utilization*. CABI Publication: Wallingford, UK and New York, USA.

Bellotti, A. C., L. Smith and S. L. Lapointe. 1999. Recent advances in cassava pest management. *Annu. Rev. Entomol.* 44:343–70.

Ceballos, H., T. Sánchez, N. Morante, M. Fregene, D. Dufour, A. M. Smith, K. Denyer, J. C. Pérez, F. Calle and C. Mestres. 2007. Discovery of an amylose-free starch mutant in cassava (*Manihot esculenta* Crantz). *J. Agr. Food Chem.* 55(18):7469–76.

Ceballos, H., T. Sánchez, K. Denyer, A. P. Tofiño, E. A. Rosero, D. Dufour, A. Smith, N. Morante, J. C. Pérez and B. Fahy. 2008. Induction and identification of a small-granule, high-amylose mutant in cassava (*Manihot esculenta* Crantz). *J. Agr. Food Chem.* 56(16):7215–22.

Ceballos, H., J. Ramirez, A. C. Bellotti, A. Jarvis and E. Alvarez. 2011. Adaptation of cassava to Changing Climates, pp. 411–25. In S. S. Yadav, B. Redden, J. S. Hatfield, G. Lotze-Campen and A. Hall (Eds), *Crop Adaptation to Climate Change*. Wiley–Blackwell: Hoboken, NJ.

Cock, J. 1985. *Cassava: New Potential for a Neglected Crop*. Westview Press: Boulder, CO.

CIAT, 1980. Cassava Program Annual Report 1979. Centro Internacional de Agricultura Tropical (CIAT), Cali, Colombia.

Du, L., M. Bokanga, B. L. Møller and B. A. Halkier. 1995. The biosynthesis of cyanogenic glucosides in roots of cassava. *Phytochemistry* 39:323–6.

El-Sharkawy, M. A., J. H. Cock, J. K. Lynam, A. D. P. Hernandez and L. F. Cadavid. 1990. Relationships between biomass, root yield and single-leaf photosynthesis in field-grown cassava. *Field Crops Res.* 25(3–4):183–201.

El-Sharkawy, M. A. 2009. Pioneering research on C_4 leaf anatomical, physiological, and agronomic characteristics of tropical monocot and dicot plant species: implications for crop water relations and productivity in comparison with C_3 cropping systems. *Photosynthetica* 47:163–83.

El-Sharkawy, M. A. 2012. Stress-tolerant cassava: the role of integrative ecophysiology-breeding research in crop improvement. *Open J. Soil Sci.* 2:162–86.

Gleadow, R. M. and I. E. Woodrow. 2002. Constraints of effectiveness of cyanogenic glycosides in herbivore defense. *J. Chem. Ecol.* 28:1301–13.

Gleadow, R. M., J. R. Evans, S. McCaffery and T. R. Cavagnaro. 2009. Growth and nutritive vaslue of cassava (*Manihot esculenta* Crantz.) are reduced when grown in elevated CO_2. *Plant Biol.* 11:76–82.

Gollin, D., D. Lagakos and M. E. Waugh. 2014. The Agricultural Productivity Gap. *Q. J. Econ.*, 129:939–93.

Henry, G. 1995. *Global Cassava Sector Constraints and Estimated Future R&D Benefits*. Centro Internacional de Agricultura Tropical (CIAT): Cali, Colombia, p. 39.

Herrera Campo, B. V., G. Hyman and A. Bellotti. 2011. Threats to cassava production: known and potential geographic distribution of four key biotic constraints. *Food Sec.* 3:329–45.

Hershey, C. H. 2008. A Global Conservation Strategy for Cassava (*Manihot esculenta*) and Wild *Manihot* Species. (https://www.croptrust.org/wp-content/uploads/2014/12/cassava-strategy.pdf)

Howeler, R. H. 2002. Cassava mineral nutrition and fertilization, pp. 115–47. In R. J. Hillocks, J. M. Thresh and A. C. Bellotti. (Eds), *Cassava: Biology, Production and Utilization*. CABI Publication: Wallingford, UK and New York, USA.

Howeler, R. H. 2011a. Chapter 16. Short- and long-term N, P and K requirements of cassava, pp. 468–93. In R. H. Howeler (Ed) *The Cassava Handbook. A Reference Manual based on the Asian Regional Cassava Training Course*. Centro Internacional de Agricultura Tropical (CIAT): Thailand.

Howeler, R. H. 2011b. Chapter 20. Soil erosion control, pp. 566–608. In R. H. Howeler (Ed) *The Cassava Handbook. A Reference Manual based on the Asian Regional Cassava Training Course*. Centro Internacional de Agricultura Tropical (CIAT): Thailand.

Howeler, R. H., E. Sieverding and S. Saif.1987. Practical aspects of mycorrhizal technology in some tropical crops and pastures. *Plant Soil* 100:249–83.

Iheanacho, O., A. Samireddypalle, L. Kaptoge, C. Fauquet, J. Atehnkeng, R. Bandyopadhyay, P. Kulakow, A. Duncan, T. Alabi and M. Blummel. 2015. Technical innovations for small-scale producers and households to process wet cassava peels into high quality animal feed ingredients and aflasafe™ substrate. *Food Chain* 5:71–90. doi:http://dx.doi.org/10.3362/2046-1887.2015.005.

Imai, K., D. F. Coleman and T. Yanagisawa. 1984. Elevated atmospheric partial pressure of carbon-dioxide and dry matter production of Cassava (*Manihot esculenta* Crantz). *Jpn. J. Crop Sci.* 53:479–85.

Jarvis, A., J. Ramirez-Villegas, B. V. Herrera Campo, C. Navarro-Racines. 2012. Is cassava the answer to African climate change adaptation? *Trop. Plant Biol.* 5:(1) 9–29.

Jennings, D. L. and C. Iglesias. 2002. Breeding for crop improvement. In R. J. Hillocks, J. M. Thresh and A. C. Bellotti (Eds), *Cassava: Biology, Production and Utilization*. CABI Publishing: New York, pp. 149–66.

Kawano, K. 2003. Thirty years of cassava breeding for productivity – biological and social factors for success. *Crop Sci.* 43:1325–35.

Kawano, K. and J. H. Cock. 2005. Breeding cassava for underprivileged: institutional, socio-economic and biological factors for success. *J. Crop Improv.*14:197–219.

Long, S. P., E. A. Ainsworth, A. Rogers and D. R. Ort. 2004. Rising atmospheric carbon dioxide: Plants face the future. *Annu. Rev. Plant Biol.* 55:591–628.

Long, S. P., E. A. Ainsworth, A. D. B. Leakey, J. Nösberger and D. R. Ort. 2006. Food for thought: lower-than-expected crop yield stimulation with rising CO_2 concentrations. *Science* 312:1918–21.

Lynam, J. K. 2008. Development paths for cassava in Africa. Global Cassava Partnership, GCP-1, held in Ghent, Belgium. July 21–25 2008. PowerPoint presentation.

McMahon, J. M., W. L. B. White and R. T. Sayre. 1995. Cyanogenesis in cassava (*Manihot esculenta* Crantz). *J. Expt. Bot.* 46:731–41.

Mkumbira, J., L. Chiwona-Karltun, U. Lagercrantz, N. Meso Mahungu, J. Saka, A. Mhone, M. Bokanga, L. Brimer, U. Gullberg and H. Rosling. 2003. Classification of cassava into 'bitter' and 'cool' in Malawi: From farmers' perception to characterisation by molecular markers. *Euphytica* 132:7–22.

Nassar, N. M. A. and R. Ortiz. 2008. Cassava genetic resources: Manipulation for crop Improvement. *Plant Breed. Rev.* 31:247–75.

Nweke, F. I., D. S. C. Spencer and J. K. Lynam. 2002. *The Cassava Transformation: Africa's Best-kept Secret*. Michigan State University Press, East Lansing, Michigan, USA p. 273.

Okogbenin, E., S. S. Kahya and M. Fregene. 2011. Use of biotechnology tools in cassava breeding, pp. 119–45. In R. H. Howeler (Ed) *The Cassava Handbook. A Reference Manual based on the Asian Regional Cassava Training Course*. Centro Internacional de Agricultura Tropical (CIAT): Thailand.

Olsen, K. M. and B. A. Schaal. 2001. Microsatellite variation in cassava (*Manihot esculenta*, Euphorbiaceae) and its wild relatives: further evidence for a southern Amazonian origin of domestication. *Am. J. Bot.* 88(1):131–42.

Reilly, K., D. Bernal, D. F. Cortes, R. Gomez-Vasquez, J. Tohme and J. R. Beeching. 2007. Towards identifying the full set of genes expressed during cassava post-harvest physiological deterioration. *Plant Mol. Biol.* 64:187–203.

Rosenthal, D. M. and D. R. Ort. 2012. Examining cassava's potential to enhance food security under climate change. *Trop. Plant Biol.* 5:30–8.

Sánchez, T., H. Ceballos, D. Dufour, D. Ortiz, N. Morante, F. Calle, T. Zum Felde and F. Davrieux. 2014. Carotenoids and dry matter prediction by NIRS and hunter color in fresh cassava roots. *Food Chem.* 151:444–51.

Sánchez, T., D. Dufour, I. X. Moreno and H. Ceballos. 2010. Comparison of pasting and gel stability of waxy and normal starches from cassava, potato, maize, and rice under thermal, chemical and mechanical stress. *J. Agr. Food Chem.* 58:5093–9.

Sánchez, T., G. Mafla, N. Morante, H. Ceballos, D. Dufour, F. Calle, X. Moreno, J. C. Pérez and D. Debouck. 2009. Screening of starch quality traits in cassava (*Manihot esculenta* Crantz). *Starch/Stärke* 61:12–19.

Stupak, M. 2008. Improving protein content in cassava storage roots. PhD dissertation ETH, Zurich. www.e-collection.library.ethz.ch.

Wheatley, C. C. and G. Chuzel. 1995. Cassava: the nature of the tuber and use as a raw material pp 734-43. In R. Macrae, R. K. Robinson and M. J. Sadler (Eds), *Encyclopedia of Food Science, Food Technology and Nutrition*. Academic Press: London.

Zeddies, J., R. P. Schaab, P. Neuenschwander and H. R. Herren. 2001. Economics of biological control of cassava mealybug in Africa. *Agr. Econ.* 24:209–19.

Chapter 9

Targeting smallholder farmers to adopt improved cassava technologies: challenges and opportunities

Kolawole Adebayo, Federal University of Agriculture, Nigeria

1 Introduction

2 Research on improving cassava production and understanding smallholder farmers and their decisions

3 Research on smallholders' support systems

4 Case studies

5 Summary

6 Future trends in research

7 Where to look for further information

8 References

1 Introduction

Cassava (*Manihot esculenta* Crantz) is a root crop, cultivated and consumed as a staple food in many regions of the developing world. The importance of cassava to the livelihoods of many millions of poor people has made the crop a significant target for research and development (R&D) interventions. The African Union New Partnership for Africa's Development Agency (NEPAD) adopted the slogan 'Cassava: A Powerful Poverty Fighter in Africa' for its Pan African Cassava Initiative. Presidential Initiative on Cassava (PIC) has been implemented in Ghana and Nigeria. These initiatives seek to explore the potential of the crop as a cheap source of food calories and are a means of bringing benefits to poor producers and consumers. The per capita consumption of cassava in the leading consuming countries is 120–255 kg/capita (Nweke, 2004). Cassava also plays multiple roles as a famine reserve, food and cash crop, industrial raw material and livestock feed. In the last decade, the crop has become more important in areas with higher densities of population and which are closer to market centres.

Current trends in world food markets have increased the potential role of cassava in the international food equation. Even though the vast majority of the cassava grown in Africa is marketed through traditional food markets, recent attempts to synchronize cassava production, processing and marketing (*value chains*) have helped to expose the

http://dx.doi.org/10.19103/AS.2016.0014.09

untapped business opportunities that exist in using cassava to transform the lives of many poor people across the continent (Adebayo et al., 2010). Market trends also indicate unmet demands for cassava products from the food, livestock, textile and pharmaceutical industries (Bennett et al., 2012).

There are several efforts at the continental and national levels, as well as in research institutes and universities, to develop the comparative advantages that Africa has in the cassava value chains (Odedina and Adebayo, 2012). Exports of cassava products, new convenience products from cassava and products made from composite flour with cassava are beginning to appear in the markets (Bennett et al., 2012). The importance of these trends is that the demand for cassava roots will continue to rise in the foreseeable future. Yet, the supply capacity is limited by the fact that cassava production is dominated by smallholder farmers characterized by low farm yields and low productivity. It is important that smallholder farmers adopt productivity-enhancing technologies so that they do not lose out on the commercial opportunities that these new demands for cassava roots predict. The achievement of these goals requires a careful understanding of the cassava technologies adoption and diffusion processes.

The literature on issues relating to the adoption of improved technologies in developing countries is voluminous (Kidd, 1968; Patel and Anthonio, 1971; Hays and Raheja, 1977; Zandstra et al., 1981; Norman et al., 1995; Franzel et al., 2000; Collinson, 2001; Adebayo, 2009; Osuntade et al., 2012). Many of these studies were premised on the assumption that if the factors influencing the adoption of these technologies are known, the reactions of farmers to the introduction of technological packages could be determined.

In studies on the concepts of adoption and diffusion of innovation, asking respondents whether or not a practice was acceptable has not been found to be useful to the understanding of the processes involved (Franzel et al., 1996, 2001). The farm and household characteristics that are examined most frequently in adoption research for their association with testing and continued use of a practice included gender, level of formal education, household size, farm size and wealth level (Ladebo, 1999; Franzel et al., 2000). Sall et al. (2000) point to government policy as an important factor in the adoption of improved technologies at the farm level. Furthermore, Hays and Raheja (1977) and Sall et al. (2000) have shown that farmers' perceptions of technology-specific characteristics significantly influence adoption decisions relating to improved farm practices. Collinson (2001) identifies deficiencies in the process of technology generation as the main cause for the low adoption by small farmers in developing countries. Sinclair (2001) reviewed a series of studies in process-based research and concluded that the involvement of farmers in the technology development process is crucial for successes in adoption. He noted, however, that in addressing longer-term sustainability issues, a predictive understanding of farmers' adoption is needed. Arising from all these precedents, models of the technology adoption process have been constructed in at least three ways. The first applies a minimalist approach which assumes that human behaviour is passive and unchanging, even in the face of significant global change. A second approach is bio-economic modelling, which confronts behavioural problems head-on and makes very explicit assumptions about how people act with respect to managing their resources. The third and most promising approach to modelling the adoption process is through constructing multiple, simultaneous regression models of human behaviour (Adebayo, 2002, 2005, 2009).

Targeting smallholder farmers to adopt improved cassava technologies requires more than a simple 'we know it works, take it' mentality that often forms the premise of traditional production-centred research. The reasons for this complexity are as follows:

a A smallholder farming system is more than just an economic unit; it is also a social system with complex dynamics between the production and consumption system and moderating forces. These dynamics are often poorly understood (Adebayo, 2002). According to Hartmann (2004), 'A successful approach cannot only be about agriculture – it has to recognise the vital role it plays in the bigger picture.'
b Cassava requires nearly a full year from planting to harvest. This implies that smallholder farmers cannot reap the benefits of the technologies adopted in a period of less than one year. Yet, some of the market dynamics that determine the level of benefits accruable from technology adoption are influenced by shorter-term market actions and reactions that are not within the control of smallholder farmers. Therefore, the tendency among smallholder farmers is to take decisions not necessarily influenced by the expected benefits of the improved technologies (Dipeolu et al., 2002).
c Smallholder farmers are not a homogeneous group. Targeting requires an understanding of the diversity among smallholder farmers and, therefore, the requirement for differences in the technology promotion strategies to support adoption (Adebayo et al., 2005, 2008).
d A static technology promotion system cannot address the needs of a dynamic client group. Innovativeness in the agricultural extension systems promoting improved cassava technologies is a necessary requirement for stimulating improved technology adoption among smallholder farmers (Adebayo et al., 2003).

This chapter will therefore explore how each of these key issues is being addressed, review some case studies of how research has been used to improve cassava productivity among smallholders in practice, and provide an insight into future trends in research on targeting smallholder farmers to adopt improved cassava technologies for yield and profitability.

2 Research on improving cassava production and understanding smallholder farmers and their decisions

2.1 Research on improving cassava production and yield

Traditional cassava development research in the African National Agricultural Research Systems (NARS) had focused almost exclusively on improving cassava production and yield, particularly looking at the development of new cassava varieties that are higher yielding and resistant (or tolerant) to key diseases such as cassava mosaic and brown streak diseases. Several programmes also developed optimal agronomic practices for achieving high yield or returns to inputs used in cassava production. These studies have established that productivity-enhancing products and practices can significantly and sustainably increase smallholder farmers' incomes, as well as improve their livelihoods (Ortiz and Hartmann, 2003). Yet, they are not adopted as extensively as they should be by the world's 1.5 billion smallholder farmers. A recent study looks at how pioneering companies and organizations have succeeded in enrolling smallholder farmers in productivity-enhancing schemes, sustainably and at scale – either as buyers of produce or as sellers of products, services or equipment (Hystra, 2015). This report, among others, exemplifies the development and use of the value chain approach in the development of the cassava sector. It has brought to the fore issues that were previously not the central focus of most cassava research for targeting smallholder farmers to adopt improved cassava technologies.

2.2 Research on supporting smallholder decision making

Several studies have indicated that the adoption of an innovative technology is influenced by social, economic, innovation-related factors; technology development process related; or even exogenous factors such as government policy and developments in the international arena (Bernet et al., 2001; Collinson, 2001; Smith et al., 1995; Agbamu, 1995, Anaglo et al., 2009). Many of these studies have been conducted on new crop varieties and pest control practices. On the other hand, the influence of market dynamics on the decision making of smallholder cassava farmers has been relatively understudied. It is clear that the commercial cultivation of cassava presents a classic *cobweb* situation in economic analysis. The cobweb theory is an economic model that explains why prices might be subject to periodic fluctuations in certain types of markets. It describes the cyclical supply and demand in a market, where the quantity produced must be determined before prices are negotiated (Akerman, 1957; Labys, 2006). The cassava supply–demand situation, with its annual imbalances in many African nations, depicts this phenomenon. Yet, very few studies have explored the influence of this economic misalignment between changes in market forces and farmers' response, such as the decision to adopt or not adopt improved cassava technologies.

2.3 Research on understanding smallholder farmers

Adebayo (2005) examined three decades of adoption and diffusion research in Nigeria and posited that smallholder farmers are not a homogenous group. While smallholder farmers growing grains such as maize and rice can adopt improved technological packages meant for larger farmers, the same is almost impossible for smallholder cassava farmers. Three factors may explain why this is so:

a The use of stem cuttings rather than seeds for propagation: Cassava is a vegetatively propagated crop using stem cuttings that are bulky to transport, and can lose viability within days of harvest if the proper conditions to initiate growth are not provided. Unlike seeds, whose bulk can be managed by simply opening a 100 kg bag and taking a smaller unit (say 1 kg) for use on a small farm, a planter designed for planting more than 60 bundles of stem cuttings on larger farms will be difficult to operate when it is dealing with one bundle on a 100 square metre farm.

b Farmers' interest in the starchy roots: The main reason most farmers grow cassava is to use or sell its starchy roots. As with stems, starchy roots are bulky and difficult to transport. Therefore, techniques or processes that are developed to move larger quantities are simply not applicable to or adoptable by smaller production units. Conversely, there may be techniques or processes better suited to small enterprises. For instance, if a technique is developed for treating fresh cassava roots to keep them fresh for more than 30 days, and which is based on treating single root units, it immediately poses a special challenge for larger producers on how to apply the technique to large volumes of roots.

c The irregular size and shape of cassava roots: The irregular sizes of cassava roots in a single harvest provide a similar challenge for application of the same solution for large and small farmers. The handling of small quantities in bags and small containers suits smallholder situations. Simply increasing the size of bag and container would not work well in larger holdings. In the same vein, application of motorized containers and conveyor belt systems in larger processing units poses cost and energy challenges for smallholders.

These factors make targeting smallholder farmers to adopt improved cassava technologies designed for larger, more commercially oriented farmers more challenging. Adebayo (2005) also found that 38 variables explain the adoption of improved technologies by smallholder farmers and placed them in six broad categories: (i) individual farmer's socio-economic factors; (ii) social system factors; (iii) extension agent and agency factors; (iv) macro-economic factors; (v) innovation characteristics; and (vi) communication factors (Table 1).

The large number and broad scope of these variables may explain why technology development and dissemination in human social systems has remained a complex venture and difficult to achieve with the precision that physical and chemical models provide. Among the 38 factors, age, education, farm size, family size and contact with extension services feature more prominently in research studies than other variables in determining the adoption of farm technologies in Nigeria. It is noteworthy that for any specific locality, type of technology and category of farmers, the set of key variables that need to be considered in the design of technology may be obtained through iterative processes like factor analysis, regression analysis and stakeholder analysis.

3 Research on smallholders' support systems

It makes business sense to work with and for smallholder farmers, to help them achieve higher incomes and aim to transform their lives in the process (Zaria and Omotayo, 1997; Adebayo et al., 2010). But a balanced and stable relationship is essential to ensure that both sides – the farmers and the organization working with them – grow and thrive (Hystra, 2015). Systemic issues requiring farmer support systems to physically adjust their foci and realign their axes to ensure a proper fix have been rather challenging for the technology generation (research) systems.

Even though smallholder farmers have been described by some scientists as *stubborn* and unwilling to accept modern, improved technologies, the same can sometimes be said of research systems with their embedded tradition of what is considered *scientific* research. Some studies in the last four decades have brought up the position of indigenous technologies in modern scientific research (Ibitoye, 2011; Fuller-Wimbush and Adebayo, 2014). Yet, it is more difficult to obtain funding for research with an exclusive focus on smallholder clients without proving commercial gains.

In Nigeria, for instance, several studies were carried out to determine the impact of the extension service under the World Bank funding arrangement (Adebayo and Idowu, 2000). Most of these studies indicate positive impact, in terms of growth in the output of major food crops and adoption of innovations by farmers (Chikwendu et al., 1995; Okwoche, et al, 1998; Adebayo et al., 1999; Ladebo, 1999; Omotayo et al, 2001). These studies show that about 60% adoption rates on improved varieties of cassava were recorded, 75% adopted fertilizer technology in sorghum and 50% adopted new maize varieties. Technologies related to pesticide use were the least adopted. The reasons

Table 1 Explanatory variables in adoption studies in Nigeria

I Socio-economic factors
Age – Okwoche et al., 1998; Akinola,1986a; Bello, 2000; Adekoya and Ajayi, 2000
Marital status – Adekoya and Ajayi, 2000
Religion – Patel and Anthonio, 1971; Adekoya and Ajayi, 2000
Gender – Adebayo, 1998; Adekoya and Ajayi, 2000
Ancestry – Adebayo, 1994
Education – Akinola, 1986a; Onu, 1991; Agbamu, 1993; Okwoche et al., 1998; Ladebo, 1999; Adekoya and Ajayi, 2000
Farm size – Patel and Anthonio, 1971; Akinola, 1986a,b; Onu, 1991; Agbamu, 1995; Okwoche et al., 1998; Ozor, 1998; Ladebo, 1999; Adekoya and Ajayi, 2000
Farming experience – Akinola, 1986b
Income – Okwoche et al., 1998; Ladebo, 1999
Resource endowment – Okwoche et al., 1998
Attitude – Onu, 1991
Diversification factor – Akinola, 1986b
II Social system factors
Family size – Patel and Anthonio, 1971; Okwoche et al.,1998
Social position – Vabi et al., 1993
Social participation – Patel and Anthonio, 1971; Onu, 1991
Membership of organizations – Akinola, 1986a; Okwoche et al.,1998
Number of previous adopters – Akinola, 1986b
Leadership – Agbamu, 1993
III Extension agent and agency factors
Extension contact –Akinola, 1986a,b; Onu, 1991; Agbamu, 1993; Okwoche et al., 1998; Adekoya and Ajayi, 2000
IV Macro-economic factors
Access to credit – Patel and Anthonio, 1971; Akinola, 1986a; Okwoche et al., 1998
Product prices – Akinola, 1986b
Quantity of inputs available for sale – Akinola, 1986b
Number of active selling points – Akinola, 1986b
Advertisement – Akinola, 1986b
V Innovation characteristics
Difficulty in use of innovation – Chukwu, 1995
Cost of innovation – Akinola, 1986b
VI Communication factors
Period of awareness –Agbamu, 1993
Cosmopoliteness –Agbamu, 1993; Ladebo, 1999

Source: Adebayo (2005).

for not adopting some of the technologies include unavailability and the cost of inputs. The arrangement under the training and visit (T&V) system, which is the main thrust of the extension management strategy under the World Bank project, precludes extension agents from getting involved in input procurement and distribution (Benor et al., 1984). The T&V system of agricultural extension is designed for a professional extension service operating regular and predictable visits for training farmers and getting trained. Involvement in input distribution is believed to be capable of diluting the extension efforts of the village extension agent. Indeed this has been one of the major problems associated with the failure of previous extension approaches in Nigeria. Under the T&V system, extension agents can only identify sources of inputs and link farmers up with the providers. The arrangement has not been quite effective because inputs are either too costly or not available to the farmers at the right time. Even subsidized inputs such as fertilizer fail to reach the farmers at the subsidized rates. This is one of the reasons for the policy inconsistency concerning agricultural subsidies in Nigeria. An impact study in Somalia (Chapman, 1987) reported similar findings.

Studies in Asia focus mainly on assessment of operations of smallholder farmers' support systems (Feder and Slade, 1983; Howell, 1982; Feder and Slade, 1984; Feder et al., 1985; Cernea et al., 1988; Staal, et al., 2008; McDermott, et al., 2010). While the results of some of these studies suggest high cost (potentially underfunded) and huge personnel build-up inherent in smallholder farmers' support systems, some point to the effectiveness of the system in terms of the regularity of visits to farmers by extension officers, the relevance of production recommendations and improved innovation adoption rates.

4 Case studies

4.1 Case 1 – Commercialization of traditional processed cassava products to maximize benefits and sustain rural livelihoods (*Fufu* Project)

This project was implemented between 1998 and 2001 in Nigeria and Ghana to develop 'best practice' tools and technologies for the value chain to commercialize a traditional processed cassava product (*fufu*) from production to consumption. It considered market demand and consumer preferences, the needs of rural processors, the private sector and market traders to develop an understanding of the changes that different approaches to commercialization are likely to have on cassava producers and processors. The project was led by the Natural Resources Institute of the University of Greenwich in collaboration with the University of Agriculture, Abeokuta, Nigeria; the Food Research Institute (FRI), Ghana; and Food and Flour Limited (FFGL), Ghana. The project developed cassava-based foods to meet the changing and growing urban demand, through the manufacture of products that are convenient, of high quality and safe. This was achieved through the development of private sector partnerships which were central to providing linkages between smallholder cassava producers and the urban market.

Activities included:

1 developing partnerships between researchers in Europe and Africa, the private sector and rural cassava processors;

2 working with small/medium-scale enterprises (SMEs) to bridge the gap between primary local processing and urban demand;
3 developing specific processed products to meet identified markets (such as dried fermented odourless *fufu*, instant unfermented *fufu* and High Quality Cassava Flour (HQCF));
4 developing, in partnership with a local fabricator, two dryers appropriate for producers in West Africa;
5 establishing private sector training centres for cassava processing for training local businesses, professionals and postgraduate students;
6 developing a food safety training manual for SME cassava processors;
7 developing improved processes for fermented cassava products that reduced the cost, increased throughput and reduced environmental pollution;
8 establishing the consumer acceptability of the new cassava products in major urban locations in West Africa so that products are developed to meet their needs; and
9 establishing the factors critical to success for small-scale cassava processors such as the scale of production, use of water and fuel, and economic/financial and social factors.

The institutional and interdisciplinary partnerships and collaborations fostered by the *Fufu* Project blossomed into one of the leading consortia for cassava research and development in Africa. In Nigeria and Ghana, the major products developed and pilot-tested (dried fermented odourless *fufu* (Nigeria), instant unfermented *fufu* (Ghana) and HQCF (Nigeria and Ghana)) have become an integral part of non-traditional value chains for new SMEs and for influencing the livelihoods of cassava farmers. The processing methods (particularly fermentation and drying) are now employed in the emerging cassava-based industries in Africa (Dipeolu et al., 2001). The description and results of this project can be found at: https://www.gov.uk/dfid-research-outputs/identification-of-an-approach-to-the-commercialisation–of-cassava-fufu-processing-in-west-africa-that-maximises-benefits-to-sustainable-rural-livelihoods-final-technical-report-project-r7495

4.2 Case 2 – Adding Value to Waste in the Cassava Processing–Goat Keeping Systems in Nigeria

This project was funded by the World Bank Development Marketplace (DM) in 2008 and comprised of five innovations:

- a simple technology: a drying platform for the cassava peels (to be used as a value-added product instead of being burnt as waste);
- a new product: clean dried cassava peels that can be sold as goat feed;
- a knowledge component: a diet designed by animal scientists and prescribed to goat farmers, that utilizes cassava peels and maximizes the growth rate and health of the goats;
- access to credit: facilitating micro-credit loans to build the drying platforms; and
- a new market mechanism: linking cassava processors and goat keepers.

The innovations provide a measurable economic benefit to both cassava processors and goat farmers. Original estimates suggest an annual increase of $384 USD and $198 USD for processors and goat farmers, respectively. An assessment of the project by

Fuller-Wimbush and Adebayo (2014) shows that the average quantity of cassava peels at processing centres in Osun and Ogun States was estimated to be 263 kg/day and 247 kg/day, respectively, showing a sufficient source of base material to utilize for animal feed. Level of education, marital status and system of goat keeping were shown to influence the perceived benefit from the use of fortified cassava peel for feeding goats. Farmers' age, level of education, marital status and farming system are the factors that influence their willingness to pay for cassava-peel-based products. The study found that the amount that farmers are willing to pay is $1 USD/25 kg of dried cassava peel (100%) and $7/25 kg of fortified cassava peel. The fortified cassava peel is made up of dried cassava peel (70%), dried cassava leaves (10%), cowpea haulm (17.5%), sulphur (1%), salt (0.5%) and bone meal (1%). These additional ingredients were appreciated by the farmers. The study shows an additional net income of $0.33 US/day for cassava processors selling only dried cassava peels and an additional $1.75 USD/day (net income) for fortified dried cassava peels. Further, recycling the cassava waste into a marketable product provides an environmental benefit by preventing the burning of cassava waste and the subsequent release of harmful toxins into the air. In its first two years (2009–11), the project exceeded the initial target of reaching 3600 cassava processors and 600 goat keepers, expanding its reach within project life to 21 additional locations and directly benefitting 6078 processors and 886 goat keepers. The expansion was a result of support from agricultural extension officers and the communities they represent. Project team leaders estimated that there is potential to reach 200 000 of the 350 000 farmers in the state of Ogun, in addition to thousands more in other states within the country. A record of this project is available at: http://documents.worldbank.org/curated/en/269631468333871352/Lessons-from-scaling-up-recycling-Cassava-waste-a-triple-win-for-development-a-case-study-from-Nigeria

The project aligns with the goals of the Federal Government of Nigeria, the World Bank and the Millennium Development Goals. It has the advantages of strong leadership and offers economic and environmental benefits, but still lacks a devoted institutional champion to advocate for the innovation and to help bring it to scale. This may be a research institute, university or government ministry with a clear strategy, budget, plans and commitment.

4.3 Case 3 – Cassava: Adding Value for Africa (C:AVA)

The C:AVA Project was initiated in 2008 with funding from the Bill & Melinda Gates Foundation to develop value chains for HQCF in Ghana, Tanzania, Uganda, Nigeria and Malawi. The goal is to improve the livelihoods and incomes of at least 90 000 smallholder households as direct beneficiaries, including that of women and disadvantaged groups. The project promoted the use of HQCF as a versatile raw material for which diverse markets exist. The project is led by the Natural Resources Institute of the University of Greenwich (NRI) working closely with the University of Agriculture, Abeokuta, Nigeria; the Food Research Institute, Ghana; the Tanzania Food and Nutrition Centre; the Africa Innovations Institute, Uganda; and Chancellor College, University of Malawi, along with a range of other partners.

The project focused on three key intervention points in the value chain: ensuring a consistent supply of raw materials; developing viable intermediaries acting as secondary processors or aggregating agents in the value chains; and driving market demand and building market share (in, for example, the bakery industry, and components of traditional foods or for use in plywood/paperboard).

Farmers and farmer/processors were supported in their production and primary processing activities through partnership with NGOs or public sector extension services. Business development and other specialists supported intermediaries to meet the requirements of end users, who were supported technically in adopting HQCF. C:AVA has met its original target of benefitting 90 000 smallholder farmers, earning over $33.3 million for them over the six years of the project. C:AVA has achieved this by providing objective research and making strategic technological improvements; linking cassava suppliers with markets; and encouraging efficient, sustainable investments from processors, end users and governments. Another significant result is the emergence of fabricators of more efficient processing equipment such as presses, graters and dryers. Even though there are widely divergent views on the level of success for HQCF as a substitute for wheat flour, there are a growing number of rural-based bakeries in Africa partially substituting wheat flour with HQCF. The project's 2014 annual report indicated that 22 160 tons of HQCF was used by end users in Nigeria, Ghana, Tanzania, Uganda and Malawi. A record of this project is available at: http://cava.nri.org.

A second phase of the project (CAVA II) began in 2014, led by the Federal University of Agriculture, Abeokuta, Nigeria. The overall goal of the CAVA II project is to facilitate systems where about 200 000 smallholder farmers would sell more than 2 million tons of fresh cassava roots to targeted value chains in Nigeria, Ghana, Tanzania, Uganda and Malawi between 2014 and 2019. CAVA II has been working across the value-added cassava chains, specifically focusing on a number of bottlenecks that farmers, processors, advisory services and other actors face in making these value chains work in favour of smallholders. The main interventions of the project include:

1 Enhancing smallholder productivity – by improving the profitability of cassava production, through the pull of market demand in specific value chains, increasing farmer yields with improved varieties, promotion of good agronomic practices and facilitation of systems where buyers of large quantities of cassava and cassava products purchase from smallholders.
2 Developing and expanding processing capacity – by promoting a range of drying technologies (sun, solar, bin and flash) that best suit smallholder supply and specific end-use models. In Uganda, Tanzania and Malawi, where artificial drying is new, CAVA II has multiplied the effects of sun drying by scaling out its use within each project country, while introducing artificial drying technologies (solar and flash). In West Africa, artificial drying is being promoted by improving the efficiency of both bin and flash dryers and decreasing the costs of producing processed cassava products using these technologies.
3 Expanding product options, market opportunities and geographies, by expanding the uptake of HQCF and other cassava products to new cassava growing and consumption areas in each country. The project is working with partners to explore and develop new market opportunities for fresh cassava roots by building and linking smallholder farmers to large markets in starch, ethanol, plywood, paperboard, milling and livestock (including aquaculture) feed sector. Efforts are also employed in raising consumer awareness and promotion for cassava-based products at local, national and international fora.

In its first two years of operation (2014 and 2015), CAVA II has facilitated systems where smallholder farmers sold 728 333 tons of fresh cassava roots to targeted value chains in Nigeria, Ghana, Tanzania, Uganda and Malawi. This is worth almost USD35 million in two years of project operation. Thirty-eight efficient flash driers are operational, mostly in Nigeria, but also in Malawi. In addition, 12 bin dryers are operational in Ghana, 71 rural entrepreneurs are engaged in cassava processing and 222 Community Processing Groups are profitably and sustainably producing HQCF. A record of this project is available at: http://cava2.unaab.edu.ng.

5 Summary

Research is essential for unearthing hidden knowledge and transforming such knowledge to practical use. But research needs to operate within a holistic environment that respects and taps into the potential of other subsystems such as extension services, the media, as well as market dynamics, and social and cultural factors to make meaningful impacts on enhanced and sustainable crop production. Improvements in productivity/yields, supporting more sustainable cultivation or improving smallholder security, are tangible objectives for which research evidence is needed, but they also require a thorough understanding of socio-economic, environmental, political and communication factors that influence the uptake of the outcome of research.

6 Future trends in research

Targeting smallholder farmers to adopt improved cassava technologies for yield and profitability will remain a desirable goal for R&D organizations and for governments. For a major crop that is currently grown mainly by resource-constrained smallholders, this goal will remain desirable in the foreseeable future. In this respect, four lines of research will be attractive. These are:

1 following a value chain approach in the development of the cassava research agenda;
2 exploring the influence of short- and long-term market dynamics on the decision making of smallholder cassava farmers;
3 categorizing smallholder farmers into socio-economic and/or ecological groups for more precise targeting for the development, promotion and adoption of improved cassava technologies;
4 addressing systemic issues such as gender inequities that influence technology adoption.

These research foci will provide the tools to direct and redirect scientific enquiries in ways that not only benefit farmers but make research outputs more desirable for adoption.

7 Where to look for further information

1 International Society for Tropical Root Crops (ISTRC) www.istrc.org
2 Global Cassava Partnership in the 21st century (GCP 21) https://ciat.cgiar.org/gcp21
3 Natural Resources Institute (NRI) of the University of Greenwich www.nri.org

4 Federal University of Agriculture Abeokuta (FUNAAB) www.unaab.edu.ng
5 Cassava: Adding Value for Africa Phase II (CAVA II) http://cava2.unaab.edu.ng
6 International Institute of Tropical Agriculture (IITA) www.iita.org/cassava
7 Food and Agriculture Organisation of the United Nations (FAO) www.fao.org/docrep

8 References

Adebayo, K. (1998). *Communication in Agriculture*. Abeokuta: Greenlinks International, Abeokuta 102p.

Adebayo, K. (2002). Simulation models of the adoption process in smallholder farming systems in Nigeria. PhD Thesis, University of Agriculture, Abeokuta.

Adebayo, K. (2005). Three decades of adoption and diffusion research in Nigeria: lessons and prospects. *Journal of Agricultural Extension* 8:65–75.

Adebayo, K. (2009). Dynamics of technology adoption in rural-based cassava processing enterprises in South-West Nigeria. *International Journal of Agricultural Economics and Rural Development* 2(1):15–25.

Adebayo, K., Fabusoro, E. and Fapojuwo, E. O. (2008). Factors determining selective adoption of the cassava grater and screw press in southwest Nigeria. *Journal of Agricultural Sciences, Science, Environment and Technology Series C* 3(1):40–52.

Adebayo, K. and Idowu, I. A. (2000). The aftermath of the withdrawal of the World Bank funding for the Ogun State Agricultural Development Programme (OGADEP) in Nigeria. *Journal of Sustainable Agriculture* 17(2/3):79–98.

Adebayo, K., Idowu, I. A., Omotayo, A. M., Apantaku, S. O. and Olunuga, B. A. (1999). An appraisal of the prospects and constraints to the sustainable funding of the Ogun State Agricultural Development Programme. *Journal of Agricultural Extension* 3:28–38.

Adebayo, K., White, J. L., Morris, M. J., Dipeolu, A. O., Ayinde, I. A., Wandschneider, T. S., Sanni, L. O., Oyewole, O. B., Tomlins, K. and Westby, A. (2003). Innovativeness and stakeholdership in the *fufu* processing systems in southwest Nigeria. *Journal of Agricultural Sciences, Science, Environment and Technology (ASSET) Series A* 3(4):15–27

Adebayo, K., Abayomi, L., Abass, A., Dziedzoave, N. T., Forsythe, L., Hillocks, R. J., Gensi, R., Gibson, R. W., Graffham, A. J., Ilona, P., Kleih, U. K., Lamboll, R. I., Mahende, G., Martin, A. M., Onumah, G. E., Orr, A. W., Posthumus, H., Sanni, L. O., Sandifolo, V. and Westby, A. (2010). Sustainable inclusion of smallholders in the emerging high quality cassava flour value chains in Africa: challenges for agricultural extension services. *Journal of Agricultural Extension* 14(1):1–10.

Agbamu, J. U. (1993). Categorisation of Prospective Adopters and the Implication for Planning Extension Communication Strategies. Paper presented at the Inaugural Annual Conference of the Agricultural Extension Society of Nigeria, ARMTI, Ilorin. 24–26 February 1993.

Agbamu, J. U. (1995). Analysis of farmers' characteristics in relation to adoption of soil management innovations in Ikorodu area of Nigeria. *Japanese Journal of Tropical Agriculture* 39(4):213–22.

Akerman, G. (1957). The cobweb theorem: a reconsideration. *The Quarterly Journal of Economics* 71(1):151–60.

Akinola, A. A. (1986a). Farmers' participation in the National Accelerated Food Production Programme in Nigeria. *Agricultural Systems* 20(4):301–12.

Akinola, A. A. (1986b). Determinants of diffusion of cocoa spraying chemicals among Nigerian cocoa farmers. *Agricultural Systems* 22(3):191–202.

Anaglo, J. N., Westby, A., Morton, J. F., Atengdem, P. B. and Adebayo, K. (2009). Assessment of farmers' decision to add value to cassava roots in the Greater Accra Region, Ghana. *Journal of Sustainable Development* 6(1).

Bennett, B., Kleih, U., Phillips, D., Naziri, D., Mahende, G., Towo, E., Kirya, M. and Jagwe, J. (2012). Linking breeding with market demand in value chain for cassava and processed cassava products findings from Uganda and Tanzania. Presented at the Global Cassava Partnership in the 21st Century (GCP21) II, Kampala, Uganda. 18–22 June 2012.

Benor, D., Harrison, J. Q. and Baxter, M. (1984). *Agricultural Extension: The Training and Visit System.* The World Bank, Washington, DC, p. 284.

Bernet, T., Ortiz, O., Estrada, R. D., Quiroz, R. and Swinton, S. M. (2001). Tailoring agricultural extension to different production contexts: a user-friendly farm-household model to improve decision making for participatory research. *Agricultural Systems* 69(3):183–98.

Cernea, M., Coulter, J. and Russel, F.(1988). *Agricultural Extension by Training and Visit: The Asian Experience.* The Word Bank, Washington, DC, p. 48.

Chapman, N. P. (1987). The Evolution of the T & V in Somalia. In Howell, J. (Ed.), *The Training and Visit System of Extension in Practice. Agricultural Administration Unit Occasional Paper 8,* London ODA.

Chikwendu, D. O., Chinaka, C. C. and Omotayo, A. M. (1995). Adoption of minisett technique of seed yam production by farmers in Eastern forest zone of Nigeria. *Discovery and Innovation* 7(4):367–75.

Collinson, M. (2001). Institutional and professional obstacles to a more effective research process for smallholder agriculture. *Agricultural Systems* 69(1–2):27–36.

Dipeolu, A. O., Adebayo, K., Ayinde, I. A., Oyewole, O. B., Sanni, L. O., Pearce, D. M., Wandschneider, T. S., White, J. L. and Westby, A. (2001). *Fufu Marketing Systems in Southwest Nigeria.* NRI Report Number R2626. Natural Resources Institute (NRI), United Kingdom and University of Agriculture Abeokuta, Nigeria, p. 55.

Dipeolu, O. A., Adebayo, K. and Ogundokun, M. O. (2002). Consumer behaviour towards Non-grain staple foods in a dynamic price and income regime in Abeokuta. Nigeria *Journal of Agriculture and Social Research* 2(1):52–66.

Feder, G. and Slade, R. H. (1983). Experiences with the monitoring and evaluation for training and visit extension in India. *World Bank Staff Working Paper No 595,* The World Bank, Washington, DC.

Feder, G. and Slade, R. H. (1984). Contact farmer selection and extension visits: the T & V in naryara India. *Quarterly Journal of International Agriculture* 23(1): 6–21.

Feder, G., Slade, R. H. and Sundaram, A. K. (1985). The training and visit system: an analysis of operations and effects. *World Bank Staff Working Paper 719.* The World Bank, Washington, DC.

Franzel, S., Arimi, H., Murithi, F. and Karanja, J. (1996). Boosting milk production and income for farm families: the adoption of Calliandra calothyrsus as a fodder tree in Embu District, Kenya. *East African Agricultural and Forestry Journal* 61(2):235–51.

Franzel, S., Coe, R., Cooper, P., Place, F. and Scherr, S. J. (2001). Assessing the adoption potential of agroforestry practices in sub-Saharan Africa. *Agricultural Systems* 69(1–2): 37–62.

Franzel, S., Ndufa, J. K., Obonyo, C. O., Bekele, T. and Coe, R. (2000). Farmer-designed agroforestry tree trials: farmers' experiences in western Kenya. AFRENA Report No. 131. International Centre for Research on Agroforestry, Nairobi.

Fuller-Wimbush, D. and Adebayo, K. (2014). Lessons of endogenous leadership in Nigeria: innovating to reduce waste and raise incomes in the cassava processing and goat keeping systems. *Development in Practice* 24(5 and 6):693–8.

Hartmann, P. (2004). *An Approach to Hunger and Poverty Reduction in Sub-Saharan Africa.* International Institute of Tropical Agriculture, Ibadan.

Hays, H. M. and Raheja, A. K. (1977). Economics of sole crop cowpea production in Nigeria at the farmer's level using improved practices. *Experimental Agriculture* 13(2):149–54.

Howell, J. (1982). Managing agricultural extension: The T. & V system in practice. *Agricultural Administration* 10:45–60.

Hystra (2015). Smallholder Farmers and Business: 15 pioneering collaborations for improved productivity and sustainability. A Report by Hybrid Strategies Consulting, July 2015, p. 176.

Ibitoye, S. J. (2011). Indigenous technology and agricultural production: The case of poultry incubator. *International Journal of Poultry Science* 10(6):493–5.

Kidd, D. W. (1968). Factors affecting farmers' response to extension in Western Nigeria. *Consortium for the Study of Rural Development (CSNRD)* 30:179.

Labys, W. C. (2006). *Modelling and Forecasting Primary Commodity Prices*. Ashgate Publishing Limited, Aldershot, p. 243.

Ladebo, O. J. (1999). Determinants of adoption of new technologies among rice farmers in Ifo LGA of Ogun State, Nigeria. *Acta Universitatis Agriculturae et Silviculturae Mendelianae Brunensis (Brno)* XLVII(3):83–7.

McDermott, J. J., Staal, S. J., Freeman, H. A., Herrero, M. and Van de Steeg, J. A. (2010). Sustaining intensification of smallholder livestock systems in the tropics. *Livestock Science*. Doi: 10.1016. livsi.2010.02.014.

Norman, D. W., Worman, F. W., Siebert, J. D. and Modiakgotla, E. (1995). The Farming Systems Approach to Development and Appropriate Technology Generation. FAO Farm Systems Management Series 10. AGSP. Food and Agricultural Organization of the United Nations, Rome, Italy.

Nweke, F. I. (2004). New Challenges in the Cassava Transformation in Nigeria and Ghana. EPTD Discussion Paper: International Food Policy Research Institute, Washington, DC, p. 103.

Odedina, S. A. and Adebayo, K. (2012). *Commercial Production of Cassava in the Humid Tropics*. Livelihoods Support and Development Centre, Abeokuta, p. 138.

Okwoche, V. A., Voh, J. P. and Ogunwale, S. A. (1998). Socio-economic characteristics influencing adoption behaviour of women cooperators and non – cooperators in Oju LGA of Benue State. *Journal of Agricultural Extension* 2:31–8.

Omotayo, A. M., Chikwendu, O. D. and Adebayo, K. (2001). Two decades of World Bank assisted extension services in Nigeria: lessons and challenges for the future. *Journal of Agricultural Education and Extension* 7(3):143–52.

Onu, D. O. (1991). Factors associated with small-scale farmers' adoption of improved soil conservation technologies under intensified agriculture in Imo State, Nigeria, PhD Dissertation, University of Nigeria, Nsukka.

Ortiz, R. and Hartmann, P. (2003). Beyond Crop Technology: The challenge of African rural development. In Vol. 2 Reference Material of the Sub-Saharan Africa Challenge Program 'Building Livelihoods through Integrated Agricultural Research for Development – Securing the Future for Africa's Children' Forum for Agricultural Research in Africa (FARA), Accra, Ghana, pp. 39–72.

Osuntade, O. B., Adebayo, K., Iposu, S. O. and Banmeke, T. O. A. (2012). Lessons from the management of microcredit for supporting cassava technology adoption in ogun state. *Nigerian Journal of Farm Management* 13(2): 51–7.

Patel, A. U. and Anthonio, Q. B. O. (1971). An analysis of selected factors that influence the adoption of improved practices among tobacco farmers in the Western State. *The Nigerian Agricultural Journal* 8(2):123–31.

Sall, S., Norman, D. and Featherstone, A. M. (2000). Quantitative assessment of improved rice variety adoption: the farmer's perspective. *Agricultural Systems* 66(2):129–44.

Smith, J., Woodworth, J. B. and Dashiell, K. E. (1995). Government policy and farm level technologies: the expansion of soybean in Nigeria. *IITA Research* 11:14–18

Staal, S. J., Pratt, A. N. and Jabbar, Md. (2008). Dairy Development for the resource Poor: A Comparison of Dairy Policies and Developments in South Asia and East Africa. FAO/ILRI. 44p.

Zandstra, H. G., Price, E. C., Litsinger, J. A. and Morris, R. A. (1981). *A Methodology for On-Farm Cropping Systems Research*. International Rice Research Institute, Los Banos, Philippines.

Zaria, M. B. and Omotayo, A. (1997). Improving Every Farmer's Access to Extension Services in Sub-Saharan Africa: Approaches and Challenges. In Samanta, R. K. and Arora, S. K. (eds), *Management of Agricultural Extension in Global Perspectives*. B. R. Publishing Co. Delta, pp. 133–7.

Chapter 10

GCP21: a global cassava partnership for the 21st century

Claude M. Fauquet and Joe Tohme, International Center for Tropical Agriculture (CIAT), Colombia

1 Introduction
2 GCP21 as a platform for the science of cassava production
3 Making connections in cassava research
4 Looking for gaps in R&D from basic science to final products
5 Improving global access to information and technology
6 Conclusion
7 Where to look for further information
8 References

1 Introduction

Cassava (*Manihot esculenta* Crantz), originally from South America, where it was domesticated several thousands of years ago, is believed to have been transported throughout the world, since the sixteenth century, by the Portuguese. Africa adopted cassava primarily for human food production. And, although its spread through the continents is fairly recent, by 1960 it was already the most important food crop (in fresh root tonnage) in sub-Saharan Africa. Today, more than 55%, or 146 million tons (mt), of the world cassava production is from Africa (FAO 2016). At the same time cassava became very important in Asia as well, but mostly for industrial purposes (e.g. feed, starch, sweeteners and alcohol). Moreover, since 1998, cassava is the food crop that has outperformed, in relative production, all other major crops in the world. If this trend continues, by 2050 cassava will be the world's third most important food crop after maize and rice, with about a billion tons produced annually.

There are several reasons why cassava has been so successful. First, it is semi-perennial – that is, its roots can stay in the ground more than two years, thereby offering much flexibility in its storage and harvesting time. Second, cassava can grow on a variety of soils and in a range of agroclimatic conditions. Third, cassava was once seen as growing 'from Timbuktu (200 mm of rainfall per year) to the slopes of Mt. Cameroon' (with more than 6000 mm of rainfall per year). And although the harvest can vary, cassava's huge adaptability was key for its wide adoption. Finally, there is now another reason for cassava

http://dx.doi.org/10.19103/AS.2016.0014.12

to be even more successful: its resilience to global climatic changes (Rosenthal et al., 2012). The plant can grow under very high temperatures, is already drought resistant, and perhaps will respond better than most crops to elevated levels of carbon dioxide in the coming decades. For all these reasons, cassava will obviously play a very important role in feeding the world by 2050, especially in Africa, whose booming population is estimated to exceed 2.4 billion by that year (United Nations 2015).

Although the maximum potential yield of cassava is unknown, there have been records of 65–100 Mt/ha reported in several places around the world. The world average yield in 2013 was 13 Mt/ha, but it was only about 10 Mt/ha in Africa. In the last 25 years some countries in South East Asia have invested heavily in cassava, where yields have increased significantly: India is now above 40 Mt/ha, Thailand at 22 Mt/ha and Vietnam above 17 Mt/ha. Clearly, cassava productivity in all of Asia is progressing quickly (FAO 2016).

Yet, cassava has always been considered a 'poor man's crop'; consequently, it has been neglected as compared with many other crops such as cereals, legumes and fruits. Investment in science and technology – even industrial development – has been well below what is needed to meet cassava's huge potential. As a result, its yields are quite low. This is especially true in Africa, where yields have stayed basically at the same level since 1961 (De Sousa et al. 2016).

Cassava research is traditionally carried out by the public sector (Consultative Council for International Agricultural Reseacrch (CGIAR) international centres, national agricultural research centres, advanced labs in developed countries and universities). On the other hand, the private sector has used cassava as a commodity but without investing in its research. For all the reasons mentioned above, there has been a real need to gather the cassava community together and shape a unique platform for exchanging information, identifying gaps in research and development (R&D), and changing the mindset about this highly important crop. This is the mandate of the Global Cassava Partnership for the twenty-first century (GCP21), created in 2003 (Fauquet et al. 2012).

2 GCP21 as a platform for the science of cassava production

A crop cannot be developed without also engaging with the science and technology in its development. Because the investment in cassava in the past 50 years has been very limited, and knowledge and technological tools are equally so.

GCP21 is a non-profit organization, currently hosted by the International Center for Tropical Agriculture (CIAT), Colombia. The mission is threefold: advocating for cassava R&D; indeontifying gaps in value chains from the field to the plate; and providing all types of information about the crop. The general mandate is to trigger activities and promote new ideas and projects but not to implement these projects. It is governed by a secretariat assisting the director of GCP21 and advised by a Technical Advisory Committee of world cassava experts. The funding comes from donations from international donors. It is the only global platform for cassava existing in the world.

It was in 2009 that GCP21 managed, with a coalition of US scientists, to obtain the first draft of the cassava genome (Prochnik et al. 2012). Although the genetic map was far from perfect, it unlocked cassava genomics, and several other genome-based projects were subsequently developed in the United States and in Asia. Perhaps the best-funded one is the so-called NextGen project (http://www.nextgencassava.org/), which aims to use genome-wide selection for breeding multiple traits simultaneously in cassava.

In 2010, GCP21 organized the first strategic meeting in Bellagio, Italy, in which a group of experts reviewed all the major issues with the crop and identified a number of required actions to remedy to some of these issues (see http://gcp21.org/strategic1.html). This roadmap (http://gcp21.org/Bellagio/FinalStatement.pdf) has proved useful in the past years, and can be used in the future as guidelines and a canvas for all individuals and organizations searching for research projects and funding possibilities on cassava. Some of the most important problems were detailed in a special issue of *Tropical Plant Biology* in 2012 (see http://rd.springer.com/ journal/12042/5/1/page/1).

Despite these activities, there are many other areas open for consideration for research:

- all para-genomic studies;
- in-depth genetic description of the cassava genome;
- production of double haploids and inbred lines;
- understanding of the role of heterosis in yield, quality and resistance;
- development of detailed and automated phenotyping studies;
- molecular understanding of the storage root starch accumulation process;
- detailed studies and diagnosis of the many pests and diseases affecting cassava;
- better knowledge of the postharvest deterioration syndrome;
- deeper understanding of the resilience of cassava to global climatic change.

In parallel, there should be the development of a range of technologies related to all these scientific ideas to improve breed, cultivate, preserve, protect and process the starchy roots to produce more and better food for the world. All these topics should be addressed through workshops, publications and projects; GCP21 aims to tackle some of these important issues.

3 Making connections in cassava research

3.1 Supporting collaboration among the cassava community

The global cassava community is very small and spreads out across more than 106 countries through many hundreds of organizations, universities, centres, administrations, private companies and non-governmental organizations. It is, therefore, absolutely necessary to strengthen relations between all these members to enhance the circulation of knowledge and thereby speed up the whole R&D cycle. It is well established that groups of scientists the world over are more productive and more effective when they coordinate their activities. In keeping with this, GCP21 has organized a series of world conferences: Brussels, Belgium (2008) (http://gcp21.org/ belgium.html); Kampala, Uganda (2012) (http://gcp21.org/kampala.html); and Nanning, China (2016) (http://gcp21.org/wcrtc.html). These conferences cover all topics, from genomics to production to market products. This large scope of interest is quite challenging; it is necessary to strengthen the community and to make the exchange of information easier. As these tri-annual conferences are more and more successful, they may need to be held more often so as to keep up with participant demand for interaction and collaborative opportunities.

Bridges and pipelines need to be built between the advanced world and the developing world. Many technologies are being developed on model plants or crops used in temperate regions that could benefit cassava. Africa will not anytime soon develop the R&D capacity that exists in North America and Europe, so it is imperative to establish North–South and South–South collaborations on a variety of topics that could be fruitful for cassava. Technologies can reduce the time it takes to transfer knowledge from one system to another. The Basic Research to Enable Agricultural Development programme (BREAD), a collaboration between the US National Science Foundation and the Bill & Melinda Gates Foundation, is an excellent example of what is required to open new doors and new avenues. GCP21, through all the meetings being organized, will facilitate this sort of collaboration and promote its development.

Equally critical is the exchange of information and knowledge between the different continents. Brazil and South East Asia, for example, are much more advanced with respect to mechanization and processing, and all the components and issues associated with these technologies. Over the past decades, Brazil has developed many types of processed food based on cassava that would be very useful in Africa with its burgeoning urbanization. Indonesia is very good at mastering the fermentation of cassava products in many ways that are not used in the rest of the world.

Finally, R&D for cassava is largely funded by a myriad of public agencies and donors (country aid services, international organizations and foundations). Each has its own agenda and develops its own projects, sometimes without consulting others and rarely by consolidating efforts. Although it is difficult to coordinate all of these initiatives, it is possible to inform them about cassava R&D needs, about new ideas and to the development of possible common projects. GCP21 invites representatives from different donors to each strategic meeting that it organizes, to make sure they have access to the information and are part of both the discussions and the elaboration of roadmaps for needs and new R&D projects. We believe that this systematic approach can improve decision-making and coordination among donors.

3.2 Promoting a 'whole system' approach

Too often, scientists and developers and donors work in isolation; this can lead to mistakes, failures or simply to a lack of impact. GCP21 is developing and promoting the practice of thinking more in systems or along value chains – that is, to look into whole systems, what is not done, what is not working or what is simply unknown. This thinking is the basis for the 'GCP21 strategic meetings', whereby a small group of experts in a chosen topic will deliberate and discuss on that subject in various dimensions and perspectives, to identify the obstacles to be removed in order to get the whole system working better and to be more fruitful.

GCP21 provides non-partisan leadership for cassava R&D, independent from national, international and private organizations. All too frequently, organizations are defending their own positions to secure funding and to give credit to their scientists. In doing so, however, they may bypass important issues or repeat failures for decades.

The very best example could be the 'War on Cassava Viruses in Africa'. Viruses have been known to exist in Africa since 1894, and more than a century later they are still the continent's most important problem for the crop. And, with the accelerated rate of international trade, the risk of spreading diseases in other continents is increasing. In 2013, a group of experts under the auspices of GCP21 met at the Rockefeller Center in Bellagio,

Italy, and developed a comprehensive roadmap to better control and contain these viral diseases (Legg et al. 2014). A subsequent meeting organized by the French scientists of CIRAD-La Réunion (http://gcp21.org/strategic5.html) analysed the feasibility of setting up a Pan-African Cassava Surveillance Network (PACSUN). They offered to use their centre from La Réunion and to establish an international cassava transit site to allow the exchange of material between Africa and the rest of the world. The activities defined by these roadmaps are now up for funding consideration at the International Fund for Agricultural Development and European agencies.

A more recent example is the Gari Revolution strategic meeting, organized by GCP21 and the CGIAR Research Program on Roots, Tubers and Bananas (RTB), at the International Institute for Tropical Agriculture, 4–6 October 2016 (http://gcp21.org/strategic6.html). Gari is the most important food produced from cassava in Africa – probably in the world; however, it is produced in conditions that are often unsanitary, ecologically unsound, and very inefficient: a minimum of 5 tons of fresh roots are required to produce 1 ton of gari, and many women and children are working in unhealthy conditions for a miserable pay. Firewood which is the only source of energy used to roast gari is devastating the remaining forests and causing poor health conditions for the women roasting gari; all the solid and liquid waste are polluting the workplace and the environment; and the poor food safety conditions to produce gari are putting millions of consumers at risk constantly, especially with ongoing, rapid urbanization. The experts at the strategic meeting dissected the whole gari process – from the roots to the plate – and came up with an array of recommended actions to improve gari processing at all steps. These recommendations will be turned into both a roadmap (to be published) and several concept notes offered to donors for funding opportunities.

4 Looking for gaps in R&D from basic science to final products

An important element of the GCP21 strategy is to look for gaps in the value chains, from the scientific process to products. If a link in a chain is weak or broken, the whole chain does not work or is inefficient. Good examples of these gaps are the targeted breeding for consumer preference traits, the increase of the dry matter content (DMC) of improved varieties and the use of cassava waste.

For decades, cassava breeders have bred for high yield and for resistance to diseases and have achieved many successes. It is, however, increasingly apparent that it is often very difficult to find in the collection of improved varieties any individual one that will fully satisfy all desired production-related characteristics as well as the particular processing and consumer traits. In the recent GCP21 Gari Revolution workshop, the link between breeding and gari quality was established. A list of priority actions aimed at developing useful quality traits and molecular markers for these traits for gari has been elaborated. This will require first a change in mindset of the scientists involved, and then implementation to breed new varieties specifically designed for the gari market. It will take time, but beginning this process is essential to significantly improve the whole gari processing chain.

DMC is recognized more and more as a key element in the overall improvement of cassava cultivation and use. Although the industry has been using this criterion to pay farmers delivering roots at the factories for decades, it has not been used for other human food or animal feed businesses. The efficacy of gari or fufu processing, and of any cassava-derived product, is directly correlated with the DMC of the fresh roots used. In other words, we could produce much more food in Africa with the same amount of cassava production, if the varieties cultivated for this use had a higher DMC. DMC in the harvested lots for the same variety can vary widely. The factors influencing the differential stability for DMC are not well known. For example, the DMC of most varieties falls dramatically at the beginning of the rainy season, but some clones are able to maintain significantly higher and more stable levels, like 32±3% (H. Ceballos, pers. comm.). As a result, scientific activities should focus on the improvement of this trait, and breeders and agronomists must work to change their mindset about the importance of DMC as a selection criterion. GCP21 will organize strategic workshops to this effect, to impact a variety of people involved in the cassava value chains (e.g. scientists, policymakers, producers, processors, traders and consumers).

Another example of identifying gaps in the food chain is the work done on the use of cassava by-products and waste. Processing cassava for food in Africa generates huge amounts of solid and liquid waste. Main wastes and by-products are the peels, the high-fibre, low-starch residues from starch extraction, and the wastewater from various processes, but especially from starch extraction. An evaluation study done in 2014 concluded that at least 14 million Mt/y of waste was generated in Nigeria alone (http://gcp21.org/Ibadan/FauquetC_Estimated_cassava_waste_in_Nigeria_141127.pdf). GCP21 organized a workshop dedicated to this topic, and concluded that cassava waste could be used to feed all sorts of animals (http://gcp21.org/strategic4.html). Subsequently, a project was developed by RTB using solid waste from gari processing centres as a partial substitute for corn in several types of feed. Animals were fed with this feed which was found to be equally nutritious and suitable at least for small ruminants, pigs and chicken; fish are now being tested (Okike et al. 2015) (https://www.youtube.com/watch?v=JwfgyHKnkLE&feature=youtu.be). Several other projects led by the private sector in Nigeria have been initiated and may reach commercial level in the near future. If this new business was effectively adopted by the feed milling industry, it would be a completely new business from cassava in Africa. It would also improve the life of the women involved in the gari processing business, and help to clean up the environment. Furthermore, millions of tons of maize per year could be used as food for humans instead of as animal feed. Another avenue for the use of liquid waste could be the development of the so-called biogas technology. In Paraná, Brazil, this technology has been in use for many years and works fantastically and economically. There is no reason that this technology could not be applied in other parts of the world for cassava processing factories? GCP21 will work with Brazilian companies to spread the information about this biogas technology.

Along these lines, GCP21 is investigating the possibility of using bio-fermentation of cassava waste to produce high-quality fish feed. This is a new technology being developed by BioTork, a start-up US company, based on the selection of microbes to feed on waste products. Preliminary tests with cassava peels have been done using algae. The tests demonstrate that it would be possible to produce a high-quality fish feed, high in protein and very rich in omega-3 fatty acids. Calculations have been done that show the potential of the application of such a technology relative to both the feed market and production cost. The development of this new technology based on cassava waste would, however, need capital investment to develop a pilot project leading to commercialization. GCP21 has been actively investigating that possibility and looking for partners to develop it.

5 Improving global access to information and technology

5.1 Importing and exporting technologies

Cassava is destined for various purposes and used in many different ways around the world. And although cassava is used as food for humans in much of Africa, in Asia it is almost exclusively used for the production of industrial products, with the exception of Indonesia. In South America, the uses are highly diverse – for human food and animal feed and also for industry. On each of these continents technologies have been developed over time to mechanize cassava production and to produce different types of food from cassava. By 2050, to feed the African continent, cassava productivity will have to increase very significantly, from about 10 Mt/ha to at least 25 Mt/ha. In addition, Africa is rapidly urbanizing: by 2050, in a number of African countries, more than 50% of their population will be concentrated in cities, with some countries reaching 75% urbanization. This huge change in African society will call for changes in the cassava food chains. Farmers will have to be more productive to respond to the growing demand in food, and to compensate for the loss in labour in the countryside. Cassava will have to be transported longer distances, and women in cities will not have the time or the physical environment to produce traditional food as it is done in villages today. All of these changes call for research on new cassava production systems, new cassava processes and new types of cassava food. GCP21 will be proactive at transferring some of these technologies from South America and Asia to Africa, through workshops and demonstrations wherever possible. GCP21 will, in particular, investigate the possibility of transferring food technologies from Brazil and Indonesia, where alternative food technologies have been well developed.

5.2 Supporting the flow and dissemination of information

In many crop development examples, it is the private sector that is organizing the whole chain – from production in the field to consumption on the plate. Private companies can align a long-term plan involving R&D in a continuous manner. But cassava value chains essentially reside in the public sector or are being used by small, medium and large industries all over the world. There is no unique system to establish very close linkages between the scientists involved in the improvement of cassava productivity and those who use cassava roots. The industry producing starch from cassava does not invest in research, nor are the more informal networks of food processors well connected to research scientists. Therefore, the only way to improve the flow of information back and forth is to link all of these different actors in a unique network, and to accelerate the circulation of information to speed up changes and increase impact. This is one of the important roles for GCP21.

Work started with the organization of triennial conferences based on cassava, inviting a range of actors from scientists to private sector representatives. It is also a matter of getting these groups to talk together, which is quite difficult because of the wide gaps in background and knowledge. GCP21 has now organized three of these international conferences in Belgium, Uganda and China (http://gcp21.org/conferences.html), and is aiming at a fourth conference in 2018. The work continued with the organization of small strategic workshops on specific topics of interest in the cassava value chains (see above) (http://gcp21.org/meetings.html). Finally, an important element is the development of a GCP21 website, where all sorts of users can access useful information to improve knowledge and contact the best people with whom to collaborate.

The GCP21 website (http://gcp21.org/index.html), currently under construction, will progressively become a unique cassava portal to access all sorts of information about cassava:

- scientific and technical information, technical databases, and statistical information about cassava production and cassava products;
- conferences, meetings, and workshops information around the world;
- presentations and posters of all past and present GCP21 conferences, workshops, and meetings;
- contact information for cassava organizations, associations, projects, companies, and donor contacts for cassava R&D;
- PowerPoint presentations, pictures, and videos about cassava and cassava products and publications.

The idea is not to have all information residing on the GCP21 website, but to provide links to access all information available in the world to speed up the flow and dissemination of information. GCP21 will promote individual interaction and participation in this communication effort, as well as to limit the cost of maintenance and updates through the use of up-to-date information and communications technologies such as specialized websites, newsletters, WhatsApp, Twitter and Facebook.

6 Conclusion

GCP21 is the first and only global cassava platform. Created in 2003, it is now coming to the fore. Because cassava will feed at least 2 billion people by 2050, it is critical that production of the crop is improved and its uses expanded worldwide. Despite the investment in cassava R&D having been very low for the past 50 years, the potential of cassava is several times that of current global yields. But many limitations to achieving this potential persist. And, although cassava is resilient to global climatic changes, it is prone to pests and diseases that will also increase in the future. A significant investment will therefore be necessary to increase yields from 13 Mt/ha to 25 Mt/ha in order to feed the booming population, particularly in Africa. To achieve this goal, the cassava community has to strengthen both in numbers and in capacity. It has to exchange far more information in order to accelerate investment, produce more and feed people better. GCP21 is an important force for reaching these goals.

7 Where to look for further information

All information about GCP21 is being loaded on the GCP21 website (http://www.gcp21.org/index.html) and the reader should go through the different pages. Of particular interest is the new series of 'Cassava Stories' launched in 2017 (http://www.gcp21.org/stories.html) and the section on 'Cassava Discover' (http://www.gcp21.org/discover.html) with sites, books, infography and statistics about the crop. Finally, all the strategic meetings and conferences organized by GCP21 are posted on 'Cassava Events' (http://www.gcp21.org/events.html) with all the posters and presentations downloadable since 2008.

8 References

De Souza, A. P., Massenburg, L. N., Jaiswal, D., Cheng, S., Shekar, R. and Long, S. P. 2016. Rooting for cassava: insights into photosynthesis and associated physiology as a route to improve yield potential. *New Phytologist* doi:10.1111/nph.14250.

FAO – Food and Agriculture Organisation of the United Nations. 2016. Statistics of production. Available from http://faostat3.fao.org/home/E

Fauquet, C. M., Taylor, N. J. and Tohme, J. 2012. The Global Cassava Partnership for the 21st Century (GCP21). *Tropical Plant Biology* 5: 4. doi:10.1007/s12042-012-9100-2.

Legg, J., Somado, E. A., Barker, I., Beach, L., Ceballos, H., Cuellar, W., Elkhoury, W., Gerling, D., Helsen, J., Hershey, C., et al. 2014. A global alliance declaring war on cassava viruses in Africa. *Food Security* 6: 231–48. doi:10.1007/s12571-014-0340-x.

Okike, I., Samireddypalle, A., Kaptoge, L., Fauquet, C., Atehnkeng, J., Bandyopadhyay, R., Kulakow, P., Duncan, A., Alabi, T. and Blummel, M. 2015. Technical innovations for small-scale producers and households to process wet cassava peels into high quality animal feed ingredients and aflasafe™ substrate. Practical Action Publishing: http://www.developmentbookshelf.com/. Available from http://dx.doi.org/10.3362/2046-1887.2015.005

Prochnik, S., Marri, P. R., Desany, B., Rabinowicz, P. D., Kodira, C., Mohiuddin, M., et al. 2012. The Cassava Genome: Current Progress, Future Directions. *Tropical Plant Biology* 5: 88. doi:10.1007/s12042-011-9088-z.

Rosenthal, D. M., Slattery, R. A., Miller, R. E., Grennan, A. K., Cavagnaro, T. R., Fauquet, C. M., Gleadow, R. M. and Ort, D. R. 2012. Cassava about-FACE: greater than expected yield stimulation of cassava (*Manihot esculenta*) by future CO2levels. *Global Change Biology* 18: 2661–75.

United Nations. 2015. World population prospects: the 2015 revision, key findings and advance tables. Working Paper No. ESA/P/WP. 241. New York, NY: United Nations.

Part 3

Sustainable intensification of cassava cultivation

Chapter 11

Integrated crop management for cassava cultivation in Asia

Tin Maung Aye, CIAT, Vietnam; and Reinhardt Howeler, CIAT Emeritus - formerly CIAT, Thailand

1 Introduction
2 Implementing GAP in cassava pre-harvesting
3 Implementing GAP for cassava crop health
4 Implementing GAP in cassava post-harvesting
5 Adoption of GAP by farmers
6 Conclusion
7 Future trends
8 References

1 Introduction

Cassava is grown in tropical and subtropical areas of Africa, Asia and Latin America. In Africa and much of Latin America cassava is used mainly for direct human consumption, after boiling or processing. In Asia cassava has now become more of an industrial crop that is processed into a wide array of products, such as starch and many starch-derived products, modified starch, animal feed and bio-ethanol. Moreover, due to very high population pressure in Asia, the crop is grown mostly by smallholder farmers on very small plots of land and is often grown in the same field year after year. As a consequence, cassava production increases have to be achieved mainly by increasing yields rather than through area expansion. Obtaining high yields requires very careful crop and soil management to prevent soil degradation caused by nutrient depletion and erosion.

The demand for cassava in Asia is expected to grow in the years to come largely because of an increasing demand for animal feed, biofuels and multiple uses of starch. The livestock and poultry feed industry is growing rapidly as more animal protein is incorporated into the Asian diet. The rapid expansion of the biofuel industry in recent years is driven by the expectation of long-term declining reserves of fossil fuels, but perhaps more significantly by new policy developments aimed in part at reducing the carbon footprint of fuels, but also because of the power of farm lobbies in major producing countries to advocate for

http://dx.doi.org/10.19103/AS.2016.0014.23

new markets for grains, especially maize. Starch is a commodity of continuing growth, and multiple crop sources can provide the raw material for most uses. Maize starch leads all sources, followed by that of cassava, with the latter having the greatest annual rate of increase. While there are certain to be significant year-to-year fluctuations, the cassava industry will see continuous growth, especially with the development of value-added cassava traits and products in Asia (Hershey, 2015). Increasing the quantity and quality of cassava products in response to growing demand will require careful and sustainable increases in cassava production. Expansion into newly cleared land is not a sustainable option, given the declining available land and negative environmental impacts. Science must focus on sustainable increased productivity per unit area to keep prices competitive in the marketplace, improve farmer income and protect the environment.

Cassava is better adapted to less favourable environments than most other major crops, especially in low-fertility and drought-prone areas. In Southeast Asia, upland soils are less fertile compared to the lowlands, and they are usually located in areas that are not supported by well-developed infrastructure, such as roads, markets, electricity, water supply and agricultural supply stores (Dierolf et al., 2001). However, most smallholder farmers and their families live and grow food crops on these marginal upland soils and in fragile environments, where manure and fertilizers are not commonly applied or are applied in insufficient amounts and at inappropriate application times. The crop is often considered the 'food of last resort' (Howeler and Aye, 2014).

Cassava extracts soil nutrients in proportion to the production that is removed from the field. In the absence of proper crop and soil management, cassava cultivation may lead to soil erosion and land degradation. Its ability to grow on already-depleted soils is one of its great attributes, but effective soil management and other sound crop management practices must also be in place to take advantage of the crop's efficiency. In addition, adverse environmental consequences of rapid commercialization of cassava are also apparent in fragile upland farming systems. The increase in cassava cultivation seen in Southeast Asia, particularly during the last decade in Cambodia, Laos and Myanmar, may have been in part a response to declining soil fertility and increasing soil erosion from other cropping systems that could not be sustained. This dynamic has had the unfortunate consequence that cassava is widely seen as a crop that causes soil depletion and erosion, even though these effects are often caused by other crop systems.

The fact that cassava is well adapted to adverse conditions is fully compatible with its cultivation in a sustainable manner, through the adoption of good agricultural practices (GAP). Under ideal conditions, maximum root yield of cassava can potentially be about 30 t/ha of dry roots, equivalent to 75–100 t/ha of fresh roots (Cock et al., 1979). However, the average cassava yields in Asia and throughout the world are much lower. The difference between average actual yields of farmers and potential yields – the so-called 'yield gap' – is still high for cassava in Asia. Some 20 years ago the economically achievable cassava yield in Asia was estimated at about 25 t/ha (Henry and Gottret, 1996), and this yield is already being achieved in a few countries. However, a recent review of the cassava research literature indicates that if all current constraints could be eliminated, the attainable yield in Asia could be as high as 46 t/ha (Howeler, 2014, unpublished). Appropriate crop management practices, including better land preparation, use of suitable high-yielding varieties and good quality planting materials, along with the correct plant populations, balanced fertilizer application, and effective weed and erosion control, will all be required to improve cassava yields.

2 Implementing GAP in cassava pre-harvesting

GAP include those practices that are technically feasible, economically viable, environmentally sustainable and socially acceptable.

The focus of GAP should be on the sustainable optimization of cultivation. This has also been defined as 'A sustainable agriculture is one that over the long term enhances environmental quality and the resource base on which agriculture depends; provides for basic human food and fibre needs; is economically viable; and enhances the quality of life for farmers and society as a whole' (American Society of Agronomy, 1989).

Key concerns for achieving sustainable cassava production include the crop's actual and perceived contributions to soil erosion and decline in soil fertility. Additionally, the environmental and economic sustainability of the cassava sector is under increasing pressure from the lack of high-yielding, adapted varieties for some agro-ecological conditions, inappropriate crop management, emerging pests and diseases, rising labour costs and the difficulty in mechanizing the smallholder production system, especially those located on sloping land.

Like for any other crop, cassava cultivation cannot focus solely on short-term profits, but must take into account long-term environmental sustainability for the overall success of the cassava value chains in Asia. Moving towards sustainable cassava cultivation will depend upon the wise use of limited natural resources (e.g. soil and water), but a goal of improved environmental quality without ensuring the viability of short-term returns to growers would also jeopardize production (Norman et al., 1997). Therefore, sustainable cassava cultivation emphasizes the efficient use of natural resources, so that production remains economically viable and socially acceptable while environmental quality is still maintained or improved.

Some common myths about cassava have contributed to reluctance by many governments to support cassava research and development, and have inhibited development and adoption of best practices:

Myth # 1: *Cassava does not need fertilizer.* It is true that cassava will tolerate low-fertility soil conditions, and still produce some yield where other crops would fail. Cassava is often cultivated on infertile soils by resource-poor farmers with little to no fertilizer or manure. In traditional systems, soil fertility may have been maintained by fallow periods, crop rotation, intercropping and other means. As pressure on land increased, these practices have tended to decline, but the alternative practices for maintaining soil fertility may not have been adopted because of the myth that cassava does not need fertilizers. However, cassava is often highly responsive to appropriate nutrient application. With application of optimum types and amounts of nutrients, the highest quantity and quality of cassava roots will be achieved.

Myth # 2: *Cassava cultivation degrades the soil.* Cassava is often cultivated on infertile or highly degraded soils; however, this is not necessarily because the crop has 'caused' this soil degradation. A relatively large amount of nutrients is removed from the soils when cassava roots are harvested from the field. Therefore, continuous planting of cassava on the same soil without adequate nutrient application will lead to soil nutrient depletion. Cassava is also often cultivated on sloping lands, again because it performs better under these conditions than alternative crops. However, due to its slow early growth and poor initial canopy formation, cassava creates conditions for soil erosion from rainfall, when cultivated without soil conservation measures, especially in light-textured soils. Therefore, cassava cultivation can result in serious soil erosion and degradation if the crop and soil are not properly managed.

The lack of promotion and adoption of GAP at the level of both agricultural officials and farmers has hampered the development of sustainable cassava systems. This chapter describes in detail the different elements of GAP. It includes every production step practiced by cassava producers, especially smallholder farmers, to increase cassava yields in profitable and sustainable ways. Further details on crop nutrient management are provided in 'Improving soil and nutrient management in cassava cultivation', 'Addressing nutritional disorders in cassava cultivation' and 'Nutrient sources and their application in cassava cultivation' by Howeler.

Major causes of low cassava yields in some parts of Southeast Asia are (Bellotti et. al., 2012; Howeler, 2001, 2014; Howeler and Aye, 2014):

- Lack of high-yielding and best-adapted varieties
- Low soil fertility and inadequate fertilizer or manure use
- Soil erosion when cassava is cultivated on sloping land
- Use of poor-quality planting materials
- Inappropriate agronomic practices, including land preparation and suboptimum plant populations
- Strong weed competition
- Pest and disease problems
- Inappropriate mixed cropping

2.1 Variety selection and production of planting materials

The use of suitable cassava varieties and good quality planting materials are essential for obtaining the stable high yield and quality of products that markets demand. Farmers in Southeast Asia grow a wide range of cassava varieties; the highest-yielding and best-adapted varieties vary from region to region, and according to end use. However, there are some varieties from China, Thailand and Vietnam that have demonstrated high yield potential across many parts of the region, though typically varieties perform best within specific agro-ecological zones. The names of the most popular released varieties and their main characteristics for industrial processing are described in 'Cassava cultivation in Asia' by Aye. Local testing for adaptation to specific agro-ecological and socio-economic conditions is essential to assure that the best varieties are used by farmers.

Cassava variety identity can only be assured when the source of the planting materials used is known and controlled, since many varieties may appear quite similar to each other. The choice of variety arises from a consensus, and sometimes compromises, among the expectations of farmers, consumers, cassava industries and traders. This can be achieved by inviting representatives of these various actors in the value chain to field days when on-farm variety trials are being harvested. This provides an opportunity to discuss the various features such as agronomic characteristics, resistance to major pests and diseases, and market requirements. For example, the variety Rayong 72 from Thailand is broadly popular because it has high yield and good plant architecture, is drought tolerant, has a relatively high starch content and is suitable for starch processing and animal feeding (after chipping and drying). It was tested among all the value chain actors before release and promotion.

Smallholder farmers usually produce their own cassava planting material. Farmers plant cuttings (also called 'stakes') taken from the stems of mother plants, for clonal propagation. This means that the new plants will have characteristics identical to those of the mother plants. But it will also increase the risk of disease or pest transmission to the next generation if the planting materials are not carefully selected to be free of those problems. Selecting quality planting materials of suitable cassava varieties is thus necessary to obtain high yield and root quality. Among many management factors, two key points stand out for obtaining the highest-quality planting materials, both of which can boost early season production and have lasting effects on yield potential: freedom from diseases and pests, and good physiological quality of the planting materials, including nutritional quality.

Good quality planting materials provide a certain degree of resilience against unfavourable conditions, such as early season lack of rain, low soil fertility, and pests and diseases. To ensure good quality planting materials the following guidelines are suggested:

1. *Variety:* Selected varieties must have good early rooting and high bud sprouting capacities. These capacities can be assessed from the percentage of sprouted plants at 15–20 days after planting.
2. *Age of the mother plant:* Planting materials should be cut from plants that are between 10 and 12 months old, when a good proportion of the stems are mature enough for rapid sprouting. Younger planting materials dry out easily and older planting materials tend to have delayed sprouting.
3. *Storing of stems:* Stems to be used as planting materials should be stored upright in the shade of trees or in buildings. Leaving the stems exposed in the open field will reduce their viability. In areas with danger of frost, the stems should be stored in trenches in the field or dug into hillsides, covered with straw and soil to protect against frost damage. Storage time should be as short as possible and preferably not more than two months after harvest.
4. *Part of the stem:* The cuttings should be taken from lignified stems that are not too old and not too young, that is, they are usually taken from the middle two-thirds of the stems of the mother plants (discarding older bottom and younger top of the plant). When the stems have sprouted or the ends have dried during storage, these parts should be eliminated before cutting the planting stakes.
5. *Size of the stems used as planting materials:* The most suitable stems are those with a diameter of the central pith equal to or less than 50% of the stem diameter. High-quality stem cuttings have a diameter between 1.5 and 2.0 cm.
6. *Length of the planting stakes:* Planting stakes should be about 20–25 cm long regardless of the planting positions used (i.e. vertical, horizontal or inclined).
7. *Number of nodes per planting stake:* The planting stakes should have between five and seven nodes. The number of nodes may vary depending on the varieties and on weather conditions during the growth of the mother plants.
8. *Cutting of stems into stakes and cutting angle:* The cutting of stems into stakes should be done with a sharp knife, machete, cleaver or lopper. Under favourable conditions, planting stakes that have been cut either transversely or bevelled may give good yields. But planting stakes that have been transversely cut will root more uniformly around the whole stem perimeter, resulting in a better root distribution. To prevent excessive drying and damage during transport, the cutting of stems into stakes is best done in the field and just before planting.

2.2 Land preparation

Land preparation is one of the most expensive and critical management operations, necessary for the success of cassava production. In general, land preparation involves ploughing, harrowing and levelling the ground to make it suitable for planting. Where cassava is traditionally grown as the first crop after clearing the land, no land preparation may be required, other than the removal of bushes, shrubs and vines, if planting is to be done by hand. When the first rains have softened the ground, farmers loosen the soil in individual planting holes with a hoe or sharp spade or planting stick, and proceed to plant the cassava stakes.

In fields where successive crops of cassava are grown, as soon as one cassava crop has been harvested, the land can be prepared again for the next planting. However, if cassava stems and leaves are left in the field, it is better to wait several weeks before these are incorporated into the soil in order to prevent their re-sprouting among the newly planted stakes. Cassava growers should understand the different tillage methods and equipment used for land preparation, and be able to select the most appropriate tools and practices for their situation. Methods of land preparation have a significant effect on cassava root yield but not on the root starch content (Jongruyasub et al., 2007).

Conservation tillage, including minimal and zero tillage, are possible solutions in land preparation to improve soil health. The aim of land preparation is to condition the soil with the best physical attributes for cassava to grow. Land preparation may produce a lighter structure, which makes planting by hand easier; it also incorporates plant residues, weeds and soil amendments such as basal fertilizers and manures. However, excessive land preparation using tractors, especially on wet soil, can lead to soil compaction or the formation of a hardpan, which reduces water infiltration and may lead to soil erosion. Depending on the crop's requirement, land topography and availability of resources, land preparation may be done manually with a hoe, animal traction or mechanized using a two-wheel or four-wheel tractor. Any land preparation method should consider the efficient use of soil and water resources, the effect on the environment and the expected economic returns on investment.

Land preparation is achieved by loosening the soil with a hoe or using bullocks or a tractor to plough, often followed by a harrow or rototiller to break up big soil clumps and smoothen out the soil surface. In most countries in Southeast Asia, however, cassava fields are very small, often located on steep slopes where the use of tractors is impossible. Farmers cultivating small areas, such as in Indonesia, Northern Vietnam and Northern Laos, often use only a hoe to loosen the whole area to be planted, or only the planting hole itself.

In areas of intermediate farm size, on flat or slightly sloping lands, such as in the Philippines, Indonesia, Southern India and the Ayeyarwady delta of Myanmar, farmers will often plough the fields with bullocks or buffaloes, usually one or two times, to loosen the soil and incorporate crop residues and weeds (Aye, 2012). In southern India and Myanmar ploughing may be followed by making mounds by hand to concentrate the most fertile topsoil for the planting of stakes. In high rainfall areas farmers may plant cassava on top of ridges to prevent waterlogging of the root system. These ridges are either made by hand, using hoes or are made with a tractor-mounted disk. Regardless of the soil type and moisture condition, most farmers in Cambodia, Thailand and South Vietnam, where slopes are not too steep, use a hired four-wheel tractor to prepare their land. In Thailand this is usually done with a three-disk plough followed by a seven-disk harrow and sometimes a ridger; however, some farmers use their own two-wheel tractor equipped with a small disk plough or a rotovator. Farmers prefer to plant cassava in well-prepared loose soil without any weeds, which facilitates vertical or inclined planting of stakes and reduces early weed competition (Howeler, 2014). Land preparation in smallholder systems is likely to change over time as technologies for small farm mechanization advance, and labour costs continue to rise. Mechanization is often used for the purpose of early land preparation to ensure timely planting.

When considering the choice of land preparation methods, it is important to have clearly defined short- and long-term objectives, and to monitor the resulting effects. Any tillage operation has more than one effect. For example, if the soil is tilled to decrease the soil ped (aggregate) sizes, weeds will be controlled and crop residues will be incorporated, but soil moisture may be lost. Tractor operators typically prefer to prepare the field in straight lines parallel to the roads or field borders, irrespective of slope direction, or they prefer driving up and down the slope rather than along the contour. These kinds of land preparation could exacerbate soil erosion, especially on sloping land with sandy soils. On sloping land, it is important to plough along the contour to reduce erosion. Conventional mouldboard or disk ploughing can also create a 'plough sole', or a compacted layer, at 15–20 cm depth, restricting root growth, drainage and nutrient penetration. Compacted layers can be broken by using a subsoiler, but this is a very energy-intensive, that is, expensive, process (Watananonta et al., 2006).

2.3 Planting methods

Three methods, that is, horizontal, inclined (slanted) and vertical, are commonly used for planting cassava stakes. In the case of the vertical or inclined method about half of the stake is pushed into the soil while the other half remains above the ground. With horizontal planting the stakes are placed in a shallow trench about 2–5 cm deep and covered with soil. With the vertical or inclined method, the cassava stakes sprout more quickly than with the horizontal method, but the vertical method tends to produce deeper roots which are more difficult to harvest in clayey soils. Planting vertically or slanted generally produces higher yields than planting horizontally (CIAT, 1979), particularly during periods of drought. Vertical planting is most suitable for sandy soils and under conditions of limited rainfall (Tongglum et al., 1992); it results in higher sprouting rates and root yields compared to horizontal planting, as shown in Table 1. Most mechanical planters are designed to plant stems horizontally. They open a furrow, drop in the stake and cover it with soil (Ospina et al., 2007). More recent designs also allow vertical planting (Fig. 1).

Figure 1 Cassava vertical planting machine.

Table 1 Effect of stake position, stake length and planting depth on cassava yield (t/ha), planted in both the rainy and dry season at Rayong Field Crops Research Center, Thailand. Data are the average of three years, 1987–1989

	Rainy season (May–August)			Early dry season (November)		
Treatments	No. of plants survived ('000/ha)	Root yield (t/ha)	Starch content (%)	No. of plants survived ('000/ha)	Root yield (t/ha)	Starch content (%)
Method of planting						
Ridge	14.6	15.0	16.6	10.7	14.7	18.6
No ridge	14.4	13.5	16.7	12.1	15.0	18.6
F-test	NS[1]	NS	NS	**	NS	NS
Stake position						
Vertical	14.9	16.0	17.0	13.0	17.7	19.0
Inclined	14.9	15.5	17.1	12.0	16.4	18.7
Horizontal	13.7	11.1	15.8	9.3	10.3	18.2
F-test	**[2]	**	**	**	**	**
Stake length (cm)						
20	14.5	14.5	16.7	10.6	14.5	18.5
25	14.4	13.5	16.7	13.0	15.4	18.9
F-test	NS	*[3]	NS	**	NS	NS
Planting depth (cm)						
5–10	14.4	13.9	16.6	9.7	13.1	18.2
15	14.6	14.4	16.7	12.7	16.2	19.0
F-test	NS	NS	NS	**	**	**

No interaction between methods and treatments in all characters.
[1] NS = not significantly different.
[2, 3] Mean within a column separated by DMRT at 0.01% and 0.05%, respectively.
** = F-test highly significantly different;
Source: Tongglum et al., 1992, 2001.

2.4 Plant density and planting arrangement

Plant density (number of plants per hectare) will determine the quantity of planting material required for cassava cultivation and will influence management practices such as weeding; it will also impact the final yield and affect the quality of the resulting planting material (e.g. high density will tend to produce thinner stems). The goal is to have enough healthy and vigorous planting material at the time of the next planting. The best plant spacing of cassava will vary according to the branching habit of the variety, the fertility of the soil, climatic conditions and also whether the crop is grown in monoculture or in association with intercrops (Leihner, 1983). This can be summarized as follows:

- Under favourable climatic conditions and in fertile soil, cassava grown in monoculture can be planted at a population of about 10 000–12 000 plants per hectare, or at a spacing of about 100×100 cm to 90×90 cm.
- Under less favourable climatic conditions or in less fertile soil, the cassava plant population under monoculture should be increased to between 12 000 and 18 000 plants per hectare, corresponding to a spacing of about 90×90 cm to 75×75 cm.
- When cassava is intercropped and grown under favourable conditions, the inter-row spacing is widened and the inter-plant spacing is reduced to obtain a cassava plant population of about 8 000–10 000 plants per hectare.
- When cassava is intercropped and grown under less favourable conditions, the cassava plant population should be increased to about 10 000–12 000 plants per hectare, with the inter-row and inter-plant spacing adjusted according to the relative importance of cassava in comparison with the associated crop(s).
- Highly branching varieties will require wider spacing than semi-branched ones, which in turn require wider spacing than non-branching varieties.
- Whether the stakes are planted in a square or in a rectangular (wider inter-row and narrower inter-plant spacing) arrangement will not greatly affect yields as long as the optimum number of plants per hectare is being maintained.

2.5 The use of intercropping

Farmers plant cassava either as a sole crop in monoculture or in a mixed intercropping system. Mixed intercropping involves several crops grown together in the same space, using a random or specifically designed distribution. Each crop may be planted and harvested at various times according to its specific maturity characteristics. Generally, smallholder farmers prefer to grow two or more crops at the same time in the same field in a mixed intercropping system. Common intercrops with cassava in Asia include upland rice (*Oryza sativa* L.), maize (*Zea mays* L.) and legume crops such as peanut (*Arachis hypogaea*), cowpea (*Vigna unguiculata*) and mung bean (*Vigna radiata*). In China, especially in subtropical areas, farmers often prefer intercropping with watermelon or sweet potatoes. In other areas with plantation crops such as rubber, coconut or cashew, cassava is often intercropped for a few years between the rows of young trees, or in the case of coconut, among either the young or the old trees (Villamayor et al., 1992). Intercropping cassava with leguminous crops such as common bean (*Phaseolus vulgaris*), cowpea, mung bean and peanut is beneficial, as these legumes will provide the necessary proteins for the family and their livestock. In addition, the legumes fix nitrogen from the atmosphere, and cassava may benefit from this symbiosis. Because of the legumes' faster growth, they cover the ground and protect the soil from the direct impact of rainfall when the cassava canopy is not yet closed, reducing soil erosion. Early maturing food crops should be selected for intercropping with cassava in order to reduce the period of competition and to avoid excessive shading of the young cassava plants.

Intercropping systems vary markedly from country to country as well as among different regions within the same country, depending on the soil and climatic conditions, especially the length of the rainy and dry seasons. Intercropping has many advantages for smallholder farmers to reduce the risk of crop failure, to provide diversity of crops and to obtain food or income at different times of the year. Numerous experiments have been conducted to determine the best intercrops for cassava, as well as the best planting arrangements and relative time of planting (Leihner, 1983; Aye and Howeler, 2012).

Probably the most intensive intercropping systems are found in the wetter zones of West Java and southern Sumatra of Indonesia. Here, cassava is often intercropped with simultaneously planted upland rice between cassava rows and maize between plants within the cassava row. Once the upland rice and maize are harvested at about four months after planting, a short-duration grain legume, such as mung bean, soya bean, cowpea or peanut, is planted in the inter-row space previously occupied by rice. If rainfall permits, a fourth intercrop such as mung bean or peanut is planted in the space previously occupied by the harvested grain legume. In East Java, on the other hand, the dry season is longer and cassava can be intercropped by only one crop, usually maize. In South Vietnam, cassava is often intercropped with maize or planted among young rubber or cashew trees, while in north Vietnam, the crop is often intercropped with peanut or black bean (cowpea). In Guangxi province of China, cassava is often intercropped with maize, peanut, sweet potato or watermelon, while in Hainan province, the crop is often interplanted among young rubber trees or bananas. In Thailand, cassava is seldom intercropped because farmers cultivate larger cassava fields and do not have enough labour to manage more intensive intercropping, but the crop is sometimes planted for a few years among young rubber or coconut trees.

Intercropping normally reduces the yield of each crop, compared to their yields in monoculture, but the overall system is typically more economically productive, as well as having the other benefits already described. Competition between the crops can be reduced by modifications of the plant spacing or planting patterns of both crops, or by adjusting the relative time of planting, and the amounts of fertilizer applied to each crop to maximize yields. There are many alternative ways of planting cassava in intercropping systems. Tongglum et al. (2001) reported that intercropping cassava with peanut seems to have had a long-term beneficial effect on soil fertility and cassava yields in Rayong, Thailand. The most commonly used intercropping systems are shown in Table 2.

When smallholder farmers adopt intercropping cassava with other crops, a relatively small area of land may be sufficient to provide the family with their basic dietary needs, such as energy, proteins, minerals and vitamins.

Table 2 Intercropping systems with cassava in Asia

Country	Associated crops
Cambodia	Upland rice, maize, cashew nut, rubber
China	Maize, watermelon, sweet potato, peanut, rubber
East Timor	Maize, peanut, vegetables, banana
India	Maize, cowpea, vegetables, coconut
Indonesia	Upland rice, maize, soya bean, cowpea, mung bean, peanut, coconut, rubber
Laos	Upland rice, maize, Job's tear, peanut
Myanmar	Maize, peanut, common bean, banana
Philippines	Maize, peanut, sweet potato
Thailand	Maize, rubber, coconut, cashew nut
Vietnam	Maize, upland rice, peanut, black bean, rubber, cashew nut, coffee, tea

Source: Aye and Howeler, 2012.

3 Implementing GAP for cassava crop health

3.1 Soil fertility and soil health

The effect of low soil fertility can be difficult to notice in cassava because the plants do not display clear deficiency symptoms of major nutrients (N, P, K). Growth and development will be slow and yield will be low if some nutrients are not present or are present in either inadequate or excessive amounts. Nutrient deficiencies can be corrected by applying balanced fertilizers. However, most smallholder farmers in Southeast Asia do not apply adequate amounts of fertilizers or manure on cassava. Cassava grows reasonably well and produces reasonable yields even in low-fertility soils that may be too infertile for other crops, or on soils that have been depleted by previous crops. However, cassava is also highly responsive to nutrient application (mineral and organic fertilizers).

High-yielding cassava varieties require more plant nutrients than traditional varieties, since the former have usually been selected, especially for responsiveness to favourable growing conditions. Application of the optimum type and amount of nutrients is critical to achieving the highest quantity and quality of cassava roots. Application of high rates of N may increase the cyanide concentration of the roots, while high rates of K may increase the starch and decrease the cyanide concentration. External nutrient sources must be used to fill the gap between the cassava crop's needs and the soil's indigenous supply. The nutrients removed when roots and/or plant tops are harvested must be replaced in the soil to maintain stable high yields. Additional nutrients may need to be added to account for removal of stems and leaves, if these are not reincorporated into the soil.

Soil organic matter (SOM) improves soil moisture and aeration and reduces nutrients from being leached down the soil profile. It can also act as a buffer against adverse environmental effects such as high temperature and drought. SOM can be maintained through mulching, which adds a protective layer of plant material on top of the soil between crop plants. Mulching can be achieved by allowing cassava stems and leaves to decompose on the soil surface, by manure application and by appropriate crop rotations or intercropping practices. Mulches can also be from decaying weeds, grass or compost, and have the following benefits:

- Protect the soil from erosion
- Reduce compaction from the impact of heavy rain
- Conserve soil moisture
- Prevent weed growth
- Maintain a more even soil temperature (reduces the extremes of either high or low temperatures)

3.2 Soil erosion control

Preventing erosion is a key aspect of maintaining soil health. Due to population pressure and scarcity of agricultural land in Asia, many smallholder farmers grow cassava on sloping lands. High rainfall and very intensive cropping on steep slopes can cause severe soil losses due to erosion, which tends to be more serious in Asia than in Africa or Latin America (Chorley, 1969; Milliman and Meade, 1983). Many cassava soils in Asia are light-textured, and are thus more susceptible to erosion, especially in Thailand, the Philippines, Vietnam and some parts of southern China. In those areas, the land-use systems should be designed to strengthen the soil's physical structure and avoid leaving the soil surface exposed to direct rainfall impact and wind at times of highest risk. Erosion results in deteriorating soil quality (physical and chemical characteristics), which in turn reduces cassava yields. The erosion process mainly removes SOM, and certain clay fractions, which provide soil water and nutrient holding capacity. If cassava is grown on steep or exposed lands, physical anti-erosion structures, such as contour hedgerows, bunds and grass strips, should be used.

Soil conservation techniques aim to reduce the amount and speed of runoff water by improving water infiltration into the soil. They also aim to protect the soil from direct rainfall impact by stimulating rapid early crop growth and canopy formation of cassava, or by including intercrops to cover the soil more quickly, while the cassava canopy develops. Contour cultivation with contour grass barriers or hedgerows also reduces the length or steepness of slopes in cassava fields by filtering out the soil sediments and depositing those above the hedgerows, this results in natural terrace formation.

Erosion control experiments using different crops grown on sloping land often show that cassava causes more soil loss by erosion than many other crops (Quintiliano et al., 1961; Margolis and Campos Filho, 1981; Putthacharoen et al., 1998). This is because cassava plants have to be widely spaced to reduce inter-plant competition, while initial plant growth is slow, leaving the soil surface between plants exposed to the direct impact of rain drops. This results in the disintegration of soil aggregates and the movement of the smaller soil particles downhill with the rainfall runoff. This is particularly serious when cassava is grown on light-textured soils that have little clay and organic matter to bind the aggregates together, or in heavier soils that have been compacted by the use of heavy machinery when the soil is too wet. This soil compaction leads to poor internal drainage, resulting in excessive overland runoff and soil erosion.

While cassava has intrinsic characteristics that may lead to serious soil erosion, research has shown that there are many simple agronomic and soil conservation practices that will markedly reduce erosion while often increasing yields. Numerous experiments, demonstration plots and Farmer Participatory Research (FPR) trials have been conducted in Thailand, Vietnam, China and Indonesia to determine the effect of these practices on cassava yields and soil loss due to erosion (Howeler, 2008, 2012j, 2014). Different sets of treatments were compared in these experiments and trials depending on the situation and preferences of farmers, with the most promising treatments repeated in many different experiments. The average relative yield and soil loss obtained for each treatment in these experiments or trials are shown in Tables 3 and 4 for Thailand and Vietnam, respectively. In both tables the relative yield and soil loss for each treatment in a specific experiment were determined in comparison with the yield and soil loss of a check treatment of either cassava monoculture or intercropped with peanut (the latter in Vietnam only), with application of a medium level of fertilizers but no hedgerows or other special cultural practices.

In both Vietnam and Thailand, one of the most effective agronomic practices, both in terms of increasing yields and decreasing erosion, was the application of fertilizers, as this markedly improved initial plant growth resulting in early canopy closure, thus reducing the area of bare soil exposed to the direct impact of raindrops. In Vietnam, the lack of fertilizer application reduced yields on average by 68% and increased erosion losses by 37%, compared with the fertilized check treatment, while in Thailand the lack of fertilizers decreased yields on average by 4% but increased soil losses by erosion (140%). Other very effective practices in Thailand include the planting of contour hedgerows of vetiver grass (*Vetiveria zizanioides*), while in Vietnam similar results were obtained with hedgerows

Table 3 Effect of various soil conservation practices on the average[1] relative cassava yield and dry soil loss due to erosion as determined from soil erosion control experiments, FPR demonstration plots and FPR trials conducted in Thailand from 1994 to 2003

	Soil conservation practices[2]	Relative cassava yield (%)	Relative dry soil loss (%)
1	With fertilizer; no hedgerows, no ridging, no intercrop (check)	100	100
2	With fertilizer; vetiver grass hedgerows, no ridging, no intercrop[4]	90 (25)	58 (25)
3	With fertilizer; lemon grass hedgerows, no ridging, no intercrop[4]	110 (14)	67 (15)
4	With fertilizer; sugarcane for chewing hedgerows, no intercrop	99 (12)	111 (14)
5	With fertilizer; *Paspalum atratum* hedgerows, no intercrop[4]	88 (7)	53 (7)
6	With fertilizer; *Panicum maximum* hedgerows, no intercrop	73 (3)	107 (4)
7	With fertilizer; *Brachiaria brizantha* hedgerows, no intercrop[3]	68 (3)	78 (2)
8	With fertilizer; *Brachiaria ruziziensis* hedgerows, no intercrop[3]	80 (2)	56 (2)
9	With fertilizer; elephant grass hedgerows, no intercrop	36 (2)	81 (2)
10	With fertilizer; *Leucaena leucocephala* hedgerows, no intercrop[3]	66 (2)	56 (2)
11	With fertilizer; *Gliricidia sepium* hedgerows, no intercrop[3]	65 (2)	48 (2)
12	With fertilizer; *Crotalaria juncea* hedgerows, no intercrop	75 (2)	89 (2)
13	With fertilizer; pigeon pea hedgerows, no intercrop	75 (2)	90 (2)
14	With fertilizer; contour ridging, no hedgerows, no intercrop[4]	108 (17)	69 (17)
15	With fertilizer; up and down ridging, no hedgerows, no intercrop	104 (20)	124 (20)
16	With fertilizer; closer spacing, no hedgerows, no intercrop[4]	116 (10)	88 (11)
17	With fertilizer; C + peanut intercrop	72 (11)	102 (12)
18	With fertilizer; C + pumpkin or squash intercrop	90 (13)	109 (15)
19	With fertilizer; C + sweet corn intercrop	97 (11)	110 (14)
20	With fertilizer; C + mung bean intercrop[3)]	74 (4)	41 (4)
21	No fertilizer; no hedgerows, no or up and down ridging	96 (9)	240 (10)

[1] Number in parentheses indicates the number of experiments/trials from which the average values were calculated.
[2] C = cassava.
[3]* = Promising soil conservation practices; [4]** = Most promising soil conservation practices;
Source: Howeler, 2008.

Table 4 Effect of various soil conservation practices on the average[1] relative cassava yield and dry soil loss due to erosion as determined from soil erosion control experiments, FPR demonstration plots and FPR trials conducted in Vietnam from 1993 to 2003

		Rel. cassava yield (%)		Rel. dry soil loss (%)	
	Soil conservation practices[2]	Cassava monoculture	Cassava + peanut	Cassava monoculture	Cassava + peanut
1	With fertilizer; no hedgerows (check)	100	–	100	–
2	With fertilizer; vetiver grass hedgerows[4]	113 (17)	115 (23)	48 (16)	51 (23)
3	With fertilizer; *Tephrosia candida* hedgerows[4]	110 (17)	105 (23)	49 (16)	64 (23)
4	With fertilizer; *Flemingia macrophylla* hedgerows[3]	103 (3)	109 (4)	51 (3)	62 (3)
5	With fertilizer; *Paspalum atratum* hedgerows[4]	112 (17)	–	50 (17)	–
6	With fertilizer; *Leucaena leucocephala* hedgerows[3]	110 (11)	–	69 (11)	–
7	With fertilizer; *Gliricidia sepium* hedgerows[3]	107 (11)	–	71 (11)	–
8	With fertilizer; pineapple hedgerows[3]	100 (8)	103 (9)	48 (8)	44 (9)
9	With fertilizer; vetiver + *Tephrosia* hedgerows	–	102 (7)	–	62 (7)
10	With fertilizer; contour ridging, no hedgerows[3]	106 (7)	–	70 (7)	–
11	With fertilizer; closer spacing, no hedgerows	122 (5)	–	103 (5)	–
12	With fertilizer; peanut intercrop, no hedgerows[3]	106 (11)	100	81 (11)	100
13	With fertilizer; maize intercrop, no hedgerows	69 (3)	–	21 (3)	–
14	No fertilizer; no hedgerows	32 (4)	92 (15)	137 (4)	202 (12)

[1] Number in parentheses indicates the number of experiments/trials from which the average values were calculated.
[2] IC = intercrop, HR = hedgerows.
[3] = Promising soil conservation practices
[4] = most promising soil conservation practices
Source: Howeler, 2008.

of paspalum grass (*Paspalum atratum*) or the legume shrub *Tephrosia candida*; the latter species grows best under subtropical conditions in northern Vietnam and southern China. These contour hedgerows, spaced between 6 and 20 m apart, depending on the slope, decreased soil losses on average by about 40–50% and either slightly decreased yields or actually increased yields by 10–13%.

Other very effective practices included the planting of cassava in alley cropping systems between contour hedgerows of *Leucaena leucocephala* or *Gliricidia sepium* (Nguyen Huu

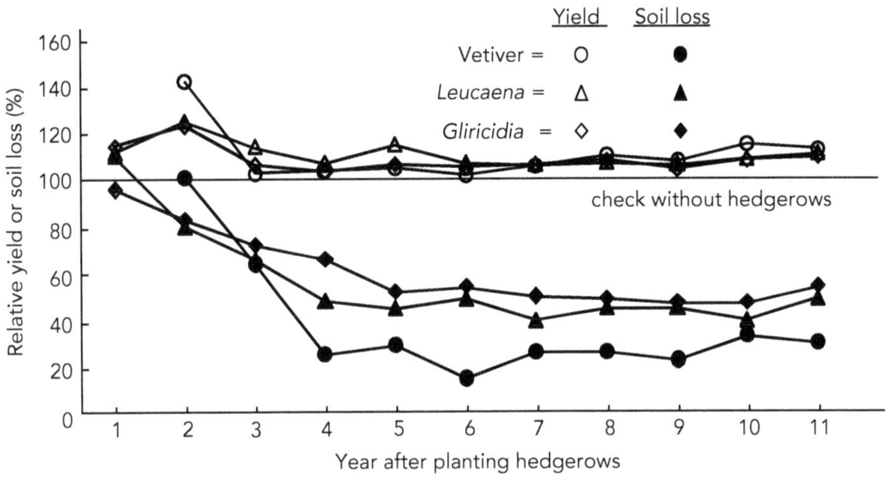

Figure 2 Trends in relative yield and relative soil loss by erosion when cassava was planted with contour hedgerows of vetiver grass, *Leucaena leucocephala* or *Gliricidia sepium* in comparison with the check without hedgerows during 11 consecutive years in Hung Loc Agric. Research Center in South Vietnam from 1997/98 to 2007/08. Source: Nguyen Huu Hy et al., 2010.

Hy et al., 2010) or with *Flemingia macrophylla* or elephant grass (*Pennisetum purpureum*) (Wani Hadi Utomo, pers. comm.; Wargiono et al., 1992, 1995, 1998, 2001; Howeler, 2014). In general, these contour hedgerows became more and more effective in reducing erosion over time, as shown in Fig. 2 (Nguyen Huu Hy et al., 2010; Howeler, 2012j, 2014). Other very effective practices are planting of cassava on contour ridges, planting at closer plant spacing and intercropping with peanut. The common practice of up and down land preparation and ridging slightly increased yields in Thailand, but also markedly increased soil loss by erosion (Howeler, 2008, 2014).

The method of land preparation can also have a significant effect on both cassava yield and soil loss by erosion, but this is not always consistent and will vary with the type of soil and the water content of the soil at the time of land preparation. Currently, many people recommend the practice of zero tillage to improve soil health and prevent soil compaction. This practice can be used for cassava production, but is most suitable in areas where cassava is planted by machine, as manual planting requires a well-prepared and loose soil in order to control weeds and facilitate the insertion of the cassava stakes. Also, most experiments have shown that the effect of zero tillage on yield and soil erosion is highly variable, but usually results in lower yields than when the soil is well prepared with a disk plough followed by harrowing and contour ridging. Table 5 shows that in one location in Thailand zero tillage resulted in low yields and high soil loss due to erosion, while in another nearby site the yield was relatively high and the soil loss was low. Long-term trials comparing zero tillage with conventional land preparation may be necessary to determine whether yields will increase after several years when soil tilth and nutrient availability may have improved as a result of zero tillage.

The data also shows that lack of fertilizer application consistently resulted in low yields and high levels of soil erosion. From these results it can be concluded that soil erosion can

Table 5 Effect of various tillage practices and fertilizer application on the average cassava yield and annual soil loss due to erosion when cassava was planted on 5–8% slope in Si Racha, Chonburi Province, and in Pluak Daeng, Rayong Province, Thailand

	Cassava yield (t/ha)		Dry soil loss (t/ha)	
Tillage treatment	Si Racha (1987–1988)	Pluak Daeng (1989–1990)	Si Racha (1987–1988)	Pluak Daeng (1989–1990)
1. Zero tillage, with fertilizer	28.5	17.0	49.8	10.7
2. Conventional tillage[1], with fertilizer	28.6	14.4	20.8	17.5
3. Conventional with contour ridging, with fertilizer	32.6	15.6	8.1	13.2
4. Conventional with up–down ridging, with fertilizer	29.4	16.1	23.6	19.8
5. Conventional tillage, no fertilizer	21.5	12.2	35.8	25.8

[1] Conventional tillage is ploughing with a three-disk plough followed by harrowing with a seven-disk harrow.
Source: Adapted from Jantawat et al., 1994.

be quite serious in cassava fields, but that many agronomic and soil conservation practices will both increase yields and reduce soil losses by erosion.

The increases in yield due to specific soil conservation practices, such as the planting of contour hedgerows, is likely to be small, ranging from 0 to 20%, as these hedgerows take up space in the field and may compete with cassava plants growing nearby. This competition can be minimized by the right choice and management of the hedgerow species. But if the crop is grown on slopes and is poorly managed, cassava yields are likely to decrease over time, due to the loss of top soil, which contains most of the organic matter, nutrients and beneficial microorganisms, as well as by gully formation, and by washing out plants, or the covering of small plants with eroded sediments.

However, contour ridging, fertilization and intercropping require more work and usually imply higher production costs. Hedgerows also require more work for establishment and maintenance and may reduce yields by occupying 10–20% of the land. Thus, farmers have to consider the trade-off between immediate costs and benefits versus long-term benefits of less erosion and improved fertility (Table 6).

3.3 Weed control

A 'weed' in a generic sense is an unwanted plant that causes damage to native or endemic vegetation or ecosystems, or to agricultural production. Weeds typically propagate prolifically. Depending on the species, they reproduce either through seeds, or by vegetative propagation (i.e. bulbs, rhizomes and root suckers, and stem fragments). Weeds compete with cassava for space, nutrients, light and water. Cassava grows slowly for the first three months and is generally not competitive with fast growing weeds during these early stages of growth. Cassava will suffer serious yield losses, or in extreme cases becomes completely eradicated, if the crop is not adequately weeded, especially during the early stages of its growth. The critical period for weeding is during the first two months after planting. Weed competition during the first 60 days after planting diminishes yields to approximately 50% of the weed-free control case (Villamayor, 1988). In less intensive cultivation systems, especially in Africa, farmers may prefer early branching varieties that

Table 6 Effect of various soil/crop management practices on erosion and yield, as well as on labour and monetary requirements, and long-term benefits in cassava-based cropping systems

Erosion control practice	Erosion control	Terrace formation	Effect on cassava yield	Labour requirement	Monetary cost	Long-term benefits	Main limitations
Minimum or zero tillage	++	–	–	+	––	+	Compaction, weeds
Mulching (carry-on)	++++	–	++	+++	+	++	Mulch availability, transport
Mulching (*in situ* production)	+++	–	++	++	+	++	Competition
Contour tillage	+++	+	+	+	+	++	
Contour ridging	+++	+	++	++	++	+	Not suitable on steep slopes
Leguminous tree hedgerows	++	++	+	+++	+	+++ [1]	Delay in benefits
Cut-and-carry grass strips	++	++	––	+++	+	+++ [1]	Competition, maintenance
Vetiver grass hedgerows	+++	+++	+	+	+	+++	Availability of planting materials
Natural grass strips	++	++	–	+	–	++	High maintenance costs
Cover cropping (live mulch)	++	–	–––	+++	++	+	Severe competition
Manure or fertilizer application	++++	–	+++	+	+++	+++	High cost
Intercropping	++	–	–	++	++	+++	Labour-intensive
Closer plant spacing	+++	–	+	+	+	++	

\+ = effective, positive or high.
– = not effective, negative or low.
[1] = value added in terms of animal feed, staking material or fuel wood.
Source: Howeler, 2012j.

more quickly develop a canopy to provide some weed control. Yield reduction varies from 40% in early branching varieties to 70% in late- or non-branching varieties (IITA, 1990).

In general, cassava should be weeded 2–3 times during the first three months or until canopy closure. A field should be weed-free at planting, and first weeding is done about one month later. If pre- or post-emergence herbicides are used, the weeding schedule will vary. Weeding can be done by hoe, animal-drawn cultivator or hand tractor, but can also be done by a tractor-mounted cultivator, or with herbicides. In light-textured soils in Thailand, farmers sometimes use a 'poor man's plough', that is, a cultivator blade pulled by hand, between cassava rows for weed control. Weed control after four months of age for cassava may not increase root production. While hand weeding can be effective, safe and minimally damaging to the crop, it is also often the most expensive method.

Chemical weed control in cassava is of increasing interest to farmers as a component of improved production technology. Recommended pre-emergence herbicides (referring to the time of weed seed germination) should be applied right after planting (Table 7). Farmers should have pre-emergence herbicides applied and activated – by rainfall or irrigation – prior to the initiation of weed seed germination. Farmers want to have an 'active' chemical barrier present in the soil solution when the target weed seeds imbibe water. This is accomplished by applying pre-emergence herbicides early so that there is ample time for rainfall to occur.

To achieve pre-planting weed-free conditions, a broad-spectrum post-emergence herbicide such as Roundup® (glyphosate) should be applied at the rate of 4–5 l/ha about ten days before land preparation, or it may be applied in no-till or minimum till situations. Post-emergence 'burn-down' herbicides (e.g. Fluazifop) may be applied as soon as weeds begin to emerge after the pre-emergence herbicide treatment. For best results, post-emergence herbicides should be sprayed only when farmers are sure of having at least three hours of sunshine after spraying. They are basically non-selective, localized contact herbicides and should be sprayed with a shield over the spray nozzle to ensure that only the weeds receive the chemical. Post-emergence herbicides should not be applied on windy days as the spray may drift and damage the cassava plants. If the cassava field is infested with difficult-to-control weeds such as *Imperata* grass (*Imperata cylindrica*) and selective herbicides (Gesapax for *Imperata*) should be used. Systemic herbicides like glyphosate, Fusilade or Sarasota are absorbed and translocated throughout the plant and should be carefully applied with manufacturers' guidelines for each of the herbicides.

Weather conditions affect herbicide performance. Herbicides should not be applied soon after heavy rainfall, or before the likelihood of rain, as the chemicals may become diluted and become less effective. If herbicides are used, farmers should be well-informed and trained about their effective use, safety and precautions. However, environmental conditions can still impact the level of control.

Weed competition can also be reduced by adequate and early application of fertilizers to speed up canopy closure, by intercropping and by planting towards the end of the rainy season when weed growth is less vigorous. To prevent the fertilizer actually stimulating weed growth, the fertilizer should not be broadcast but be band applied near the cassava stake or plant. However, some agricultural practices can contribute to weed multiplication and propagation. For example, the use of cattle manure which has not properly decomposed can contribute to serious weed seed dispersal. Timely weeding and integrated weed control measures are encouraged for sustainable cassava production.

Table 7 Various herbicides used for the control of weeds in cassava

Technical name	Commercial name	Selectivity for cassava	Time of application	Dosage of CP/ha [1]	Type of weeds controlled
Diuron	Karmex	intermediate	Pre	2.0–3.0 kg	broadleaved
Alachlor	Lazo	high	Pre	3.0–4.0 lit	grasses
Fluometuron	Cotoran	intermediate	Pre	4.0–5.0 lit	broadleaved
Oxifluorfen	Goal	intermediate	Pre	2.0–4.0 lit	broadleaved/ grasses
Metribuzin	Sencor	intermediate	Pre	1.0–1.5 lit	grasses
Linuron	Afalon	intermediate	Pre	2.0–3.0 kg	broadleaved/ grasses
Trifluralin	Treflan	high	Ibp	2.5–3.5 lit	broadleaved/ grasses
Metolachlor	Dual	high	Pre	3.0–4.0 lit	grasses
	Karmex + Lazo	intermediate	Pre	1.0–1.5 + 1.5–2.0	broadleaved/ grasses
	Cotoran + Lazo	intermediate	Pre	1.0–2.5 + 1.5–2.0	broadleaved/ grasses
	Goal + Lazo	intermediate	Pre	1.0–2.0 + 1.5–2.0	broadleaved/ grasses
	Afalon + Lazo	intermediate	Pre	1.0–1.5 + 1.5–2.0	broadleaved/ grasses
	Karmex + Dual	intermediate	Pre	1.0–1.5 + 1.5–2.0	broadleaved/ grasses
	Cotoran + Dual	intermediate	Pre	1.0–2.5 + 1.5–2.0	broadleaved/ grasses
	Goal + Dual	intermediate	Pre	1.0–2.0 + 1.5–2.0	broadleaved/ grasses
	Afalon + Dual	intermediate	Pre	1.0–1.5 + 1.5–2.0	broadleaved/ grasses
Glyphosate	Roundup	not select.	Post	2.0–3.0 lit	broadleaved/ grasses
Glufosinate	Basta	not select.	Post	1.0–3.0 lit	broadleaved/ grasses
Fluazifop	Fusilade	high	Post	1.0–3.0 lit	grasses
Paraquat	Gramoxone	not select.	Post	2.0–3.0 lit.	broadleaved/ grasses

Ibp = Incorporated before planting; Pre = pre-emergence; Post = post-emergence; CP = commercial product
[1]Lower dosage for use in light-texture soils and higher dosage in heavy-texture soils.
Source: Calle, 2002.

3.4 Pest and disease control

Pests and diseases can threaten the sustainable production of cassava. Traditionally, farmers in Asia have paid little attention to the issue, since cassava in the past had not been severely affected by pests or diseases. However, cassava has begun to suffer from various pest and disease problems that can markedly reduce cassava yields (Bellotti et al., 2012), and effective preventative control has become necessary (see also 'Diseases affecting cassava' by Legg and 'Integrated pest/disease management in cassava cultivation' by Wykhuys).

Cassava used for industrial processing requires year-round production of roots, a practice which has aggravated pest and disease problems as there have to be cassava plants in the field throughout the year at different stages of development. The most important pests now found in Asia include whiteflies, mealybugs, red spider mites, scale insects, white grubs, termites and several pests attacking dried cassava during storage. The most important cassava diseases in Asia are Indian cassava mosaic disease and Sri Lankan cassava mosaic disease (SLCMD), cassava witches' broom disease, cassava bacterial blight (CBB) caused by *Xanthomonas axonopodis* pv *manihotis* (*Xam*), root rots (*Phytophthora* spp.), cassava anthracnose (*Glomerella manihotis*), brown leaf spot (*Cercospora henningsii*) and white leaf spot (*Phaeoramularia manihotis*).

Since the accidental introduction of the cassava mealybug *Phenacoccus manihoti* to Thailand in 2008, this pest has caused serious reductions in cassava growth and yields. Due to the lack of effective native biological control this cassava mealybug rapidly spread throughout Thailand and then into neighbouring countries, particularly to the cassava growing regions of Cambodia and Laos. Whitefly (*Aleurodicus dispersus*) and red spider mite (*Tetranychus* sp.) problems have also become more serious in Southeast Asia's cassava growing regions.

Concurrently, the previously unknown witches' broom disease has also spread in major cassava growing areas of Cambodia, Laos, Thailand, the Philippines and Vietnam with the first reported infestation in Thailand in early 1993 (Graziosi et. al., 2016). Witches' broom disease is transmitted mainly through the use of infected planting material of cassava; however, as there has been limited study on the disease in the past, there are no effective measures to control the disease once the plants are infected.

Recently, the outbreak of SLCMD in the Rattanak Kiri province of northeastern Cambodia (Wang et al., 2016) has caused a potentially serious threat to cassava production in Asia. The causal virus is transmitted by infected planting material as well as by the whitefly *Bemisia tabaci*, which has been found in both Vietnam and Thailand. CBB, root rots, cassava anthracnose, brown leaf spot and white leaf spot are also often seen on cassava in Asia. However, because cassava is a long-season crop that is exposed to these problems over a long period, it is hardly ever economic or effective to control pests and diseases through single control measures such as the application of pesticides; the use of resistant varieties should be the best method, especially for diseases. Integrated pest management is the most effective approach to manage pests and diseases, and may include a combination of resistant varieties and cultural practices, including mechanical, biological and chemical methods.

4 Implementing GAP in cassava post-harvesting

Harvesting cassava is time-consuming and physically arduous, especially under tropical weather conditions. The crop is traditionally harvested by cutting off the top (stems and leaves) about 20 cm from the soil surface and then pulling on the remaining stump of the

Figure 3 Cassava harvesting tool.

stem until the roots emerge from the soil. This may require some digging around the roots with a hoe, shovel or other tools. Some farmers still use a harvesting stick for harvesting cassava, but most use metal harvesting tools. A simple tool, invented by farmers in Thailand, consists of a metal plate with a notched cut out V-shape, which is welded onto a metal cylinder, and then attached to a wooden or metal pole (Fig. 3). The notched V of the metal plate is pulled around the stump and the attached pole is pulled up to function as a lever to more easily lift the stump with the attached roots out of the ground; normally this does not require additional digging of the soil. There are many variations of this harvesting tool, but all make use of the same concept – the power of leverage. In larger farms and commercial plantations, or in very hard clay soils, cassava is now often harvested by tractor-mounted harvesting tools that dig under the roots and lift the root clumps out of the soil and onto the soil surface. This is currently mainly practised in heavy clay soils in eastern Thailand, as these soils become too hard in the dry season to harvest the roots by hand. It is also practiced by a large commercial plantation in eastern Cambodia.

5 Adoption of GAP by farmers

While many management practices to control problems such as soil erosion have been recommended by researchers and extension agents, not enough of these practices have been adopted by farmers. This is mainly because most of the recommended practices require either additional labour or money, and benefits are usually accrued over the long-term, while most poor farmers are in desperate need of immediate income to feed their families.

Cramb and Nelson (1998) reported the results of a modelling exercise to predict the long-term effect of planting contour hedgerows in a relatively eroded soil in the Philippines on the long-term yield of maize and on net present value (NPV). In this example, the model predicts that when maize is grown in open fields without hedgerows, yields will decline markedly during the first few years. With hedgerows, yields will be lower initially, as hedgerows occupy space in the field, but maize yields with hedgerows will overtake those without hedgerows after two years and remain fairly constant at 2–3 t/ha for the next 25 years. However, the NPV for planting maize without hedgerows was higher than planting with hedgerows for the first five years. The NPV for the first two years was very low due to the high initial cost of establishing the hedgerows and the cost of maintenance and lower maize yields obtained. Thus, the farmer will not receive economic benefits from planting hedgerows until after five years. It is only after 10–15 years that farmers will reap substantial economic benefits from these soil conservation practices, but that is too long for most farmers with a short planning horizon, or with immediate needs for adequate income. This example shows the main dilemma in promoting soil conservation practices: most recommended practices were selected by researchers because they are effective in controlling erosion, but few consider whether poor farmers can actually bear the economic burden of adopting these practices. If they cannot, governments may have to provide some incentives, since part of the benefits of better erosion control are reaped off-site by people living downstream or in the cities – benefits sometimes referred to as *ecosystem services*.

Another problem in the transfer of soil conservation technologies is that many soil erosion control trials were conducted on experiment stations under optimum and uniform conditions. These conditions seldom correspond with those faced by farmers living in mountainous areas with heterogeneous soils, topography and climates, and with economic opportunities that vary markedly from place to place depending on distance to roads and markets.

Many practices that seemed very effective in controlling erosion, and may have economic benefits under the conditions of the experiment station, may be rejected by farmers simply because they are not effective or not appropriate under the farmer's specific biophysical and socio-economic conditions. For that reason it is more effective to present farmers with a range of options, from which they can select those that they consider useful, and let them try out some of these options in simple trials on their own fields. In this way farmers can observe and decide which is the most effective and useful practice for their own conditions. This Farmer Participatory Research (FPR) methodology is particularly useful for developing and disseminating technologies such as erosion control practices that are highly site-specific and where there are many trade-offs between costs and benefits. Only farmers themselves can decide about the costs they can bear and risks they can take now in order to obtain benefits sometime in the future.

FPR has been conducted in over 100 villages in the cassava growing regions of China, Indonesia, Thailand and Vietnam from 1994 to 2003 (Howeler, 2008, 2012k, 2014). Table 8 shows an example of a simple FPR erosion control trial conducted by six farmers having adjacent plots on a uniform slope of 35–45%.

During the third year of cropping, some erosion control practices, such as intercropping with peanut, application of fertilizers and contour hedgerows of vetiver grass or *Tephrosia candida* reduced soil loss to about one-third, while doubling gross and net income. These were the practices most farmers selected as most useful for their particular conditions. Farmers selected a combination of practices, such as new high-yielding varieties, better fertilization and intercropping with peanut, which increased income, in combination with contour ridging and contour hedgerows that mainly reduced erosion, so as to obtain both short-term and long-term benefits.

Table 8 Effect of various crop management treatments on the yield of cassava and intercropped peanut as well as the gross and net income and soil loss due to erosion in an FPR erosion control trial conducted by six farmers in Kieu Tung Village of Thanh Ba District, Phu Tho Province, Vietnam, in 1997 (3rd year)

			Yield (t/ha)		Gross income[2]	Product costs	Net income	
Treatment[1]	Slope (%)	Dry soil loss (t/ha)	Cassava	Peanut[1]	(mil. dong/ha)			Farmers ranking
C monocult., with fertilizer, no hedgerows	40.5	106.1	19.17	–	9.58	3.72	5.86	6
C + P, no fertilizer, no hedgerows	45.0	103.9	13.08	0.70	10.04	5.13	4.91	5
C + P, with fertilizer, no hedgerows	42.7	64.8	19.23	0.97	14.47	5.95	8.52	–
C + P, with fertilizer, *Tephrosia* hedgerows	39.7	40.1	14.67	0.85	11.58	5.95	5.63	3
C + P, with fertilizer, pineapple hedgerows	32.2	32.2	19.39	0.97	14.55	5.95	8.60	2
C + P, with fertilizer, vetiver hedgerows	37.7	32.0	23.71	0.85	16.10	5.95	10.15	1
C monocult, with fert., *Tephrosia* hedgerows	40.0	32.5	23.33	–	11.66	4.54	7.12	4

[1] Fertilizers = 60 kg N+40 kg P_2O_5, + 120 kg K_2O/ha; all plots received 10 t/ha pig manure.
[2] Prices: cassava (C) dong 500/kg fresh roots
peanut (P) 5000/kg dry pods
US$1 = approx. 13 000 dong
Source: Howeler, 2010c, Howeler et al., 2007.

Table 9 Extent of adoption (% of households)[1] of new technologies by participating and non-participating farmers (n = 767) in the cassava project in Thailand and Vietnam in 2003

	Thailand			Vietnam			Full sample		
Technologies	Partic. n = 109	Non-partic. n = 308	Total n = 417	Partic. n = 126	Non-partic. n = 224	Total n = 350	Partic. n = 235	Non-partic. n = 532	Total n = 767
New varieties									
100% improved varieties	100	88.0	91.1	50.0	38.8	42.9	73.2	67.3	69.1
75% improved varieties	0	11.7	8.6	5.6	6.7	6.3	3.0	9.6	7.6
50% improved varieties	0	0.3	0.2	26.2	18.3	21.1	14.0	7.9	9.8
25% improved varieties	0	0	0	4.0	5.4	4.9	2.1	2.3	2.2
No improved varieties	0	0	0	14.3	30.8	24.9	7.7	13.0	11.3
	100	100	100***[2]	100	100	100***[2]	100	100	100***[2]
Soil conservation practices									
Contour ridging	52	22	30***	35	31	33	43	26	31***
Hedgerows	60	10	23***	50	12	25***	54	11	24***
No soil conservation	21	72	59***	23	58	45***	22	67	53***
Intercropping	28	8	13***	79	49	59***	55	25	34***
Fertilization									
Chemical fertilizers	98	86	89***	85	86	86	91	86	87**
Farmyard or green manure	55	25	33***	74	60	65**	65	40	48***
No fertilizer	0	13	9***	12	8	9	6	11	9*

[1] Percentages may total more than 100% as households can adopt more than one type of technology simultaneously.
Significant differences between participants and non-participants. * $P<=0.10$; ** $P<=0.05$; *** $P<=0.01$.
[2] Level of significance refers to differences between participants and non-participants in terms of the category distribution, not to the level of adoption.
Source: Dalton et al., 2007.

Table 9 shows the results of a survey about the extent of adoption of various technologies that farmers in Thailand and Vietnam had tested in their FPR trials conducted between 1994 and 2003. The survey was conducted in four project sites each in Thailand and Vietnam as part of focus group discussions with farmers that had participated in the FPR trials and those that lived in villages nearby that had not participated in the project. The data shows that farmers mainly adopted the new varieties, followed by improved fertilizer use, while fewer farmers adopted soil conservation practices and significant numbers of farmers (only in Vietnam) adopted intercropping.

Among soil conservation practices, more farmers adopted contour ridging than contour hedgerows. There were highly significant differences in adoption rates between participating and non-participating farmers, indicating that FPR trials are highly conducive to the adoption of new technologies, especially of those technologies where the immediate benefits are not as clear. In 2015, 12 years after the project ended, many farmers in north Vietnam were still planting and promoting the planting of contour hedgerows of *Paspalum atratum* in their cassava fields, which serve both to reduce erosion on sloping land and as fodder to feed their cattle and water buffaloes.

6 Conclusion

Many of the current cassava cultivation systems have been developed and used for many decades. The world's annual cassava production has increased by an estimated 100 million tonnes since 2000, driven in Asia mainly by demand for starch and dried cassava for use in livestock feed, biofuels and other industrial applications. In Africa demand has been driven by expanding urban markets for cassava food products as a result of population increases and urbanization (FAO, 2013). Farmers have shown that these cultivation systems have some degree of sustainability. However, there is increasing pressure to produce still more cassava to accommodate a growing world population. At the same time, the adverse environmental consequences of rapid commercialization of cassava have also become apparent in some fragile upland farming systems. Furthermore, climate change is adding complexity to the situation by placing greater demands for new technologies that respond to various climate change challenges for cassava in Asia.

Cassava is more resilient than most crops in the face of multiple biotic and abiotic constraints since it can endure a number of pest and disease attacks or even periods of severe drought; however, it is very sensitive to excess water or flooding. Since the crop has no specific maturity period, there is no period of growth (after initial establishment) during which it is especially vulnerable to environmental stresses (Hershey et al., 2012). There is, therefore, great potential for further production increases in many tropical areas. Jarvis et al. (2012) reported that cassava is a highly resilient crop capable of resisting even the most severe consequences of climate change in the tropics and subtropics. Furthermore, cassava fresh root yields can potentially reach 75–80 tonnes per hectare (Cock et al., 1979), compared to the current world average yield of just 11.2 tonnes (FAOSTAT, 2016). The opportunities for increasing sustainable cassava production can only be achieved by using the best-adapted cassava varieties and implementing GAP in cassava cultivation. Cassava production also needs to be profitable with key focus on smallholder farmers. Farm labour scarcity and rising labour costs are some of the factors most responsible for driving various adaptations, including mechanization, in the cassava cultivation system. Ensuring the availability of a wide range of expertise to develop and implement sustainable cassava cultivation strategies is necessary.

7 Future trends

Despite the importance of cassava in food security and income generation in tropical and subtropical countries, most governments give low priority to cassava-related programmes and policies to direct cassava research and implement good agronomic practices for sustainable cassava cultivation. During the past several decades, international research organizations, especially CIAT and IITA, have been collaborating with many of the national agricultural research and extension institutions to conduct research on cassava breeding and evaluation of cassava germplasm, agronomic practices and soil management, cassava cropping systems, and utilization and marketing tools. However, there is still a strong need to further strengthen cassava research holistically. For instance, there could be better collaboration between cassava breeders and agronomist to develop goals and strategies to optimize the response of new varieties to improved agronomic practices. In addition, there exist great opportunities to achieve more widespread adoption of improved varieties and practices by researchers working together with extensionists to help farmers conduct their own research to develop the best practices for their own conditions.

Regarding future priority research areas, more emphasis should be placed on developing better-integrated methods of effective and profitable cassava intensification, the effective use of inputs and the conservation of natural resources in cassava cultivation systems with the primary goal of improving smallholder farmers' livelihoods while protecting the environment. There is still much to learn about the implementation of GAP in sustainable cassava cultivation. The current recommended agronomic practices will need to be adapted to the site-specific conditions of farmers' fields and farmers' requirements.

8 References

American Society of Agronomy. 1989. Decision reached on sustainable agriculture. Agronomy News. January, Madison, Wisconsin, p. 15.

Aye, T. M. 2012. Cassava agronomy: Land preparation, time and method of planting and harvest, plant spacing and weed control. In Howeler, R. H. (Ed.), *The Cassava Handbook – A Reference Manual based on the Asian Regional Cassava Training Course held in Thailand*. Cali, Colombia: Centro Internacional de Agricultura Tropical (CIAT), pp. 588–612.

Aye, T. M. and Howeler, R. H. 2012. Cassava agronomy: Intercropping systems. In Howeler, R. H. (Ed.), *The Cassava Handbook – A Reference Manual based on the Asian Regional Cassava Training Course held in Thailand*. Cali, Colombia: Centro Internacional de Agricultura Tropical (CIAT), pp. 613–25.

Bellotti, A. C., Herrera, C. L., Hernandez, M. P., Arias, B., Guerrero, J. M. and Melo, E. L. 2012. Cassava pests in Latin America, Africa, Asia. In Howeler, R. H. (Ed.), *The Cassava Handbook – A Reference Manual Based on the Asian Regional Cassava Training Course held in Thailand*. Cali, Colombia: Centro Internacional de Agricultura Tropical (CIAT), pp. 199–257.

Calle, F. 2002. Control de malezas en el cultivo de la yuca (Weed control in cassava). In Ospina, B. and Ceballos, H. (Ed.), *La Yuca en el Tercer Milenio. Sistemas Modernos de Produccion, Processamiento, Utilizacion y Commercializacion*. Cali, Colombia: CIAT, pp. 126–8.

Centro Internacional de Agricultura Tropical (CIAT). 1979. Annual Report for 1978. Cali, Colombia: CIAT, Cassava Program, p. A55–76.

Chorley, R. J. 1969. *Water, Earth and Man: A Synthesis of Hydrology, Geomorphology and Socio-economic Geography*. London, UK: Metheun Co. Ltd., p. 588.
Cock, J. H., Franklin, D., Sandoval, G. and Juri, P. 1979. The ideal cassava plant for maximum yield. *Crop Sci*. 19: 271–9.
Cramb, R. A. and Nelson, R. A. 1998. Investigating constraints to the adoption of recommended soil conservation technology in the Philippines. In Penning de Vries, F. W. T., Agus, F. and Kerr, J. (Eds), *Soil Erosion at Multiple Scales. Principles and Methods for Assessing Causes and Impacts*. Oxon, UK: CABI Publ., pp. 99–120.
Dalton, T. J., Lilja, N. K., Johnson, N. and Howeler, R. 2007. Impact of participatory natural resource management research in cassava-based cropping systems in Vietnam and Thailand. In Waibel, H. and Zilberman, D. (Eds), *International Research on Natural Resource Management. Advances in Impact Assessment*. Wallingford, Oxfordshire, UK: CABI, pp. 91–117.
Dierolf, T. S., Fairhurst, T. H. and Mutert, E. W. 2001. *Soil Fertility Kit: A Toolkit for Acid, Upland Soil Fertility Management in Southeast Asia*. Singapore: GTZ, FAO, PT Jasa Katom and PPI/PPIC, p. 136.
FAOSTAT (The Statistics Division of FAO). 2016. www.faostat.fao.org.
Food and Agriculture Organization (FAO). 2013. Save and Grow: Cassava. *A Guide to Sustainable Production Intensification*. (prepared by Howeler, R. H., Lutaladio, N. and Thomas, G.). Rome, Italy: FAO, p. 129.
Graziosi, I., Minato, N., Alvarez, E., Ngo, D. T., Hoat, T. X., Aye, T. M., Pardo, J. M., Wongtiem, P. and Wyckhuys, K. A. G. 2016. Emerging pests and diseases of South-east Asian cassava: a comprehensive evaluation of geographic priorities, management options and research needs. *Pest Manag Sci*. 72: 1071–89.
Henry, G. and Gottret, V. 1996. *Global Cassava Trends. Reassessing the Crop's Future*. CIAT Working Document No. 157. Cali, Colombia: CIAT, p. 45.
Hershey, C. 2015. Redesigning research and capacity building to drive income growth for smallholder cassava producers. In Howeler, R. H. (Ed.), Sustainable Cassava Production in Asia for Multiple Uses and for Multiple Markets. Proceedings of 9th Regional Workshop, held in Nanning, Guangxi, China. 27 November to 3 December 2011, pp. 3–12.
Hershey, C. H., Álvarez, E., Aye, T. M., Becerra, L. A., Bellotti, A., Ceballos, H., Fahrney, K., Howeler, R., Lefroy, R., Ospina, B. and Parsa, S. 2012. Eco-efficient interventions to support cassava's multiple roles in improving the lives of smallholders. In Hershey, C. H. and Neate, P. (Ed), *Eco-efficiency: From Vision to Reality*. Cali, Colombia: Centro Internacional de Agricultura Tropical (CIAT).pp. 135-60.
Howeler, R. H. 2001. Cassava agronomy research in Asia: Has it benefited cassava farmers? In Howeler, R. H. and Tan, S. L. (Eds), Cassava's Potential in Asia in the 21st Century: Present Situation and Future Research and Development Needs. Proceedings of 6th Regional Workshop, held in Ho Chi Minh city, Vietnam. 21-25 February 2000, pp. 345–82.
Howeler, R. H. 2008. Results, achievements and impact of the Nippon Foundation Cassava Project. In Howeler, R. H. (Ed.), Integrated Cassava-based Cropping Systems in Asia. Proceedings of the Workshop on the Nippon Foundation Cassava Project in Thailand, Vietnam and China, held in Thai Nguyen, Vietnam. 27-31 October 2003, pp. 161–223.
Howeler, R. H. 2010c. Technology adoption and impact as a result of the Nippon Foundation Cassava Project in Thailand, Vietnam and China. In Howeler, R. H. (Ed.), A New Future for Cassava in Asia: Its Use as Food, Feed and Fuel to Benefit the Poor. Proceedings of 8th Regional Workshop, held in Vientiane, Lao PDR. 20-24 October 2008, pp. 357–400.
Howeler, R. H. 2012j. Soil erosion control. In Howeler, R. H. (Ed.), *The Cassava Handbook. A Reference Manual Based on the Asian Regional Cassava Training Course*, held in Thailand, pp. 524–55.
Howeler, R. H. 2012k. Farmer participation in research and extension: The key to achieving adoption of more sustainable cassava production practices in Asia. In, Howeler, R. H. (Ed.). *The Cassava Handbook. A Reference Manual Based on the Asian Regional Cassava Training Course*, held in Thailand, pp. 556–87.
Howeler, R. H. 2014. Sustainable Soil and Crop Management of Cassava in Asia – A Reference Manual. Cali, Colombia: Centro Internacional de Agricultura Tropical (CIAT). p. 280.

Howeler, R. H. and Aye, T. M. 2014. *Sustainable Management of Cassava in Asia – From Research to Practice*. Cali, Colombia: Centro Internacional de Agricultura Tropical (CIAT). p. 148.

Howeler, R. H., Watananonta, W. and Ngoan, T. N. 2007. Farmer participation in research and extension: The key to achieving adoption of more sustainable production practices on sloping land in Asia and their impact on farmers' income. In de Graaff, J., Cameron, J., Sombatpanit, S., Píeri, Ch. and Woodhill, J. (Eds), *Monitoring and Evaluation of Soil Conservation and Watershed Development Projects*. Enfield, NH., USA: Science Publishers, Inc. Chapter 24, pp. 435–76.

IITA, 1990. *Cassava in Tropical Africa: A Reference Manual*. Ibadan, Nigeria: IITA, p. 7–8.

Jantawat, S., Tongglum, A., Putthacharoen, S., Poolsanguan, P. and Howeler, R. H. 1994. Conservation practices for sustainable production of cassava on sloping land. Sustaining environmental quality: The erosion control challenge. In Proceedings of 25th Conference of Intern. Erosion Control Assoc. pp. 521–6.

Jarvis, A., Ramírez-Villegas, J., Herrera-Campo, B. and Navarro-Racines, C. 2012. Is cassava the answer to African climate change adaptation? *Trop. Plant Biol.* 5(1): 9–29.

Jongruaysub, S., Namwong, P., Tiensiriroek, A., Laochaikarm, C., Joodkong, A., Katong, S., Watananonta, W. and Howeler, R. H. 2007. Minimum tillage for cassava in Thailand. In Howeler, R. H. (Ed.), Cassava Research and Development in Asia: Exploring New Opportunities for an Ancient Crop. Proceedings of 7th Regional Workshop, held in Bangkok, Thailand. 28 October-1 November, 2002. pp. 251–64.

Leihner, D. 1983. *Management and Evaluation of Intercropping Systems with Cassava*. Cali, Colombia: Centro Internacional de Agricultura Tropical (CIAT). p. 70.

Margolis, E. and Campos Filho, O. R. 1981. Determinação dos fatores da equação universal de perdas de solo num podzólico vermelho amarelo de Glória do Goitá (Determination of factors in the universal soil loss equation for a red-yellow podzolic soil in Gloria do Goita). In Anais do 3rd Encontro Nacional de Pesquisa Sobre Conservação do Solo, held in Rengife, Pernambuco, Brazil. 28 July-1 August, 1980, pp. 239–50.

Milliman, J. D. and Meade, R. H. 1983. World-wide delivery of river sediments to the ocean. *J. Geol.* 91: 1–21.

Nguyen Huu Hy, Nguyen The Dang, and Tong Quoc An. 2010. Soil fertility maintenance and erosion control research in Vietnam. In Howeler, R. H. (Ed.), A New Future for Cassava in Asia. Its Use as Food, Feed and Fuel to Benefit the Poor. Proceedings of 8th Regional Workshop, held in Vientiane, Lao PDR. 20-24 October 2008, pp. 263–74.

Norman, D., Janke, R., Freyenberger, S., Schurle, B. and Kok, H. 1997. Defining and implementing sustainable agriculture, *Kansas Sustainable Agriculture Series, Paper #1*. Manhattan, KS, USA: Kansas State University.

Ospina, B., Cadavid, L. F., Garcia, M. and Alcalde, C. 2007. Mechanization of cassava production in Colombia. In Howeler, R. H. (Ed.), *Cassava Research and Development in Asia: Exploring New Opportunities for an Ancient Crop*. Proceedings of 7th Regional Workshop, held in Bangkok, Thailand, 28 October to 1 November 2002, pp. 277–87.

Putthacharoen, S., Howeler, R. H., Jantawat, S. and Vichukit, V. 1998. Nutrient uptake and soil erosion losses in cassava and six other crops in a Psamment in eastern Thailand. *Field Crop. Res.* 57: 113–26.

Quintiliano, J., Margues, A., Bertoni, J. and Barreto, G. B. 1961. Perdas por erosão no estado de São Paulo (Erosion losses in Sao Paulo State). *Brigantia* 20(2): 1143–82.

Tongglum, A., Vichukit, V., Jantawat, S., Sittibusaya, C., Tiraporn, C., Sinthuprama, S. and Howeler, R. H. 1992. Recent progress in cassava agronomy research in Thailand. In Howeler, R. H. (Ed.), Cassava Breeding, Agronomy and Utilization Research in Asia. Proceedings of 3rd Regional Workshop, held in Malang, Indonesia. 22-27 October 1990, pp. 199–223.

Tongglum, A., Suriyapan, P. and Howeler, R. H. 2001. Cassava agronomy research and adoption of improved practices in Thailand – Major achievements during the past 35 years. In Howeler, R. H. and Tan, S. L. (Eds.), Cassava's Potential in Asia in the 21st Century: Present Situation and Future Research and Development Needs. Proceedings of 6th Regional Workshop, held in Ho Chi Minh city, Vietnam. 21-25 February 2000, pp. 228–58.

Villamayor, F. G. 1988. Agronomic research on cassava in the Philippines. In Howeler, R. H. and Kawano, K. (Eds), Cassava Breeding and Agronomy Research in Asia. Proceedings of 2nd Regional Workshop, held in Rayong, Thailand. 26-28 October, 1987, pp. 261–96.

Villamayor, F. G., Dingal, A. G., Evangelio, F. A., Ladera, J. C., Medellin, A. C., Sagise Jr., G. E. and Burgos, G. B. 1992. Recent progress in cassava agronomy research in the Philippines. In Howeler, R. H. (Ed.), Cassava Breeding, Agronomy and Utilization Research in Asia. Proceedings of 3rd Regional Workshop, held in Malang, Indonesia. 22-27 October 1990, pp. 245–59.

Wang, H. L., Cui, X. Y., Wang, X. W. and Liu, S. S. 2016. First report of Sri Lankan cassava mosaic virus infecting cassava in Cambodia. *Plant Dis.* 100(5): 1029.

Wargiono, J., Guritno, B. and Hendroatmodjo, K. 1992. Recent progress in cassava agronomy research in Indonesia. In Howeler, R. H. (Ed.), Cassava Breeding, Agronomy and Utilization Research in Asia. Proceedings of 3rd Regional Workshop, held in Malang, Indonesia. 22-27 October 1990. pp. 185–98.

Wargiono, J., Guritno, B., Sugito, Y. and Widodo, Y. 1995. Recent progress in cassava agronomy research in Indonesia. In Howeler, R. H. (Ed.), Cassava Breeding, Agronomy Research and Technology Transfer in Asia. Proceedings of 4th Regional Workshop, held in Trivandrum, Kerala, India. 2-6 November 1993, pp. 147–74.

Wargiono, J., Kushartoyo, S. H. and Guritno, B. 1998. Recent progress in cassava agronomy research in Indonesia. In Howeler, R. H. (Ed.), Cassava Breeding, Agronomy and Farmer Participatory Research in Asia. Proceedings of 5th Regional Workshop, held in Danzhou, Hainan, China. 3-8 November 1996, pp. 307–30.

Wargiono, J., Widodo, Y. and Utomo, W. H. 2001. Cassava agronomy research and adoption of improved practices in Indonesia – Major achievements during the past 20 years. In Howeler, R. H. and Tan, S. L. (Eds), Cassava's Potential in Asia in the 21st Century: Present Situation and Future Research and Development Needs. Proceedings of 6th Regional Workshop, held in Ho Chi Minh city, Vietnam. 21-25 February 2000, pp. 259–78.

Watananonta, W., Tangsakul, S., Katong, S., Pretprapi, P., Jantawat, S., Samuthong, N. and Howeler, R. H. 2006. Effect of methods of land preparation on the yields of four cassava cultivars in Thailand. In Tan, S. L. (Ed.), Proceedings of 2nd International Symposium on Sweetpotato and Cassava: Innovative Technologies for Commercialization. International Society of Horticultural Sciences (ISHS). *Acta Hort.* 703: 225–32.

Chapter 12

Seed systems management in cassava cultivation

James George and Sarojini Amma Sunitha, ICAR-Central Tuber Crops Research Institute, India

1 Introduction
2 General description of cassava seed
3 Formal and informal seed systems
4 Quality and standards for planting material
5 Production of quality planting material
6 Rapid multiplication techniques
7 Formal seed system
8 Case study: mini-stake-based rapid multiplication at ICAR-CTCRI, India
9 An integrated approach for the future
10 Future trends and conclusion
11 Where to look for further information
12 References

1 Introduction

Seed is one of the most crucial elements in the livelihoods of agricultural communities. It is the repository of the genetic potential of crop species and their varieties resulting from continuous improvement and selection over time. Seeds are at the heart of agricultural production and determine what farmers grow and what they harvest.

A sustainable seed system will ensure that high-quality seeds of farmer-demanded varieties and crops are produced and fully available on time and affordable to farmers and other stakeholders. Production increases brought about by the use of adapted varieties increases farmers' income when market linkages exist. Food security is heavily dependent on the seed security of the farming community. However, in many developing countries, farmers have not yet been able to fully benefit from the advantages of using quality seed due to a combination of factors, including inefficient seed production, non-existent or ineffective distribution and quality assurance systems, as well as bottlenecks caused by a

http://dx.doi.org/10.19103/AS.2016.0014.24

lack of good seed policy or key issues such as access to credit for inputs. Furthermore, the uncertainties from fluctuating food prices and climate change create additional challenges.

Cassava (*Manihot esculenta* Crantz) is the fourth most important food crop in the world, after rice, wheat and maize. Its high productivity under adverse conditions, year-round availability, and its wide adaptability to various farming and food systems make it an ideal food security crop and versatile industrial raw material. For commercial plantings, cassava is cloned by means of stem cuttings (stakes). For breeding and research purposes, it is also propagated through botanical seeds. Plants from botanical seed segregate widely for many traits, which is unsuitable for commercial production.

Availability of high-quality cuttings for planting is one of the most important problems in cassava production systems. Efficient seed systems are generally lacking throughout cassava production areas. Seed availability and quality can be especially an issue during the dry season or in dry zones, and where pests and diseases cause degeneration of planting material from one cycle to the next (Dickson, 2003). Planting material is typically produced and used on-farm but sometimes also traded with neighbours, or marketed through middlemen or in rural markets. Cassava's low traditional multiplication rate (about 1:5–10) is generally satisfactory in stable production systems, but limits expansion of area or adoption and diffusion of new varieties (IITA, 1998). After crop harvest, stems are often stored from a few days up to a few months. In most areas, only minimal efforts are made to preserve cassava stem cuttings at a high level of quality after harvest of the roots (Eke-okoro et al., 2005).

The large majority of cassava seed is produced informally. Private company interest in the multiplication, distribution and marketing of cassava varieties has been generally low; however, there are several emerging cases of private investment. Constraints to private seed sector development are: low or fluctuating demand; the high cost of a year-long period of producing cassava planting materials; the large area required to produce substantial quantities of stems; the bulkiness of the planting materials, which has implications for management and transportation; and limited storage period compared to grain crops. This combination of constraints discourages both the public sector and private companies from investing in production, distribution and marketing of cassava stem cuttings to farmers.

This chapter describes the production and management of high-quality cassava seed under both informal and formal seed systems, and under both traditional and different types of rapid propagation methods.

2 General description of cassava seed

Each viable bud distributed on the cassava stem is capable of developing into individual plants. These buds, on mature lignified stems, are the fundamental basis of cassava seed. The stem cuttings, or stakes, of 15–20 cm, the typical length used for planting, will typically have 10–12 buds. Stem length determines the quantity of seed produced, which in turn is influenced by the growth habit of varieties. Erect less-branching types can give rise to a greater number of stakes (up to 30), and facilitate the harvest, storage and transport of stems. The reproductive rate of highly branched types can be as low as 1:3–5 (FAO, 2010). Plant height may vary from 1 to 4 m depending upon the variety, growth habit, climate, soil conditions, management, etc. The normal rate of multiplication of planting material is about 1:5–10 at harvest (George, 2006). Planting material from the central part of the main stem, after discarding the green tender portion at the top as well as the basal woody portion, has been found to be ideal for obtaining higher root yield (Ravindran et al., 2006).

3 Formal and informal seed systems

Farmers, particularly small farmers, practice or participate in multiple kinds of seed systems to produce and obtain the seed they need. These systems can be broadly divided into two types: a formal seed system and a local system. The local system is also sometimes called the informal, traditional or farmer seed system.

The formal seed system is easier to characterize, as it is a deliberately constructed system that involves a chain of activities leading to clear products: certified seed of verified varieties. The chain usually starts with plant breeding and selection, resulting in new varieties, and leading to formal variety release, promotion and maintenance. Guiding principles in the formal system are to maintain varietal identity and purity and to produce seed of optimal physical, physiological and sanitary quality. Certified seed marketing and distribution take place through a limited number of officially recognized seed outlets, usually for financial sale (Louwaars, 1994).

In a local or informal seed system, activities tend to be integrated and locally organized. Normally the local system embraces most of the other ways in which farmers themselves produce, disseminate and access seed: directly from their own harvest; through exchange and barter among friends, neighbours and relatives; and through local markets. What characterizes the local system most is its flexibility, and it encompasses a wider range of seed system variations. Varieties may be pure landraces, mixed races (multi-clonal), or hybrids resulting from breeding programmes. In addition, the seed is of variable quality (Almekinders and Louwaars, 1999). The same general steps or processes take place in the local system as in the formal sector, but they take place as integral parts of farmers' production systems (where generally the main product is the roots) rather than as discrete activities.

The principal actors for the informal sector include farmers, traders, cooperatives, processors and other business people. The formal sector includes universities, national and international agricultural research institutes, regulatory bodies, small and medium enterprises, national and multinational seed companies, agro-input agencies, financial service providers, etc.

There is growing interest and policy research in informal seed systems, as informal channels provide 80–90% of the materials (across crops) that farmers sow in their fields worldwide (Cooper,1993). A wide range of crops, including cassava, move primarily through the informal seed system. Local markets are the major venue for much of the grains and legumes, in contrast to the social networks (neighbours, relatives, friends, etc.) which are the standard conduits for moving planting materials of vegetatively propagated crops such as cassava and sweet potato (Sperling et al., 2013).

Better understanding of informal markets is an important precursor to strengthening them, as such markets have unrealized potential to deliver more and higher quality seed of both modern and local varieties. Support for informal seed markets could usefully feature in rural livelihood and social protection programmes, but this will require basic shifts in interventions and further refinements in market analysis.

The basic process of growing cassava for multiplication of stems for use as seed is the same as that for root production. Unlike other root and tuber crops such as yams and aroids, in cassava it is the stems left after harvest rather than the roots or tubers that form the planting material for the next generation. The fact that the main commercial product (roots) does not compete for seed needs, as in the grains, legumes and some clonal crops, is a great advantage for cassava.

Informal systems can be improved relatively easily, through managing a separate area within production fields, specifically for seed production – typically about 10% of the area that will be planted in the next growing season (taking into account that the normal rate of multiplication of planting material is about 10-fold at harvest). This on-farm seed production area should have adequate management conditions and periodic inspections as close as possible to that of a more formal seed nursery to ensure quality planting material. Emphasis may be given to the production of maximum number of quality stems, rather than roots, but roots are also likely to be harvested and marketed.

4 Quality and standards for planting material

4.1 Degradation of planting material

Degradation of planting material, which results in declining yields from one planting cycle to the next (Seesahai and Ousman, 2012), is a major concern in many vegetatively propagated crops, including cassava, and may be caused by any one or more of the following:

- Systemic diseases, caused by viruses, bacteria or phytoplasma.
- Low soil fertility and nutrient imbalance.
- Soil salinity.
- Inappropriate crop cycle, where the crop is planted in the off-season, not coinciding with rainfall availability and the drought affects the vegetative growth.
- The socio-economic situation of the farmer, which impacts the overall management of the crop.
- Disorganized systems of supply, which result in staggered harvesting and replanting.
- Low storage potential of stakes related to nutritional status of mother plants and physiological quality of the stakes. There appear to be genetic differences for storage potential of stems, but this is not well-documented by research.

4.2 Pests and diseases transmitted through planting material and their management

Pests and pathogens carried from one generation to the next through propagules are a severe problem in vegetatively propagated crops. This section focuses on their effects on seed quality and their management for improved seed. These problems are rarely universal, but rather affect specific regions under specific conditions of growth and management.

Sucking pests frequently infest stems, or move from leaves to stems, which then become the vehicle for transmission to the next crop cycle. Infested cuttings are the most important vehicle for the cassava green mite (*Mononychellus tanajoa*), scale insects (*Aonidomytilus albus, Saissetia miranda*) and mealybugs (*Phenacoccus herreni* and, *P. manihoti*). The eggs and larvae of other insects such as thrips (*Frankliniella williamsi, Corynothrips stenopterus* and *Caliothrips masculinus*) and lace bug (*Vatiga* sp.) adhere to the surface of stems and spread via infested cuttings.

Stems free of scales and mites should be collected if possible. As a prophylactic measure, the stems should be sprayed with or dipped in a suitable pesticide during storage.

The insects most commonly found within the stem are the stem borers from different species of Coleoptera (*Coelosternus* sp. and *Lagochirus* sp.), Lepidoptera (particularly *Chilomima* sp.) and Hymenoptera fruit flies (*Anastrepha* sp.). The surface or subterranean cutworms that feed on the stem (*Agrotis ipsilon* and *Prodenia eridania*) are often carried inadvertently from one place to another through infested planting material. In addition, the galleries that they make in the stem facilitate access of microorganisms that cause stem rot, but on the other hand, also provide an easy way to identify infested planting material.

Systemic pathogens include vascular (viruses, bacteria and/or phytoplasma) and cortical or epidermal (different fungi) causal agents that invade the host systematically, often without leaving any visible signs in the mature portion of the stem. Without proper precautions there is every chance that these systemic pathogens can be disseminated through stem cuttings. The most important diseases propagated by infected planting materials are discussed in the following sections.

4.2.1 Cassava mosaic disease (CMD)

The disease is prevalent in Africa, India and Sri Lanka (Fig. 1). It spreads mainly through indiscriminate use of infected planting material. Since the causal virus is also vectored by the whitefly *Bemisia tabaci*, clean plantations may later become infected if there are neighbouring fields as sources of inoculum. CMD might be effectively managed solely through the use of disease-free planting material if there is a very effective seed system in which all farmers participate. However, this is very difficult for cassava, and therefore it is recommended that clean seed strategies be combined with moderate to high levels

Figure 1 Cassava mosaic disease.

of host plant resistance to manage the disease. Regular field inspection to rogue out the infected plants, even those with low disease incidence, must be followed while multiplying the planting materials.

4.2.2 Cassava brown streak disease (CBSD)

The disease is currently reported only in East Africa but is moving slowly westward, posing an eventual threat across sub-Saharan Africa and even to Asia and Latin America. The fact that infected plants may not show leaf symptoms makes this disease very challenging from a seed systems perspective. The virus is vectored by whiteflies (*Bemisia* sp.) and is also transmitted through infected cuttings. Harvesting plants and observing the cut roots is the best visual method to select clean planting material. Regular inspection to remove the suspected plants by uprooting and control of vectors can be followed in endemic areas.

4.2.3 Cassava bacterial blight (CBB)

The disease is found globally and caused by *Xanthomona saxonopodis* pv. *Manihotis* (Fig. 2). It is systemic and is favoured by cool wet conditions. Clean planting material may more likely be collected from symptomless plants and from the lignified portion of stems up to 1 m from the base. Parts of the stem with vascular browning should be discarded. Regular disinfection of tools; crop rotation and fallowing; removing infected debris and weeds; dry heating of stems (50–52°C for 20–30 minutes); foliar spraying with Agrimycin, copper sulphate or Bordeaux mixture; and use of resistant varieties are recommended to manage the disease.

4.2.4 Super-elongation disease (SED)

A fungal disease induced by *Sphaceloma manihoticola* (Teleomorph: *Elsinoe brasiliensis*) that is widespread in the Americas. Symptoms include an exaggerated elongation of stem

Figure 2 Cassava bacterial blight.

internodes, an easily recognized diagnostic feature. The disease spreads readily through infected cuttings, and plants with symptoms should be discarded as seed sources.

4.2.5 Root rots

Several species of *Phytophthora* cause root rot and the infected stems are the most important dissemination mechanism. The *Diplodia manihotis* fungus causes root and stem rots in Africa, Latin America and the Caribbean. The disease mainly spreads through water, contaminated tools and plant debris with fungal inoculum, and infested plantations should be avoided as sources of seed.

4.2.6 Cassava frog skin disease (CFSD)

Unknown aetiology although suspected causal agents including various viruses and a phytoplasma are implicated. CFSD is an economically important root disease affecting cassava in Colombia, Brazil, Venezuela, Peru, Costa Rica and Panama. CFSD is a challenging disease to manage in seed systems because of a general lack of foliar symptoms. In fact, because the disease restricts photosynthate partitioning to the roots, the aerial part of the plant is often more vigorous, and therefore might be more appealing to farmers as a source of seed. Cassava growers often do not realize until harvest that the plants have been infected and roots damaged. However, it is known from wide experience in Latin America that it is possible to attain good control of the disease if plants are harvested, and roots observed for CFSD symptoms, prior to cutting stems for seed for the next generation.

Routine diagnostic methods have been standardized for detection of ICMV and SLCMV in planting material, which includes serological methods like enzyme linked immunosorbent assay (ELISA), dot immunobinding assay (DIBA) and tissue immunoblot assay (TBIA) using monoclonal antibodies SCR58 and 60 (Makeshkumar and Nair, 2001). Nucleic-acid-based diagnostic methods include: polymerase chain reaction (PCR) (and their different types like RT-PCR, multiplex PCR) using specific primers and nucleic acid spot hybridization (NASH) using a coat-protein-gene-specific probe. Among all these techniques, ELISA, DIBA, multiplex PCR and NASHG are being extensively used at CTCRI for routine diagnosis of cassava mosaic disease. For the phytoplasma agent causing cassava witches broom disease, RT-PCR techniques can effectively identify infected plants.

A main challenge with molecular diagnostics to detect and monitor seed health is the large number of samples that typically need to be managed. It is rarely possible to test every plant, especially in a large-scale seed system. Typically, visual symptoms remain the primary focus of phytosanitary assessment, and molecular diagnostics are applied on the basis of random sampling throughout a field. If positive samples are found, additional targeted testing can be done to determine the extent of infection and whether measures can be taken to obtain clean seed.

4.3 Physiological quality of planting material

Physiological quality relates to the ability of the stake to sprout easily, including after a normal period of storage, and produce a vigorous, high-yielding plant. Several criteria define physiological quality for cassava seed (FAO, 2010):

- High dry matter content, that is, good carbohydrate reserves for stake storage and for nutrition of the newly growing buds after planting.

- Lignified xylem vessels and a reduction of medullary rays and parenchyma cells.
- High levels of N, P, K, Ca and Mg. Cassava plants grown with low levels of nitrogen, phosphorus and potassium produce stakes that are also low in those nutrients, starch, reducing sugars, and total sugars, and in turn have a lower rate of sprouting, followed by fewer stems and lower root yields (Lopez and El Sharkawy, 1995).
- Stake viability is directly proportional to its moisture content.
- Cut sections dehydrate quickly. When stored under high temperatures and low relative humidity, 20–50% moisture is lost within a matter of days, causing up to 50% or more reduction in viability.
- Latex flow at the point of a fresh cut is a good indicator of stake quality and viability.
- Plants affected by pests and pathogens often have less vigorous growth, thereby affecting the overall physiological quality of stakes (in addition to the risk of pest or disease transmission).

5 Production of quality planting material

Optimized production and preparation of stakes are very important for better cultivation of cassava. The conventional cassava multiplication method, based on stem pieces of 15–20 cm, is the easiest and most widely used (Fig. 3 and 4). However it has the key disadvantage (among others) of a low multiplication ratio (1:5–10 per year). On the other hand, rapid multiplication techniques have multiplication ratios of 1:60–100 or even up to 1:many thousands per year for some systems. This section first describes the conventional method based on mature stems and standard planting stakes, followed by the description of three broad types of rapid propagation to accelerate propagation rate and reduce phytosanitary risks.

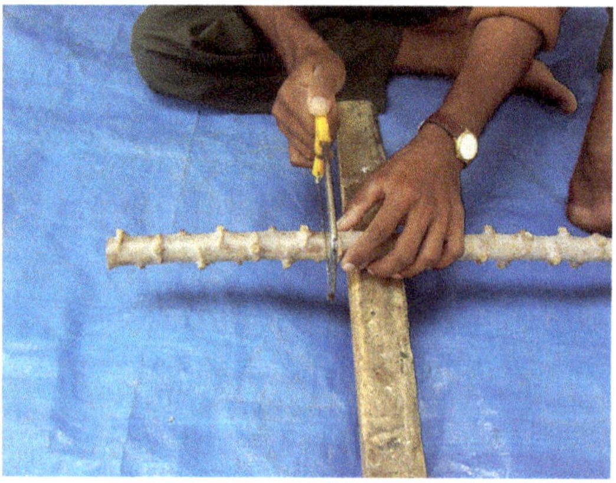

Figure 3 Normal sett preparation.

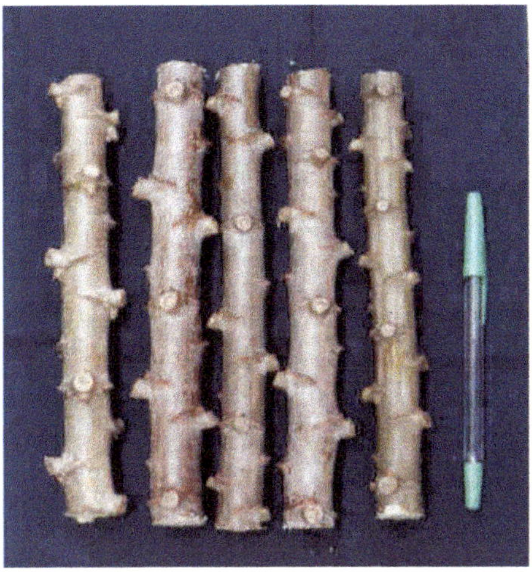

Figure 4 Setts of 20 cm length with 10–12 buds.

5.1 Selection of a variety

A variety to be used as planting material should be in high demand among the local community, recommended by a research institution, and released as a State/National variety by the Government or a certified seed production agency. Projected farmer demand will determine the choice of materials to multiply. However, newly released varieties with potential high demand for food consumption and industrial use should be included in order to meet expected demand as farmers learn about them and would like to try them.

Two types of varieties are generally bred and released by researchers: varieties for industrial use and varieties for consumption. For industrial purposes, the objective is usually to breed for yield, starch content and quality. For culinary purposes, the consideration is generally for yield, cooking quality, and storability. Both types of variety should have tolerance to predominant pests and diseases, and adaptation to the climate and soil conditions where grown. Both types of variety are amenable to the same seed system methods and technologies. However, since varieties for industrial use may tend to be managed at higher levels of inputs and technologies, seed systems will also often be more advanced for industrial varieties.

5.2 Selection of planting material

Only healthy planting materials should be used. The guidelines for selecting good-quality planting materials are:

- Plants that are mature – about 7–12 months old. Proper maturity (neither too young nor too old) is indicated by good lignification and emergence of latex from cut ends.
- Vigorous plants with robust stems (2–3 cm girth) and with lush foliage.
- Avoid wounding or bruising stems. Wounds are potential entry sites for pathogens.
- It is better to have a smooth circular cut rather than an irregular cut for uniform callus formation and root initiation.

- Avoid plants with pests and diseases. Many cassava pests and diseases are stem-borne and are spread through distribution and planting of infested or infected cuttings. Treat cuttings infected with cassava mealybugs and cassava green mites with a recommended pesticide before storage and distribution/planting (e.g. dimethoate 0.05%). Among the diseases, cassava mosaic spreads through planting material and hence roguing out of infected plants and field sanitation are important.

5.3 Land preparation and planting

In situations where seed is produced on-farm, the land dedicated to growing plants for seed will need to be about 10% of area anticipated for planting for root production in the following season (precise area will depend on the specific multiplication rate expected). Where seed is produced for trade or sale, the area planted will depend on expected market demand. Land for cassava seed production should meet the following requirements (FAO, 2012):

- Fertile and well drained.
- Free from cassava residue and volunteer plants of a previous crop. The land should not be planted with cassava at least for the previous season.
- Free from shaded conditions.
- Free from drainage from other cassava fields.
- If produced for sale or trade – accessible to beneficiaries or seed buyers for easy transportation.
- Isolated from other cassava fields. For cassava, isolation of seed production fields from commercial production is not necessary to assure seed purity, but may be advisable if there are chances of spread of pests or pathogens into the seed plots, or if the risk of mixing of certified and uncertified seed, or of varieties, is a factor based on proximity of different fields.

5.4 Field maintenance and inspection

Land for cassava seed multiplication should be prepared early to enable planting with the early rains. Early planting enables the crop to establish while there is still adequate moisture. Other land preparation practices are similar as for plants for root production.

The use of two stakes planted per hill, as used by some farmers in order to increase the rate of multiplication. However, this practice can considerably reduce the number of high-quality planting materials (Mohankumar et al., 1980). Thus it is recommended to plant a single stake only per mound to avoid excessive competition. The fields should be weeded and fertilized with manures and fertilizers at rates somewhat above those recommended rates for root production for the respective location. These higher rates may contribute more to foliage growth, and hence to seed quality, than to increased root yield. Nonetheless, soil fertility management will be quite specific for variety and for location, and should be discussed with local extension agents or other experts.

Attack of pests and diseases transmitted through stakes as described above must be checked and eliminated to obtain vigorous stems. Routine inspections, at fortnightly intervals, should be carried out to ensure that infected plants (e.g. with mosaic) are uprooted and destroyed outside the field. For pests such as scales (Fig. 5) and mealybugs,

Figure 5 Stems attacked by scales.

need-based pesticide spray may be applied to the growing plants, even if such would not be warranted on a crop destined for root production only.

Good planting materials should be true-to-type to the variety being multiplied, with no admixtures. During field inspections off-types must be identified, uprooted and destroyed. This inspection will typically be based on visual characteristics such as leaf and stem colours, branching habit and leaf shape. Molecular fingerprinting tools are available, but currently their cost is typically not justified in routine production of certified seed. It is expected that costs will continue to decline and molecular ID techniques will become routine.

5.5 Stem harvesting and collecting

Stems for use as planting material are collected from mature plants of 8–18 months of age, depending on the variety and on growing conditions. Since cassava has no physiological maturity, it is preferable to maintain the planting materials in the nursery fields rather than harvest them too early, entailing the need to store for 2–3 months. Also, requirements for field management, such as rotation to other crops, or preparation for the next cassava nursery, will influence the flexibility in time of harvest of planting material.

Planting materials should be selected from plants with high root yield. This simple practice can markedly increase production, especially of traditional varieties that may have degenerated due to slow accumulation of pests and pathogens over a long period of time, even if they are moderately resistant (FAO, 2013). Uprooting prior to harvest of stems, and discarding of lower-yielding plants, can be one of the most effective practices to improve seed quality. Nonetheless, it adds a cost and requirement for expertise that must be factored into planning and management. In addition, uprooting prior to stem harvest will not be possible in ratoon cropping systems (see Section 5.8).

There is no dormancy in cassava and fresh stems can be planted immediately after harvest. Plants should be cut 20–25 cm above the ground. The younger the plant, the closer to the basal, more lignified part of the stem selected for the cutting should be. One practical way of knowing whether a stem is sufficiently mature is to determine the relationship between the diameter of the pith and the full stem in a transversal cut (Fig. 6). If the diameter of the pith is equal to or less than 50% of the diameter of the stem, it is sufficiently mature to be used for propagation (Lozano et al., 1977). Stakes are prepared by discarding one-third of the total length of the stem from the top and the thick, very lignified section at the bottom (at least 5 cm). It was observed that survival of stakes and root yield were significantly higher when the lower part of the stem was used as planting material as compared to the immature upper part (Mohankumar and Potty, 1999). This is likely due to the early vigour advantage that more mature stakes provide as a result of having higher carbohydrate reserves.

During harvest, the plants should be screened again to eliminate any stems showing evidence of pest or disease attack.

The immature branches are cut and discarded and mature stems meeting the quality standards may be cut and tied together to form bundles of convenient size for storage and transport. Each bundle should be properly labelled with name of variety, date of collection, location, etc., before storing, packing or transporting. On an average, each stem can yield 5–7 stakes. Depending on the growth habit, the number varies from 3 to 12 stakes. In most well-managed cassava plantations, each plant will have an average of about 2 stems, therefore with a yield of about 10–14 stakes per plant. The number of stems that can be harvested depends on the variety, soil fertility and management of weeds, pests and diseases. At and after harvesting, care should be taken to avoid bruising the stems since bruised buds may not sprout. At and after harvesting, care should be taken to avoid bruising the stems since bruised buds may not sprout.

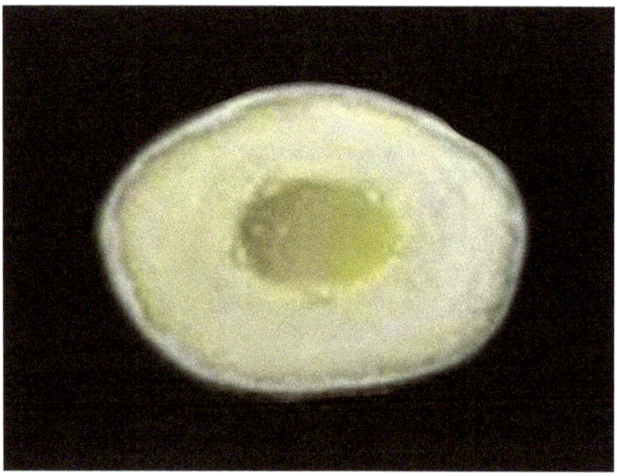

Figure 6 Cross section of cassava stem showing the ideal relationship between pith thickness and diameter.

5.6 Storage of planting material

Once the clean stems are harvested, they should be properly stored to ensure the viability of buds at a later stage of planting. Fresh stakes from mature stems are ideal for obtaining best results. However, many producers have to store the stems before they are used for cultivation due to prolonged drought, excess moisture in the field, or field management considerations. Long storage causes loss of moisture and nutrients, and exposure to many pests. Uncut stems must be stored in shady, well-ventilated areas, avoiding wet conditions. Farmers usually store the stems vertically under shade of trees for up to about 2–3 months (Fig. 7). Stems produced through mini-stake techniques (see Section 6.1) also should be stored in a well-ventilated storehouse with diffused light. Long stems (>1 m) are better preserved than short ones. Application of chlorpyrifos at the base of stems at the beginning of storage and at 60 days intervals protect them from termite infestation. To control fungal attack, cassava stems can be treated with a combination of systemic and protectant fungicides. As a prophylactic measure, spray the stems with 0.05% dimethoate to control pest attack during storage.

Although stems are not considered to express dormancy, storing them for a period of 15 days has been observed to give maximum sprouting (96%) as compared to planting fresh stakes (90%). Storage beyond 60 days resulted in the loss of viability of planting material and sprouting was less than 75% (Nair et al., 2008). Planting materials dehydrate and lose significant amounts of carbohydrate depending on the duration and conditions of storage. Thus physiological and pathological factors may both cause deterioration of planting material in storage, which account for loss of viability and sprouting vigour and resulting in poor yield of the subsequent crop (Ravi et al., 2008).

Figure 7 Storage of cassava stems.

Stems stored in vertical position gave better sprouting when compared to horizontal storage (CTCRI, 1983). The stems can be kept vertical by providing a support structure, for example inside a storage shed or under a shade roof, against a tree, or by supporting bundles against each other in a free-standing stack. A low-cost and farmer-friendly method for storing cassava planting materials is a *zero energy cool chamber* (ZECC). Cassava stems can be safely stored for over three months in a ZECC under hot, dry weather conditions, which is practised for example in some parts of Andhra Pradesh, India. The ZECC is a room built of double brick walls with sand sandwiched in between, and kept moistened. The structure has an elevated thatched roof with open ventilation above walls, on all sides. A single ZECC can hold cassava stems adequate for one hectare (Ravi et al., 2008). Stems selected as planting materials could be safely stored in a ZECC for 3–4 months. Significant differences were observed in viability of stems from the ZECC, which remained fresh and green as compared to the stems stored under tree shade.

Results from a survey in Ghana (Osei et al., 2009) showed that long-term storage (over 8 weeks) of stems is not possible under farmers' conditions because of pest and disease attack, dehydration and the sub-optimal quality of planting material used. Vertical storage was found to be the most common method. The various storage methods identified are: 1) stems are tied into bundles and stored in vertical position under a tree or other shade with the lower end inserted in soil; 2) stems are stored horizontally under a tree or other shade; 3) some of the roots are harvested such that the stems continue to receive nutrients and water from the remaining roots; 4) stems are cut into planting stakes (about 20 cm), put in sawdust under shade and covered by a perforated polybag; or 5) individual stems are inserted upright into the soil without tying them into bundles.

Improving storage conditions such as avoiding direct sunlight, ensuring that the buds face upwards and the use of long and mature stems under the vertical storage method will go a long way toward improving quality of planting materials in storage under farmers' conditions (Osei et al., 2009). Further, covering the stems with foliage or straw during storage reduces dehydration. Stems can be cut into stakes of suitable size just prior to planting in order to minimize dehydration of stakes.

5.7 Transport of planting material

In addition to the constraint imposed by low multiplication rate, cassava stem cuttings are bulky, difficult to transport and highly perishable. It is often necessary to transport stems from one place to another. The large volume of cassava stems makes the use of trucks/lorries obligatory for mass transportation of stems over long distances (IITA, 2002). The transportation of cassava cuttings should ideally be to points as close as possible to the source, on good roads to minimize node damage from desiccation or mechanical injury, and to reduce costs. Nevertheless, rural roads in cassava-growing areas are frequently in very poor condition. The transportation shock canal can be ameliorated by using suitable packing materials. The ideal packaging will protect the stems from damage during handling, transport and storage, with containers of uniform size that can be conveniently stacked and handled, and easily account for quality and appropriate weight.

Transporting long stems often becomes difficult. Hence the stems need to be cut into suitable sizes and packed properly to avoid desiccation and maintain viability after transport. Studies conducted at ICAR-CTCRI revealed that packing mini-stakes in gunny bags or cloth bags for transportation for up to five days does not adversely affect sprouting. For long-term transport of normal stem cuttings of 20 cm length gunny bags are better, while packing in polythene causes fungal attack of stems (Sunitha and Ravindran, 2012). When long stems are transported, putting banana leaves or gunny bags in between the bundles minimizes damage of stems.

5.8 Ratoon cropping for seed production

Cassava fields for seed production have the strong advantage that they may be ratoon-cropped. That is, after stem harvest at the end of the first growth cycle, plants are left in the field to re-grow. Several shoots sprout from the stumps after cutting, often more than sprout from the original stake in the first season. These should be thinned subsequently to 2–3 per stump, which will mature into stems.

While ratooning, fertilizers should be applied to boost growth and the field should be kept weed free. Another set of stems can be harvested 9–11 months after ratooning. In theory, the ratooning process can be repeated indefinitely as long as there is no build-up of diseases and pests. However, practical experience indicates that only one ratoon crop is typically recommendable. Also, ratooning is not recommended where disease pressure is high, such as the cassava mosaic disease (pan-Africa and South Asia), cassava frog skin disease (Latin America), or cassava brown streak disease (East Africa). In such cases, the crop has to be uprooted after each season to ensure that the plant is healthy for further multiplication.

6 Rapid multiplication techniques

A low multiplication ratio is normally not a constraint in stable cassava systems. However, if area is expanding rapidly, or new varieties are introduced, there is often a need to employ methods of rapid multiplication. Many methods have been developed for increasing the multiplication rate of cassava. Most are based on some variation of three basic methods which include mini-stakes, rooted shoots or tissue culture methods.

There have been many hypothetical multiplication rates presented. However, in the real world, these rates are hardly ever achieved. In seed system planning, it is important to use realistic multiplication rates, based on results obtained under real-life conditions. Exaggerated expectations of rapid propagation have been a major reason for ill-advised investment in cassava seed systems in past years. Often, the use of rapid propagation systems is most appropriate at the very early stages of the seed system, such as to progress from small quantities of breeder seed to larger quantities of foundation seed. At the level of production of certified seed, rapid propagation systems often are overwhelmed by the complexity of managing very large numbers very quickly, and without a consistent continuing demand to support the investment. While it is beyond the scope of this chapter to provide extensive detail on various rapid propagation systems, the fundamentals are described in the following section.

6.1 Mini-stake system

The mini-stake system is the simplest of rapid multiplication systems. It is based on using a shorter-than-normal stem piece in order to increase the number of propagules from each plant. The minimum requirement is that the stem piece used must have at least one node, or bud. Otherwise, no sprouting will occur. The minimum practical size of the stake depends on the moisture in the soil at planting time, storage period of the planting material, sprouting capacity of varieties, and overall physiological and nutritional quality of the planting material. One- or two-node mini-stakes have been used successfully for rapid multiplication where moisture availability is not a constraint (Ceballos et al., 2010). As an alternative to field planting, two-node mini-stakes can be grown at high density in moist chambers where they sprout, and are later transplanted to the field. These mini-stakes may be used in two basic ways: 1) for field planting or transplanting, to directly produce plants that will be the source of seed or 2) to produce multiple shoots that will be harvested and rooted, and then transplanted to the field (described further in the next section). Stems produced through the mini-stake techniques should be cut to final size close to the time planting since they will dry out quickly after cut.

The mini-stake system is the only rapid propagation system that is somewhat widely used both in formal and informal seed systems. It requires no special facilities and only a basic training that, in most cases, both farmers and technicians can easily absorb and apply.

6.2 Rooted shoots

There are two basic techniques to produce rooted shoots, which are then transplanted to the field to produce plants for commercial seed production. The first involves successively cutting emerging shoots from mature stem pieces, which may vary from mini-stakes to stakes of normal length. These shoots are harvested about every 3 weeks at 15–20 cm long. The lower sections are immediately immersed in water until well rooted, and then they are transferred to soil for hardening. The facilities needed for an efficient hardening are simply a cool or fresh environment under shade, where extreme temperatures can be avoided and adequate humidity provided. Shade-net houses are ideal.

The second generalized method of rooted shoot multiplication involves cutting an individual petiole and leaf bade, along with the bud at the base of the petiole and a small section of stem, such that the bud and the petiole remain firmly connected to each other. The entire unit is placed in a soil or synthetic rooting medium. The bud will sprout and produce roots. These can be directly transplanted to the field, or shoots may be cut and rooted successively, as in the rooted cutting technique described above.

While the rooted cutting methods are relatively simple, there are many details of the management that need to be finely tuned for each specific set of conditions, in order to achieve good success, for example, with relation to control of light, temperature and humidity. Rooting hormones may accelerate rooting. There has been a broad adoption of rooted shoots in the early stages of formal seed systems, but there are few cases where the methods have been sustained in the long term.

6.3 *In vitro* systems

In vitro systems have the potential for very high multiplication rates. Just as importantly, when combined with appropriate methods and diagnostic tools, they can greatly reduce or even eliminate the transmission of disease-causing microorganisms from one generation to another. Tissue culture approaches such as the use of meristems and somatic embryogenesis have been used since the 1970s (Fregene et al., 2002). Many researchers advocated and demonstrated the potential of meristem culture and micro-propagation of cassava genotypes to maximize its regeneration and multiplication to

Figure 8 Micro-propagation in Cassava.

produce a large number of disease-free planting material within a short period (Kartha et al., 1974; Nair, 1990; Zok et al., 1992; Acedo and Labana, 2008; Shiji et al., 2014a). Usually a small meristem tip is free from viruses, but follow-up testing and confirmation is always required. Hence meristem tip culture is an efficient tool for the elimination of CMV and other viruses, and regeneration of disease-free planting material. The synthetic MS medium is very efficient for successful cassava meristem culture.

The critical period for the hardening process, after the plants are taken from *in vitro* conditions, is one week. Hardening starts with the transfer of the small plant from the *in vitro* condition to a container (plastic bags or seedling trays) with a mixture of soil and sand that has ideally been sterilized at 100°C. A high moisture condition is required for the first week after transplanting. Different alternatives have been proposed ranging from the use of plastic disposable coffee cups with small holes at the base and placed inverted over the plant so that it is completely covered, to the use of moist chambers (Fregene et al., 2002; Segovia et al., 2002). After the first week, the plants are gradually exposed to conditions with lower humidity and higher temperature. Hardened plantlets can be transplanted in the field within a period of two months (Shiji et al., 2014b). The potential multiplication rate of planting material through micro-propagation techniques over a period of one year is estimated to be in the range of 16 000–17 000 in place of 1:10 in traditional methods of propagation in cassava (Fig. 8) (Shiji et al., 2014a). Nonetheless, as previously noted, there have been few successful experiences of using micro-propagation on a large scale over several years. It is important to understand the implications of scaling up on the actual achieved multiplication rate and the economics of the system.

In vitro systems involve a number of basic strategies of varying degrees of sophistication and efficacy. Some of the key methods are:

- Culture of individual meristems, which develop into rooted plantlets for hardening and transplanting to the field. The method allows moderately high rates of multiplication each cycle – typically 1: several hundreds, and several cycles can be repeated in a year, by use of immature mother plants of 3–4 months, or even younger. Substantial investment in facilities and laboratory expertise are required. There are a range of levels of sophistication that may be applied to multiplication through meristem culture. For example, at CIAT these techniques have even been applied at the village level at a pilot study level, but have not been further adopted for commercial use.
- Culture of individual meristems (as above), but with one or more cycles of sub-culturing *in vitro* to increase the multiplication rate. Sub-culturing involves cutting

individual nodes of plantlets, and re-establishing those nodes in new sterile media to re-start the process of rooting and plantlet growth. Each cycle of growth and cutting new nodes requires about six weeks. In a hypothetical scenario of 100% establishment success, one can see that this system (multiplication rate of 1:6 every 6 weeks) will produce on the order of 10 000 000 plantlets in one year, starting from an individual *in vitro* plantlet. It is evident that the planning and investment for such rapid scaling up needs to be done with extreme care, and in fact is rarely possible, nor warranted, in cassava seed systems.

6.3.1 Somatic embryogenesis

Somatic embryogenesis relies on isolating cells from leaf mesophyll, developing friable callus, and from that, generating rooted plants from each somatic embryo. Alternatively, methods have been developed for synthetic seeds, where somatic embryos are encapsulated in a polymer matrix, stored for short periods of time, germinated in media, hardened in the screen house/greenhouse, and then transplanted to the field. Somatic embryogenesis involves sophisticated facilities and training, and is not yet used in commercial seed systems for cassava.

7 Formal seed system

7.1 Classes of planting materials under formal systems

Well-organized and documented seed programmes have been largely lacking in cassava. Most cassava seed material is produced by the informal system by the farmers themselves, with no special management. Globally, few farmers will have the access to such certified seed. The formal systems to produce and distribute certified seed material in cassava are largely confined to research institutions and extension agencies. Recently however, well-funded projects in both East and West Africa are developing sustainable models for formal cassava seed systems involving public-private partnerships (www.rtb.cgiar.org/blog/2016/04/22/new-project-to-build-commercially-sustainable-cassava-seed-system-in-nigeria; http://meda.org/market-systems-our-projects/market-systems-current-projects/74-tanzania-cassava-seed-system).

Just like cereals and legumes, the classes of planting materials for cassava under a formal system are:

- *Breeder seed:* Breeder seed represents the variety released and maintained at a high level of purity by the breeder.
- *Basic/foundation seed:* This is the progeny of breeder seed (or in the case of cassava, the clonal derivatives of breeder seed). Its production is normally under the control of respective institutions and under the direct supervision of a national certification agency.
- *Certified seed:* This is produced through the multiplication of basic/foundation seed by research institutions, national seed companies, or certified private entrepreneurs, for distributing directly to farmers or through traders or service providers.

7.2 Stages of multiplication of certified seed

A large-scale multiplication of certified seed may not be possible at the research institute level alone. Instead, multiplication will typically need to take place at various levels in a chain – from a primary site in the vicinity of research institutions, to a secondary site or to even a tertiary site. In all the stages, the relevant seed agencies, research institutes and trained personnel for assuring the quality of planting material need to be involved.

- *Production at primary sites:* At the primary stage, production sites are either at or close to the research station for easy supervision by research scientists.
- *Production at secondary sites:* At the secondary multiplication stage, production sites are managed by institutions such as foundation seed producing institutions, extension workers, departments of the Ministries of Agriculture, or non-governmental organizations (NGOs), supported by breeders/researchers through training about the varietal characteristics and multiplication system of the varieties.
- *Production at the tertiary site:* Under tertiary site production, fields are managed by small-scale farmers, perhaps under the auspices of NGOs, and under the supervision of a certification scheme.

7.3 Quality standards of planting material

FAO (2012) has prescribed certain norms for the quality standards for selection of healthy planting material of cassava as follows.

7.3.1 Age of the planting material

Stems should not be stored longer than three months under optimal conditions. Storage should preferably be less than one month. When stems are cut into planting stakes, the exudation of latex from the cortex is an indication of good stem condition.

7.3.2 Number of nodes per cutting

Each stem node has an axillary bud. Theoretically, one plant can be obtained from each node. It has been found that cuttings with 1–3 nodes have low percentages of sprouting in the field. Longer stakes, with 8–10 buds, have a better chance of conserving their potential viability, but require more planting material per unit area. Ideal stakes should, therefore, have from 5 to 7 nodes and be about 20 cm long.

7.3.3 Pith thickness

The pith diameter should be, as a general rule, 50% or less of the total diameter of the stem.

7.3.4 Visual inspection of stems/stakes

A visual inspection of the stems can detect physical damage during storage and/or transport as well as presence of symptoms or signs of diseases and pests. Good-quality planting material should have less than 5% of stems showing these types of problems.

7.3.5 Sprouting capacity

Quick and vigorous sprouting is a critical objective of any scheme to produce cassava planting material. Therefore, a sample of stakes can be tested for sprouting capacity by placing them in plastic bags with soil and adding water. Sprouting percentage and vigour can be evaluated within ten days.

8 Case study: mini-stake-based rapid multiplication at ICAR-CTCRI, India

ICAR-CTCRI has developed quality standards for the production of healthy planting materials in cassava (George, 2006, 2011), and the mini-stake system is presented here as a case study for rapid propagation.

8.1 Overview

ICAR-CTCRI has standardized the mini-stake (or mini-sett) technique for rapid multiplication of cassava planting material (George et al., 2004) (Table 1). In the traditional system of planting, cassava stakes of 20 cm length with 10–12 buds are used as planting material. However, out of the 10 or 12 buds, only two sprouts on the opposite sides are retained and allowed to grow, while the rest of the sprouts are clipped off. This avoids competition among the shoots. In the mini-stake technique, the objective is to utilize the capability of every bud to sprout and grow as an individual plant and thus enhance the multiplication ratio. It is based on the concept that once the bud sprouts, the developing roots would start drawing nutrients from the soil rather than from the mother planting material. Therefore the size of planting material is less important as far as sprouting and vigour are concerned. Based on this concept, the mini-stake technique was standardized for the rapid production of quality planting materials in cassava. The technology has been provided to the farmers in the cassava-growing states of India, whereby farmers have been enabled to multiply planting materials of high-yielding varieties on their own.

8.2 Preparation of mini-stakes

For producing mini-stakes, the first step is selection of mature, disease-free stems (preferably those obtained originally from meristem culture). Two-node mini-stakes have been standardized as the optimum size for rapid production of planting materials in cassava.

Mini-stakes are prepared from healthy stems by making two-node cuttings using a sharp hacksaw. The top one-third portion is usually discarded in the traditional system; however, in the mini-stake technique, the whole stem is used. When tip cuttings are taken, it is advisable to place them in water immediately to prevent dehydration. The stem just below the growing tip is very tender with prominent axillary buds. Hence from this portion, cuttings with four nodes are taken instead of two as the latter may dry out very fast. During the course of mini-stake preparation, sufficient care should be taken such that no damage is caused to the axillary buds.

Table 1 Quality standards developed by ICAR-CTCRI for cassava seed production from mini-stakes

Management component	Description
Soil	Well-drained loamy soil is best suited. The optimum soil pH is between 5 and 6.
Climate	Cassava tolerates high temperature, but growth declines and poor storage root formation occurs below 18°C. It can be successfully grown under a wide range of rainfall, between 800 and 3000 mm.
Water source/Structures	In the case of poor rainfall area, irrigating the crop is essential. Ground water or surface well could be utilized and sprinkler or drip irrigation method could be adopted.
Skilled manpower	Three person-years per ha
Mother block	A clean and well protected area with ideal soil and environmental conditions should be set apart exclusively for the multiplication of planting materials. This should be isolated from the general cassava plots at a distance not less than 400 m.
Methodology for production of planting materials	• Healthy, mature and disease-free stems from previous harvest, stored for a maximum of one month, are normally used as mother seed (planting material). • For disease-free and quality planting material production, use meristem-derived cassava stems. • Such stems are cut into mini-stakes of two nodes and planted in nursery bed under a shade-net house. • Within a month, they are to be transplanted in the prepared main field on ridges at a spacing of 45 × 45 cm. accommodating 49 400 mini-stakes in a hectare of land. • Apply FYM @ 12.5 t ha^{-1} at the time of land preparation and after establishment of mini-stakes in the field. • Apply NPK @ 100:50:100 ha^{-1}. Both N and K may be applied in two equal splits, the first soon after establishment of the transplanted mini-stakes and the second, one month later, while the entire P should be applied as basal. • On an average about 50 000 quality planting materials (stems) could be obtained from a hectare.
Good nursery practices for production of planting materials	• Nursery bed should be under a net house of 35% shade, facilitating optimum light required for the growth of the crop • The shade-net house may preferably be of 30 × 7 m dimension or of any convenient size. The sides of the net house may also be covered so as to prevent the entry of insects and pests, however, one or two swing doors could be provided. • The bed should be of about one metre width and of any convenient length. • They are planted in nursery bed of sand:soil media, of equal proportion. • No fertilization is required; however, dry FYM or any other organic manure may be added to the nursery bed, if need be. • Planting in nursery bed is done at a closer distance, end to end; taking care they do not touch each other. • Micro sprinklers may be used for irrigating the nursery bed. • Nursery area of 165 m^2 would be sufficient for planting mini-stakes in one hectare of land. • In about a month's time, when two or three fully opened leaves appear, the mini-stakes may be transplanted
Important diseases	
Cassava mosaic disease (CMD)	
Causal organism	Indian Cassava Mosaic Gemini Virus and Sri Lankan Cassava Mosaic Virus

(Continued)

Table 1 (*Continued*)

Management component	Description
Detection and diagnosis	• On the leaves, chlorotic areas intermixing with green tissues give mosaic pattern. • Symptoms are seen even in freshly emerging leaves. • In severe cases leaves are reduced in size, twisted and distorted resulting in reduced chlorophyll content and photosynthetic area. • Virus indexing for CMD may be done for effective diagnosis.
Management	• CMD free planting materials must be used. • Resistant variety like Sree Padmanabha may be used. • Rogue and destroy infected plants
Tuber rot	
Causal organism	*Phytophthora palmivora* var. *palmivora*
Detection and diagnosis	• Brown discolouration of internal tissues. • Decomposes/rots and emits foul smell
Management	• Keep the field clean, remove and destroy infected plants and tubers • Deep ploughing (24") of the field using 'sub-soiler' in case of hard pan underneath the top soil • Ridge planking and proper drainage facility • Soil treatment with *Trichoderma viride*, prior to planting.
Important pests	
Spider mites (*Tetranychus telarius, T. neocaledonicus, T. cinnabarinus and T. neocaledonicus*)	
Detection and diagnosis	• Elongated streaks, chlorosis, rusted leathery appearance, withering and drying of leaves.
Management	• Spray dimethoate or monocrotophos 0.05% during severe infestation (January to April) at monthly interval or spray water
Scale insects (*Aonidomytilus albus*)	
Detection and diagnosis	• Attack mostly stems in field as well as in storage • In serious infestation, stems become weak and dry.
Management	• While storing, healthy stems should be kept in vertical position under shade, facilitating easy aeration and diffused light. • As a prophylactic measure, mini-stakes may be dipped in dimethoate 0.05% at the time of planting
White fly (*Bemisia tabaci*)	
Detection and diagnosis	• Visual observation. It is the vector for CMD.
Management	• Spraying the crop with dimethoate (0.03%) at monthly intervals
Spiralling white fly (*Aleurodicus dispersus*)	
Detection and diagnosis	• Eggs deposited in spiral manner and covered with white waxy material • Adults also covered with powdery substance. Develops sooty mould on the infested leaf.
Management	Apply neem-based insecticides.

Management component	Description	
Standards for healthy planting materials.	**Cassava stakes**	
	Purity	100% (true-to-type)
	Stem diameter	2–3 cm
	No. of nodes/20 cm length	>5
	Moisture	>60%
	Free from pest and diseases	
	Cassava mini-stakes	
	Purity	100% (true-to-type)
	Source	Meristem-derived stems
	Size of mini-stake	2 nodes
	No. of leaves	2–3 fully opened
	No. of roots	>2
	Free from diseases and pests	
Size and contents of label/tag to be used	Label size: 10 × 7.5 cm **Label contents:** • Serial no: • Name of variety: • Place grown: • Season/month of harvest: • Nuclear seed (white label), Foundation seed (red label), Certified seed (blue label): • Name and address of the supplier: • Best planted before:	

(Source: George, 2006, 2011)

The following are different types of mini-stakes derived from a cassava plant (Fig. 9):

1 Tip portion mini-stake
2 Four-node mini-stake, from the top tender portion
3 Two-node mini-stake, from the mature (woody) portion.

The woody stock at the base is discarded. About 60–70 mini-stakes could be prepared from an individual plant.

8.3 Planting in a shade-net house

The shade-net house should be ideally located on a flat field and the soil should be well drained, devoid of stones and pebbles, preferably near a water source. Since the mini-stakes ought to be free of any diseases and pests, it is essential to grow them in a protected environment. The structural framework comprises of GI-B class pipe with a flat roof and drooping sides. The shed is covered with shade-net that provides 35% shade, the optimum to regulate and limit the plant growth at the desired level, since higher shade would prompt etiolation. The roof as well as the sides of the net house should be totally covered so as to ensure protection from insects and pests. There should be provision for one or two doors.

Figure 9 Mini-setts of cassava.

8.4 Preparation of nursery

Nursery beds are prepared within the shade-net house. Prior to making beds, it is desirable to dig the soil well, break clods and form fine tilth. To set the beds, mark out a convenient length. The width of the nursery bed should be selected in such a way that the middle portion of the bed is easily reached from either side, which will be about 1 m. Make raised beds (20 cm height) of soil and sand, mixed in the ratio of 50:50. It is advisable to line the sides of the bed with cement hollow bricks or a similar material, to maintain soil in place during irrigation and nursery operations. A nursery area of 125 sq m would be sufficient for raising mini-stakes required for planting one hectare, taking into account rejecting 20–30% as sub-optimum plantlets.

8.5 Planting mini-stakes in the nursery

Furrows are made across the width of the bed, for example with a small hand hoe, about 5 cm deep and 5 cm between rows. Two-node cuttings are then planted in the furrow, end to end horizontally, with the buds facing either side Fig. 10). Mini-stakes planted in the furrow are then covered with a fine layer of soil:sand mixture.

Tip portion mini-stakes and four-node mini-stakes from the top tender portion of the stem must be planted vertically in the nursery rather than in horizontal position. This is because the chances of the tender mini-stakes decaying due to excess moisture are very high, if horizontally embedded in the soil. For those mini-stakes that are vertically planted, a spacing of 5 × 5 cm could be adopted.

8.6 Nursery management of mini-stakes

Light irrigation is given soon after planting and then another after four or five hours. From the next day onwards, intermittent irrigations are given for about two weeks. Irrigation can be provided through a network of micro sprinklers fitted on ¾″ PVC pipes, connected to a water tap. The mini-stakes start sprouting in about a week's time. Chemical fertilizers normally should not be applied since the plantlets are very fragile and sensitive to nutrient toxicities, and can be transplanted within a month's time. If the soil used in the nursery

Figure 10 Mini-setts horizontally placed for sprouting.

is not so fertile, a portion of well-decomposed farm yard manure or vermin compost (sand:soil:FYM-1:1:1) can be added. It is advisable to spray dimethoate (0.05 %) at fortnightly intervals as a prophylactic measure against pests, especially virus vectors like the whitefly, which is responsible for spreading cassava mosaic disease (CMD). Routine examination of the mini-stake nursery is essential to detect incidence of CMD. If present in the source material, the sprouting mini-stakes would likely express mosaic symptoms in the young leaves. If an infected mini-stake is found, it should be immediately pulled out and destroyed.

8.7 Transplanting mini-stakes to the main field

The mini-stakes will be ready for transplanting in about three to four weeks after planting. Two to three fully opened leaves stage is the optimum time for transplanting (Fig. 11). Uprooting mini-stakes from nursery beds can be done with the help of a small spade or

Figure 11 Mini-sett ready for field planting.

khurpi, taking maximum possible care not to break roots. Prior to uprooting, the main field should be properly prepared. It should be thoroughly ploughed and brought to a fine tilth. About 12.5 t ha^{-1} of dry farmyard manure is spread in the field. Preparation includes forming ridges at 45 cm apart for upright varieties and 60 cm apart for more branched types, and 30 cm high.

Uprooted mini-stakes are then carefully transplanted on the prepared ridge. About 50 000 mini-stakes could be transplanted in 1 ha of land. To avoid drying out and risk of mortality, it is essential that the transplanting of mini-stakes must coincide with rainfall, or by providing irrigation. Other management practices are the same as for normal planting of cassava in the field.

8.8 Harvest

At harvest at maturity, depending on the duration of the variety, about 60 000 cassava stems and 75–80 tons of roots can be obtained from a hectare by adopting mini-stake techniques, whereas, in the traditional production system, about 24 000 stems and 30 t/ha of roots could be obtained. If the mini-stake technique is adopted for further multiplication, then mini-stakes for planting 50–60 ha could be obtained in the next generation. Multiplication ratio in cassava planting material by this process is enhanced to 1:70 from the traditional 1:10 in one year.

9 An integrated approach for the future

Reports show that most smallholder farmers source their planting material from the informal system; they plant, sell and exchange a wide range of varieties that fall outside the production and distribution functions of the formal sector. Strategies for scaling the seed system require an integrated approach, which implies coordinated action between the formal and informal seed sectors. Most seed system functions cut across the formal and informal sectors. Informal seed channels exist everywhere, no matter how remote the farming community. They are responsive and durable. Two aspects are important for developing strategies for scaling the availability of seed in integrated seed systems: (1) how to tie together decentralized units in a coordinated way to achieve scale and (2) how best to start decentralized production at scale.

The Great Lake Cassava Initiative (GLCI) of East Africa is a decentralized organizational network for cassava, in which 1.35 million cassava farmers are linked. Networks unite decentralized producers with higher-level structures, which in turn shape clear divisions of labour (NARS, NGOs, farmers groups, entrepreneurs, private companies, etc.), provide up-to-date training and technical information and sometimes stimulate key marketing opportunities. For GLCI, the scaling had to take place through mobilizing multiple units spread over large areas geographically; this occurred not just to scale up the availability of planting material, but also to control and isolate the cassava brown streak disease (Walsh, 2012).

Like potato, the rapidly changing status of cassava as a potential cash crop could radically change opportunities for introducing more structured seed systems in the near future. There is considerable evidence that integrated seed systems can achieve important scaling opportunities that the formal seed system alone cannot. The integrated seed system has the potential to move public sector varieties more widely, extending geographic and social coverage. In vegetatively propagated crops, an integrated seed system would be the main driver for seed system impact and scaling up, as a formal sector complement is nearly non-existent (Sperling et al., 2013).

10 Future trends and conclusion

Cassava is important as a food security crop and varieties need to be developed that will supply the specific value chains of commodities such as starch, high-quality cassava flour, chips, sweeteners and fuel ethanol. A sound cassava seed system that will provide for these chains is required in all the producing countries. Governments need to give cassava its appropriate importance, especially where it is a food security crop. This will enable governments, donors and NGOs to include cassava in the inputs support programmes, and will stimulate use of improved varieties and establishment of sustainable cassava seed systems through private companies, NGOs and farmers' associations.

Policy decisions need to be taken at the operational level of research institutes to determine the relevance of future investments of research on technology for rapid multiplication of cassava stems or for exploring other options for ensuring that yields of roots can be increased and loss of fresh roots minimized. The technical and economic viability of alternative stem multiplication technologies need further research. All producing countries need cassava research to provide tangible results, giving importance especially to food and livelihood security, along with its strategic importance for responding to the climate change challenge and for industrial development.

11 Where to look for further information

ICAR-CTCRI has done commendable works on quality planting material production in cassava which have been documented in several reports. Those who are interested for further reading may check the website of ICAR-CTCRI. Further, the websites of CIAT, Columbia and IITA, Nigeria would also provide further information in the matter since similar works have been done at those centres too in the past.

12 References

Acedo, V. Z. and Labana, C. U. 2008.Rapid propagation of released Philippine cassava varieties through tissue culture.*J. Root Crops*, 34: 108–14.

Almekinders, C. and Louwaars, N. 1999. *Farmer's Seed Production: New Approaches and Practices.* Intermediate Technology Publications Ltd: London.

Ceballos, H., Okogbenin, E., Perez, C. J., López-Valle, L.A. B. and Debouck, D. 2010. Cassava. In Bradshaw, J. E. (Ed.), *Root and Tuber Crops, Handbook of Plant Breeding.*Springer, pp. 53–96.

CTCRI. 1983. Annual Report, Central Tuber Crops Research Institute, Trivandrum, India.

Dickson, D. 2003. South-South collaboration picks up steam. Editorial. Science and Development Network. 17 November. http://www.scidev.net/en/editorials/southsouth-collaboration-picks-up-steam.html.

Ekwe, K.C. 2012. *Cassava Stem Multiplication Technology: AViable Option for Industry Development, Knowledge for Development.* CTA:Wageningen, Netherlands.

Eke-okoro, O.N., Ekwe, K.C. and Nwosu, K.I. 2005. *Cassava Stem and Root Production: A Practical Manual*. National Root Crops Research Institute (NRCRI) Press:Umudike, Nigeria, p. 54.

FAO.2010. *Quality Declared Planting Material: Protocols and Standards for Vegetatively Propagated Crops*. Cassava (Hernan Ceballos and Fernando Calle, International Centre for Tropical Agriculture (CIAT), Cali, Colombia). http://www.fao.org.

FAO. 2012. *Seed Production and Training Manual* (Lambert Delimini, Fao/Germany cooperative programme for Ministry of Agriculture, Forestry and Food security) – fao - 05 December 2012. www.fao.org.

FAO. 2013.Save and Grow: Cassava A Guide to Sustainable Production Intensification. p. 140. http://www.fao.org.

Fregene, M., Tohme, J., Roca, W., Chavarriaga, P., Escobar, R. and Ceballos, H. 2002. Biotecnologia de yuca. In Ceballos, H. and Ospina, B. (Eds), *La Yucaen el TercerMilenio*. CIAT:Cali, Colombia.

George, J., Nair, S.S. and Sreekumari, M. T. 2004. *Rapid Multiplication of Planting Materials in Tuber Crops*. Central Tuber Crops Research Institute:Thiruvananthapuram, Kerala, India, p. 36.

George, J. 2006. Planting material production in cassava, yams and aroids. In George, J., Byju, G. and Edison, S. (Eds), *Quality Planting Material Production in Tropical Tuber Crops*, pp. 39–54.

George, J. 2011. Propagation techniques in tuber crops for production of disease free and quality planting materials. In Singh, B. P., Dua, V. K. and Singh, B. (Eds), *Production of Disease Free Quality Planting Material Propagated Through Tubers and Rhizomes*, pp. 85–92.

IITA. 1998. IITA (International Institute of Tropical Agriculture) Annual Report and Research Highlights for 1987–88. IITA:Ibadan, Nigeria.

IITA. 2002. Eastern Africa Root Crops Research Network (EARRNET) News. International Institute of Tropical Agriculture (IITA):Nigeria.

Kartha, K.K., Gamborg, O.L., Constable, F. and Shyluk, J.P. 1974. Regeneration of cassava plants from apical meristems. *Plant Sci. Lett.* 2: 107–13.

Lopez, J. and El Sharkawy, M.A. 1995. Increasing crop productivity in cassava by fertilizing production of planting material. *Field Crops Res.* 44: 151–7.

Louwaars, N. 1994. *Seed Supply Systems in the Tropics: International Course on Seed Production and Seed Technology*. International Agriculture Centre: Wageningen, The Netherlands.

Lozano, J.C., Toro, J.C., Castro, M. A. and Bellotti, A.C. 1977. *Production of Cassava Planting Material*. CIAT:Cali, USA.

Makeshkumar, T. and Nair, R. R. 2001. Serological and nucleic acid based detection of ICMD. In Fifth International Scientific Meeting of Cassava Biotechnology Network (CBN-V) held at St. Louis, USA during 4–9 November 2001. http://www.danforthcenter.org/media/video/cbnv/session8/S8-12.htm.

Mohankumar, C.R., Nair, G.M. and Mandal, R.C. 1980.Effect of plant population and the production of quality planting material in cassava. Proceedings of Nat. Seminar on Tuber Crops Prod. Tech. TNAU, Coimbatore.

Mohankumar, C.R. and Potty, V.P. 1999. Influence of planting material the yield cassava and sweet potato. In Balagopalan, C., Nayar, T. V.R., Sundaresan, S., Premkumar, T. and Lakshmi, K. R. (Eds), *Tropical Tuber Crops*, pp. 150–6.

Nair, G.M., Ravindran, C.S. and Nedunchezhiyan, M. 2008. Agro techniques and quality planting material production in sweet potato and cassava. In Nedunchezhiyan, M. (Ed.), *Advance Techniques in Quality Planting Material Production and Commercial Cultivation of Tropical Tuber Crops*. Jyoti Graphics:Bhubaneswar, Odisha, pp. 25–33.

Nair, N.G. 1990. Performance of viruses free cassava developed through meristem tip culture. *J. Root Crops*, 16: 123–31.

Osei, M.K., Taah, K. J., Berchie, J. N. and Osei, C. K. 2009. A survey of cassava (ManihotesculentaCrantz) planting materials in storage: A case study in two communities in the Ejisu District of Ashanti Region, Ghana. *J. Agron.*, 8: 137–40.

Ravi, V., George, J. and Ramesh, V. 2008. Storage of planting materials of tropical tuber crops. In Nedunchezhiyan, M. (Ed.), *Advance Techniques in Quality Planting Material Production and*

Commercial Cultivation of Tropical Tuber Crops. Jyoti Graphics:Bhubaneswar, Odisha, pp. 113–18.
Ravindran, C.S., Nair, G.M. and Ravi, V. 2006. Tips for better production in tuber crops. *Kissan World*, 33(11): 17–19.
Seesahai, A. and Ousman, M. 2012. Cassava production: a farmer's guide to selecting high quality planting material. www.cardi.org/CASSAVA-PRODUCTION MODULE 2 – RIO CLARO.
Segovia, R. J., Bedoya, A., Treviño, W., Ceballos, H., Gálvez, G. and Ospina, B. 2002. Metodologia para el Endurecimientode vitroplantas de yuca. In Ceballos, H. and Ospina, B. (Eds), *La Yucaen el Tercer Milenio*. CIAT: Cali, Colombia.
Shiji, R., George, J., Sunitha, S. and Muthuraj, R. 2014a. Micropropagation for rapid multiplication of planting material in cassava (*Manihotesculenta* Crantz). *J. Root Crops*, 40(1): 23–30.
Shiji, R., George, J., Sunitha, S., Vandana, A. and Muthuraj, R. 2014b. Effect of NAA and IBA on in vitro regeneration and hardening in cassava (*Manihotesculenta* Crantz). *J. Root Crops*, 40(2): 12–20.
Sperling, L., Boettiger, S. and Ian, B. 2013. Integrating Seed Systems. apxc AG Partner Xchange.
Sunitha, S. and Ravindran, C.S. 2012. Effect of packing materials used for transportation on viability of cassava setts. *J. Root Crops*. 38(2): 180–2.
Walsh, S. 2012. Seed system innovations in the Great Lake Cassava initiative. Report prepared for the Bill and Melinda Gates Foundation, July 2012.
Zok., S., Nyechemeing, L., Tambong, J. and Watoch, T. 1992. Rapid seed stock multiplication of improved clones of cassava through shoot tip culture in Cameroon. Proceedings Cassava biotechnology Network, 17: 96–8.

Chapter 13

Cassava cultivation and soil productivity

Reinhardt Howeler, CIAT Emeritus - formerly CIAT, Thailand

1 Introduction
2 Nutrient removal by the cassava crop
3 The effect of cassava on soil productivity
4 Nutrient losses in eroded sediments and runoff
5 Diagnosis of nutritional problems
6 Conclusions
7 References

1 Introduction

Cassava (*Manihot esculenta* Crantz) has become one of the most important food crops in the world, still far behind rice, wheat and maize, but ahead of potato in terms of total area planted. Over the past 20 years, the area planted to cassava has increased at a faster rate than that of any other food crop, while total production increased faster than any other crop except for maize (Table 1).

Cassava has become increasingly more important as a staple food crop in sub-Saharan Africa (SSA), because it can grow reasonably well in infertile light-textured soils, where many other crops would fail. This is because cassava is very tolerant of acid soils with low pH and high levels of exchangeable Al; it will also grow well in soils with low levels of available phosphorus (P). For that reason, it is often grown as the last crop of the production cycle in slash-and-burn systems, just before the field reverts back to bush fallow to recuperate its productivity. Rapidly increasing human populations, especially in Africa, have increased land pressure to produce enough food. The fallow period has steadily been shortened and the production cycles lengthened, resulting in a downward trend in soil fertility due to soil nutrient removal in the harvested products. Because of this gradual depletion of soil nutrients, many more-demanding crops like maize and rice may no longer produce well, leaving cassava as an alternative food crop. Moreover, due to climate change, the rainfall patterns in many areas in SSA are changing, with shorter and less predictable wet seasons and longer dry periods, which often cause crop failures except for cassava. Cassava, once established, is very drought tolerant and plants will rarely die because of drought.

However, in many parts of the world, cassava has the reputation to degrade the soil, either by excessive absorption of nutrients and their removal in the harvested products

http://dx.doi.org/10.19103/AS.2016.0014.25

Table 1 Area, production and yield of the ten most important food crops in the world in 1994 and 2014, as well as the per cent change over that 20-year period

	Area harvested (million ha)			Total production (million tonnes)			Yield (t/ha)		
Crop	1994	2014	% change	1994	2014	% change	1994	2014	% change
Wheat	214.7	220.4	+3	525.4	729.0	+39	2.45	3.31	+35
Maize	138.0	184.8	+34	569.0	1037.8	+82	4.12	5.62	+36
Paddy rice	147.3	162.7	+10	538.9	741.5	+38	3.66	4.56	+25
Barley	72.3	49.4	−32	161.3	144.5	−10	2.23	2.92	+31
Sorghum	44.1	45.0	+2	60.2	68.9	−15	1.36	1.53	+13
Millet	38.4	31.4	−18	28.4	28.4	0	0.74	0.90	+22
Cassava	16.8	23.9	+42	164.6	268.3	+63	9.81	11.24	+15
Potato	18.1	19.1	+6	271.2	381.7	+41	15.02	19.99	+33
Oats	19.1	9.6	−50	33.9	22.7	−33	1.78	2.37	+33
Sweet potato	9.0	8.4	−7	125.5	106.6	−15	14.00	12.76	−9

Source: FAOSTAT, December 2016.

or by causing serious erosion when planted on slopes. This belief is mainly based on the observation that cassava is often seen growing on seriously degraded soils or on eroded slopes. In fact, there are few scientific studies comparing the effects of different crops on soil productivity. Cong Doan Sat and Deturck (1998) compared the physical and chemical properties of a Haplic Acrisol (Ultisol) in the eastern regions of South Vietnam that had been under forest, rubber, sugarcane, cashew or cassava for many years. They reported that soils that had been under cassava had the lowest clay content, aggregate stability and water retention, as well as the second lowest infiltration rate, and third highest bulk density, indicating a physical degradation of the soil due to continuous cassava production. Moreover, cassava soils had also suffered chemical deterioration, as indicated by low levels of organic C, total N, CEC, and exchangeable K and Mg. Available P levels in soils under cassava were higher than in those under forest or cashew, but lower than those under rubber or sugarcane, indicating that some source of P had been applied to cassava as well as to rubber and sugarcane. The question remains whether cassava cultivation on these soils is the cause or the result of the physical and chemical degradation, that is, does cassava cultivation cause soil degradation, or is cassava generally grown on those soils that are already degraded because of its exceptional ability to still produce something on those soils where other crops would not? It is well known that cassava has an ability to grow well on poor soils (Cock and Howeler, 1978; Howeler, 1991a). However, like any other crop, cassava absorbs nutrients from the soil; at harvest all or some of these nutrients are removed from the field, resulting in nutrient depletion and fertility decline. In addition, soil/crop management, such as land preparation and weeding, can lead to soil compaction or to soil erosion, which results in soil loss and nutrient losses in eroded sediments and runoff.

This chapter begins with a discussion of the rate at which cassava crops remove nutrients from the soil, and the factors affecting nutrient removal. Section 3 then considers the effect of cassava cultivation on soil productivity, including discussion of the claim that under certain conditions cassava production can actually improve soil fertility with respect to the availability of N and P. Section 4 reviews nutrient losses in eroded sediments and runoff in soils used for cassava cultivation. Finally, Section 5 discusses how the cassava farmer can diagnose nutritional problems in his or her crop.

2 Nutrient removal by the cassava crop

Since nutrient removal is mainly a function of yield, it is more practical to calculate nutrient removal per tonne of fresh roots harvested. Table 2 shows the average removal in the roots as well as in the total plants, as calculated from many reports in the literature (Howeler, 1981, 2014b). These data indicate that if only cassava roots are harvested (as in Thailand), the crop removes mainly K, less N and very little P. However, when farmers harvest roots, stems and leaves (as in Vietnam), the removal of N and P doubles, while that of K increases about 40%; in that case the removal of Ca and Mg also becomes significant, especially if fallen leaves are also collected. An average yield of 15 t fresh roots/ha and the removal of all plant parts from the field would remove per ha about 93 kg N, 12 kg P, 81 kg K, 42 kg Ca and 13 kg Mg. This corresponds reasonably well with data from Thai Phien and Nguyen Cong Vinh (1998) who reported losses of 62–153 kg N, 16–34 kg P and 47–102 kg K.

Since cassava is usually grown on very acid and infertile soils, people oftentimes believe that cassava absorbs large amounts of nutrient from the soil, causing serious soil degradation. In fact, comparing the nutrient absorption of cassava with those of other crops it was found that at high yields cassava extracts per hectare about the same amounts of N and P but higher amounts of K from the soil as compared to many other crops (Table 3). However,

Table 2 Average fresh and dry root yield, as well as the amount of nutrients removed when cassava roots or the whole plants are harvested, based on data from the literature[1]

	Yield (t/ha)		Nutrient removal				
Plant part	fresh	dry	N	P	K	Ca	Mg
			kg/ha				
Roots	28.9	11.4	67.1	11.2	88.1	13.5	7.9
Whole plant		19.0	179.5	22.7	156.1	81.8	25.8
			kg/t fresh roots				
Roots	28.9	11.4	2.32	0.39	3.05	0.47	0.27
Whole plant		19.0	6.22	0.79	5.41	2.83	0.89
			kg/t dry roots				
Roots	28.9	11.4	5.87	0.98	7.71	1.18	0.69
Whole plant		19.0	15.70	1.99	13.66	7.16	2.26

[1] Data are average of 15 data sets which have yields reported in both fresh and dry weight.
Source: Howeler, 2004, 2014b.

Table 3 Average nutrient removal in the harvested products of cassava and various other crops, expressed in both kg/ha and kg/t dry matter (DM) produced, as reported in the literature

	Yield (t/ha)		(kg/ha)			(kg/t DM produced)		
Crop/plant part	fresh	dry[1]	N	P	K	N	P	K
Cassava/fresh roots	28.9	11.4	67	11.2	88	5.9	0.98	7.7
Sweet potato/fresh roots	25.2	5.0	61	13.3	97	12.0	2.63	19.2
Maize/dry grain	6.5	5.6	96	17.4	26	17.3	3.13	4.7
Rice/dry grain	4.6	4.0	60	7.5	13	17.1	2.40	4.1
Wheat/dry grain	2.7	2.3	56	12.0	13	24.1	5.17	5.6
Sorghum/dry grain	3.6	3.1	134	29.0	29	43.3	9.40	9.4
Beans/dry grain	1.1	0.9	37	3.6	22	39.6	3.83	23.4
Soya/dry grain	1.0	0.9	60	15.3	67	69.8	17.79	77.9
Groundnut/dry pod	1.5	1.3	105	6.5	35	81.4	5.04	27.1
Sugarcane/fresh cane	75.2	19.6	43	20.2	96	2.3	0.91	4.4
Tobacco/dry leaves	2.5	2.1	52	6.1	105	24.8	2.90	50.0

[1] Assuming cassava fresh roots to have 39% DM, grain 86%, sweet potato 20%, sugarcane 26%, and dry tobacco leaves 84%.
Source: Adapted from Howeler, 1991b, 2014b.

Table 4 Dry matter and nutrient distribution in 12-month-old cassava cv. MVen 77, grown with and without fertilization in Carimagua, Colombia in 1983/84

	DM	N	P	K	Ca	Mg	S	B	Cu	Fe	Mn	Zn
	(t/ha))	(kg/ha)										
Unfertilized												
-tops	5.1	69.1	7.4	33.6	37.4	16.2	8.2	0.07	0.03	0.45	0.33	0.26
-roots	10.7	30.3	7.5	54.9	5.4	6.5	3.3	0.08	0.02	0.38	0.02	0.10
-fallen leaves	1.5	23.7	1.5	4.0	24.7	4.0	2.5	0.04	0.01	–	0.37	0.18
Total	17.4	123.1	16.4	92.5	67.5	26.7	14.0	0.19	0.06	–	0.72	0.54
Fertilized[1]												
-tops	6.9	99.9	11.7	74.3	55.0	15.3	9.6	0.08	0.03	0.78	0.57	0.30
-roots	14.0	67.3	16.8	102.1	15.5	8.4	7.0	0.07	0.03	0.90	0.06	0.17
-fallen leaves	1.9	30.5	2.0	7.1	31.9	4.7	2.6	0.05	0.02	–	0.46	0.19
Total	22.7	197.7	30.5	183.5	102.4	28.4	19.3	0.20	0.08	–	1.09	0.66

[1] 1 t/ha of 10–20-20 and 5 kg Zn/ha were applied in fertilized plots.
Source: Howeler, 1985a, 2014b.

when the nutrient extraction is calculated per tonne of dry matter (DM) in the harvested products, cassava actually extracts much less N and P and similar amounts of K per tonne of dry roots as other crops.

Table 4 shows how nutrients are distributed at time of harvest among roots, tops (stems with attached leaves) and fallen leaves. If farmers remove from the field not only the roots but also stems, leaves and fallen leaves, they will remove substantial additional amounts of N, Ca and Mg, since 75% of N, 92% of Ca and 76% of Mg were found in the plant tops and fallen leaves, and only 25, 8 and 24%, respectively, in the roots. In case of P, about equal parts were found in roots and tops, while for K about 60% was found in the roots and only 40% in tops and fallen leaves. Thus, if only roots are removed, the ratio of N, P and K removed (expressed in terms of N, P_2O_5 and K_2O) is 1.8:1.0:3.8 or about 2:1:4, while if all plant parts are removed, this will be 3.3:1.0:2.9 or about 3:1:3.

However, when the data in the literature on the nutrients in the roots and in the whole plant are plotted against yield (Fig. 1) it is clear that the nutrient removal is not proportional to yield since plants with a high root yield also tend to have high nutrient concentrations in the roots as well as in leaves and stems (Howeler, 1985a, 2014b). Thus, if we base our nutrient removal calculations for a particular situation on the average removal data shown in Table 2 we would overestimate nutrient removal if yields are low and underestimate removal if yields are high. For example, at an average fresh root yield of 15 t/ha, Fig. 1 shows that

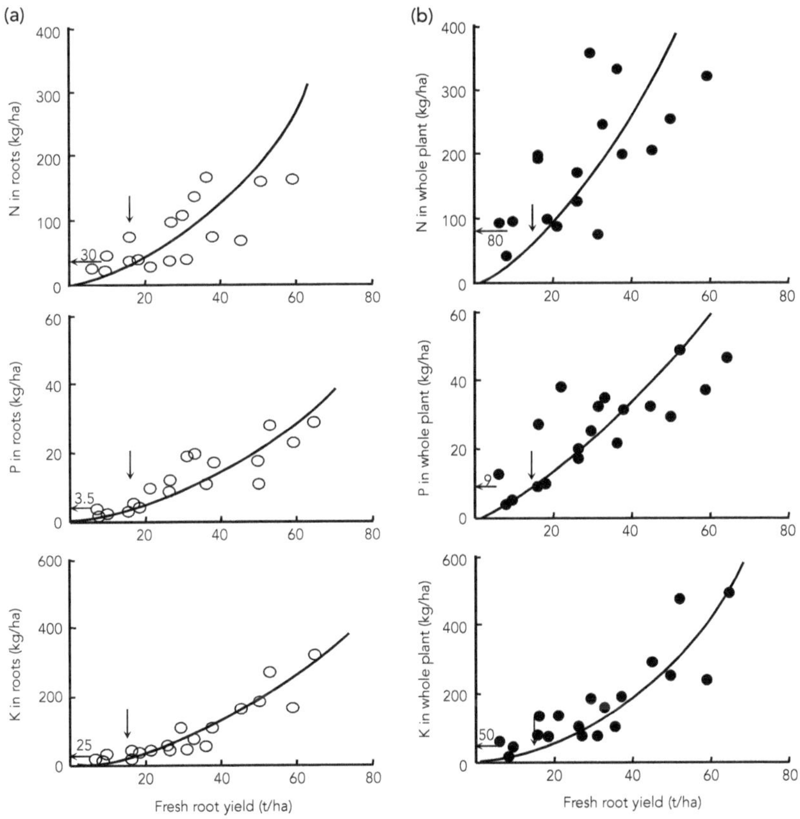

Figure 1 Relation between the amounts of N, P, and K in cassava roots (a) or in the whole plant (b) and fresh root yield, as reported by various sources in the literature. Arrows indicate the approximate nutrient content corresponding to a fresh root yield of 15 t/ha. Source: Howeler, 2002, 2004, 2014b.

the removal in the root harvest is only about 30 kg N, 3.5 kg of P and 25 kg of K/ha, rather than the 34.9 kg of N, 5.8 kg of P and 45.8 kg of K, as predicted from the average values in Table 2. But, when yields increase from 15 to 30 t/ha, K removal by the roots does not double from 25 to 50 kg/ha, but increases to about 65 kg/ha. Thus, when farmers want to increase their yields, they will need to apply more chemical fertilizers or manures, especially those fertilizers that are high in N and K and low in P.

3 The effect of cassava on soil productivity

3.1 Effect of cassava cropping on soil fertility and yields

When cassava is grown for the first time in a new field, either after slashing and burning of the fallow vegetation or after incorporation of grasslands or previous crop residues, the crop may produce well and there may not be any response to the application of chemical fertilizers or manures. This does not mean that cassava does not need and does not respond to fertilizer applications. Thousands of fertilizer experiments conducted by FAO throughout the world between 1961 and 1977 (FAO, 1980) indicate that cassava is as responsive to fertilizer applications as other crops that traditionally are fertilized, and that fertilizer application to cassava can be highly economic. But since soil nutrients are removed in the harvested products or may be lost in eroded sediments or runoff, it may take several cropping cycles to reduce the soil nutrient concentrations to below their critical levels, in order to obtain a significant fertilizer response. Without application of fertilizers or manures, cassava yields will almost invariably decline when the crop is grown continuously for many years on the same field.

Several researchers in Africa have reported that after one or more years of cassava production without application of fertilizers or manure, soil fertility had declined and cassava yields had decreased.

Table 5 Average cassava fresh root yield (t/ha) of variety BEN86052, in response to the application of various combinations of N, P and K fertilizers during three consecutive cropping cycles on the same plots in 15 farmers' fields in Hayakpa, southern Benin

Treatment[1]	1999/2000	2000/2001	2001/2002	Mean
$N_0P_0K_0$	19.6	16.0	12.9	16.0
$N_0P_0K_1$	21.4	16.7	16.7	18.3
$N_0P_1K_1$	20.4	18.6	18.9	19.3
$N_1P_0K_1$	20.9	16.1	16.9	18.0
$N_1P_1K_0$	21.5	17.8	16.6	18.5
$N_1P_1K_1$	21.0	19.8	21.9	20.9
Mean	20.7	17.5	17.3	

[1] N_1 = 60 kg N, P_1 = 16 kg P and K_1 = 138 kg K/ha.
Source: Carsky and Toukourou, 2005.

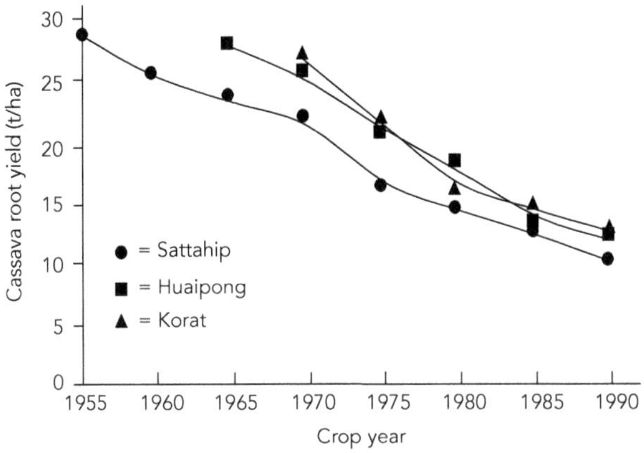

Figure 2 Decline in fresh root yields due to continuous cassava cultivation without fertilizers in three soil series in Thailand. Source: Sittibusaya, 1993; Howeler, 2012b.

While there may not have been any response to fertilizer application in the first year, there was often a response in the second year, and with increasing levels of response in subsequent years. Table 5 is an example of this situation in southern Benin in West Africa (Carsky and Toukourou, 2005).

During the first year there was no significant response to application of N, P or K; during the second year there was a highly significant response to P and a significant response to K, while in the third year there was a highly significant response to both P and K. Without fertilizer application yields declined from 19.6 t/ha in the first year to 12.9 t/ha in the third year, while with the complete NPK application yields increased slightly to 21.9 t/ha. The exchangeable K content of the soil also decreased from 0.13 me/100 g before the first planting to 0.10 me/100 g in the third year (at 6 MAP); it remained at or slightly above 0.13 me/100 g in all treatments where K had been applied. The level of available P had also decreased, from 6.1 ppm (Bray 1) before the first planting to 3.0 ppm without NPK, and to 3.4 ppm with complete NPK in the third year.

Another good example of the effect of continuous cropping of cassava on yields when no fertilizers or manures are applied is shown in Fig. 2. In this case, cassava was grown for 20–30 years without fertilizers in different farmers' fields on three soil series in Thailand. In all three areas, cassava yields were originally quite high, but they decreased markedly when the crop was grown continuously for many years.

3.2 Can cassava production actually improve soil fertility?

Cassava is one of the most important food crops in many countries in sub-Saharan Africa, but farmers seldom apply any chemical fertilizers or manures to the crop (Fermont et al., 2008, 2010; Nweke, 1996). Traditionally, soil fertility is maintained through slash-and-burn systems, in which after several years of cropping, the land is returned to bush fallow for 10–20 years to let the soil rest and replenish the nutrients that were lost during the cropping cycles. After this fallow period the bush vegetation is usually burnt to start another 3–4 year cropping cycle. However, because of rapid population growth and the consequent increase in land pressure, the fallow cycle has steadily been shortened while the cropping cycle and intensity have increased. In many densely populated countries in East and West Africa the fallow cycle has now been reduced to 1–2 years, or has actually been eliminated entirely. Instead, farmers are trying to maintain the fertility of their land through the planting of 'improved' fallows of fast-growing leguminous trees, alley-cropping systems, or by cereal-legume crop rotations. Farmers in Ghana and Benin in West Africa also use extensive cassava cropping as a 'substitute fallow' as they have observed that this system seems to restore soil productivity. The maize crop planted after cassava grows better than maize planted after maize, or even after many legume crops. Cassava cropping in many countries in West Africa is called 'jachère manioc', which literally means 'cassava fallow' (Saïdou, 2006; Saïdou et al., 2004; Adjei-Nsiah et al., 2007). Similarly, the bush-fallow cycle in many countries in East Africa has nearly disappeared, and many farmers are now rotating maize with cassava so that cassava can 'rest the soil' and improve soil fertility for the following maize crop (Fermont et al., 2008).

In an experiment in Ghana, yields of maize planted after either cassava or mucuna (*Mucuna pruriens*) indeed produced higher yields than when maize followed cowpea, maize, or a speargrass (*Imperata cylindrica*) fallow (Fig. 3).

The beneficial effect of cassava was attributed to the relatively high nutrient concentrations, especially N, of the fallen leaves and in the crop residues incorporated into the soil after the harvest of the roots. Since cassava has a relatively deep root system, part of these nutrients may have been taken up from the subsoil and are then deposited onto the soil surface where they are available for rapid uptake by the following maize crop. Also, cassava may have stimulated the native VA-mycorrhizal (VAM) population of the soil, resulting in an increased VAM spore population, which in turn increased the P-uptake by the following maize crop (Saïdou et al., 2004). Finally, once well established, cassava provides a dense canopy cover during a fairly long time, which shades out most weeds. This results in less weed competition for the following maize crop, especially when N was applied to the maize. More research on these aspects will be required to confirm these initial observations. It seems, however, that under certain conditions cassava production can actually improve soil fertility with respect to the availability of N and P, while K will probably be required from other sources to compensate for the substantial removal of this nutrient during the cassava root harvest.

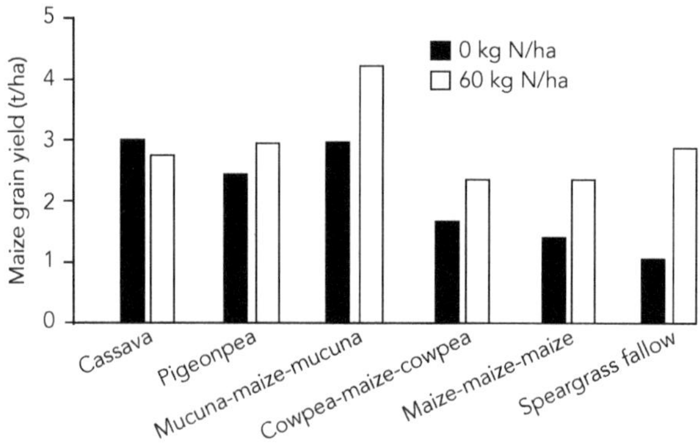

Figure 3 Effect of previous crop or crop sequence and N rate on maize grain yield produced on researcher managed plots in Ghana. Source: Adjei-Nsiah et al., 2007.

4 Nutrient losses in eroded sediments and runoff

When soil particles are dislodged by the impact of raindrops or by the scouring action of overland flow, and move down-slope with runoff, the field loses not only the most fertile part of the soil, that is, the topsoil, but also the associated organic matter, manures, fertilizers and beneficial micro-organisms, such as mycorrhizal fungi. Moreover, clay particles, once dislodged, are quickly carried down-slope, resulting in a preferential loss of clay and a lightening of soil texture. This may be the reason why soils used for a long time for cassava cultivation were found to be much lower in clay, organic C and CEC than those used for forest, rubber and cashew (Cong Doan Sat and Deturck, 1998). In addition, applied fertilizer particles can be dislodged and removed, or the water-soluble constituents can be lost with runoff water. In general it was found that eroded sediments are much higher in nutrients than the soil in the original site. This enrichment is due to preferential losses of organic matter, clay, earthworm castings and plant debris laying on the soil, or dissolved manures and fertilizers (Rupenthal et al., 1997). Thus, erosion does not only reduce the soil depth available for root growth and for uptake of nutrients and water, but it also leaves the remaining soil less fertile, while often exposing highly infertile subsoil. This has a detrimental effect on productivity, as can be seen in Fig. 4, where cassava yields on eroded soil in Colombia were about half of those on adjacent non-eroded soil.

Table 6 shows nutrient losses in eroded sediments from cassava plots with different treatments in Thailand and Colombia. The losses of P, K, Ca and Mg are in terms of available and exchangeable nutrients rather than total nutrients and are therefore much lower than if they had been expressed as losses of total nutrients.

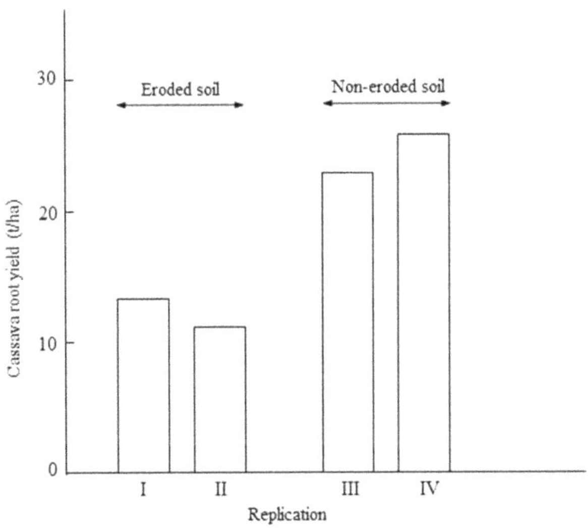

Figure 4 The average yield of 18 cassava varieties planted in two replications on an eroded slope and two replications on an adjacent non-eroded flat area in 1983/84 in Mondomo, Cauca, Colombia. Source: Howeler, 1986, 2014b.

Table 6 Nutrients in sediments eroded from cassava plots with various treatments in Thailand and Colombia

Location and treatments	Dry soil loss (t/ha/year)	(kg/ha/year) N[1]	P[2]	K[2]	Mg[2]
Cassava on 7% slope in Sriracha, Thailand[3]	71.4	37.1	2.18	5.15	5.35
Cassava on 5% slope in Pluak Daeng, Thailand[4]	53.2	22.3	1.25	3.27	–
Cassava planted on 7–13% slope in Quilichao, Colombia[5]	5.1	11.5	0.16	0.45	0.45
Cassava with leguminous cover crops in Quilichao[5]	10.6	24.0	0.24	0.97	0.81
Cassava with grass hedgerows in Quilichao, Colombia[5]	2.7	5.8	0.06	0.22	0.24
Cassava planted on 12–20% slope in Mondomo, Colombia[5]	5.2	13.3	1.09	0.45	0.36
Cassava with leguminous cover crops in Mondomo[5]	2.7	6.5	0.04	0.24	0.20
Cassava with grass hedgerows in Mondomo, Colombia[5]	1.5	3.5	0.02	0.13	0.10

[1] Total N.
[2] Available P, and exchangeable K and Mg.
[3] Putthacharoen et al., 1998.
[4] Tongglum et al., 2001.
[5] Ruppenthal et al., 1997.

Nutrient losses in Colombia were much lower than in Thailand because of lower erosion losses. Nevertheless, Ruppenthal et al. (1997) reported that in both Quilichao and Mondomo the sediments were enriched with Ca, Mg, K, P and sand compared with the original soil, with average enrichment ratios of 1.30, 1.08, 1.13, 1.11 and 1.16, respectively, while there was little enrichment of OM and N, and a slight impoverishment of silt and clay. The same authors reported that in cassava plots in Quilichao and Mondomo, on average 35% of lost P, 15% of K and 37% of Mg were found in the sediments and the remainder in the runoff. Thus, a considerable amount of nutrients were also lost in runoff, especially K. This not only results in a serious loss of nutrients from the field, reducing productivity, but may also result in nutrient pollution and eutrophication of waterways and lakes downstream.

5 Diagnosis of nutritional problems

Cassava yields are often lower than optimum because of the use of unsuitable varieties, pest and disease problems, unsuitable climatic conditions, poor crop management (especially inadequate weeding), and soil-related problems such as nutritional deficiencies or toxicities, soil erosion and soil compaction. Among these various problems, the elimination of soil management constraints is usually the most important in increasing cassava yields. Among the various soil management constraints, the improvement of soil fertility will have the greatest effect (Henry and Gottret,1996; Fermont et al., 2009; Howeler, 2014a (unpublished)). Nutritional problems in cassava are very widespread in many tropical countries but often go unnoticed, since the most common deficiency symptoms of N, P and K are not very clearly visible in cassava.

Before a nutritional problem can be corrected it is essential to diagnose the nature of the problem, that is, whether it is a deficiency or toxicity, and which element or elements are involved. Nutritional problems are generally diagnosed by visual observation of symptoms, by soil or plant tissue analyses, or by observation of the plants' response to specific nutrient applications. The remainder of this chapter will address these methods for diagnosing nutritional problems in cassava crops. Chapters 34 and 35 then provide more detailed discussion of nutritional disorders and strategies for addressing them.

5.1 Symptoms of deficiencies and toxicities

Symptoms of deficiencies or toxicities in cassava have been described and photographed by Asher et al. (1980), Lozano et al. (1981), Howeler and Fernandez (1985), Susan John et al. (2006), Alvarez et al. (2012), Howeler (2012a, 2014b) and Howeler and Aye (2014). These symptoms will be described in detail for each element in Chapter 34.

Symptoms of nutritional disorders in cassava, especially the deficiencies of N, P and K, are often not readily recognized and farmers may be unaware that their crop is suffering from nutritional stresses leading to reduced yields. When grown under nutritional stress the cassava plant does not readily translocate nutrients from the lower to the upper leaves, but instead, reduces the rate of new leaf production (De Tafur et al., 1997). Thus, cassava plants suffering from moderate N, P or K deficiencies tend to be short and weak but without showing clear deficiency symptoms. This makes the diagnosis of nutritional disorders rather difficult and in some cases the diagnosis can only be made after soil and/or plant tissue analysis. Varieties also differ to some extent in the expression of symptoms as well as in their nutrient accumulation and distribution patterns. Thus, some varieties show symptoms more clearly than others or the symptom expression may be quite different in different varieties.

5.2 Soil analyses

Soil analyses have the advantage over tissue analyses that the diagnosed problem can be corrected before the crop is planted. However, soil analyses determine only the 'available' fraction of the nutrient present in the soil, and the determination of this fraction depends on the extractant and methodology used in the laboratory. There is not enough standardization of methodologies among laboratories and no one method is optimal for all soils. Thus, in the interpretation of the results it is important to consider the extractant and methodology used in the laboratory. In any case, the 'amount' of available nutrient is only an index figure, which has to be correlated with plant response, but should not be interpreted as equivalent to so many kilos of available nutrient per hectare. Soil and tissue analyses results are interpreted by comparing the obtained values with so-called 'critical levels' or 'critical ranges' determined for each element for each crop. Critical levels of deficiency are often defined as the concentration of a particular nutrient in the soil or plant tissue below which its application has a significant effect in increasing yield, and above which no significant response to its application is expected. The critical level is often taken as the level that corresponds with 90% or 95% of the maximum yield as calculated from the curve that relates yield with nutrient concentration. Thus, in Fig. 5 relative yields of cassava were plotted against the available P and the exchangeable K concentration of the soil, both extracted with Bray-II solution. From these curves, critical levels of 4 ppm P and 0.17 me K/100 g dry soil were calculated. Thus, P and K applications to cassava are

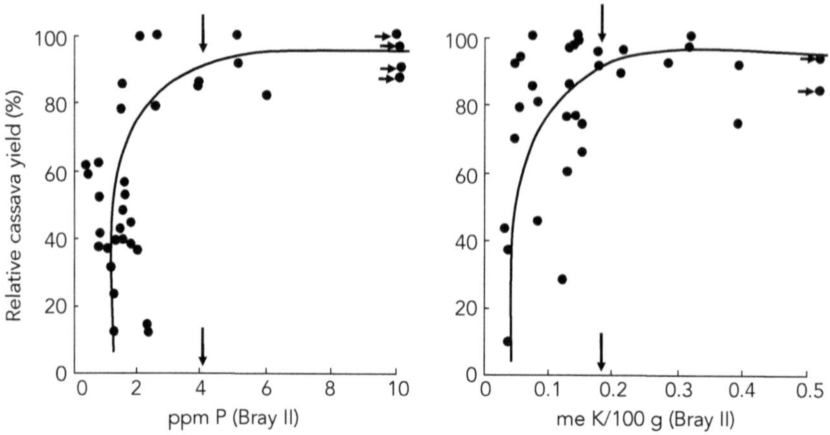

Figure 5 Relation between the relative cassava yield of P or K check plots and the available P (left) or exchangeable K (right) content of soils in 24 NPK trials in Colombia. Vertical arrows indicate the critical levels. Source: Howeler, 1985b; Howeler and Cadavid, 1990.

Table 7 Approximate classification of soil chemical characteristics according to the nutritional requirements of cassava

Soil parameter	Very low	Low	Medium	High	Very high
pH[1]	<3.5	3.5–4.5	4.5–7.0	7–8	>8
Organic matter[2] (%)	<1.0	1.0–2.0	2.0–4.0	>4.0	
Al saturation[3] (%)			<75	75–85	>85
Salinity (mS/cm)			<0.5	0.5–1.0	>1.0
Na saturation (%)			<2	2–10	>10
P[4] (ppm)	<2	2–4	4–15	>15	
K[4] (me/100 g)	<0.10	0.10–0.15	0.15–0.25	>0.25	
Ca[4] (me/100 g)	<0.25	0.25–1.00	1.0–5.0	>5.0	
Mg[4] (me/100 g)	<0.2	0.2–0.4	0.4–1.0	>1.0	
S[4] (ppm)	<20	20–40	40–70	>70	
B[5] (ppm)	<0.2	0.2–0.5	0.5–1.0	1–2	>2
Cu[5] (ppm)	<0.1	0.1–0.3	0.3–1.0	1–5	>5
Mn[5] (ppm)	<5	5–10	10–100	100–250	>250
Fe[5] (ppm)	<1	1–10	10–100	>100	
Zn[5] (ppm)	<0.5	0.5–1.0	1.0–5.0	5–50	>50

[1] pH in H_2O. 1:1.
[2] OM = Walkley and Black method.
[3] Al saturation = 100 × Al/(Al + Ca + Mg + K) in me/100 g.
[4] P in Bray II; K, Ca, Mg and Na in 1 N NH_4-acetate; S in Ca-phosphate.
[5] B in hot water; Cu, Mn, Fe, and Zn in 0.05 N HCl + 0.025 N H_2SO_4.
Source: Howeler, 1996a,b, 2014b.

only recommended if the Bray-II extractable P and K concentrations are below 4 ppm and 0.17 me/100 g, respectively. Critical levels according to various methods of soil analyses, as reported in the literature, have been previously summarized by Howeler (2012a, 2014b). Critical ranges of key nutrients and other soil parameters for cassava are shown in Table 7.

5.3 Tissue analyses

For plant tissue analysis the total concentration of nutrient in that tissue is determined. This is essentially independent of methodology, and plant tissue analyses therefore vary little among laboratories. Also, tissue analyses determine the amounts of each nutrient the plant has actually absorbed and are thus a good indicator of the plant's nutritional status and the soil's nutrient-supplying power. However, plant tissue analyses often diagnose a nutritional problem too late to correct the problem in the current crop; they may only allow preventing the same problem in the next crop.

Since nutrient concentrations in the plant vary among the different tissues and change during the growth cycle, it is very important to standardize the sampling of plant tissue for diagnosing nutritional disorders. It is recommended to sample the blades (without petioles) of the youngest fully expanded leaves (YFEL) at three to four months after planting. The leaf samples should be dried immediately in an oven at 110° C for about

Table 8 Nutrient concentrations in youngest fully expanded leaf (YFEL) blades of cassava at 3–4 MAP, corresponding to various nutritional states of the plants; data are average values of various greenhouse and field experiments

	Nutritional states[1]					
Nutrient	Very deficient	Deficient	Low	Sufficient	High	Toxic
N (%)	<4.0	4.1–4.8	4.8–5.1	5.1–5.8	>5.8	–[2]
P (%)	<0.25	0.25–0.36	0.36–0.38	0.38–0.50	>0.50	–
K (%)	<0.85	0.85–1.26	1.26–1.42	1.42–1.88	1.88–2.40	>2.40
Ca (%)	<0.25	0.25–0.41	0.41–0.50	0.50–0.72	0.72–0.88	>0.88
Mg (%)	<0.15	0.15–0.22	0.22–0.24	0.24–0.29	>0.29	–
S (%)	<0.20	0.20–0.27	0.27–0.30	0.30–0.36	>0.36	–
B (ppm)	<7	7–15	15–18	18–28	28–64	>64
Cu (ppm)	<1.5	1.5–4.8	4.8–6.0	6–10	10–15	>15
Fe (ppm)	<100	100–110	110–120	120–140	140–200	>200
Mn (ppm)	<30	30–40	40–50	50–150	150–250	>250
Zn (ppm)	<25	25–32	32–35	35–57	57–120	>120

[1] Very deficient = <40% maximum yield
Deficient = 40–80% maximum yield
Low = 80–90% maximum yield
Sufficient = 90–100% maximum yield
High = 100–90% maximum yield
Toxic = <90 maximum yield
[2] – = no data available.
Source: Howeler, 1996a,b, 2014b.

24 hours, or quickly sun-dried in a hot and dry location; the dry leaves are then ground and analysed (Howeler, 1983).

Table 8 shows the nutrient concentrations of YFEL-blades corresponding to very deficient, deficient, low, sufficient, high, and toxic levels. In general, fertilization is recommended if the nutrient concentration is below the sufficiency range. Critical levels of nutrient deficiencies, as reported in the literature (Howeler, 1981, 2002, 2014b), generally fall somewhere within the sufficiency range. These values are a useful guide for interpreting tissue analyses results, but may vary somewhat for different varieties, climatic conditions and soils (Leong, 2016).

6 Conclusions

Research has shown that:

1 Cassava does not extract more nutrients from the soil than other crops, except when yields are very high and/or all plant parts are removed from the field.
2 When only roots are harvested, nutrients are removed in the ratio (N-P_2O_5-K_2O) of about 2:1:4; when all plant parts including fallen leaves are harvested, nutrients are removed in the ratio of about 3:1:3. In the latter case, large amounts of Ca and Mg are also removed.
3 Nutrient removal in eroded soil and runoff water can be substantial, especially K in runoff and sediments, and N in sediments, but nutrient losses from erosion are generally lower than those due to crop removal.
4 Symptoms of N, P and K deficiencies are often not clearly expressed in cassava due to their low mobility in the phloem; in case of any of these deficiencies plants remain small and weak and yields will be reduced.
5 Because of the lack of clear deficiency symptoms of N P and K, farmers may not be aware that their low yields are due to any nutrient deficiencies and are thus not able to diagnose and correct the problem.

7 References

Adjei-Nsiah, S., T. W. Kuyper, C. Leewis, M. K. Abekoe and K. E. Giller. 2007. Evaluating sustainable and profitable cropping sequences with cassava and four legume crops: Effects on soil fertility and maize yields in the forest/savannah transitional agroecological zone of Ghana. *Field Crops Research* 103: 87–97.

Alvarez, E., A. Bellotti, L. Calvert, B. Areas, L. F. Cadavid, B. Pineda, G. Lano and M. Cuervo. 2012. Practical Handbook for Managing Cassava Diseases, Pests and Nutritional Disorders. CIAT, CLAYUCA, CTA. Cali, Colombia.115 p.

Asher, C. J., D. G. Edwards and R. H. Howeler. 1980. Nutritional Disorders of Cassava *(Manihot esculenta* Crantz*)*. University of Queensland, St. Lucia, Qld., Australia. 48 p.

Carsky, R. J. and M. A. Toukourou. 2005. Identification of nutrients limiting cassava yield maintenance on a sedimentary soil in southern Benin, West Africa. *Nutrient Cycling in Agroecosystems* 71: 151–62.

Cock, J. H. and R. H. Howeler. 1978. The ability of cassava to grow on poor soils. In: Crop Tolerance to Suboptimal Land Conditions. ASA, CSSA, SSSA. Madison WI, USA. pp. 145–53.

Cong Doan Sat and P. Deturck. 1998. Cassava soils and nutrient management in South Vietnam. In: R. H. Howeler (ed.), *Cassava Breeding, Agronomy and Farmer Participatory Research in Asia*. Proc. 5th Regional Workshop, held in Danzhou, Hainan, China. Nov. 3–8, 1996, pp. 257–67.

De Tafur, S. M., M. A. El-Sharkawy and L. F. Cadavid. 1997. Response of cassava (*Manihot esculenta* Crantz) to water stress and fertilization. *Photosynthetica* 34 (2): 233–9.

Fermont, A. M., P. J. A.van Asten and K. E. Giller. 2008. Increasing land pressure in East Africa: The changing role of cassava and consequences for sustainability of farming systems. *Agriculture, Ecosystems and Environment* 128: 239–50.

Fermont, A. M., P. J. A. van Asten, P. Tittonell, M. T. van Wijk and K. E. Giller. 2009. Closing the yield gap: an analysis from small-holder farms in East Africa. *Field Crops Research* 112: 24–36.

Fermont, A. M., P. Tittonell, Y. Bacuma, P. Ntawauruhunga and K. E. Giller. 2010. Towards understanding the factors that govern fertilizer response in cassava: lessons from East Africa. Nutrient Cycling in Agroecosystems 86: 133–151.

Food and Agriculture Organization (FAO). 1980. Review of data on responses of tropical crops to fertilizers, 1961–77. FAO, Rome, Italy. 101 p.

Food and Agriculture Organization (FAO). 2016. www.faostat3.fao.org

Henry, G. and V. Gottret. 1996. Global Cassava Trends. Reassessing the Crop's Future. CIAT Working Document No. 157. CIAT, Cali Colombia. 45 p.

Howeler, R. H. 1981. Mineral Nutrition and Fertilization of Cassava. Series 09EC-4, Centro Internacional de Agricultura Tropical (CIAT), Cali, Colombia. 52 p.

Howeler, R. H. 1983. Análisis del Tejido Vegetal en el Diagnóstico de Problemas Nutricionales: Algunos Cultivos Tropicales (Plant tissue analysis to diagnose nutritional problems in some tropical crops). Centro Internacional de Agricultura Tropical (CIAT), Cali, Colombia. 28 p.

Howeler, R. H. 1985a. Mineral nutrition and fertilization of cassava. In: Cassava; Research, Production and Utilization. UNDP-CIAT Cassava Program, Cali, Colombia, pp. 249–320.

Howeler, R. H. 1985b. Potassium nutrition of cassava. In: W. D. Bishop et al. (eds), *Potassium in Agriculture*. Intern. Symposium, held in Atlanta, GA, USA. July 7–10, 1985. ASA-CSSA-SSSA, Madison, WI, USA, pp. 819–41.

Howeler, R. H. 1986. El control de la erosión con prácticas agronómicas sencillas (Erosion control through simple agronomic practices). *Suelos Ecuatoriales* 16: 70–84.

Howeler, R. H. 1991a. Identifying plants adaptable to low pH conditions. In: R. J. Wright et al. (eds), *Plant-Soil Interactions at Low pH*. Kluwer Academic Publisher, Netherlands, pp. 885–904.

Howeler, R. H. 1991b. Long-term effect of cassava cultivation on soil productivity. *Field Crops Research* 26, 1–18.

Howeler, R. H. 1996a. Diagnosis of nutritional disorders and soil fertility maintenance of cassava. In: G. T. Kurup et al. (eds), *Tropical Tuber Crops: Problems, Prospects and Future Strategies*. Oxford and IBH Publishing Co. Pvt. Ltd., New Delhi, India, pp. 181–93.

Howeler, R. H. 1996b. Mineral nutrition of cassava. In: E. T Craswell, C. J. Asher and J. N. O'Sullivan (eds), *Mineral Nutrient Disorders of Root Crops in the Pacific. Proc.* Workshop, held in Nuku'alofa, Kingdom of Tonga. April 17–20, 1995. ACIAR Proc. no. 5, Canberra, Australia, pp. 110–16.

Howeler, R. H. 2002. Cassava mineral nutrition and fertilization. In: R. J. Hillocks, M. J. Thresh and A. C. Bellotti (eds), *Cassava: Biology, Production and Utilization*. CABI Publishing, Wallingford, UK, pp. 115–47.

Howeler, R. H. 2004. Nutrient inputs and losses in cassava-based cropping systems – examples from Vietnam and Thailand. In: R. W. Simmons, A. D. Noble and R. D. B. Lefroy (eds), *Nutrient Balances for Sustainable Agricultural Production and Natural Resource Management in SE Asia*. Proc. Intern. Workshop, held in Bangkok, Thailand. Feb 20–2, 2001. 30 p. Selected Papers and Presentations on CD, IWMI-CIAT.

Howeler, R. H. 2012a. Diagnosis of nutritional problems of cassava. In: R. H. Howeler (ed.), *The Cassava Handbook*. A Reference Manual Based on the Asian Regional Cassava Training Course, held in Thailand, pp. 305–20.

Howeler, R. H. 2012b. Effect of cassava production on soil fertility and the long-term fertilizer requirements to maintain high yields. In: R. H. Howeler (ed.), *The Cassava Handbook*. A

Reference Manual Based on the Asian Regional Cassava Training Course, held in Thailand, pp. 411–28.

Howeler, R. H. 2014a. Literature review on yield gaps of cassava in Asia. (unpublished).

Howeler, R. H. 2014b. Sustainable Soil and Crop Management of Cassava in Asia – A Reference Manual. CIAT Publication No. 389. CIAT, Cali, Colombia. 280 p.

Howeler, R. H. and F. Fernandez. 1985. Nutritional Disorders of the Cassava Plant. Study Guide. CIAT, Cali, Colombia. 36 p.

Howeler, R. H. and L. F. Cadavid. 1990. Short- and long-term fertility trials in Colombia to determine the nutrient requirements of cassava. *Fertilizer Research* 26: 61–80.

Howeler, R. and T. M. Aye. 2014. Sustainable Management of Cassava in Asia. From Research to Practice. CIAT, The Nippon Foundation. Cali, Colombia. 148 p.

Leong, G. J. M. 2016. Mineral nutrient calibration for cassava (*Manihot esculenta* Crantz) grown in dry-tropical north-eastern Queensland, Australia: with focus on phosphorus and boron. MSc thesis, Univ. Queensland, St. Lucia, Australia. 138 p.

Lozano, J. C., A. Bellotti, J. A. Reyes, R. Howeler, D. Leihner and J. Doll. 1981. Field Problems in Cassava. CIAT Series No. 07EC-1. Cali, Colombia. 206 p.

Nweke, F. I. 1996. Cassava: A cash crop in Africa. COSCA Working Paper No. 14. Collaborative Study of Cassava in Africa. International Institute of Tropical Agriculture (IITA), Ibadan, Nigeria.

Putthacharoen, S., R. H. Howeler, S. Jantawat and V. Vichukit. 1998. Nutrient uptake and soil erosion losses in cassava and six other crops in a Psamment in eastern Thailand. *Field Crops Research* 57: 113–26.

Ruppenthal, M., D. E. Leihner, N. Steinmuller and M. A. El-Sharkawy. 1997. Losses of organic matter and nutrients by water erosion in cassava-based cropping systems. *Experimental Agriculture* 33: 487–98.

Saïdou, A. 2006. Converging strategies by farmers and scientists to improve soil fertility and enhance crop production in Benin. PhD dissertation. Wageningen Univer., the Netherlands. 223 p.

Saïdou, A., T. W. Kuyper, D. K. Kossou, R. Tossou and P. Richards. 2004. Sustainable soil fertility management in Benin: learning from farmers. *NJAS-Wageningen Journal of Life Sciences* 52: 349–469.

Sittibusaya, C. 1993. Progress report of soil research on fertilization of field crops, 1992. Annual Cassava Program Review, held in Rayong, Thailand. Jan 19–20, 1993. (in Thai).

Susan John, K., G. Suja, S. Edison and C. S. Ravindran. 2006. Nutritional Disorders in Tropical Tuber Crops. Central Tuber Crops Research Institute (CTCRI), Indian Council of Agricultural Research, Thiruvananthapuram, Kerala, India. 74 p.

Thai Phien and Nguyen Cong Vinh. 1998. Nutrient management for cassava-based cropping systems in northern Vietnam. In: R. H. Howeler (ed.), *Cassava Breeding, Agronomy and Farmer Participatory Research in Asia*. Proc. 5th Regional Workshop, held in Danzhou, Hainan, China. Nov. 3–8, 1996, pp. 268–79.

Tongglum, A., P. Suriyapan and R. H. Howeler. 2001. Cassava agronomy research and adoption of improved practices in Thailand – Major achievements during the past 35 years. In: R. H. Howeler and S. L. Tan (eds), *Cassava's Potential in Asia in the 21st Century: Present Situation and Future Research and Development Needs*. Proc. 6th Regional Workshop, held in Ho Chi Minh city, Vietnam. Feb 21–5, 2000, pp. 228–58.

Chapter 14

Addressing nutritional disorders in cassava cultivation

Reinhardt Howeler, CIAT Emeritus - formerly CIAT, Thailand

1 Introduction
2 Nitrogen and phosphorus deficiencies
3 Potassium, calcium, magnesium and sulphur deficiencies
4 Micronutrients and soil pH
5 Short- and long-term nitrogen, phosphorus and potassium (NPK) requirements of cassava: an overview
6 Soil and nutrient management for cassava cultivation in Latin America
7 Soil and nutrient management for cassava cultivation in sub-Saharan Africa
8 Soil and nutrient management for cassava cultivation in Asia
9 Conclusions
10 References

1 Introduction

Cassava is drought-tolerant and grows reasonably well in relatively infertile soils where other crops would fail. It is becoming an ever more important crop, and over the past 20–30 years the cassava-planted areas have increased faster than those of any other food crop.

In the *ex ante* impact assessment by Henry and Gottret (1996), conducted in the mid-1990s, the elimination of soil-related constraints in Asia was estimated to increase yields up to 43%, while elimination of crop management constraints could increase yields by 26%; for Africa this was 29% and 32%, and for Latin America this was 22% and 20%, respectively. In all three continents, soil-related constraints were found to be the most, or the second most, important among the various categories of constraints. Using a slightly different methodology, Howeler (2014a) estimated that elimination of soil-related constraints in Asia could increase cassava yields by 38% from current levels while improved crop management could increase yields up to 29%. No similar data were published for Africa or Latin America, but the well-documented lack of fertilizer use in many cassava-growing areas of both continents would probably have resulted in even lower levels of current soil

http://dx.doi.org/10.19103/AS.2016.0014.34

fertility than in the mid-1990s. This has clearly been the case in Africa (Fermont et al., 2008; Pypers et al., 2012).

In farmer-managed trials conducted in 49 farms in four villages in western Kenya and four villages in Uganda, it was found that the average cassava yield under current farmer practice was only 8.6 t/ha. By changing to the use of improved practices of crop establishment (1 m x 1 m plant spacing, no intercropping and early planting at the start of the rainy season) the average yield increased to 10.1 t/ha. By also changing to new improved varieties from IITA, yields further increased to 13.6 t/ha. But by also applying NPK fertilizers (100-50-100 kg/ha of N-P_2O_5-K_2O) yields further increased to 20.8 t/ha. It was concluded that poor soil fertility was the most important constraint to cassava production in these parts of Kenya and Uganda (Fermont et al., 2009). This was further reinforced by the strong response to fertilization and by the fact that 62% of the interviewed farmers indicated that they considered poor soil fertility as a constraint and 22% considered it as the most important constraint.

In order to supply the plants with the nutrients required for optimal production, it is important to know the plant's nutritional requirements, diagnose correctly any deficiency or toxicity, and use adequate measures to correct these problems. This chapter describes these points for each nutrient separately. It also provides detailed discussion of the soil and nutrient management strategies most suitable for the three main growing regions: Latin America, sub-Saharan Africa and Asia.

2 Nitrogen and phosphorus deficiencies

2.1 Nitrogen deficiency

Some varieties of cassava do not show any symptoms of N deficiency, but the plants remain small and weak while root yields are markedly reduced. Unless farmers have well-fertilized plants for comparison, they may never know that their crop is suffering from N deficiency. Some other varieties, however, show clear symptoms of N deficiency: plants are uniformly chlorotic and leaves have a uniform light green or yellowish colour. N deficiency symptoms first appear in the bottom leaves but rapidly spread throughout the plant, leading to a generalized chlorosis. N-deficient leaves are smaller and may have fewer lobes and shorter petioles than normal leaves.

In nutrient solution trials, Forno (1977) observed only slight N deficiency symptoms in cassava, while sorghum, maize and cotton showed severe symptoms. However, like for other crop plants, cassava growth was severely reduced, which corresponds with observations in the field.

The 'critical level' for N deficiency in youngest fully expanded leaf (YFEL) blades at 3–4 months after planting is about 5.3% N (Howeler, 2012b), while the 'sufficiency range' is about 5.1–5.8%. The critical level is defined as that concentration corresponding to 95% of maximum yield, while the sufficiency range is defined as the concentration corresponding to 90–100% of maximum yield.

To control N deficiency in cassava, it is recommended to apply 50–100 kg N/ha in the form of urea or as a compound fertilizer during the first 2–3 months after planting. In light-textured soils, N may be lost through leaching; in that case, it is recommended to apply N in two applications – at planting and at two months. Nitrogen can also be applied in the form of animal manure (5–10 t/ha) or by incorporation or mulching of leguminous intercrop residues, green manures or cover crops. The latter could supply as much as 200 kg N/ha in the above-ground biomass (Tongglum et al., 1992).

Nitrogen (N) deficiency is commonly observed when cassava is grown on light-textured low organic matter soils or in highly acid soils with a low rate of N mineralization. Significant responses to N application have been reported mainly in light-textured soils of the Colombian north coast, the east and north-eastern part of Thailand, on Bohol island of the Philippines, in southern China, in Kerala state of India, on Java island of Indonesia as well as in very acid soils of the Eastern Plains and the Andean region of Colombia. These will be discussed in more detail below.

Many investigators (Vijayan and Aiyer, 1969; Acosta and Perez, 1954; Obigbesan and Fayemi, 1976; Fox et al., 1975) found that cassava responded negatively to application of high levels of N in terms of economic yields. These high levels stimulate excessive top growth, resulting in a reduction of root production. Krochmal and Samuels (1970) reported a root yield reduction of 41% and top growth increase of 11% due to high N applications. Also, high levels of N application stimulate production of N-containing compounds such as protein and HCN (Obigbesan and Agboola, 1977; Kabeerathumma et al., 1990), and may also result in a decrease in starch content. It should be noted that there are likely to be strong varietal differences for response to nitrogen, in terms of root versus top growth, but there are limited studies. Breeders should select varieties under the nitrogen levels recommended for the target region so that variety response to nitrogen will maximize economic returns.

On the other hand, high levels of N applications may be necessary for cassava forage production since the regular cutting of tops removes large amounts of N. Figure 1 shows

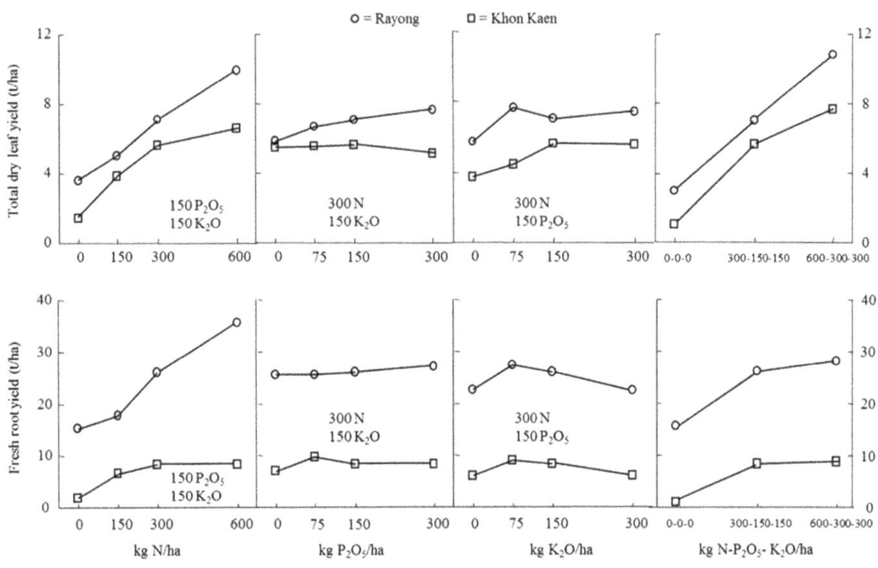

Figure 1 Effect of application of various combinations of N, P and K on the total dry cassava leaf yield from five cuts of young tops and on the final fresh root yield in Rayong and Khon Kaen Field Crops Research Centers in Thailand in 2003/04. Data are average values of varieties Rayong 5 and Rayong 72. Source: Howeler, 2012g.

the response to N, P and K applications in two locations in Thailand, in terms of both total dry forage and fresh root production. In both locations there was a nearly linear response up to 600 kg N/ha, while there was no response to the application of P and a minor response to K. With these high levels of applied N, P and K, even after five cuts of tops during a 12 month crop cycle, the root yields at one location were above 25 t/ha, while up to a total of 11 t/ha of dry forage containing 1.9 t/ha of protein were produced by the frequent cutting of top growth.

2.2 Phosphorus deficiency

Phosphorus-deficient cassava plants are generally short and with thin stems, small and narrow leaf blades and short petioles. During drought periods, upper leaves of P-deficient plants tend to droop down from the petioles. The leaves are generally dark green in colour while one or two lower leaves may be dark yellow to orange, and in some varieties purplish with necrotic white spots. These lower leaves often drop off, leaving the plant without any recognizable symptoms. The critical level for P deficiency in YFEL blades is about 0.41% P (Howeler and Cadavid, 1990) and the sufficiency range is calculated to be 0.38–0.50% P. The critical level for available P in the soil is about 4–6 ppm Bray II-extractable P (Howeler, 1990). In some soils having only 2–4 ppm available P there was still no response to P application due to a highly efficient mycorrhizal population (Howeler et al., 1987), which enabled the plants to absorb soil P from a greater soil volume (Howeler, 2012f). It has been clearly shown (Yost and Fox, 1979; Zaag van der et al., 1979; Howeler et al., 1982a, b) that cassava is extremely dependent on an effective mycorrhizal association for absorption of P from the soil. In soils with a low or ineffective mycorrhizal population, cassava growth and production can be increased by soil inoculation with a highly effective strain of mycorrhiza. In the presence of an effective mycorrhizal population cassava is very tolerant of low levels of available P. While maize and soya bean have a critical soil P level of 14–15 ppm, cassava required only 8 ppm of Bray I-extractable P (Kang et al., 1980).

Pellet and El-Sharkawy (1993a, b) studied the effect of P application (0, 50 and 100 kg P/ha) on the physiological processes of four cassava varieties during two consecutive crop cycles in CIAT-Quilichao, Colombia. The soil in the original cycle had a pH of 4.35 and Bray II-extractable P of 2.49 ppm. They found that soil available P increased with increasing P application rates, reaching a maximum at 2–4 months after planting (MAP), after which there was some decline. During the second year, the available P without P application remained low at about 3 ppm, while with the second year's application of P the available soil P increased markedly, reaching a maximum at 4 MAP of about 10 ppm in plots with 50 kg applied P, and 23 ppm in plots with 100 kg P/ha. The increase in available P was observed only in the top 20 cm of soil with no P movement down the profile. Without NPK application total biomass production and root yields were low, but plants had diverted photosynthates into production of flowers and fruits; however, with the application of N and K (each at 100 kg/ha) but without P there was little flowering. P application also markedly increased the branching of the varieties CM 523-7 and MCol 1684, as well as the number of storage roots of all varieties, reaching a maximum at 4-6 MAP, after which this remained rather constant. CM 489-1 had the greatest number of storage roots, especially in the second year and with the highest application of 100 kg P/ha. Two varieties, CM 523-7 and CM 489-1, had relatively high root yields at low P as well as a good response to P application; MCol 1684 did not respond to P application, and

Table 1 Differential dry root yield (t/ha) response of four cassava varieties to the application of fertilizer during two years of cropping at CIAT-Quilichao in Colombia

N-P-K applied (kg/ha)	CM 523-7		CM 489-1		MCol 1684		CMC 40	
	1989	1990	1989	1990	1989	1990	1989	1990
0-0-0	9.6	5.3	10.6	4.9	7.7	5.4	7.2	4.2
100-0-100	11.5	11.6	13.0	9.6	12.9	9.8	8.2	6.0
100-50-100	13.4	14.1	16.2	13.1	12.2	10.6	11.5	9.6
100-100-100	16.1	14.3	17.3	14.0	11.5	11.5	11.0	10.4
LSD 5%[1]	2.8	2.7						

[1]LSD for fertilizer levels within varieties.
Source: Pellet and El-Sharkawy, 1993a.

CMC 40 had low yields at both low and high P (Table 1). This indicates that varieties show a differential adaptation to low P soils and that germplasm can be selected that produces high yields under both low- and high P conditions.

The average P uptake rates per month increased markedly with P application, but they were not significantly different between applications of 50 and 100 kg P/ha. The four varieties differed significantly in P uptake rates in the order of CM 523-7 > CM 489-1 > MCol 1684 > CMC40, but did not differ in uptake rate per unit root length, indicating that root length or density and P availability were the two factors affecting P uptake rates. P uptake rates increased with time depending on P availability and the density of fine roots; the latter were concentrated in the top 20 cm of soil and there was little P uptake from deeper down the profile. The lack of significant response to P by MCol 1684 may be due to root sink limitations as most photosynthates were channelled into top growth production resulting in a low harvest index and low P-use efficiency.

Phosphorus deficiency is often the most limiting factor for cassava grown on acid-infertile Oxisols, Ultisols and Inceptisols in Latin America, but is less common in Asia and Africa.

To control P deficiency, it is recommended to band-apply 25–50 kg P/ha near the stake using highly soluble P sources such as single or triple superphosphate (TSP), mono- or di-ammonium phosphate or as compound fertilizers. These should be band-applied close to the stakes in order to reduce P fixation by the soil as well as control weed growth. Alternatively, P can be applied by broadcasting and incorporating less soluble sources such as basic slag, rock phosphates or thermo-phosphate. Because of their lower costs these are attractive sources of P in very acid soils. Leong (2016) observed that broadcasting of even soluble P sources may be more effective than band application in non-P fixing soils in order to distribute the P as close as possible to the whole fibrous root system.

While P is the main limiting element for cassava in many virgin soils, especially in Latin America, the application of P has a long residual effect. Thus, continuous P application builds up the available P content of the soil, improving its productivity. Once the available P level has increased above the critical level, further P applications can be greatly reduced or entirely eliminated.

Alternative methods of P application such as foliar applications or stake treatments have been tried in Carimagua and in Quilichao in Colombia. In both locations, these methods were found to be ineffective and yields increased significantly only when P sources were applied to the soil. Fractionation of P had no beneficial effect and it is recommended to apply all P at, or shortly after, planting to enhance early growth and plant vigour, which in turn reduces weed growth and erosion.

3 Potassium, calcium, magnesium and sulphur deficiencies

3.1 Potassium deficiency

Potassium-deficient plants are generally shorter and more branched compared to normal plants of the same variety and with a tendency towards a prostrate growth habit. In many varieties, the upper internodes are very short and prematurely lignified, giving rise to a 'zigzagging' of the upper stem. In some varieties, the upper leaves are small and chlorotic, while in others a few lower leaves are yellow with black spots and border necrosis. During periods of drought, leaf borders may curl upward, while during wet periods there may be severe dieback of shoot tips due to K deficiency-induced anthracnose (*Colletotrichum* spp.) and cassava bacterial blight (CBB) (*Xanthomonas axonopodis* pv *manihotis*). In many cases, however, there are no clearly recognizable symptoms except that the plants are shorter and have smaller leaves than those well-supplied with K. The critical level for K deficiency in YFEL blades at 3–4 MAP is about 1.5% K, while the sufficiency range is 1.4–1.9% K. The critical level for exchangeable K in the soil was found to be 0.15–0.18 me/100 g (Howeler, 1985a, b; 2012a; 2014b; Howeler and Cadavid, 1990; Juo, 1985).

Cassava removes large amounts of potassium (K) in the root harvest, and long-term fertility trials have shown that K deficiency becomes the most limiting nutritional factor if cassava is grown continuously without adequate K fertilization. Significant short-term responses to K have been obtained mainly in low-fertility Oxisols or Ultisols, such as in the Eastern Plains of Colombia (Howeler and Cadavid, 1990), and in northern Vietnam (Nguyen Huu Hy et al., 2010), Lao PDR (Aye et al., 2010) and in one of two varieties in Hainan, China (Tian Yinong et al., 1995). Significant long-term responses to K have been obtained in volcanic ash soils of Colombia (Howeler and Cadavid, 1990), and in Ultisols of Malaysia (Chan, 1980), Thailand (Nakviroj et al., 2007) and India (Kabeerathumma et al., 1990), as well as in Vietnam, Indonesia and China (Howeler, 2012c). In much of sub-Saharan Africa cassava is planted after cutting and burning the bush fallow; in that case, there is usually no short-term response to K fertilization as sufficient K is supplied by the ash (Juo, 1985).

Potassium deficiency in cassava can be controlled with the application of 50–100 kg K_2O/ha as potassium chloride or sulphate (Ngongi et al., 1977). The former is generally cheaper and thus more commonly used. Potassium can also be applied as a compound fertilizer or in the form of wood ash or animal manures. In soils where K deficiency is not a serious problem, compound fertilizers with an N-P_2O_5-K_2O ratio of about 2:1:2 or 2:1:3 are recommended in order to supply enough K to prevent its exhaustion in the soil. Most K fertilizers are highly soluble and should be band-applied near the stake during the first two MAP. A basal application at 30 days after planting has given highest overall yields (Howeler, 1985a). In light-textured soils K fertilizers should be applied in two smaller doses to prevent losses by leaching.

3.2 Calcium deficiency

Since Ca is not very mobile in the phloem, there is little translocation from the older to the younger tissues. Thus, for normal growth the plant requires a continuous supply of Ca. When this supply is inadequate, symptoms of Ca deficiency develop in the growing points of both shoots and roots (Asher et al., 1980). The upper leaves are deformed with leaf tips burned and curling either upward or downward. The growing points of fibrous roots die back resulting in excessive root branching. In the field, Ca deficiency is mainly characterized by slight deformation and burning of leaf tips in the upper part of the plant, but these symptoms cannot be seen in all varieties. Triploid cassava plants seem to be particularly susceptible to Ca deficiency in southern India (Susan John et al., 2005). The critical concentration for Ca deficiency in YFEL blades at 3–4 MAP was found to be 0.56% Ca, and the sufficiency range was calculated to be about 0.50–0.72% Ca. However, Ca concentrations in YFEL blades can vary among varieties. Being a phloem-immobile element, Ca concentrations tend to be higher in the lower part of the plant. A critical level of exchangeable Ca in the soil was found to be about 0.25 me/100 g.

Calcium (Ca) deficiency symptoms are easily produced in nutrient solution culture, but are seldom seen in the field. Still, in very acid soils with high levels of aluminium (Al) and low levels of Ca, cassava does respond to liming, which is mainly a response to the application of Ca and/or Mg. Thus, significant Ca or lime responses have been reported in the very acid soils of the Eastern Plains of Colombia (Spain et al., 1975; CIAT, 1985; Howeler, 1991a, 2012d, 2014b) and on peat soils of Malaysia (Tan and Chan, 1995).

Calcium deficiency is generally controlled by application of 200–400 kg Ca/ha in the form of calcitic or dolomitic limestone, as calcium oxide or hydroxide, or as gypsum (CIAT, 1985; Howeler, 2014b). Gypsum is a more soluble source, which supplies Ca as a nutrient but without affecting soil pH or exchangeable Al. All these Ca sources should be broadcast and incorporated before planting. High applications of lime may have detrimental effects by inducing micronutrient deficiencies, particularly that of zinc (Zn) (CIAT, 1976; Howeler, 2012d, 2014b).

3.3 Magnesium deficiency

Magnesium deficiency is characterized by interveinal chlorosis of the lower leaves (fishbone pattern), which starts out as slight yellowing of leaf margins and may eventually develop into necrosis of leaf tips and margins. Symptoms appear first at the lower leaves and progressively move up the plant. Critical concentrations for Mg deficiency in YFEL blades at 3–4 MAP were found to be 0.25% Mg and the sufficiency range was calculated as 0.24–0.29% Mg. No critical levels for soil exchangeable Mg have been reported, but Mg deficiency symptoms and a response to Mg application were found in soils with less than 0.2 me/100 g (Kang, 1984; Howeler, 2012a, 2014b).

Magnesium deficiency symptoms are often found in acid-infertile soils such as Oxisols, Ultisols and certain Inceptisols. They have been observed in the Eastern Plains and Andean zone of Colombia, in southern India, in the northeast of Thailand, in northern Vietnam and in central Lao PDR.

Mg deficiency can be controlled by the application of 40–60 kg Mg/ha in the form of magnesium oxide, dolomitic limestone or magnesium sulphate. The first two sources are relatively insoluble and should be broadcast and incorporated before planting. Magnesium sulphate is a soluble source, which may be band-applied near the stake after planting, but may be more effective when incorporated into the soil (Howeler,

2012d, 2014b). It has no effect on soil pH or exchangeable Al, but is also a source of sulphur (S).

3.4 Sulphur deficiency

Sulphur deficiency is characterized by a uniform chlorosis or yellowing of leaves (similar to N deficiency) in the upper and middle part of the plant. Eventually the whole plant may become chlorotic. Leaves are small and plant height may be reduced but leaves are not deformed. The critical concentration for S deficiency in YFEL blades was found to be about 0.31% S, while the sufficiency range was calculated to be 0.30–0.36% S. Critical levels in the soil have not been determined, but S responses were obtained in soils with 25–30 ppm phosphate-extractable S (Ngongi et al., 1977; Howeler, 2002, 2014b).

Sulphur deficiency in cassava is easily produced in nutrient solutions, but is not often found in the field. Sulphur deficiency symptoms and responses to S application have been found in Oxisols of the Eastern Plains of Colombia and in Ultisols in southern India (Mohan Kumar and Nair, 1985).

Sulphur deficiency can be controlled by the application of 10–20 kg S/ha as elemental sulphur or as sulphates of ammonium, potassium, calcium or magnesium. The latter four sources are relatively soluble and can be band-applied near the stake at planting, while elemental sulphur should be broadcast and incorporated before planting (Howeler, 2002). However, Mohan Kumar and Nair (1985) recommended the application of 50 kg S/ha in the form of S-containing fertilizers, while Susan John et al. (2016) suggested applications of 10–20 kg S/ha in the form of elemental S or S-containing fertilizers in acid soils of Kerala, India.

4 Micronutrients and soil pH

4.1 Micronutrient deficiencies and toxicities

Micronutrients are absorbed by the plant in very small quantities, but are a basic component of many enzymes and thus play an essential role in most metabolic processes. In cassava, there are few reports of micronutrient deficiencies, but they may be more common than is generally recognized. Cassava is especially susceptible to Zn deficiency and symptoms of severe Zn deficiency have been observed in acid soils in Colombia, Brazil, Mexico, Malaysia, Thailand and Nigeria, as well as in alkaline and/or calcareous soils in Colombia, Cuba, Mexico and Thailand. In Brazil, there were few responses to application of micronutrients, with the main response being to the application of Zn applied at 5 kg Zn/ha (Gomes and Da Silva, 2006).

Symptoms of Zn deficiency appear as interveinal chlorotic spots or lines on young leaves. When very severe the whole leaf becomes pale green to white, leaf lobes become smaller and tend to point outward away from the stem. Oftentimes, lower leaves show necrotic spotting (Asher et al., 1980).

On acid soils, Zn deficiency can be controlled by incorporation of ZnO or band application of $ZnSO_4.7H_2O$ at the rate of 10–20 kg Zn/ha. Also effective are foliar applications with 1–2% $ZnSO_4.7H_2O$, or stake treatments in a 2–4% $ZnSO_4.7H_2O$ solution during 15 minutes. In alkaline soils, minor element applications to the soil are not very effective and stake treatments or foliar applications are recommended. In a high pH soil at CIAT-Palmira a simple stake dip for 15 minutes in 4% $ZnSO_4.7H_2O$ solution increased the average root yield of 20 cassava varieties from 11.5 to 25.0 t/ha. Untreated plants of many varieties died because of Zn deficiency, while the stake treatment resulted in normal growth (CIAT, 1985; Howeler, 2012d, 2014b).

Copper (Cu) deficiency results in reduced plant height, chlorosis and curling of upper leaves and necrosis of leaf tips. Lower petioles tend to be long and droopy. Severe Cu deficiency has been reported only in peat soils of Malaysia (Chew, 1971). A basal application of 2.5 kg Cu/ha as $CuSO_4.5H_2O$ increased yield from 4 to 12 t/ha (Chew et al., 1978).

Iron (Fe)-deficient plants have smaller but normal-shaped upper leaves that are light green, yellowish or white in colour. When severe, even the upper petioles are white. Iron deficiency has been observed in calcareous soils of the Yucatan peninsula of Mexico, and in northern Colombia, in Yogyakarta Province of Indonesia and in various places in central Thailand. A practical solution is probably either a stake treatment with $FeSO_4.7H_2O$ or foliar applications with sulphates or chelates.

Manganese (Mn) deficiency is characterized by interveinal chlorosis (fishbone pattern) of middle leaves, similar to Mg deficiency, but generally not present in lower leaves. When severe the whole leaf turns uniformly yellow, very similar to Fe deficiency or salinity. Manganese deficiency has been observed on alkaline soils in the Cauca Valley of Colombia as well as along the coast in N.E. Brazil. Stake treatments or foliar applications with Mn sulphate or chelates are probably the most practical solutions.

Boron (B) deficiency in field-grown cassava is characterized by chlorotic small spots on middle or lower leaves. However, in nutrient solution B-deficient plants have a small deformed growing point with very short internodes and small deformed dark green upper leaves. Sometimes the petioles or stem exude a brown gummy substance, followed by appearance of brown lesions. The root system is short and stubby (Asher et al., 1980). Symptoms of B deficiency have been observed both in the acid soils of Carimagua and Quilichao in Colombia and on alkaline soils at CIAT-Palmira. Applications of 1–2 kg B/ha, band-applied as Borax at time of planting, eliminated these symptoms, increased plant height, increased B levels in the leaves from 3 to 40 ppm, but had no significant effect on yield. Thus, cassava appears quite tolerant of low levels of available B in the soil. B deficiency symptoms were found to be most common during periods of water stress (Leong, 2016).

B toxicity has not been observed under natural conditions but is easily induced by excessive applications of B to the soil or in stake treatments. Being a phloem-immobile element, B toxicity causes necrosis of lower leaves, but the element is not translocated to the growing point; thus, plants generally recuperate.

4.2 Soil acidity and alkalinity

Large areas of the tropics are unproductive because the soils are too acid for most crops and the lack of adequate roads makes transport of lime prohibitively expensive. In these areas cassava is often the staple food because this crop is highly tolerant of low pH and high levels of aluminium (Al) and manganese (Mn) (Spain et al., 1975; Gunatilaka, 1977; CIAT, 1978; Islam et al., 1980; Howeler 1991a, 2012d).

Symptoms of Al toxicity in the field are seldom observed, except that plants are small and lack sufficient vigour. Under severe Al toxicity conditions in nutrient solutions, lower leaves may have interveinal chlorosis and necrotic spots. High levels of Al have an especially detrimental effect on root growth, which in turn affects nutrient and water absorption.

Plants suffering from Mn toxicity have droopy yellow bottom leaves with brown and black spots along the veins. These leaves may later fall off, leaving the plant without recognizable symptoms. Mn toxicity occurs only in very acid soils high in Mn and mainly in areas of poor drainage, which enhances the solubility of Mn due to reduction processes. Severe symptoms of Mn toxicity leading to low root yields have been observed in light-textured soils that had been excessively compacted by frequent tractor use in Thailand (Silpamaneephan, 1994). Compared with other crops cassava is relatively tolerant of high levels of Mn. Among 13 plant species studied only three species were more tolerant (Edward and Asher, unpublished). Among cassava varieties considerable differences in tolerance were also observed. Mn toxicity in cassava has been reported only in acid Ultisols and Inceptisols in Quilichao, Colombia. Application of lime in acid soils decreases the concentration of both Al and Mn, reducing their toxic effects.

A lime application experiment in Carimagua, Colombia, showed that increasing rates of lime application increased soil pH and decreased exchangeable Al. Application of 6 t/ha of lime reduced the Al saturation from 85 to 20%, where % Al saturation = exchangeable Al/Σ(exch Al, Ca, Mg and K). The experiment showed that maize, rice, *Phaseolus* beans and sorghum produced very poorly without lime and required 6 t/ha of lime to reach maximum yields. On the other hand, cowpea and cassava still produced 40% of maximum yield without lime and close to maximum yield with only 2 t lime/ha (Spain et al., 1975; Howeler, 2012d). Among cassava varieties there are differences in their tolerance to acid soils and highly tolerant varieties should be selected for acid soil regions (Howeler, 1991a).

Since cassava is very susceptible to Zn deficiency, application of lime can actually have a detrimental effect by reducing the availability of Zn and other micronutrients. Thus, for maximum beneficial effect liming often has to be accompanied with Zn applications (Howeler, 2012d, 2014b).

While cassava is very tolerant of acid soils, the crop is quite susceptible to salinity, alkalinity and high pH. Islam et al. (1980) showed that in nutrient solutions cassava had optimum growth at pH 5.5–7.0, but top growth declined markedly above pH 7.5–8.0 (Howeler, 2012d). The species was among the most tolerant of low pH and most susceptible to high pH. In natural soils, high pH is generally associated with high levels of salts (salinity) and sodium (Na) (alkalinity), poor drainage and minor element deficiencies. The crop usually suffers from a combination of all these conditions. Also, salinity–alkalinity problems often occur in spots in the field, giving rise to extremely heterogeneous soils and highly variable plant growth. While there are significant differences among varieties, root yields generally decline markedly above pH 8.0, above 2.5% Na saturation and above 0.5–0.7 milliSiemens/centimetre (mS/cm) of electrical conductivity. Yield reductions are probably due to the combined effect of all three factors. In comparison, many other crops will tolerate up to 15% Na saturation or 4 mS/cm electrical conductivity (CIAT, 1977; Howeler, 2012d)

Problems of soil salinity–alkalinity are very costly to resolve. Soil amendments such as gypsum or elemental sulphur are expensive and are only effective when combined with good drainage and adequate, good quality irrigation water. The best solution is to use only irrigation water of low salt content and select crops adapted to high pH and salinity. For cassava, the selection of tolerant varieties is the only practical solution, together with stake or foliar applications of minor elements, especially Zn. To date, however, there has been little effort by breeders to select for high pH conditions, due to the large predominance of cassava production in low to neutral pH soils.

5 Short- and long-term nitrogen, phosphorus and potassium (NPK) requirements of cassava: an overview

Since cassava grows reasonably well in very infertile soils where other crops may not produce at all, many people believe that cassava does not respond to fertilizers or manure applications. This belief is also based on the lack of response to fertilizer applications in some trials that were conducted for 1–2 years. These results could be due to the fact that the experiments were conducted on a rather fertile soil with no specific nutritional problems, or the amount or type of fertilizers applied were not adequate to satisfy the crop's particular nutritional requirements. However, thousands of fertilizer trials conducted by FAO throughout the world between 1961 and 1977 indicate that cassava is as responsive to fertilizer application as other crops that are traditionally fertilized, and that fertilizer application to cassava can be highly economic, as shown in Table 2 (FAO, 1980). These fertilizer experiments conducted by FAO were mostly short-term trials. These indicate that in West Africa (Ghana) cassava responded mainly to K, in Latin America (Brazil) to P and in Asia (Indonesia) to N, followed by K and P (Richards, 1979).

However, with rapidly increasing populations around the world, and the resulting pressure on land, as well as climatic changes due to global warming, it is increasingly more likely that cassava will be grown on the same field for many consecutive cropping cycles. In that case, the harvesting and removal of cassava roots and other parts of the plant will inevitably result in a decrease in the nutrient supplying power of the soil, mainly that of K. Thus, when cassava is grown continuously for many years on the same field, K is likely to become the most limiting nutrient and adequate levels of applied K will be necessary, either in the form of chemical fertilizers or organic manures. This is especially true for light-textured acid soils with low levels of exchangeable bases, mostly Oxisols, Ultisols and Inceptisols.

6 Soil and nutrient management for cassava cultivation in Latin America

In Latin America, most fertilizer experiments with cassava have been conducted in Brazil and Colombia. In 39 short-term NPK trials conducted in nine states of Brazil from 1950 to 1983, the main limiting nutrient was P in 25 trials, K in nine and N in only six trials (Gomes and Howeler, 1980). More recent research on fertilizer use for cassava in Brazil indicates little response to N application in most soils, especially in the northeast of the country (Gomes and Da Silva, 2006). However, some of the most dramatic responses to N application have been obtained on the sandy soils of Jaguaruna in Santa Catarina State of southern Brazil. Moraes et al. (1981) reported a nearly linear response of two varieties up to levels of 150 kg N/ha. In this location yields increased from 10 to 35 t/ha by N application in a soil with 89% sand and 0.7% organic matter (OM). For both varieties highest yields were obtained with a fractionated application with one-third applied at 30, 60 and 90 days. Santana et al. (1975) reported little difference among sources of N such as FYM, urea, $(NH_4)_2SO_4$ and $Ca(NO_3)_2$, while Santos and Tupinambá (1981) found that the time of N application and fractionation also had little effect.

Table 2 The average highest per cent yield increase due to fertilization and value/cost ratio (VCR) for cassava as compared to other crops in various countries

			Average best	
Country	Crop	No. of trials	% Response	VCR
Brazil	Cassava	66	112	5.6
1970–76	Maize	510	83	2.7
	Cotton	490	84	3.7
	Beans	391	91	4.9
	Rice	385	77	4.8
	Soya bean	124	102	2.4
	Sugarcane	105	67	3.6
Colombia	Cassava	16	124	5.9
1962–70	Maize	102	95	6.4
	Beans	47	67	6.0
	Forage	41	154	1.7
	Potatoes	33	266	11.4
	Wheat	15	73	4.4
Ghana	Cassava	134	71	19.9
1961–75	Maize	775	121	9.6
	Groundnut	134	52	18.7
	Cotton	92	82	18.3
	Cowpeas	61	65	15.9
Nigeria	Cassava	28	53	11.3
1961–77	Maize	478	64	5.1
	Yams	348	43	22.6
	Rice	277	42	13. 8
Indonesia	Cassava	56	176	4.2
1969–76	Rice	378	63	3.1
	Sorghum	312	217	2.5
	Groundnut	135	60	4.9
	Soya bean	117	60	3.1

Source: FAO, 1980.

Marked responses to P application were reported in the very acid and infertile soils of the Campo Cerrado and the Amazon Basin of Brazil. In most soils the crop responded to applications of 50–100 kg P_2O_5/ha. K responses were mainly obtained in sandy soils of Brazil, especially after more than three years of continuous cassava cultivation; in other soils the crop did not respond much to K application and there was little relationship between the response and levels of soil exchangeable K (Gomes and Da Silva, 2006).

Similarly, in 22 short-term NPK trials conducted from 1980 to 1982 in four zones in Colombia, there were significant responses to the application of P in 12 locations, to K in six and to N in only four locations (Howeler and Cadavid, 1990). In most sites with acid-infertile soils in Colombia, the crop responds to levels of 50–100 kg P_2O_5/ha, although much higher initial P applications are sometimes required for maximum yields. These high application rates should not be continued as P tends to build-up in the soil and may cause reductions in the availability or uptake of other nutrients, especially Zn. In the Eastern Plains (Llanos Orientales) in Colombia, on very acid and infertile Oxisols, cassava responded mainly to applications of P followed by N and then K (Howeler and Ballesteros, 1987; Howeler and Cadavid, 1990). At the Carimagua station, located in the savannahs of the Llanos Orientales, fresh root yields increased from about 7 t/ha without P to 23 t/ha with the application of 200 kg P_2O_5/ha. Banded TSP was the most effective source, followed by broadcasted basic slag, while banded simple superphosphate (SSP) and broadcasted rock phosphates were slightly less effective (Howeler, 1985a, 2012c). At the same location, the crop showed a significant response to application of 100–150 kg N/ha but showed no response to application of K during the first year of cropping. However, in subsequent years in the same plots, yields without K application markedly decreased and the response to K increased, as shown in Fig. 2 (Howeler, 1985a, 2012c).

The crop also showed a marked response to application of N, P and K in the very acid and low P volcanic ash soils of CIAT-Quilichao station, and in farmers' fields, both in Cauca Department of Colombia. In short-term trials conducted in eight locations in this department, cassava showed a significant response to N in one location, to P in seven locations and to K in two locations (Howeler and Cadavid, 1990; Pellet and El-Sharkawy,

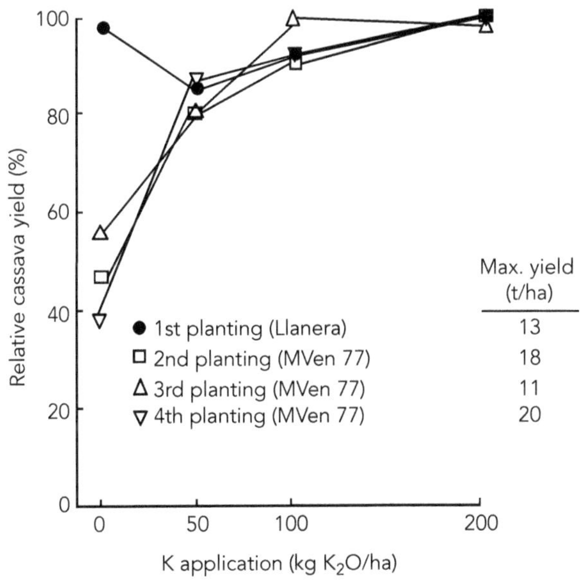

Figure 2 The effect of annual application of four rates of applied K on the relative yield of cassava cv. MVen 77 during four consecutive cropping cycles in Carimagua, Colombia. Source: Howeler, 1985a.

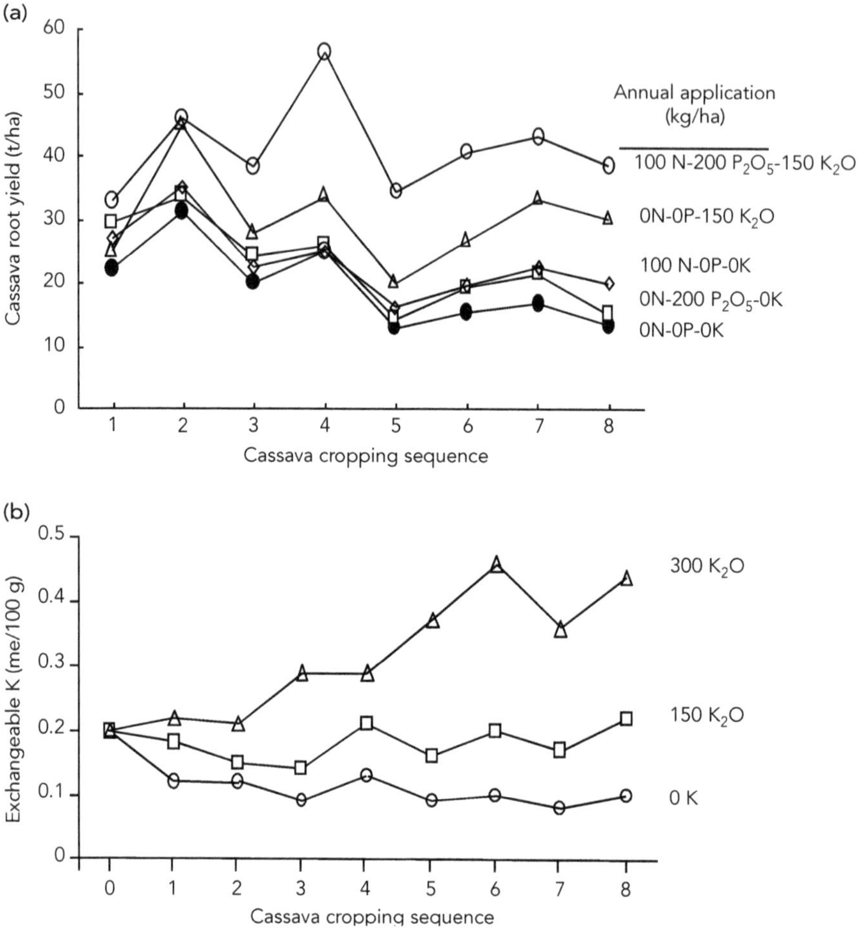

Figure 3 Effect of various levels of annual applications of N, P and K on cassava root yield (a), and on the exchangeable K content of the soil (b) during eight consecutive cropping cycles in a long-term NPK trial conducted at CIAT-Quilichao, Colombia. Source: Howeler and Cadavid, 1990.

1993a, b). However, when the crop was grown for several years in the same plots in CIAT-Quilichao, the responses to all three nutrients increased, particularly in the case of K, which is removed in largest quantities in each root harvest (Howeler, 1985b, 1991b). Figure 3 shows the results of an experiment conducted for eight consecutive years in CIAT-Quilichao. With the application of 150 kg K_2O/ha the exchangeable K level of the soil could be maintained above the critical level of 0.15 me K/100 g, and yields could be maintained at 25–30 t/ha, while without K the root yields decreased from an initial level of 20–30 t/ha to only 15–20 t/ha. With the application of 300 kg K_2O/ha the exchangeable K level increased year after year and root yields increased to 40–50 t/ha.

In contrast, in short-term trials in four sites on the Atlantic Coast of Colombia there was only a significant response to N in one site (Media Luna) and no responses to application of

P or K in any site (Howeler and Cadavid, 1990). Subsequently, a long-term experiment was conducted for eight consecutive years in the sandy, low OM soils of Media Luna to study the effect of annual application of NPK fertilizers (50 kg N, 21 P and 41 K/ha), mulching with 12 t/ha of dry *Panicum maximum* grass, and with either conventional or zero tillage. The application of NPK fertilizers significantly increased root and top biomass and reduced the HCN content of the roots, especially if no mulch was applied. Conversely, the positive effect of annual application of mulch on cassava yields was highly significant, especially in the absence of fertilizers. Mulching also significantly reduced soil temperatures and increased soil organic carbon, K, P, Ca and Mg, and prevented a drop in pH. Application of NPK fertilizers resulted in a build-up of soil P but no build-up of exchangeable K (Cadavid et al., 1998).

In Puerto Rico, Fox et al. (1975) reported a significant N response up to 160 kg N/ha in one location but no response in another site. In the latter site a greater proportion of absorbed N (38%) was found in the storage roots, resulting in a high nitrogen-use efficiency.

7 Soil and nutrient management for cassava cultivation in sub-Saharan Africa

In Africa, relatively few fertilizer experiments with cassava have been conducted. This is because few farmers apply any fertilizer to their crops, and particularly to cassava, as there is a general belief that cassava does not respond to fertilizer applications. Average fertilizer use in Africa is currently about 9 kg/ha, as compared to 73 kg/ha in Latin America and 135 kg/ha in Asia (Kelly, 2006; Fermont et al., 2010). Lack of response may have been true 40–50 years ago but is probably not the case now. In the past most crops were grown in relatively fertile soil, resulting from the burning of regrown vegetation after having left the land fallow for 10-20 years. This system of slash-and-burn agriculture is no longer possible in much of Africa due to increasing population pressure. Much shorter fallow periods and increased cropping intensity of the land has led to nutrient depletion as a result of nutrient off-take in the harvested products (Fermont et al., 2008, 2010).

During much of the past century, cassava in sub-Saharan Africa was generally planted after cutting and burning the bush fallow. In that case there was usually no response to fertilization as sufficient nutrients, especially K, were supplied by the ash. Obigbesan (1973) summarized the early research on the effect of K fertilization on cassava yields and quality parameters, mostly done in Nigeria during the 1960s and early 1970s. Most of these experiments were conducted in the savannah and forest zones of western Nigeria, as well as the forest zone of south-eastern Nigeria. Obigbesan and Agboola (1977) obtained responses to the application of 30–50 kg N/ha in the savannah and forest zones of south-western Nigeria. Obigbesan (1977) reported a significant K response on three soils of western Nigeria, but only in some years and with some varieties. Increasing the rate of K application tended to increase the root starch content up to 60 kg K_2O/ha, above which it decreased. K application also decreased the HCN content of roots. On strongly acid soils of eastern Nigeria, and on slightly acid soils with 0.23 me K/100 g in the south-western part of the country, Okeke (unpublished) observed that a significant K response was usually obtained at 3–4 years after land clearing. Kang and Okeke (1984) obtained no or a negative response to application of N during three crop cycles in the forest zone of south-western Nigeria, while there was a significant response to the application of K only in the second and third year of cropping. They reported that K deficiency was the main nutrient-limiting cassava yields on an Alfisol in the forest zone of south-western Nigeria. A significant K response was obtained when the exchangeable K content of the soil was < 0.15 me/100 g and the K concentration in YFEL blades was <1.1%. Similar results were obtained by Juo (1985) in central and western Nigeria, where cassava is generally grown on very acid Ultisols or non-acid Alfisols; usually no K response was obtained until the exchangeable K level had dropped below 0.15 me/100 g. Umeh et al. (2015) also reported a strong K response in cassava intercropped with soya bean in south-eastern Nigeria, but only in the absence of N fertilization; with application of N there was only a slight response to K due to excessive top growth. In contrast, on an Alfisol at IITA in Ibadan, Nigeria, Olasantan et al. (1997) obtained a response mainly to application of N as the soil was high in exchangeable K. They also reported that cassava growth was strongly affected by intercropped maize, but that cassava growth accelerated again after the maize harvest, depending on the nutrient availability of the soil, especially that of N.

Application of NPK fertilizers is sometimes believed to increase the severity of diseases and pests. Ogbe et al. (1993) reported that in trials conducted at NRCRI in south-eastern Nigeria, increasing the rates of application of N increased the severity of cassava mosaic disease (CMD), but that applications of P or K did not; the combined application of 60 kg N, 10 kg P and 50 kg K/ha, however, did not increase CMD severity. Njoku and Arene (1980) found that N application also increased the severity of cassava anthracnose disease (CAD), while Odurukwe and Arena (1980) reported that N application had no effect on Cassava Bacterial Blight (CBB). In contrast, K applications were found to reduce severity of CAD as well as CBB, but did not have a similar effect on CMD.

In southern Benin, Carsky and Toukourou (2005) conducted NPK trials on 15 farmers' fields and did not obtain a significant response to N application during three consecutive years of cropping, and only a minor response during the third year; they also obtained no significant response to P and K in the first year, but a significant or highly significant response in the second and third year.

In Madagascar, Roche et al. (1957) and Cours et al. (1961) found that K was the main limiting nutrient, and applications of 110 kg K_2O/ha were recommended (Anonymous, 1952, 1953).

Takyi (1972) did not observe a K response in Ghana. On a tropical forest Ochrosol in Ghana, where cassava was rotated with either maize or maize followed by groundnut, Ofori (1973) obtained a significant response to N only in the first cassava crop, to P in the first and second crop and to K in the first and third crop.

Fermont et al. (2010) reported the results obtained in 2006 of 50 on-farm and two on-station nutrient omission trials, one each in Kenya and Uganda. The on-farm trials were conducted in four sites in West Kenya and in two sites in Uganda. There was a significant response to N in all five sites in western Kenya and in 2 of 3 sites in Uganda; a significant response to P in 4 of 5 sites in Kenya and in 2 of 3 sites in Uganda; and to K in only 3 of 5 sites in Kenya and in none of the three sites in Uganda. Thus, it appears that in Africa there is some response to application of N and P in certain areas, but that the main response is to K, especially if cassava is grown for several years in the same field.

Collaborating with farmers and extension workers in the Bas-Congo region of western D.R. Congo, Pypers et al. (2012) found that burning the bush vegetation or merely cutting and burying the vegetation had the same effect on the yields of the following cassava crop, while yields decreased when the natural vegetation was removed from the field before planting. However, in the absence of the natural vegetation they obtained a very good root yield response to the application of low (283 kg/ha), medium (850 kg/ha) and high (1417 kg/ha) rates of application of 17-17-17 ($N-P_2O_5-K_2O$) fertilizers. Furthermore, by the use of two naturally growing green manures, that is, either *Tithonia diversifolia* or *Chromolaena odorata*, buried in the soil before planting cassava without chemical fertilizers, they obtained the same or significantly higher cassava yields as compared to burying the natural vegetation. But when these green manures were combined with either low or medium rates of 17-17-17 fertilizers, yields further increased, reaching 35–40 t/ha as compared to yields of 21 t/ha with the slash-and-burn system. Despite the high price of chemical fertilizers, the use of these green manures combined with fertilizers was highly economic. From discussions with farmers about the results of these trials it became clear that profitability is the main driver for adoption, but risk and uncertainty also influence farmers' decision-making.

Similar trials with farmers' participation, using the integrated soil fertility management approach, were also conducted in the Sud-Kivu Province of eastern D.R. Congo by Pypers et al. (2011) and in the Ivory Coast by Gnahoua (2016) (Gnahoua et al., 2016). In both cases the experiments included treatments of farmers' local variety and practices, change to improved varieties of cassava and intercrops, change of plant spacing arrangements, and the use of manures and chemical fertilizers. In general, the application of fertilizers and manures produced the greatest increases in cassava and intercrop yields, but the plant spacing arrangement and choice and management of intercrops had a major effect on profitability.

8 Soil and nutrient management for cassava cultivation in Asia

In Asia many short- and long-term fertilizer trials have been conducted over the years. In nearly 100 short-term NPK cassava experiments conducted in Thailand in the early 1980s, the crop responded mainly to N, followed by K and P (Hagens and Sittibusaya, 1990). The response to N was well correlated with soil OM content (Sittibusaya et al., 1988). Some short-term P responses were observed in Alfisols in eastern Java as well as in Lampung Province of Indonesia, in northern Vietnam, southern China and in the Plain of Jars in Xieng Khouang Province of Lao PDR (Howeler, 2014b). Taufiq et al. (2015) reported a significant response to K application at two sites in Lampung Province of Indonesia, with maximum economic efficiency at either 30 or 60 kg K_2O/ha.

In a long-term fertilizer trial conducted at Central Tuber Crops Research Institute (CTCRI) in Kerala, India, N was the most limiting nutrient during the first 4–5 years, but in later years K became more important (Kabeerathumma et al., 1990). Figure 4 shows that after ten years of continuous cassava cropping the yields without K application had decreased from 22 t/ha in the first year to about 6 t/ha in the tenth year. In the treatment without K the exchangeable K in the soil had decreased from an initial level of 0.17 to only 0.07 me/100 g, indicating a clear depletion of the K status of the soil due to repeated cassava cropping and harvests. With annual K application of 100 kg K_2O/ha the exchangeable K level increased to 0.23 me/100 g and yields had increased to 28 t/ha.

In Asia, at least 19 long-term NPK experiments have been conducted with a duration ranging from 4 to over 30 years of continuous cropping on the same plots with annual applications of the same treatments. Details of some of these experiments have been

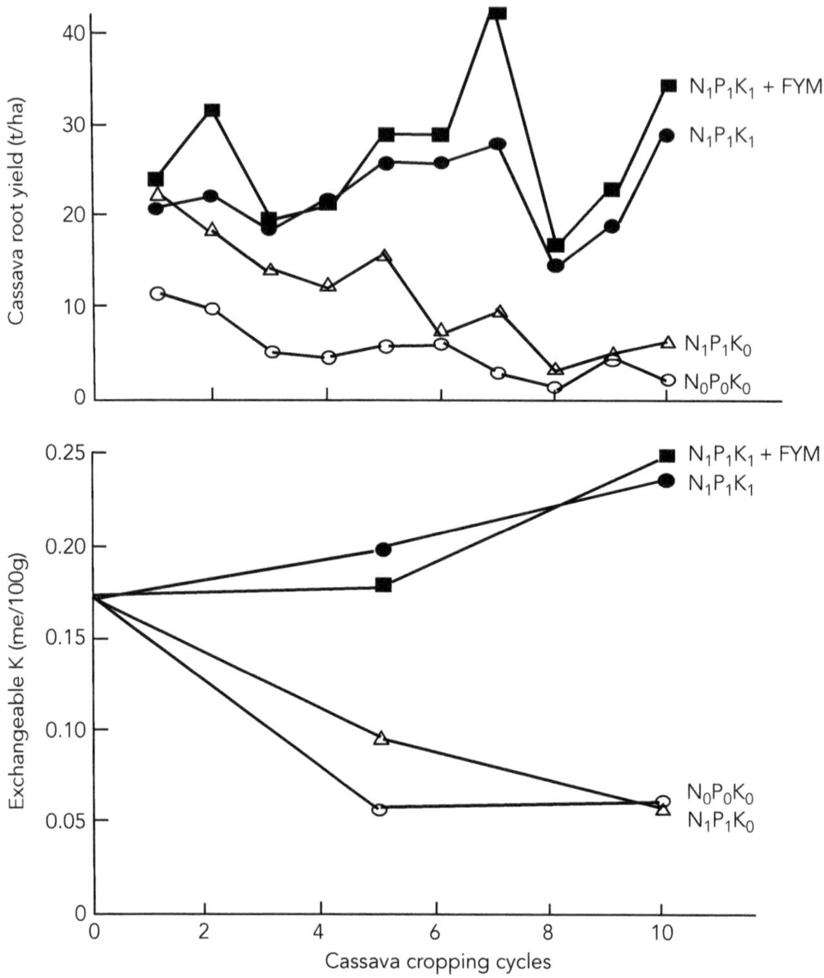

Figure 4 Cassava yield (top) and the exchangeable K content of the soil (bottom) during 10 years of continuous cropping with various NPK treatments in Trivandrum, Kerala, India. $N_1P_1K_1$ = 100 kg/ha each of N, P_2O_5 and K_2O. Source: Kabeerathumma et al., 1990.

described by Howeler (2012c, 2014b). Most of these experiments used an incomplete factorial design with 12 treatments consisting of four levels of each nutrient, with $N_0P_0K_0$ being the absolute check, $N_1P_1K_1$ corresponding to a low level, $N_2P_2K_2$ an intermediate level and $N_3P_3K_3$ a very high level of N, P and K. In most experiments cassava was grown in monoculture, but in some cassava was intercropped with one or more companion crops.

An example of the results of one experiment conducted for 23 consecutive years at Hung Loc Agricultural Research Center in Dong Nai Province of southern Vietnam is shown in Table 3. During the first four cropping cycles there was no significant response to the application of any of the three macro-nutrients, but the responses to application of K, and to a lesser extent to N and P, became more marked over the years. During the 23rd year of cropping, the yield without K was only 34%, while those without N or P were about 64% of that obtained with intermediate levels of all three nutrients ($N_2P_2K_2$). Starting in the 5th year, the responses to N, P and K were either significant or highly significant in most years, even in a soil that was initially considered rather fertile. In this and many other experiments, K became the most limiting nutrient after several years of continuous cassava cultivation on the same land.

Table 4 shows the effect of the various fertilizer treatments on the average fresh root yield, starch content, and the gross and net income during the 23rd year of cropping. It can be seen that after 23 years the main response was to the application of K, but with significant responses also to the application of intermediate levels of N and P. Without

Table 3 Effect of application of various combinations of N, P and K on the average root yields of two varieties[1] during 23 years of continuous cropping at Hung Loc Agricultural Research Center in Dong Nai province of South Vietnam from 1990/91 to 2012/13

	Cassava fresh root yield (t/ha)						
Treatments [2]	Year 1	Av Yr 1-4	Av Yr 5-8	Av Yr 9-12	Av Yr 13-16	Av Yr 17-20	Year 23
$N_0P_0K_0$	12.6	14.4	15.9	12.3	9.4	7.8	7.5
$N_0P_2K_2$	13.7	17.8	21.8	18.9	20.1	14.7	12.8
$N_1P_2K_2$	14.1	18.2	23.7	23.5	23.8	18.7	15.0
$N_2P_2K_2$	14.0	17.8	24.9	25.8	26.1	22.7	19.6
$N_3P_2K_2$	14.4	17.4	24.9	26.5	27.4	23.5	17.5
$N_2P_0K_2$	14.2	17.7	23.0	22.5	20.3	17.5	12.5
$N_2P_1K_2$	14.3	17.7	22.8	24.4	24.3	20.5	13.6
$N_2P_3K_2$	14.5	18.3	24.7	25.9	27.6	24.3	18.6
$N_2P_2K_0$	14.5	16.0	19.5	14.8	8.9	6.6	6.6
$N_2P_2K_1$	14.6	17.5	24.0	22.5	21.3	18.7	15.1
$N_2P_2K_3$	14.0	17.9	24.1	25.3	25.9	21.7	17.8
$N_3P_3K_3$	14.7	18.2	26.2	28.3	30.4	26.8	19.6
	Relative yields (%)						
N_0[3]	98	100	88	73	77	64	65
P_0	102	99	92	87	78	77	64
K_0	104	90	78	57	34	29	34

[1]Two local varieties, HL23and HL24 during the first 4 years, HL 23 and KM60 in the 5th year, and SM 937-26 and KM60 in the 6th and following years.

[2]N_0 = 0N; N_1 = 40 kg N/ha; N_2 = 80 kg N/ha; N_3 = 160 kg N/ha
P_0 = 0P; P_1 = 20 kg P_2O_5/ha; P_2 = 40 kg P_2O_5/ha; P_3 = 80 kg P_2O_5/ha
K_0 = 0K; K_1 = 40 kg K_2O/ha; K_2 = 80 kg K_2O/ha; K_3 = 160 kg K_2O/ha

[3]Relative yield of $N_0 = N_0P_2K_2/N_2P_2K_2$; $P_0 = N_2P_0K_2/N_2P_2K_2$; $K_0 = N_2P_2K_0/N_2P_2K_2$

Source: Nguyen Huu Hy, personal communication; Howeler, 2014b.

Table 4 Effect of annual application of various amounts of N, P and K fertilizer on the average root yield and starch content of two varieties, as well as the gross and net income obtained during the 23rd consecutive year of cassava cropping at Hung Loc Agricultural Research Center in Dong Nai, southern Vietnam, in 2012/13

Treatment[1]	Average root yield (t/ha)	Average starch content (%)	Gross income[2]	Fertilizer costs[3]	Total production costs[3]	Net income
			('000 dong/ha)			
$N_0P_0K_0$	7.5	22.4	7,505	0	8,130	−625
$N_0P_2K_2$	12.8	23.3	12,792	2,424	10,854	1,938
$N_1P_2K_2$	15.0	24.4	14,983	3,293	11,723	3,260
$N_2P_2K_2$	19.6	24.9	19,565	4,163	12,593	6,972
$N_3P_2K_2$	17.5	25.5	17,500	5,902	14,332	3,168
$N_2P_0K_2$	12.5	24.5	12,495	3,339	11,769	726
$N_2P_1K_2$	13.6	24.4	13,643	3,751	12,181	1,462
$N_2P_3K_2$	18.6	25.3	18,617	4,986	13,416	5,201
$N_2P_2K_0$	6.6	22.2	6,588	2,563	10,993	−4,405
$N_2P_2K_1$	15.1	24.8	15,147	3,363	11,793	3,354
$N_2P_2K_3$	17.8	25.1	17,755	5,763	14,193	3,562
$N_3P_3K_3$	19.6	24.8	19,637	8,325	16,755	2,882

[1] N_0 = 0 N; N_1 = 40 kg N/ha; N_2 = 80 kg N/ha; N_3 = 160 kg N/ha
P_0 = 0 P; P_1 = 20 kg P_2O_5/ha; P_2 = 40 kg P_2O_5/ha; P_3 = 80 kg P_2O_5/ha
K_0 = 0 K; K_1 = 40 kg K_2O/ha; K_2 = 80 kg K_2O/ha; K_3 = 160 kg K_2O/ha

[2] Prices:
cassava dong 1,000/kg fresh roots (includes the harvest of the roots by the buyer)

[3] Costs:

land preparation dong	2,000,000/ha
planting material preparation	450,000/ha
cassava planting	900,000/ha
weeding	4,500,000/ha
fertilizer application	300,000/ha
Basudin 2H (pesticide)	280,000/ha
urea (46% N)	10,000/kg
SSP (17% P_2O_5)	3,500/kg
KCl (60% K_2O)	12,000/kg
labour	120,000/person-day

US$1 = 21,000 Vietnamese dong

Source: Nguyen Huu Hy, pers. comm.

application of NPK fertilizers, yields had decreased from an initial level of 12.6 t/ha to only 7.5 t/ha and with application of only N and P (without K) this had further dropped to 6.6 t/ha. But, with the intermediate level of 80 kg N, 40 P_2O_5 and 80 K_2O/ha, yields had increased from 14 t/ha to 19.6 t/ha. The higher rate of NPK ($N_3P_3K_3$) did not further increase yields. Considering the price of fresh roots and the cost of labour and materials, the highest net income was obtained with the intermediate level of all three nutrients, while without application of K the net income was highly negative.

Table 5 shows a summary of the results of the first and the last year of all 19 long-term NPK experiments conducted in Asia, indicating the fresh root yields obtained in the absolute check and the intermediate level of $N_2P_2K_2$, which usually produced the highest net income. The table also shows the relative yields in the first and last year of each experiment in the absence of N, P or K, as compared to the intermediate level of all three nutrients combined.

During the first year of the various experiments, the yield increase due to fertilization varied from −11% to +283% and was highly dependent on the original fertility of the soil. N was the main limiting nutrient in 11 of the 15 experiments (shown in bold), K in four and P in none of the experiments. In contrast, during the final year of the various experiments, the yield increase due to fertilization varied from 1% to 2,061%. K was the main limiting nutrient in nine experiments, N in six and P in none of the experiments. It is clear that in practically all cases, the longer the cassava is grown on the same fields without fertilization, or without the right balance of N, P and K, the greater the yield reduction by nutrient depletion, and the more likely it is that K is the most limiting nutrient, followed by N and very seldom by P, at least in Asia.

Currently, many farmers in some countries are fertilizing their crop, but in many others, especially in Africa and Latin America, this is not the practice. A detailed survey of all cassava-growing provinces of Thailand, conducted in 2010, indicates that 75% of the total cassava area was being fertilized. On average farmers applied about 33 kg each of N, P_2O_5 and K_2O (OAE, 2010). With the average yield of about 19 t/ha the roots remove about 35 kg of N, 9 kg of P_2O_5 and 40 kg of K_2O/ha in the root harvest. While this level of fertilization will replace most of the removed nutrients, the NPK balance is too high in P and too low in N and K, especially if cassava had been planted in the same field for many years. Most research in Thailand shows that yields can be maintained with the application of about 65–90 kg of N, 30–40 P_2O_5 and 80–110 K_2O/ha, or 450–600 kg/ha of 15-7-18 fertilizer, as long as the crop residues are returned to the soil (FCRI, 2004).

In some other countries, such as Lao PDR and Cambodia, very few farmers apply any fertilizer, that is, only 1.7% in Lao PDR (Boupha et al., 2010) and 38% in Cambodia, and only in the eastern part of the country (El Sotheary, 2010; Ung Sopheap et al., 2015); most of that is only urea or di-ammonium phosphate, which do not contain any K. In the north-western part of Cambodia only about 1% of cassava growers apply fertilizers. They still obtain very high yields because they grow the crop on soils that have only recently been cleared of forest. It is expected that these yields will soon start to decrease unless farmers start using fertilizers with the right N, P and K balance, as has already happened in neighbouring Thailand.

To convince farmers of the need to apply adequate amounts of fertilizer with the right balance of N, P and K, it is essential that researchers conduct long-term NPK trials on their experiment stations. While these may not show any response during the first year or first few years of cropping, these trials will eventually show a marked response to fertilization. These experiments are usually very convincing to visiting farmers and government officials, and show clearly the economic benefit of adequate fertilization, and the need to apply the right kind and amount of fertilizers with the right NPK balance. This should be followed by simple fertilizer trials on farmers' fields, preferably conducted by the farmers themselves with help from researchers and/or extensionists in so-called Farmer Participatory Research trials.

Table 5 Cassava root yield response to annual application of various rates of N, P and K, as well as the relative response to each nutrient during the first and the last year of cropping in 19 long-term fertility experiments conducted in Asia

		First year						Last year					
	# of	Yield (t/ha)		% yield	Relative yield (%)[1]			Yield (t/ha)		% yield	Relative yield (%)		
Location	years	$N_0P_0K_0$	$N_2P_2K_2$ [2]	increase	$N_0P_2K_2$	$N_2P_0K_2$	$N_2P_2K_0$	$N_0P_0K_0$	$N_2P_2K_2$	increase	$N_0P_2K_2$	$N_2P_0K_2$	$N_2P_2K_0$
Philippines													
Bohol	4	25.4	30.3	19	**98**	103	107	7.5	20.4	174	58	84	**34**
Negros Occidental	4	29.6	40.1	35	**78**	97	96	8.2	10.4	27	94	144	**87**
ViSCA, Leyte [3]	6	6.5	5.8	−11	90	134	**73**	17.1	17.3	1	**88**	110	98
Indonesia													
Yogyakarta [4]	4	13.9	16.8	21	**71**	95	76	6.2	10.9	77	**60**	100	88
Jatikerto, E. Java	8	28.2	37.0	31	**75**	79	82	3.1	11.3	268	**31**	72	81
Umas Jaya, Lampung	10	22.0	23.2	5	**89**	110	117	11.1	15.0	35	111	93	**84**
Tamanbogo, Lampung [5]	16	16.1	22.7	41	**77**	112	78	3.6	13.2	264	64	57	**26**
Tamanbogo, Lampung [6]	8	9.1	22.7	148	66	88	**18**	2.9	12.2	322	58	80	**26**
China													
UCRI, Guangzhou	4	13.7	20.1	46	**77**	95	94	5.6	18.8	235	**57**	70	63
GSCRI, Nanning	8	22.0	26.3	20	**71**	95	98	12.9	18.6	44	**70**	82	85
CATAS	16	11.8	19.7	67	**81**	99	95	7.2	15.1	109	**41**	85	64
India													
CTCRI, Kerala	13	11.3	20.0	77	**68**	86	112	1.0	22.3	2,061	24	42	**7**

Malaysia													
Serdang	10	33.2	30.5	−8	130	113	**90**	20.7	50.9	146	69	73	**57**
Vietnam													
TNU, Thai Nguyen	17	6.8	26.1	283	60	81	**25**	4.4	21.8	389	67	71	**16**
Hung Loc, Dong Nai	23	12.6	14.0	11	**98**	102	104	7.5	19.6	161	65	64	**34**
Thailand													
Rayong 1st experiment	10	21.6	21.2	−2	–	96	**91**	8.7	18.3	110	–	96	**51**
Rayong 2nd experiment	21	20.3	19.5	−4	–	116	**103**	20.7	41.1	99	–	96	**65**
Khon Kaen	30	28.2	29.6	5	–	**104**	**104**	2.5	31.9	1,176	–	77	**9**
Banmai Samrong	31	20.1	24.8	23	–	**82**	83	21.7	26.9	24	–	**68**	90

[1] Relative yield of $N_0 = N_0P_2K_2/N_2P_2K_2$; $P_0 = N_2P_0K_2/N_2P_2K_2$; $K_0 = N_2P_2K_0/N_2P_2K_2 \times 100$.
[2] In some experiments only one rate of N, P and K were used; in some trials the yields are the average of two varieties planted in each plot.
[3] Cassava under mature coconut trees.
[4] Cassava intercropped with rice, maize and mungbean.
[5] Cassava intercropped with rice and maize.
[6] Cassava monoculture, but planted after 8 years of intercropping with rice and maize.
Source: Howeler, 2014a (unpublished).

9 Conclusions

Research has shown that:

1. In many cassava varieties the symptoms of deficiency of N, P or K are not clearly expressed, but in case of deficiency the plants are smaller and weaker than normal. The problem may need to be diagnosed through soil or plant tissue analyses, or through the conducting of nutrient omission trials.
2. In many low-fertility soils, or when cassava has been grown for several years in the same fields, the crop will be highly responsive to the application of well-balanced chemical fertilizers, and this can be highly economic.
3. Besides chemical fertilizers, soil fertility can also be maintained and cassava yields increased by incorporation of organic sources such as manures, ash, cassava residues, intercrop residues, mulches, weeds and prunings of alley crop species. In many cases highest yields are obtained when the right type of chemical fertilizer is combined with the application of organic sources.
4. In most cassava soils in Asia, the crop responds initially mainly to N and K applications, but after several years of continuous cassava production on the same field the crop will respond mainly to K. In many soils in Latin America cassava responds mainly to P, while in Africa the response is mainly to K, with occasional responses to N and P.

10 References

Acosta, J. R. and G. J. Perez. 1954. Abonamiento en yuca (Cassava fertilization). *Suelo Tico* (Costa Rica) 7(31): 300–8.

Anonymous. 1952. Le manioc (Cassava). *Recherche Agronomique de Madagascar* 1: 49–52.

Anonymous. 1953. Essais de fumure du manioc (Fertilizer experiments with cassava). Recherche Agronomique de Madagascar. *Compte Rendu 2*: 85–8.

Asher, C. J., D. G. Edwards and R. H. Howeler. 1980. Nutritional Disorders of Cassava *(Manihot esculenta* Crantz*)*. University of Queensland, St. Lucia, Qld., Australia. 48 p.

Aye, T. M., V. Phimphachanhvongsod and P. Kanthavong. 2010. The Nippon Foundation Cassava Project in Lao PDR. In: R. H. Howeler (ed.), *A New Future for Cassava in Asia: Its Use as Food, Feed and Fuel to Benefit the Poor*. Proc. 8th Regional Workshop, held in Vientiane, Lao PDR. Oct 20–24, 2008, pp. 425–34.

Boupha, T., S. Sacklokham, P. Chanhming and T. M. Aye. 2010. Results of the cassava survey in Lao PDR. In: R. H. Howeler (ed.), *A New Future for Cassava in Asia: Its Use as Food, Feed and Fuel to Benefit the Poor*. Proc. 8th Regional Workshop, held in Vientiane, Lao PDR. Oct 20–24, 2008, pp. 490–507.

Cadavid, L. F., M. A. El-Sharkawy, A. Acosta and T. Sanchez. 1998. Long-term effects of mulch, fertilization and tillage on cassava grown in sandy soils of northern Colombia. *Field Crops Research* 57: 45–56.

Carsky, R. J. and M. A. Toukourou. 2005. Identification of nutrients limiting cassava yield maintenance on a sedimentary soil in southern Benin, West Africa. *Nutrient Cycling in Agroecosystems* 71: 151–62.

Centro Internacional de Agricultura Tropical (CIAT). 1976. Annual Report for 1975. Cassava Program, CIAT, Cali, Colombia. 57p.

Centro Internacional de Agricultura Tropical (CIAT). 1977. Annual Report for 1976. Cassava Program, CIAT, Cali, Colombia. 76p.
Centro Internacional de Agricultura Tropical (CIAT). 1978. Annual Report for 1977. CIAT, Cali, Colombia. 386p.
Centro Internacional de Agricultura Tropical (CIAT). 1985. Cassava Program. Annual Report for 1982 and 1983. CIAT, Cali, Colombia. 521p.
Chan, S. K. 1980. Long-term fertility considerations in cassava production. In: E. J., Weber, J. C. Toro and M. Graham (eds), Workshop on Cassava Cultural Practices, held in Salvador, Bahia, Brazil. March 18–21, 1980. IDRC-151e, pp. 82–92.
Chew, W. Y. 1971. The performance of tapioca, sweet potato and ginger on peat at the Federal Experiment Station, Jalan Kebun, Selangor. Agronomy Branch, Division of Agriculture, Kuala Lumpur, Malaysia, 7 p.
Chew, W. Y., K. Ramli and K. T. Joseph. 1978. Copper deficiency of cassava (*Manihot esculenta* Crantz) on Malaysian peat soil. *MARDI Research Bulletin* 6(2): 208–13.
Cours, G., J. Fritz and G. Ramahadimby. 1961. El diagnóstico felodérmico de la mandioca (The pheloderm diagnosis of cassava). *Fertilité* 12: 3–20.
El Sotheary. 2010. Results of the cassava survey in Cambodia. In: R. H. Howeler (ed.), *A New Future for Cassava in Asia: Its Use as Food, Feed and Fuel to Benefit the Poor*. Proc. 8th Regional Workshop, held in Vientiane, Lao PDR. Oct 20–24, 2008, pp. 470–89.
Fermont, A. M., P. J.A. van Asten and K. E. Giller. 2008. Increasing land pressure in East Africa: The changing role of cassava and consequences for sustainability of farming systems. *Agriculture, Ecosystems and Environment* 128: 239–50.
Fermont, A. M., P. J. A. van Asten, P. Tittonell, M. T. van Wijk and K. E. Giller. 2009. Closing the yield gap: an analysis from small-holder farms in East Africa. *Field Crops Research* 112: 24–36.
Fermont, A. M., P. Tittonell, Y. Bacuma, P. Ntawauruhunga and K. E. Giller. 2010. Towards understanding the factors that govern fertilizer response in cassava: lessons from East Africa. Nutrient Cycling in Agroecosystems 86:133–51.
Field Crops Research Institute (FCRI). 2004. Field Crops Production: Cassava, pp. 267–80. (in Thai).
Food and Agriculture Organization (FAO). 1980. Review of data on responses of tropical crops to fertilizers, 1961–1977. FAO, Rome, Italy, 101p.
Forno, D. A. 1977. The mineral nutrition of cassava (*Manihot esculenta* Crantz) with particular reference to nitrogen. PhD thesis. University of Queensland, St. Lucia, Qld, Australia.
Fox, R. H., H. Talleyrand and T. W. Scott. 1975. Effect of nitrogen fertilization on yields and nitrogen content of cassava, Llanera cultivar. *Journal of Agriculture* (University of Puerto Rico) 56: 115–24.
Gnahoua, G. J. B. 2016. Sustainable intensification of smallholder cassava production on poor Ferralsols. PhD dissertation. Ghent University, Ghent, Belgium. 161p.
Gnahoua, G. J. B., D. J. B. Ettien, B. N'Zue, C. Ebah, B. Kone, S. De Neve and P. Boeckx. 2016. Intensification pathways for improvement of smallholder cassava production systems in southern Cote d'Ivoir. *Experimental Agriculture*. Doi 10.1017/S0014479716000041.
Gomes, J. de C. and R. H. Howeler. 1980. Cassava production in low fertility soils. In: R. J. Weber, J. C. Toro and M. Graham (eds), Proc. Workshop on Cassava Cultural Practices, held in Salvador, Ba., Brazil. March 18–21, 1980. IDRC, Ottawa, Canada, pp. 93–102.
Gomes, J. de C. and J. Da Silva. 2006. Correção da Acidez e Adubação. Liiro Agronomia. Capito 9. EMBRAPA, CNPMF, Cruz das Almas, Brazil, pp. 215–47.
Gunatilaka, A. 1977. Effects of aluminium concentration on the growth of corn, soybean, and four tropical root crops. MSc thesis. University of Queensland, St. Lucia, Qld, Australia.
Hagens, P. and C. Sittibusaya. 1990. Short and long term aspects of fertilizer applications on cassava in Thailand. In: R. H. Howeler (ed.), Proc. 8th Symposium International Society of Tropical Root Crops, held in Bangkok, Thailand. Oct. 30–Nov. 5, 1988, pp. 244–59.
Henry, G. and V. Gottret. 1996. Global Cassava Trends. Reassessing the Crop's Future. CIAT Working Document No. 157. CIAT, Cali Colombia. 45p.

Howeler, R. H. 1985a. Mineral nutrition and fertilization of cassava. In: Cassava; Research, Production and Utilization. UNDP-CIAT Cassava Program, Cali, Colombia, pp. 249–320.

Howeler, R. H. 1985b. Potassium nutrition of cassava. In: W. D. Bishop et al. (eds), Potassium in Agriculture. Intern. Symposium, held in Atlanta, GA, USA. July 7-10, 1985. ASA-CSSA-SSSA, Madison, WI, USA, pp. 819–41.

Howeler, R. H. 1990. Phosphorus requirements and management of tropical root and tuber crops. In: Proc. Symp. on Phosphorus Requirements for Sustainable Agriculture in Asia and Oceania, held at IRRI, Los Baños, Philippines. March 6–10, 1989, pp. 427–44.

Howeler, R. H. 1991a. Identifying plants adaptable to low pH conditions. In: R. J. Wright et al. (eds), *Plant-Soil Interactions at Low pH*. Kluwer Academic Publisher, Netherlands, pp. 885–904.

Howeler, R. H. 1991b. Long-term effect of cassava cultivation on soil productivity. *Field Crops Research* 26: 1–18.

Howeler, R. H. 2002. Cassava mineral nutrition and fertilization. In: R. J. Hillocks, M. J. Thresh and A. C. Bellotti (eds), *Cassava: Biology, Production and Utilization*. CABI Publishing, Wallingford, UK, pp. 115–47.

Howeler, R. H. 2012a. Diagnosis of nutritional problems of cassava. In: R. H. Howeler (ed.), *The Cassava Handbook*. A Reference Manual Based on the Asian Regional Cassava Training Course, held in Thailand, pp. 305–20.

Howeler, R. H. 2012b. Effect of cassava production on soil fertility and the long-term fertilizer requirements to maintain high yields. In: R. H. Howeler (ed.), *The Cassava Handbook*. A Reference Manual Based on the Asian Regional Cassava Training Course, held in Thailand, pp. 411–28.

Howeler, R. H. 2012c. Short- and long-term N, P and K requirements of cassava. In: R. H. Howeler (ed.), *The Cassava Handbook*. A Reference Manual Based on the Asian Regional Cassava Training Course, held in Thailand, pp. 429–54.

Howeler, R. H. 2012d. Secondary and micronutrient requirements of cassava and the use of soil amendments. In: R. H. Howeler (ed.), *The Cassava Handbook*. A Reference Manual Based on the Asian Regional Cassava Training Course, held in Thailand, pp. 455–68.

Howeler, R. H. 2012f. Importance of mycorrhiza for phosphorus absorption by cassava. In: R. H. Howeler (ed.), *The Cassava Handbook*. A Reference Manual Based on the Asian Regional Cassava Training Course, held in Thailand, pp. 497–523.

Howeler, R. H. 2012g. Cassava leaf production for animal feeding. In: R. H. Howeler (ed.), *The Cassava Handbook*. A Reference Manual Based on the Asian Regional Cassava Training Course, held in Thailand, pp. 626–48.

Howeler, R. H. 2014a. Literature review on yield gaps of cassava in Asia. (unpublished).

Howeler, R. H. 2014b. Sustainable Soil and Crop Management of Cassava in Asia – A Reference Manual. CIAT Publication No. 389. CIAT, Cali, Colombia. 280 p.

Howeler, R. H., L. F. Cadavid and E. Burckhardt. 1982a. Cassava response to VA mycorrhizal inoculation and phosphorus application in greenhouse and field experiments. *Plant and Soil* 69: 327–39.

Howeler, R. H., D. G. Edwards and C. J. Asher. 1982b. The effect of soil sterilization and mycorrhizal inoculation on the growth, nutrient uptake and critical P concentration of cassava. In: 5th Symposium. Intern. Society Tropical Root Crops (ISTRC), held in Manila, Philippines. 1979, pp. 519–37.

Howeler, R. H. and D. Ballesteros. 1987. El Cultivo de la Yuca en los Llanos Orientales de Colombia. Variedades y Prácticas Agronómicas. Serie Boletines Técnicos, Programa de Yuca, Número 1. CIAT, Cali, Colombia. 29p.

Howeler, R. H., E. Sieverding and S. Saif. 1987. Practical aspects of mycorrhizal technology in some tropical crops and pastures. *Plant and Soil* 100: 249–83.

Howeler, R. H., and L. F. Cadavid. 1990. Short- and long-term fertility trials in Colombia to determine the nutrient requirements of cassava. *Fertilizer Research* 26: 61–80.

Islam, A. K. M. S., D. G. Edwards and C. J. Asher. 1980. pH optima for crop growth: Results of flowing culture experiment with six species. *Plant and Soil* 54(3): 339–57.

Juo, A. S. R. 1985. Potassium response in root and tuber crops. In: Potassium in the Agricultural Systems of the Humid Tropics. 19th Colloqium of the Intern. Potash Institute, held in Bangkok, Thailand, pp. 277–88.

Kabeerathumma, S., B. Mohankumar, C. R. Mohankumar, G. M Nair, M. Prabhakar and N. G. Pillai. 1990. Long range effect of continuous cropping and manuring on cassava production and fertility status. In: R. H. Howeler (ed.), Proc. 8th Symposium International Society of Tropical Root Crops, held in Bangkok, Thailand. Oct 30–Nov 5, 1988, pp. 259–69.

Kang, B. T. 1984. Potassium and magnesium responses of cassava grown in Ultisol in southern Nigeria. *Fertilizer Research* 5: 403–10.

Kang, B. T. and J. E. Okeke. 1984. Nitrogen and potassium responses of two cassava varieties grown on an Alfisol in southern Nigeria. In: Proc. 6th Symposium International Society of Tropical Root Crops, held in Lima, Peru. Feb. 21–26, 1983, pp. 231–7.

Kang, B. T., R. Islam, F. E. Sanders and A. Ayanaba. 1980. Effect of phosphate fertilization and inoculation with VA-mycorrhizal fungi on performance of cassava (*Manihot esculenta* Cranz) grown on an Alfisol. *Field Crops Research* 3: 83–94.

Kelly, V. A. 2006. Factors Affecting Demand for Fertilizers in Sub-Saharan Africa. Agriculture and Rural Development Discussion Paper 23. The World Bank, Washington DC. USA.

Krochmal, A. and G. Samuels. 1970. The influence of NPK levels on the growth and tuber development of cassava in tanks. *CEIBA* 16: 35–43.

Leong, G. J. M. 2016. Mineral nutrient calibration for cassava (*Manihot esculenta* Crantz) grown in dry-tropical north-eastern Queensland, Australia: with focus on phosphorus and boron. MSc thesis, Univ. Queensland, St. Lucia, Australia. 138p.

Mohan Kumar, B. and P. G. Nair. 1985. Lime, sulphur and zinc in cassava production. Technical Bulletin No. 2. CTCRI, Kerala, India.

Moraes, O. De, E. Mondardo, J. Vizzotto and M. O. Machado. 1981. Adubação quimica e calagem da mandioca (Fertilization and liming of cassava). Boletim Técnico No.8. Empresa Catarinense de Pesquisa Agropecuâria. Florianopolis, Santa Catarina, Brazil. 20 p.

Nakviroj, C., K. Paisancharoen, O. Boonseng, C. Wongwiwatchai and S. Roongruang. 2007. Cassava long-term fertility experiments in Thailand. In: R. H. Howeler (ed.), *Cassava Research and Development in Asia: Exploring New Opportunities for an Ancient Crop*. Proc. 7th Regional Workshop, held in Bangkok, Thailand. Oct 28–Nov 1, 2002, pp. 212–23.

Nguyen Huu Hy, Nguyen The Dang and Tong Quoc An. 2010. Soil fertility maintenance and erosion control research in Vietnam. In: R. H. Howeler (ed.), *A New Future for Cassava in Asia. Its Use as Food, Feed and Fuel to Benefit the Poor*. Proc. 8th Regional Workshop, held in Vientiane, Lao PDR. Oct 20–24, 2008, pp. 263–74.

Ngongi, A. G. N., R. H. Howeler and H. A. MacDonald. 1977. Effect of potassium and sulphur on growth, yield and composition of cassava. In: J. Cock, R. MacIntyre and M. Graham (eds), Proc. 4th Symp. International Society of Tropical Root Crops, held at CIAT, Cali, Colombia. Aug 1–7, 1976. IDRC, Ottawa, Canada, pp. 107–113.

Njoku, B. O. and O. B. Arene. 1980. The influence of potassium, nitrogen, phosphorus and magnesium fertilization on anthracnose disease and tuber yield of cassava. Potassium Workshop, held at IITA, Ibadan, Nigeria. October 8–10, 1980, pp. 193–8.

Obigbesan, G. O. 1973. The influence of potassium nutrition on the yield and chemical composition of some tropical root and tuber crops. In: 10th Coloquium International Potash Institute, held in Abidjan, Ivory Coast, pp. 439–51.

Obigbesan, G. O. 1977. Investigations on Nigerian root and tuber crops. Response of cassava cultivars to potassium fertilizers in Western Nigeria. *Journal of Agricultural Science* 89: 23–7.

Obigbesan, G. O. and A. A. Agboola. 1977. An evaluation of the yield and quality of some Nigerian cassava varieties as affected by age and nitrogen application. In: C. L. A. Leakey (ed.), Proc. 3rd Symp. ISTRC, held at IITA, Ibadan, Nigeria. Dec 2–9, 1973, pp. 321–6.

Obigbesan, G. O. and A. A. A. Fayemi. 1976. Investigations on Nigerian root and tuber crops: Influence of nitrogen fertilization on the yield and chemical composition of two cassava cultivars (*Manihot esculenta*). *Journal of Agricultural Science* 86: 401–6.

Odurukwe, S. O. and O. B. Arene. 1980. Effect of N, P, K fertilizers on cassava bacterial blight and root yield of cassava. *Tropical Pest Management* 26: 391–5.

Office of Agricultural Economics (OAE). 2010. Report of Cassava Survey. Year 2010. Bangkok, Thailand. (in Thai).

Ofori, C. S. 1973. Decline in fertility status of a tropical forest ochrosol under continuous cropping. *Experimental Agriculture* 9: 15–22.

Ogbe, F. O., A. C. Ohiri and E. C. Nnodu. 1993. Effect of NPK fertilization on symptom severity of African cassava mosaic virus. *International Journal of Pest Management* 39(1): 80–3.

Olasantan, F. O., H. C. Ezumah and E. O. Lucas. 1997. Response of cassava and maize to fertilizer application, and a comparison of the factors affecting their growth during intercropping. *Nutrient Cycling in Agroecosystems* 46: 215–23.

Pellet, D. and M. A. El-Sharkawy. 1993a. Cassava varietal response to phosphorus fertilization. I. Yield, biomass and gas exchange. *Field Crops Research* 35: 1–11.

Pellet, D. and M. A. El-Sharkawy. 1993b. Cassava varietal response to phosphorus fertilization. II. Phosphorus uptake and use efficiency. *Field Crops Research* 35: 13–20.

Pypers, P., J. M. Sanginga, B. Kasereka, M. Walangululu and B. Vanlauwe. 2011. Increased productivity through integrated soil fertility management in cassava-legume intercropping systems in the highlands of Sud-Kivu, D. R. Congo. *Field Crops Research* 120: 76–85.

Pypers, P., W. Bimponda, J. P. Lodi-Lama, B. Lele, R. Mulumba, C. Kachaka, P. Boeckx, R. Merckx and B. Vanlauwe. 2012. Combining mineral fertilizer and green manure for increased, profitable cassava production. *Agronomy Journal* 104(1): 1–10.

Richards, I. R. 1979. Response of tropical crops to fertilizer under farmers conditions - Analysis of results of the FAO Fertilizer Programme. *Phosphorus in Agriculture* 76: 147–56.

Roche, P., J. Velly and B. Joliet. 1957. Essai de determination des seuils de carence en potasse dans le sol et dans les plantes (Methods to determine the level of K deficiency in soil and in plants). *Revue de la Potasse* 1957: 1–5.

Santana, A. M., J. E. B. Carvalho and I. O. Borges. 1975. Competição de fontes de nitrogênio em solos para mandioca. In: Universidade Federal da Bahia, Escola de Agronomia, Projecto de Mandioca. Cruz das Almas: UFBA: Brascan Nordeste, pp. 39–55.

Santos, Z. G. and E. A. Tupinambá. 1981. Resultados do experimento de adubação mineral sobre a produção de mandioca (*Manihot esculenta* Crantz) no Estado de Sergipe. EMBRAPA-UEPAE Aracaju.

Silpamaneephan, W. 1994. Effect of land preparation on soil physical characteristics, germination and yield of cassava. MSc thesis. Kasetsart University, Bangkok, Thailand. 78 p.

Sittibusaya, C., C. Narkaviroj and D. Tunmaphirom. 1988. Cassava soils research in Thailand. In: R. H. Howeler and K. Kawano (eds), *Cassava Breeding and Agronomy Research in Asia*. Proc. Regional Workshop, held in Rayong, Thailand. Oct 26–28, 1987, pp. 145–56.

Spain, J. M., C. A. Francis, R. H. Howeler and F. Calvo. 1975. Differential species and varietal tolerance to soil acidity in tropical crops and pastures. In: E. Bornemisza and A. Alvarado (eds), *Soil Management in Tropical America*. North Carolina State University, Raleigh, NC, USA, pp. 308–29.

Susan John, K., C. S. Ravindran and James George. 2005. Long Term Fertilizer Experiments. Three Decades Experience in Cassava. CTCRI, Thiruvananthapuram, Kerala, India. 83p.

Susan John, K., James George, S. U. Shanida Beegum and Y. S. Shivay. 2016. Soil fertility and nutrient management in tropical tuber crops – An overview. *Indian Journal of Agronomy* 61(3) 263–73.

Takyi, S. K. 1972. Effects of potassium, lime and spacing on yields of cassava (*Manihot esculenta* Crantz). *Ghana Journal of Agricultural Science* 5(1): 39–42.

Tan, S. L. and S. K. Chan. 1995. Recent progress in cassava varietal improvement and agronomy research in Malaysia. In: R. H. Howeler (ed.), *Cassava Breeding, Agronomy Research and Technology Transfer in Asia*. Proc. 4th Regional Workshop, held in Trivandrum, Kerala, India. Nov. 2–6, 1993, pp. 337–54.

Taufiq, A., Subandi and H. Suyamto. 2015. Response of cassava (*Manihot esculenta* Crantz) to potassium application on acidic dryland in Indonesia. e-ifc no. 42: 18/34-26/34.

Tian Yinong, Lee Jun, Zhang Weite and Fang Baiping. 1995. Recent progress in cassava agronomy research in China. In: R. H. Howeler (ed.), *Cassava Breeding, Agronomy Research and Technology*

Transfer in Asia. Proc. 4th Regional Workshop, held in Trivandrum, Kerala, India. Nov 2–6, 1993, pp. 195–216.

Tongglum, A., V. Vichukit, S. Jantawat, C. Sittibusaya, C. Tiraporn, S. Sinthuprama and R. H. Howeler. 1992. Recent progress in cassava agronomy research in Thailand. In: R. H. Howeler (ed.), *Cassava Breeding, Agronomy and Utilization Research in Asia*. Proc. 3rd Regional Workshop, held in Malang, Indonesia. Oct. 22–27, 1990, pp. 199–223.

Umeh, S. I., C. C. Onyeonagu and B. U. Umeh. 2015. Potassium nutrition and translocation in cassava (*Manihot esculenta* Crantz) intercropped with soybean. *American Journal of Experimental Agriculture* 5(4): 281–6.

Ung Sopheap, A. Patanothai, T. M. Aye and R. Howeler. 2015. Nutrient balances and farmers' perception on cassava cultivation in Kampong Cham province of Cambodia. In: R. H. Howeler (ed.), *Sustainable Cassava Production in Asia for Multiple Uses and for Multiple Markets*. Proc. 9th Regional Workshop, held in Nanning, Guangxi, China PR. Nov 27–Dec 3, 2011, pp. 200–14.

Vijayan, M. R. and R. S. Aiyer. 1969. Effect of nitrogen and phosphorus on the yield and quality of cassava. *Agricultural Research Journal of Kerala* 7(2): 84–90.

Yost, R. S. and R. L. Fox. 1979. Contribution of mycorrhizae to P nutrition of crops growing on an Oxisol. *Agronomy Journal* 71: 903–8.

Zaag van der, P., R. L. Fox, R. S. Pena and R. S. Yost. 1979. Phosphorus nutrition of cassava, including mycorrhizal effects on P, K, S, Zn and Ca uptake. *Field Crops Research* 2: 253–63.

Chapter 15

Nutrient sources and their application in cassava cultivation

Reinhardt Howeler, CIAT Emeritus - formerly CIAT, Thailand

1 Introduction
2 Organic and inorganic nutrient sources
3 When and where to apply chemical fertilizers, manures and soil amendments
4 Determination of fertilizer rates and nitrogen, phosphorus and potassium (NPK) balance
5 Site-specific nutrient management
6 Fertilizer requirements for balanced nutrition
7 Genetic variation in nutrient use efficiency (NUE) and tolerance to low soil fertility
8 Conclusions
9 References

1 Introduction

Cassava is drought-tolerant and well-adapted to grow on low fertility soils. It is becoming an ever more important crop, used both for direct human consumption as well as for industrial processing into animal feed, starch and many starch-derived products, as well as bio-fuel. However, production could be further increased if important soil-related constraints are addressed.

This chapter covers the choice between organic and inorganic nutrient sources, the determination of fertilizer rates and nutrient balance and site-specific nutrient management. It ends by considering the genetic variation among cassava plants in nutrient use efficiency (NUE) and tolerance to low soil fertility.

2 Organic and inorganic nutrient sources

Once a farmer knows how much N, P or K (or any secondary or micronutrients) to apply, he or she has to decide about the best source of these nutrients. These sources are usually

http://dx.doi.org/10.19103/AS.2016.0014.35

divided into either organic sources, such as animal manures, compost, wood ash, crop residues (especially those from crops that can fix N from the air), or inorganic sources such as chemical fertilizers, lime or rock phosphates.

Table 1 shows the average nutrient and moisture contents of these various sources. In sandy and low-OM soils it is often recommended to apply organic manures or incorporate green manures, which increase the soil's nutrient- and water-holding capacity, improve soil structure and supply organic matter as well as small amounts of a large number of plant nutrients. Organic manures are a good source of nutrients for small areas and where easily and cheaply available. However, for larger extensions the amounts required are often not available or its application becomes impractical unless fields are close to large animal operations. In that case, chemical fertilizers are equally good sources of nutrients and are cheaper to transport and apply, but require more careful monitoring of the plants' nutrient status to prevent imbalances or deficiencies, especially of micronutrients.

Table 1 shows that many chemical fertilizers have very high concentrations of mainly N, P or K, but may not have any other essential plant nutrients, such as Ca, Mg, S and micronutrients. On the other hand, organic nutrient sources contain many different nutrients but only in small amounts; moreover, these manures tend to have a high moisture content, which further reduces their nutrient contents per kg applied. Thus, a compound fertilizer like 15-15-15 contains about 10–20 times more N, P and K than most of the organic manures. To obtain high cassava yields it is recommended to apply most of the crop's N, P and K requirements in the form of chemical fertilizers while also applying relatively small amounts (5 t/ha) of animal manures in order to supply also the essential secondary and micronutrients as well as organic matter, which will improve soil structure, water holding capacity and aggregate stability.

To speed up decomposition, the organic sources should be broadcast and incorporated into the soil before planting in order to achieve better contact with the soil; this, however, is likely to stimulate weed growth throughout the field. It may be even more of a problem with the use of cattle manure, which often contains a lot of weed seeds. Thus, both chemical fertilizers and organic sources have their advantages and disadvantages. Chemical fertilizers are easier to apply, have known nutrient contents and result in a quick response to the specific nutrient(s) most limiting yield; this may account for most of the root yield response, while the animal manure may increase yield another 20–30% and improve soil health. This has been shown by several experiments (see Fig. 4 of Chapter 34) (Kabeerathumma et al., 1990; Howeler, 2002, 2012e, 2014b).

Table 2 shows the long-term effect of applying chemical fertilizer or farmyard manure (FYM) separately or in combination on the yield of cassava during 13 years of continuous cropping in southern India. Without fertilizers or manure yields gradually dropped from 11.3 t/ha in the first year to 1.0 t/ha during the 13th year of cropping, while with the annual application of 100 kg/ha each of N, P_2O_5 and K_2O as chemical fertilizers yields increased from 20 t/ha in the first year to 22.3 t/ha in the last year. Meanwhile, the annual application of 12.5 t/ha of FYM alone did not increase yields compared to the check in the first year, but maintained that yield over the 13-year period. However, the combination of chemical fertilizer plus FYM not only increased yields another 18% over that obtained with fertilizer in the first year, but actually increased yields further, reaching 29 t/ha (30% higher than with fertilizer alone) during the last year (Susan John et al., 2005).

Very similar results were obtained in Indonesia where the application of different chemical fertilizers was compared with cattle manure or compost alone or in combination

Table 1 Moisture and nutrient content of commonly used organic and inorganic fertilizers

Source of nutrients	Moisture (%)	N	P	K	Ca	Mg	S
		(% of dry matter)					
Organic nutrient sources							
Cattle manure	68.2	1.85	0.81	1.69	1.54	0.62	0.29
Pig manure	60.0	2.04	1.38	1.38	–	–	–
Chicken manure	43.0	2.91	1.37	1.54	4.56	0.83	–
Sheep manure	–	3.00	0.62	2.68	1.72	0.86	0.43
Human manure	–	1.20	0.06	0.21	–	–	–
City/rural compost	–	1.16	0.37	0.90	–	–	–
Rice straw compost	73.7	1.07	0.19	0.69	–	–	–
Peanut stems + leaves (compost)	58.6	0.81	0.10	0.38	–	–	–
Water hyacinth	–	2.00	1.00	2.30	–	–	–
Wood ash	–	–	0.87	4.17	23.2	2.10	0.40
Inorganic nutrient sources							
Urea	0	46	0	0	0	0	0
Ammonium sulphate	0	21	0	0	0	0	24
Ammonium nitrate	0	33	0	0	0	0	0
Mono-ammonium phosphate (MAP)	0	11	21	0	0	0	0
Di-ammonium phosphate (DAP)	0	18	20	0	0	0	0
Triple superphosphate (TSP)	0	0	20	0	14	0	0
Single superphosphate	0	0	8	0	19	0	11
Basic slag	0	0	6	0	37	1	0
Potassium chloride	0	0	0	50	0	0	0
Potassium sulphate	0	0	0	42	0	0	18
Calcium sulphate (approx.)	0	0	0	0	12	0	10
Magnesium sulphate	0	0	0	0	0	10	13
Magnesium oxide	0	0	0	0	0	32	0
Calcitic lime (approx.)	0	0	0	0	30	0	0
Dolomitic lime (approx.)	0	0	0	0	24	12	0
Elemental sulphur	0	0	0	0	0	0	100
15-15-15	0	15	6.6	12.5	0	0	0
10-20-20	0	10	8.7	16.7	0	0	0
10-30-10	0	10	13.1	8.3	0	0	0
15-7-18	0	15	3.1	15.0	0	0	0

Source: Howeler, 2004, 2007, 2014b.

Table 2 The effect of annual applications of chemical fertilizers with or without farmyard manure and wood ash on the fresh root yield of cassava, variety H-1687, during 13 years of continuous cropping at the Central Tuber Crops Research Institute in Thiruvananthapuram, Kerala, India

	Cassava fresh root yield (t/ha)					
Treatments[1]	Year 1	Av. Yr. 1–4	Av. Yr. 5–8	Av. Yr. 9–12	Year 13	Av. Yr. 1–13
$N_0P_0K_0$	11.3	7.6	4.0	2.5	1.0	4.5
$N_1P_1K_1$	20.0	20.6	23.5	25.1	22.3	23.0
FYM	11.6	8.2	10.9	11.9	11.6	10.4
FYM + $N_1P_1K_1$	23.7	24.0	29.2	31.2	29.0	28.2
Wood ash	–	9.7	11.6	11.9	5.3	12.3
Wood ash + FYM	–	10.4	16.3	15.9	12.3	14.1

$N_1P_1K_1$ = 100 kg/ha of N, P_2O_5 and K_2O; FYM = 12.5 t/ha cattle manure = 148 kg N, 14 P_2O_5 and 129 K_2O/ha; Wood ash = 2.65 t/ha of wood ash = 230 kg K_2O/ha.

Source: Adapted from Susan John et al., 2005.

with the chemical fertilizers. In this case, the application of chemical fertilizers (135 kg N/ha) increased yields from 11.0 to 35.6 t/ha; 10 t/ha of compost alone increased yields only to 22.7 t/ha, while the combination of 135 kg/ha of N with 5 t/ha of compost increased the yield to 39.3 t/ha (Utomo et al., 2010; Howeler, 2012e, 2014b).

Similar trials conducted in Thai Nguyen in North Vietnam and in Nanning, China, produced similar results (Howeler, 2012e, 2014b). In all these cases, the application of chemical fertilizers supplied most of the macronutrients (N, P and K), resulting in 70–80% of the yield increase, while the additional application of organic manures or compost resulted in an additional yield increase (20–30%) by supplying small quantities of both macro- and micronutrients as well as organic matter to improve the soil's long-term health.

The application of 4–5 t/ha of animal manure or compost may be necessary when both roots and top growth are removed from the field at time of harvest. But when all crop residues (except stems used for planting material) are reincorporated into the soil before the next cropping cycle, the use of animal manures may not be necessary to maintain soil health. Figure 1 shows that in Khon Kaen, Thailand, cassava yields of 30–40 t/ha could be maintained during more than 25 years of continuous cassava cropping as long as the application of adequate amounts of N, P and K (100-50-100 kg/ha N-P_2O_5-K_2O) in chemical fertilizers were combined with the return of crop residues to the soil, or with the incorporation of 12.5 t/ha of municipal compost (Nakviroj et al., 2007; Howeler, 2012b).

In the absence of chemical fertilizers and manures, farmers can produce reasonable crop yields by slash-and-burn methods, still used widely in Africa and in parts of Latin America and Asia, as well as by the use of green manures, alley cropping, intercropping or by incorporating or mulching the biomass of certain weeds that have a high nutrient content, such as *Tithonia diversifolia* or *Chromolaena odorata*, which can be found along roadsides or in fallow fields (Pypers et al., 2012). The advantages and disadvantages of these methods have been described in detail by Howeler (2002, 2012e, 2014b) and Howeler and Aye (2014).

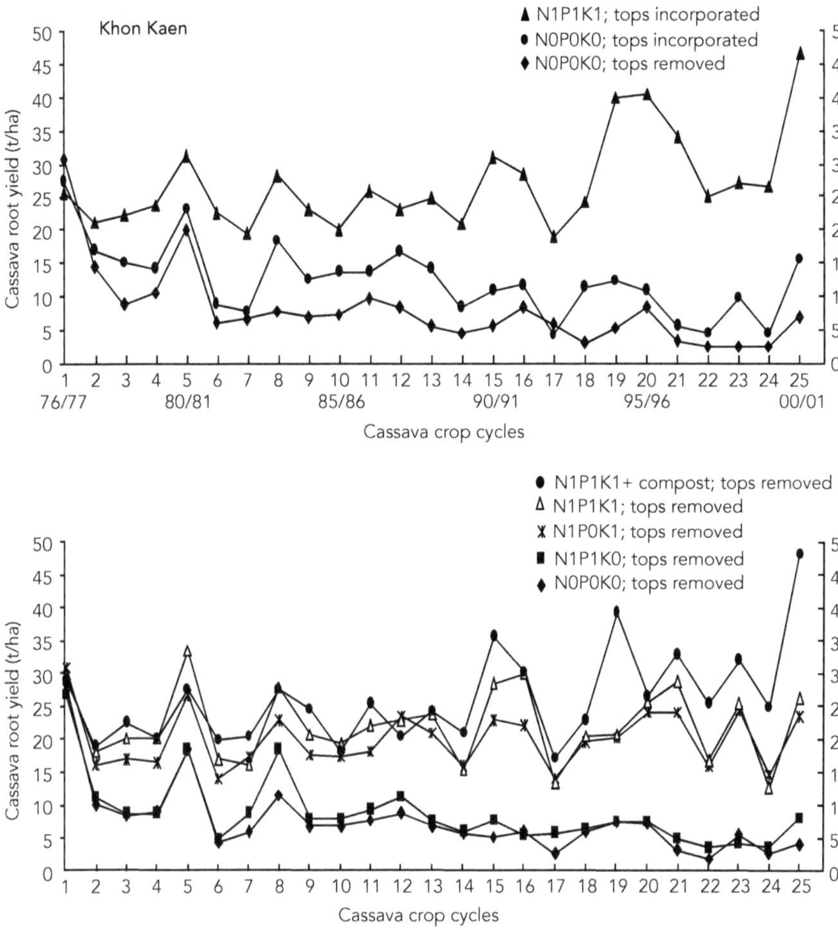

Figure 1 Effect of annual fertilizer application and crop residue management on cassava yield during 25 consecutive cropping cycles in Khon Kaen, Thailand. Source: Chumpol Nakviroj and Kobkiet Paisancharoen, personal communication.

3 When and where to apply chemical fertilizers, manures and soil amendments

Animal manures and compost as well as soil amendments such as lime, rock phosphates or elemental sulphur need to be decomposed or dissolved before the nutrients they contain can be taken up by the plant. These processes are enhanced by good contact with the soil. For that reason these nutrient sources are generally broadcast and incorporated to a depth of 15–20 cm several weeks before planting the crop. Most chemical fertilizers, however, are rather soluble and nutrients may be lost by volatilization or leaching or are washed away by runoff and soil erosion, whether applied on the soil surface or incorporated into the soil; the latter may also stimulate weed growth. For that reason it is recommended to band apply most chemical fertilizers close to, but not in direct contact with the planted stake. If stakes are planted horizontally the fertilizer may be applied in a continuous band alongside the planted stakes (5–10 cm from the stakes) or in the same furrow as long as the fertilizer is covered with some soil before the stakes are planted. Stakes should never be in direct contact with chemical fertilizers as this may inhibit sprouting. Alternatively, the fertilizer is applied in short (20–30 cm) bands next to, and about 5–10 cm away from, the stakes once the young plants have emerged from the ground. Huang Jie et al. (2015) obtained slightly higher yields when horizontally planted stakes all faced in the same direction and fertilizers were applied in short bands about 3 cm deep and 15–20 cm from the rooting end of each stake. In case of vertical or slanted planting of stakes, the fertilizers can either be applied right after planting or at 15–30 days after planting, placing the fertilizers in a short band, 3–5 cm deep and 5–10 cm from each stake.

When single element fertilizers, such as urea, single or triple superphosphate, or potassium chloride are used, all of the P fertilizer should be applied at planting or at 15 days after planting. The young plants need P for rapid early growth and before the roots have had time to become infected with VA mycorrhizae, which will enhance P uptake from the soil. The N and K fertilizers are best applied fractionated, i.e. one-half at or shortly after planting, together with the P, and the other half at 60 days after planting in order to reduce losses by leaching or by runoff and erosion. When using a compound fertilizer, all three nutrients can be applied together at planting, or, preferably, at 15–30 days after planting.

In order to increase the efficiency of fertilizer use, it is important to know how much fertilizer to apply for maximum efficiency, to apply the right type of fertilizer, at the right time, and in the right way.

4 Determination of fertilizer rates and nitrogen, phosphorus and potassium (NPK) balance

Farmers who want to increase their cassava yields often ask how much fertilizer they should apply and what kind. Ideally they should have their soils analysed so experienced researchers or extension workers can make a reasonable recommendation. Unfortunately, few countries have good soil or plant tissue testing facilities, and even if they do, farmers usually do not have the money to pay for this service. In that case, it is recommended that researchers and/or extension workers help farmers conduct simple fertilizer trials on their own fields and at time of harvest organize a field day for other farmers to see and discuss the results of these trials. Neighbouring farmers will then have some idea what fertilizers will produce the highest yields or the highest net income. This could be so-called 'nutrient omission trials', in which the plots have treatments such as: $N_0P_0K_0$. $N_0P_1K_1$, $N_1P_0K_1$, $N_1P_1K_0$ and $N_1P_1K_1$, where N_0, P_0 and K_0 are without application of N, P or K, respectively, while N_1 = 75 kg N/ha, P_1 = 35 kg P_2O_5/ha and K_1 = 90 kg K_2O/ha if crop residues are reincorporated before the next planting and only roots are removed from the field. If plant tops are also removed from the field the rates of N_1, P_1 and K_1 should increase to 120 kg N, 60 P_2O_5 and 120 K_2O/ha, respectively. The plot with the lowest yield among the three plots with the missing element will be the most limiting nutrient. This would be the most important nutrient to apply, while the remaining one or two nutrients may be of less importance and their application can be reduced or eliminated. However, after 2–3 years of cassava cropping on the same field, the rates of N can usually remain at about 80–120 kg/ha, while the application of P can be markedly reduced or eliminated, and that of K increased to 100–120 kg K_2O/ha. If only compound fertilizers are available the best would be to apply about 500 kg/ha of 15-7-18 (N-P_2O_5-K_2O) in the first few years and about 600 kg/ha of 14-4-24 after more than 4–5 years of continuous cassava cropping. An alternative would be to apply about 300 kg/ha of 15-15-15 fertilizer at planting and apply additional urea and KCl at two months after planting to increase the N and K levels to the recommended rates.

In case soil analyses data are available, these can indicate whether the levels of each nutrient are above or below the critical levels or whether they are in the low, sufficient or high range. The nutrients that are below the critical level or in the low range should be applied at the rates of about 75 kg N, 35 kg P_2O_5 and 90 kg K_2O/ha, or about 500 kg/ha of a compound fertilizer like 15-7-18 if the expected yields are around 20 t fresh roots/ha. For those nutrients that are far below the critical level or in the very low range, these rates should be further increased, up to as high as 200 kg of N, 100 kg P_2O_5 and 200 kg K_2O/ha, and preferably with the addition of about 5 t/ha of animal manure.

To be more precise, several authors have suggested calculating the requirement for each nutrient to be applied as fertilizers or manures based on the nutrient extraction by the crop at an expected yield level, minus the amount of the available nutrient in the soil, and divided by the utilization efficiency of the applied nutrients. Not all the applied nutrients will be taken up by the plant as some may be lost by leaching, volatilization, fixation, immobilization or erosion. For example, in highly P-fixing soils, the efficiency of applied P may be only 10–30%, while the efficiency of applied N and K is normally between 50 and 70%. Guerrero (1980) as quoted by Cadavid (2012) suggested the following formula:

$$NF = (WRC - S_N) \times 100/E$$

where NF = need for fertilizer application (kg/ha)
WRC = weighted nutritional requirement of the crop (kg/ha)
S_N = availability of the nutrient in the soil (kg/ha)
E = efficiency of the fertilizer

To calculate the nutritional requirement of the crop (WRC) one has to estimate the desired yield level. Cadavid (2011, 2012) reported the nutrient removal by the whole plant per tonne of **fresh** roots harvested as 4.42 kg N, 0.67 kg P, 3.58 kg K, 1.36 kg Ca, 0.82 kg Mg and 0.42 kg S. Thus, with a fresh root yield of 30 t/ha, the plants will remove 132.6 kg N, 20.1 kg P, 107.4 kg K, 40.8 kg Ca, 24.6 kg Mg and 12.6 kg S/ha. These nutrient requirements need to be met by the soil alone or by the soil plus applied fertilizers.

The amount of available nutrients in the soil is calculated from the results of soil analyses, which can be expressed in various units such as ppm available P, me K/100 g dry soil (= cmol/kg), and % organic matter, % organic C or total N. These units need to be converted to kg/ha assuming that dry soil in one hectare to about 20 cm depth weighs about 2 million kg, or 2.6 million kg if the specific gravity of the dry soil is 1.3 g/cm^3. Analyses of the soil in Carimagua in the Eastern Plains of Colombia indicate that the soil had 4.56% OM, 3.0 ppm Bray II-extractable P and 0.08 me exchangeable K/100 g. With 4.56% OM the organic N content is estimated at 4.56/20 = 0.228%. If the annual rate of mineralization is 2.5%, the available N content would be 5.7×10^{-5} kg N/kg soil or 148.2 kg N/ha. Thus, Cadavid (2012) calculated the available soil nutrient contents to be 148.2 kg N, 7.8 kg P and 81.1 kg K/ha.

Finally, using the approximate efficiency data for the NPK nutrients, i.e. 70% for N, 20% for P and 70% for K, in the formula mentioned above, the need for fertilizers (NF) was calculated to be: 22.0 kg N, 61.2 kg of P and 37.6 kg of K/ha. Thus, in this particular case of a very acidic, highly P-fixing soil, the recommendation is to apply 50 kg of N, 60 kg of P and 50 kg of K/ha, as well as 500 kg/ha of dolomitic lime. However, if cassava would be grown for several years on the same field, the need for P fertilizer would probably decline, while the need for N, and especially for K fertilizers, would increase, as shown for the third year in the same soil in Fig. 2 of Chapter 34.

5 Site-specific nutrient management

Recently, a team of cassava researchers at CTCRI in Kerala, India, developed a new method of more precise fertilizer requirement determination called 'Site-Specific Nutrient Management' (SSNM) (Byju et al., 2012). The objective of this methodology is not only to increase the yield of cassava but also to maximize the agronomic efficiency of the applied nutrients and prevent the excessive application of some nutrients that may cause pollution of the environment. The SSNM methodology has been extensively tested in the three main production zones of cassava in southern India, in comparison with the current farmer fertilizer practices (FFP). It was found that the use of SSNM markedly increased cassava yields by the careful adjustment of the amounts of applied N, P and K fertilizers to the requirements of the crop and the local soil fertility and climatic conditions, so as to reduce losses and increase farmers' profits.

The need for a more site-specific fertilizer recommendation was prompted by the realization that there is large spatial and temporal variability of soil fertility in the cassava growing areas and that the use of blanket fertilizer recommendations often result in excessive application of some nutrients and inadequate application of others, leading to low yields and high fertilizer costs. Thus, different soils may require a different balance of N, P and K as well as a different amount of each nutrient. To determine the optimum amount of each nutrient, they used an existing computer model, called Quantitative Evaluation of Fertility of Tropical Soils (QUEFTS), originally developed by Janssen et al. (1990) and validated for rice by Witt et al. (1999) and for wheat by Pathak et al. (2003). Byju and co-workers used data from many fertilizer experiments that had been conducted between 1975 and 2002 in different cassava-growing areas of India, to validate the parameters in the QUEFTS model for cassava, and adjust them to the specific relationship between yield and nutrient uptake in different environments. From the data of these experiments they determined that the mean nutrient removal by the crop (above- and below-ground) per tonne of **dry** cassava roots produced was 13.4 kg N, 2.3 kg P and 14.5 kg K, slightly lower for N and higher for P and K than those reported by Howeler (2004). They also determined the **potential nutrient supply** in each location, i.e. the maximum amount of the nutrient supplied by the soil plus fertilizers under optimum growth conditions. For every location they needed to determine the indigenous supply of N, P and K (in kg/ha) from the soil. The **indigenous nutrient supply** (INuS) is defined as the amount of nutrient under study that had been taken up from the soil by the crop at maturity (harvest time) under optimal conditions when all the other nutrients had been adequately supplied. This can be estimated from the total nutrients in the plants in a nutrient omission trial in each site. For instance, INS is the amount of N taken up by the plants in the treatment $N_0P_1K_1$ where no N is supplied but with adequate amounts of applied P and K, while IPS and IKS are the total amounts of P and

Table 3 The average indigenous supply of N, P and K (kg/ha) as related to the soil chemical properties in the two main cassava-growing regions in India

Location	Regression equations	Number of observations	R^2
Kerala	INS = 188.84 Org C – 6.2265	35	0.78
	IPS = 0.3302 Bray-I P + 8.3511	21	0.73
	IKS = 0.7398 Exch K – 9.9405	22	0.83
Tamil Nadu	INS = 221.94 Org C + 4.8519	23	0.86
	IPS = 0.6067 Olsen P + 1.084	15	0.82
	IKS = 0.2499 Exch K + 29.051	15	0.81

Note: Organic C in %; Bray-I or Olsen P and Exchangeable K both in kg/ha, where the weight of dry soil in 1 ha down to 15 cm depth is estimated at 2.24 million kg.
Source: Byju et al., 2012.

K taken up from plots $N_1P_0K_1$ and $N_1P_1K_0$, respectively. The values for INS, IPS and IKS varied greatly among the various sites where these NPK trials had been conducted, depending on the original soil fertility, climate and management conditions. The INuS values can also be determined from soil analyses data, but the relationship between soil fertility parameters and INS, IPS and IKS will vary greatly among sites. Table 3 shows the average indigenous supply of N, P and K as they relate to the soil chemical properties in the two main cassava-growing regions of India. In addition, they determined the mean value for the **internal nutrient efficiency**, i.e. the kg dry root weight increase per kg of nutrient applied. For instance, $\mathbf{IE_N = (Y_1–Y_0)/F}$ where Y_1 is the dry root yield (kg/ha) with N application, Y_0 is the dry root yield without N application, and F is the amount of N applied (kg/ha). The values of Y_1 and Y_0 are the dry root yields in plots $N_1P_1K_1$ and $N_0P_1K_1$, respectively. Similarly, the $IE_P = (Y_1–Y_0)/F$ where Y_1 is the dry root yield (kg/ha) with P application, Y_0 is the dry root yield without P application and F is the amount of P applied (kg/ha). Furthermore, the **nutrient recovery efficiency** is the total plant uptake of N, P or K (kg/ha) in fertilized plots minus the N, P or K uptake in unfertilized plots per kg of applied nutrient. Thus, $\mathbf{RE_N = (U_1–U_0)/F}$ where U_1 is the total N uptake by plants in plots $N_1P_1K_1$ and U_0 is the total N uptake by plants in plots $N_0P_1K_1$, while F is the amount of N applied.

With the use of the SSNM model it is possible to determine the optimum rate of N, P and K application to reach a certain yield goal. To do this it is necessary to first determine the **potential yield** (Ymax) of cassava in each site. In the data set used to calibrate the model, the potential yields in the four cassava-growing regions of India ranged from 17.5 to 24.5 t **dry** roots/ha. The NPK uptake at different yield potentials showed that the relation between yield and nutrient (N, P and K) uptake is linear at lower yield targets, indicating that plant growth was mainly limited by NPK uptake. At greater yield targets that are closer to the yield potential, there was a great reduction in the internal nutrient recovery efficiency. Thus, it is clear that maximizing the nutrient recovery efficiency by balanced NPK application will give more profit to farmers than aiming for greater yields.

The model calibrated for cassava suggested a linear increase in dry root yield with N, P and K uptakes of 17.6, 2.2 and 15.6 kg of applied N, P and K, respectively, per 1000 kg of dry root yield. The ratio of N, P and K in total plant dry matter was found to be 7.6:1.0:6.9. An example of the use of the QUEFTS model for one site in Tamil Nadu state of India is shown in Table 4.

Table 4 Predicted dry cassava root yield by the QUEFTS model and the actual yield in Salem, Tamil Nadu, India

Nutrient	Indigenous supply (kg/ha)	NPK fertilizer requirement (kg/ha)	Predicted nutrient uptake (kg/ha)	Predicted dry root yield (t/ha)	Measured dry root yield (t/ha)
N	134.2	120	185	10.5	11.4
P	15.7	30	24		
K	123.4	93	164		

The dry root yield potential was set at 24.5 t/ha
Source: Byju et al., 2012.

Once the QUEFTS model was calibrated for cassava, the model could be used to determine the most efficient combinations of applied N, P and K to reach a certain yield target. The model was validated in ten sites in each of the three main cassava states of India: Kerala, Tamil Nadu and Andhra Pradesh, where the average cassava **fresh** root yields are 36.3, 41.2 and 20.0 t/ha, respectively. However, within each state, yields vary markedly among farmers' fields. Ten on-farm trials were conducted for two successive seasons (2010 and 2011) in the main cassava growing districts of each of the three states. The experimental design was a nutrient omission trial, but also included one treatment using the SSNM recommendation and one treatment with the farmer fertilizer practice (FFP), i.e. the fertilizer practice the farmer normally uses. The SSNM recommendation had been calculated to maximize the internal N, P and K efficiency to achieve a particular yield goal using the yield data, indigenous nutrient supply and nutrient uptake data from the nutrient omission plots. Both the SSNM and FFP fertilizer application rates varied from site to site and between the two years. The **fresh** root yield under SSNM, averaged over all 30 sites and two years was 37.6 t/ha, while that under FFP was 30.6 t/ha. The SSNM treatment reduced the cost of fertilizers on average by $10/ha as compared to the FFP treatment, while the SSNM treatment increased the gross return above fertilizer cost by $254/ha.

While SSNM is definitely more efficient in fertilizer use and may increase yield, it is not practical to make site-specific recommendations for each cassava field, as soil and plant tissue testing facilities are not always available and the service is too expensive for farmers. The short-term aim is to produce zonal NPK recommendation maps of the main cassava-growing areas showing zones with the same SSNM fertilizer recommendations based mainly on existing soil test data, yield data and general crop management practices (rainfed or irrigated), and for a certain target yield (Byju et al., 2016a). Specific SSNM-based fertilizer recommendations are now available for each zone. In addition, custom-made compound fertilizers, made according to the specific SSNM recommendations, can be made available for each zone. Besides the recommended N, P and K balance in these fertilizers, they will also be fortified with Mg, Zn and B according to the soil fertility conditions in each zone (Byju et al., 2016a,b). An example of the map made for Kerala state in India is shown in Fig. 2.

An alternative approach is to produce an SSNM fertilizer recommendation chart that shows recommendations for the application rates of N, P and K for cassava based on soil testing results, for particular cassava-growing areas and different management practices and yield targets, as shown in Table 5.

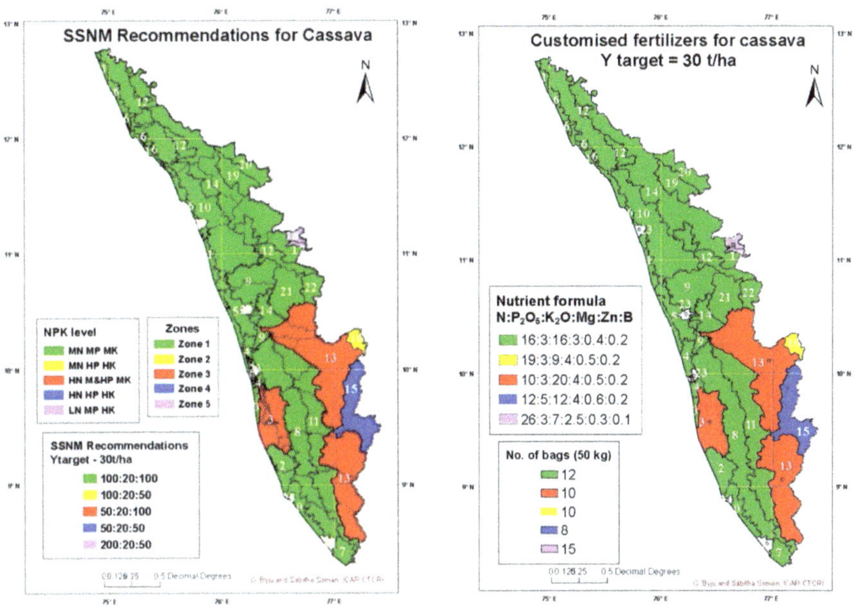

Figure 2 SSNM recommendation maps and customized fertilizer ratios for cassava in Kerala state of India. Source: Byju et al., 2016a.

Table 5 SSNM recommendation chart for cassava cultivation in major growing environments in Kerala and Tamil Nadu states of India based on soil test data and the modified QUEFTS computer model

	Yield target (t/ha)				Yield target (t/ha)				Yield target (t/ha)		
	20	30	40		20	30	40		20	30	40
Org. C (%)	N rate of application (kg/ha)			Avail. P (kg/ha)	P rate of application (kg/ha)			Exch. K (kg/ha)	K rate of application (kg/ha)		
Kerala (rainfed)											
<0.4	100	200	–	<10	20	30	–	<100	100	200	–
0.4-0.8	50	100	200	10–20	10	20	30	100–200	50	100	200
0.8–1.2	25	50	100	>20	10	20	30	200–300	25	50	100
>1.2	15	30	50					>300	15	30	50
Tamil Nadu (irrigated)											
<0.3	160	240	–	<15	–	40	60	<200	–	160	240
0.3–0.6	80	160	240	15–22	–	20	40	200–400	–	80	160
0.6–1.0	40	80	160	>22	–	15	20	400–600	–	40	80
>1.0	20	40	80					>600	–	20	40

Note: Available P by Bray I in Kerala and by Olsen in Tamil Nadu. Recommended rates are in kg/ha of N, P_2O_5 and K_2O.
Source: Byju et al., 2016b.

6 Fertilizer requirements for balanced nutrition

A similar approach as the one described above was reported by Ezui et al. (2016) for validating the QUEFTS model for cassava in West Africa by calculating the N, P and K application rates for a **balanced nutrition**. Using the calibrated QUEFTS model, the amount of N, P and K required for obtaining a target yield under balanced nutrition can be calculated, and its use will optimize the **nutrient use efficiency** within the plant, resulting in higher yields at lower fertilizer application rates, thus lowering the cost of fertilizers, increasing farmers' income and preventing nutrient loss to the environment, which may cause pollution.

Unlike in India, fertilizer use for cassava in Africa is minimal and few fertilizer experiments have been conducted in the past that can be used for validating the QUEFTS model. For that reason, Ezui et al. (2016) conducted three NPK experiments, one in southern Togo, one in southern Ghana and one in northern Ghana. These experiments were basically nutrient omission trials with ten treatments and four replications, using the variety Gbazekoute (TME 419) in Togo and Afisiafi (TME 771) in Ghana. At harvest time, they determined the fresh and dry root and top yields as well as the N, P and K concentrations within the roots and the tops (above-ground growth). From this they could calculate the harvest index (HI) as well as the N, P and K contents in the roots and tops, in order to calculate the internal nutrient supply under different levels of N, P and K applied. The data from these three experiments, called **Set 1**, were used to calculate the various parameters in the QUEFTS model for balanced nutrition. In addition, they conducted one experiment in three sites in southern Togo and one experiment in two sites in northern Ghana, with two trials in each location, for a total of ten trials. These trials, called **Set 2**, had four replications of five treatments of different combinations of N, P and K application rates, including an absolute check of 0N, 0P and 0K. In these Set 2 trials the fresh and dry yields of roots and tops were again determined in each treatment but no chemical analyses were conducted on the roots and tops to determine their nutrient concentrations. Using the parameters for the QUEFTS model determined from the Set 1 experiments, they were able to calculate the simulated dry root yields in the various Set 2 experiments and compare these with the observed yields.

A basic prerequisite for balanced nutrition assessment is the conversion of kg of N, P and K, both in the plants and those supplied by the soil and fertilizers, into **crop nutrient equivalent** (CNE) under the assumption that balanced nutrition is achieved when the supplies of these three nutrients, expressed in CNE, become equal to each other. The relation between root DM yield and nutrient uptake is indicated by the **physiological nutrient use efficiency** (PhE) of each nutrient, which varies between the minimum and maximum observed PhE. The minimum PhE (PhE_{min}) is generally obtained at high nutrient application rates when the uptake of that nutrient and the internal concentration of that nutrient are high, but the NUE, i.e. the root dry matter production per kg total absorbed nutrient (in roots plus tops), is low, resulting in maximal **accumulation**; conversely, the maximum PhE (PhE_{max}) corresponds to a low uptake of the nutrient resulting in maximal **dilution** of the nutrient within the plant and maximum dry matter production per kg of absorbed nutrient. According to Sattari et al. (2014), PhE_{min} and PhE_{max} are highly dependent on the crop or variety's HI, the ratio of the weight of the economic plant component (storage roots in the case of cassava) over the weight of the whole plant (total biomass including tops and roots). Hence, as a first approximation, PhE_{min} and PhE_{max} were calculated using the following equations:

$$PhE_{max} = 1000 \times HI/(HI \times C_{min}.roots + (1 - HI) \times C_{min}.tops)$$

$$PhE_{min} = 1000 \times HI/(HI \times C_{max}.roots + (1 - HI) \times C_{max}.tops),$$

where HI is the harvest index and C_{min}.roots, C_{min}.tops and C_{max}.roots, C_{max}.tops are the lowest and highest concentrations (g/kg) of N, P or K in the roots and tops, respectively, as determined in the Set 1 experiments. These were as follows:

In roots: C_{min} = 2.5 g N, 0.8 g P and 2.8 g K; C_{max} = 6.6 g N, 1.5 g P and 11.0 g K/kg DM roots

In tops: C_{min} = 7.9 g N, 0.9 g P and 3.4 g K; C_{max} = 17.9 g N, 2.8 g P and 18.8 g K/kg DM tops

The average of PhE_{max} and PhE_{min}, called PhE_{med}, is used to calculate the factors needed to convert kg of N, P and K into CNE units. Under balanced nutrition, 1 kilo of CNE (1 kCNE) of a given nutrient is the quantity of that nutrient that has the same effect on root DM yield as 1 kg N. The conversion factor (CFP) to convert kg P into kCNE units of P is obtained as:

$$CFP = PhE_{med} \text{ of } N/PhE_{med} \text{ of } P$$

and the conversion factor (CFK) to convert kg K into kCNE units of K:

$$CFK = PhE_{med} \text{ of } N/PhE_{med} \text{ of } K.$$

Thus, 1kCNE of P (kCNEP) = CFP kg of P, and 1 kCNE of K (kCNEK) = CFK kg of K.

Table 6 shows these parameters as obtained for the two different varieties and at two different harvest indices each, as obtained from data in Set 1 and Set 2.

Table 6 Harvest index, physiological nutrient use efficiency for maximum accumulation (PhE_{min}) and maximum dilution (PhE_{max}), and the conversion factors for P (CFP) and K (CFK) used in model calculations for two varieties (Gbazekoute and Afisiafi) in Set 1 and Set 2 experiments

		PhE_{min}			PhE_{max}			PhE_{med}				
Varieties	HI	N	P	K	N	P	K	N	P	K	CFP	CFK
Gbazekoute-Set 1	0.50	41	232	34	96	589	160	69	411	97	0.167	0.706
Gbazekoute-Set 2	0.55	47	262	38	112	653	178	80	458	108	0.174	0.736
Afisiafi-Set 1	0.65	61	329	47	148	782	214	105	556	131	0.188	0.801
Afisiafi-Set 2	0.70	70	365	53	170	848	233	120	607	143	0.198	0.839

Source: Ezui et al., 2016.

The Total Available supply of N, P and K, denominated as TAN, TAP and TAK, can come from the soil, denominated as SAN, SAP and SAK, as well as from nutrient inputs like fertilizers. Without external nutrient applications, TAN, TAP and TAK are equal to SAN, SAP and SAK. This is generally an unbalanced nutrient situation. In a balanced nutrient situation, TAN = TAP = TAK when expressed in CNE. To achieve this balanced situation the soil supply needs to be augmented with required nutrient inputs (I) from fertilizer. This is done by raising the smallest amounts of available nutrients among TAN, TAP and TAK to the level of the largest amount until TAN = TAP = TAK in units of CNE. Thus, as an example, if the TAN, TAP and TAK are equal to 75, 25 and 40 kCNE, respectively, we need to increase TAP by 50 kCNE and TAK by 35 kCNE by application of P and K fertilizers, respectively, to attain the balanced nutrition with TAN = TAP = TAK = 75kCNE. The sum of these nutrients (TAN + TAP + TAK), denoted as ΣA, is equal to 225 kCNE. Once a balanced situation has been achieved, i.e. TAN = TAP = TAK, identical quantities (in CNE) are added until maximum dry root yields (Y_{max}) are approached.

The nutrient supply from the fertilizer is calculated from the amount of applied fertilizer (I) multiplied by the **maximum recovery fraction (MRF)** of that nutrient that is taken up by the plant, i.e.

I_N x MRFN, I_P x MRFP, and I_K x MRFK. Thus, TAN = SAN + I_N x MRFN; TAP = SAP + I_P x MRFP; and TAK = SAK + I_K x MRFK.

Table 7 shows the N, P and K supply from the soil as well as the maximum recovery fractions of the nutrients supplied by the fertilizer.

Plotting root DM yield against ΣA (expressed in CNE) we obtain different curves for each variety depending on their HI (Fig. 3). This is because PhE_{max} and PhE_{min} will vary with different HIs (see Table 6). The curves have three sections: at low yield levels the yields are

Table 7 Soil supply of available N, P and K (SAN, SAP and SAK in kg/ha) and the maximum recovery fractions (MRFN, MRFP and MRFK) of all ten experimental sites in Togo and Ghana

Data Set	Country	Site	SAN	SAP	SAK	MRFN	MRFP	MRFK
Set 1	Togo	Davie	177	24	70	0.69	0.44	1.05
	Ghana	Kumasi	94	21	65	0.33	0.15	0.10
		Nyankpala	86	18	104	0.49	0.03	0.33
	Average					**0.50**	**0.21**	**0.49**
Set 2	Togo	Gbave	170	23	67	0.95	0.60	0.95
		Davie Tekpo	250	34	99	0.95	0.60	0.95
		Sevekpota black soil	186	25	74	0.69	0.44	0.80
		Sevekpota white soil	122	17	48	0.81	0.51	0.80
		Sevekpota red soil	147	20	58	0.69	0.44	0.80
	Ghana	Gbanlahi	74	15	89	0.69	0.21	0.46
		Savelugu	113	24	136	0.64	0.20	0.43

Source: Ezui et al., 2016.

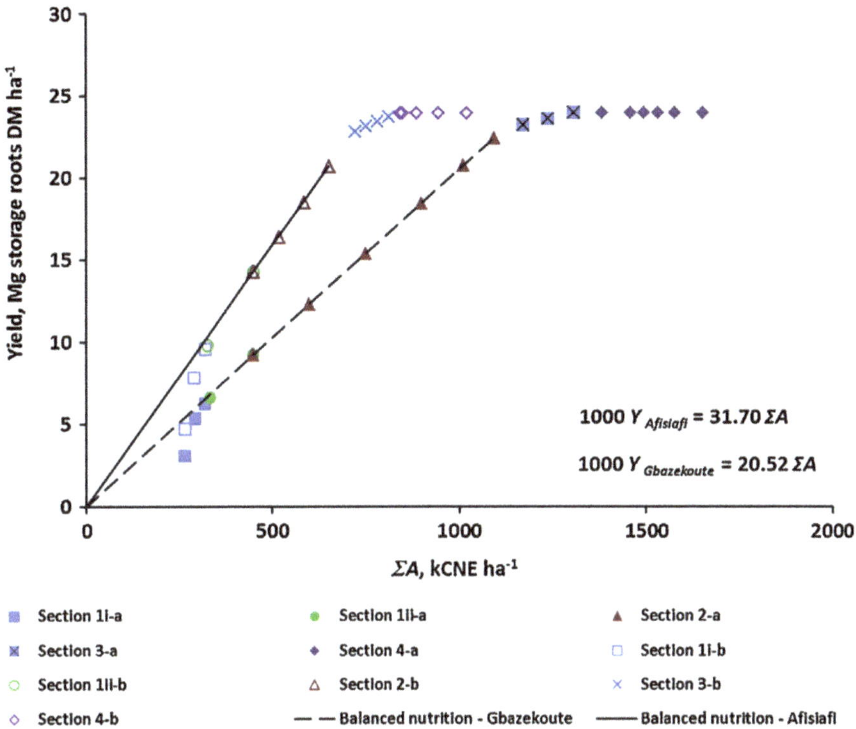

Figure 3 Simulated relations of cassava dry root yield to the sum of available N, P and K (ΣA) expressed in CNE for varieties Gbazekoute and Afisiafi. The linear part of the regression between the yield (Y) and ΣA expresses the situation of balanced nutrition for each variety. Sections 1i-a, 1ii-a, 2-a, 3-a and 4-a refer to Gbazekoute and sections 1i-b, 1ii-b, 2-b, 3-b and 4-b refer to Afisiafi. Source: Ezui et al., 2016.

determined mainly by the nutrient supply of the soil and yields rise rapidly with increasing supply. At intermediate yield levels, yields increase at a constant rate with increasing balanced nutrient supply from both the soil and fertilizers. The highest NUE is achieved during this stage. The rate of yield increase starts to slow when the maximum yield (Ymax) is being approached resulting in decreasing NUE. The slope of the linear regression of 1000Y over ΣA is taken as the NUE of the variety for a specific HI, i.e. kg DM root yield per kCNE. The highest slope of the regression between 1000Y and ΣA (in kCNE) was 31.70 kg DM root per kCNE for variety Afisiafi used in Ghana and 20.52 kg DM root per kCNE for variety Gbazekoute used in Togo. This is due to the higher HI of Afisiafi as compared with Gbazekoute (see Table 6), which makes the variety Afisiafi more nutrient use efficient: it produces more root DM per kg absorbed nutrients.

Once the parameters used in the QUEFTS model are established, it is possible to calculate the additional requirements of N, P and K to obtain balanced nutrition for specific target yields. These required N, P and K inputs will vary among sites because of differences in the amount of nutrients supplied by the soil. Table 8 shows the additional N, P and K requirements to achieve balanced cassava nutrition for several root DM yield targets at different sites in Togo and Ghana.

Table 8 Additional plant nutrient requirements to achieve balanced nutrition for different target yields of variety Gbazekoute in the Togo sites and Afisiafi in the Ghana sites

Country	Site and soil type	Target yield (t/ha dry roots)[1]	Additional nutrients required (kg/ha)		
			N	P	K
Togo	Davie – Rhodic ferralsol	8	0	0	22
		12	18	8	67
		16	83	19	113
	Gbave – Rhodic ferralsol	8	0	0	15
		12	0	6	56
		16	54	16	98
	Davie Tekpo – Rhodic ferralsol	8	0	0	0
		12	0	0	24
		16	0	5	66
	Sevekpota black soil – Alfisol	8	0	0	8
		12	0	4	49
		16	38	14	91
	Sevekpota white soil – Alfisol	8	0	2	34
		12	46	12	75
		16	102	22	117
	Sevekpota red soil – Alfisol	8	0	0	24
		12	21	9	65
		16	77	19	107
Ghana	Kubasi – Ferric acrisol	8	0	0	3
		12	34	3	37
		16	76	11	71
	Nyankpala – Gleyi-ferric lixisol	8	0	0	0
		12	42	6	0
		16	84	14	32
	Gbanlahi – Gleyi-ferric lixisol	8	0	0	0
		12	37	7	4
		16	74	14	35
	Savelegu – Gleyi-ferric lixisol	8	0	0	0
		12	0	0	0
		16	35	5	0

[1] 8, 12 and 16 t/ha dry roots is equivalent to about 22, 33 and 44 t/ha of fresh roots.
Source: Ezui et al., 2016.

From these data it is clear that the soils in all six sites in Togo are most deficient in K; N is the most limiting nutrient in the three lixisols in northern Ghana, while both N and K are limiting in the Ferric acrisol of Kumasi in southern Ghana. In Togo, at the lowest target yield of 8 t/ha dry roots the application of only K is necessary to obtain balanced nutrition, while at the higher target yields all three nutrients are required in most sites with fairly high applications of N and K and much lower rates of P for the highest target yield of 16 t/ha. In Ghana, the models predicted that the lowest yield of 8 t/ha could be achieved without NPK fertilizer application, but at higher target yields there will be a need for application of all three nutrients, except in Savelegu where only N and P are required and only at the highest target yield according to the model.

In both countries national recommendations of blanket fertilizer rates for cassava are used, i.e. 76 kg N, 13 P and 25 K/ha (this is 76-30-30 in terms of N-P_2O_5-K_2O) in Togo, and 68 kg N, 20 P and 57 K/ha (68-46-68 in terms of N-P_2O_5-K_2O) in Ghana. These rates are recommended irrespective of major differences in soil fertility characteristics. Evaluation of these blanket fertilizer rates did not result in balanced nutrition on most study sites, and resulted in higher costs of fertilizers and lower yields as compared to site-specific and balanced application rates. This is a key reason that very few farmers apply any fertilizer to cassava. On the other hand, it is currently impractical to make fertilizer recommendations for individual farms, but it may be possible to make recommendations of balanced fertilization on the basis of major soil types and agro-ecological zones (Ezui et al., 2016), similar to the site-specific nutrient management maps produced by Byju et al. in India, as shown in Fig. 2.

As cheaper and more sensitive sensors and diagnostics, along with meta-data analysis, become available in the future it will be possible to make site-specific fertilizer recommendations for individual farms, and even for individual fields within each farm. Moreover, the models should also be able to calculate the fertilizer recommendations needed to maximize economic returns based on the current prices of fresh roots and the cost of fertilizers.

7 Genetic variation in nutrient use efficiency (NUE) and tolerance to low soil fertility

Ezui et al. (2016) showed that the short variety Afisiafi had a high HI and was more nutrient use efficient than the taller Gbazekoute, which had a much lower HI (Table 6 and Fig. 3). This suggests the hypothesis that short varieties may have a higher NUE than taller varieties, in terms of nutrient use per kg of roots produced. This was confirmed by El-Sharkawy and De Tafur (2010), who compared the photosynthesis, growth, productivity and NUE of seven tall- and eight short-stemmed cassava varieties grown for two crop cycles in a highly P-fixing Inceptisol, under rainfed conditions at the CIAT-Quilichao research station in Colombia. The tall varieties ranged in height between 2.3 and 2.6 m, while the short ones ranged from 1.2 to 1.7 m. All stakes were planted at a moderate density of 10 000 plants/ha (1.0 x 1.0 m), to accommodate the taller varieties. Plants received a uniform fertilizer application of 50 kg N, 44 P and 84 K/ha (equivalent to 500 kg/ha of 10-20-20 (N-P_2O_5-K_2O)). Six fully bordered plants were harvested at 2, 4, 6 and 10 MAP, to determine total dry biomass, dry root yield and nutrient uptake in the roots and tops (above-ground biomass without fallen leaves). They also determined at regular intervals the upper canopy leaf photosynthetic rate (P_N), seasonal stomatal conductance (g_s) and seasonal leaf area index (LAI).

Table 9A shows the average dry matter yields of roots and total biomass (roots and tops), as well as the N, P, K, Ca and Mg contents of the total biomass, while Table 9B shows the total nutrient uptake per tonne dry roots and the resulting nutrient use efficiencies.

From these data it is clear that the tall varieties in this set of materials have a much higher leaf area index and lower HI than the short ones; these tall plants take up more nutrients which mainly end up in the top growth. In these experiments the total dry biomass and root yields

Table 9 Comparison between tall and short varieties of cassava in terms of leaf area index, root and total biomass yields at 10 months after planting, and the nutrient uptake in the total biomass as well as the efficiency of these nutrients to produce one tonne of dry storage roots. Data are the average of seven tall and eight short varieties over two crop cycles in CIAT-Quilichao, Colombia

A

	Leaf area index (LAI)	Dry root yield (t/ha)	Dry total biomass (t/ha)	Harvest index (HI)	Total nutrient uptake in roots and tops (kg/ha)				
					N	P	K	Ca	Mg
Tall	2.56	12.8	18. 9	0.7	116	20	97	37	22
Short	1.72	9.9	13.0	0.8	76	11	61	23	15

B

	Total nutrient uptake (kg/tonne dry roots)					Nutrient Use Efficiency (NUE) (kg dry root/kg total nutrient uptake)				
	N	P	K	Ca	Mg	N	P	K	Ca	Mg
Tall	9.1	1.6	7.6	2.9	1.7	110	640	132	346	581
Short	7.7	1.1	6.2	2.3	1.5	130	899	162	430	659
Short/tall						1.18	1.40	1.23	1.24	1.13

Source: Adapted from El-Sharkawy and De Tafur, 2010.

were higher for the tall varieties because of the rather wide plant spacing used. However, the NUE was poorer as these tall varieties required more nutrients to produce a tonne of dry roots than the short ones, as shown in Table 9B. Farmers could expect to get higher yields with short varieties if these are planted at a closer spacing, such as 0.8 x 0.8 m, or even closer. These short varieties, besides having a higher NUE also have earlier root bulking, which is important under seasonally dry or semi-arid conditions, and to provide earlier food supply or market opportunities. They also had higher seasonal leaf photosynthetic rates and longer leaf life than the taller varieties. Thus, although cassava breeding programs have to consider many other attributes as well, they should preferably use shorter varieties as parental material, providing that the branching habit is suitable for farmers.

The NUE shown in Table 9B is **internal** use efficiency, i.e. the amount of a certain nutrient taken up by the whole plant to produce one tonne of dry storage roots. These nutrients are taken up from both the soil's native nutrients and applied fertilizers or manure. Not all nutrients are taken up from these sources in equal proportions. N and K are rather mobile within the soil and are taken up in rather large proportions (maybe 50% of available N and K), while P is very immobile and moves only slowly by diffusion in the soil solution. Only a small proportion of available P in the soil, on the order of 20% (see Table 7), enters the roots, especially in case of cassava, which has a very course fibrous root system. But due to the almost universal infection of cassava roots by VA-mycorrhizae – a fungus naturally present in almost all soils, cassava is actually quite efficient in P absorption and the crop can therefore grow well in many low-P soils without the need for high applications of this nutrient (Howeler, 2012f). Table 9A shows that cassava takes up much less P than either N or K, so much of the applied P will remain in the soil and this may result in a build-up to high P levels, which inhibits further mycorrhizal infection of the root system.

Very similar results had been obtained in a previous experiment in the same location using five varieties each of short-, medium- and tall-stemmed cassava varieties during two consecutive cropping cycles. There were significant differences in leaf area index (LAI), dry biomass and nutrient uptake in the above-ground growth, both within and among groups, but with higher values for tall, followed by medium and short varieties. However, there was earlier root bulking and more biomass partitioning to the roots in the short- and medium-height varieties resulting in significantly higher NUE for root production of these *versus* the tall varieties (El-Sharkawy et al., 1998).

The same group of researchers also studied the effect of water stress and fertilization (50 kg N, 100 P_2O_5 and 100 K_2O/ha) on cassava photosynthetic rates, LAI, dry matter production and root yields, using two cassava varieties, MCol 1684 and CM 507-37. Water stress was induced at 4 MAP until harvest at 12 MAP by covering the soil with plastic. They reported that there was strong water absorption by roots down to 90 cm soil depth during the first 24 days of water stress, especially in the fertilized plots; and more water absorption down to >160 cm depth at 160 days of stress, indicating the deep rooting ability of cassava during periods of drought. Water stress reduced the LAI, but less so in the fertilized plants. Photosynthetic rates (P_N) decreased markedly in water stressed plants, but young leaves continued to produce photosynthates under water stress and store starch, especially when fertilized. With fertilizer there was no significant decrease in DM production due to drought; fertilizer application resulted in significantly higher yields, but without fertilizers, yields were actually higher for the water-stressed than non-stressed plants. This was significant only for variety MCol 1684 (Table 10) (De Tafur et al., 1997).

In a somewhat similar experiment at the same location, El-Sharkaway and Cadavid (2002) planted four clones with early (2–6 MAP), mid-season (4–8 MAP) and late-season (6–12 MAP) water stress, as well as a control treatment without water stress, during two cropping cycles. Table 11 shows that water stress during any stage of growth significantly reduced shoot biomass, but the average dry root yields were not affected by water stress

Table 10 The effect of water stress and fertilizer application on seasonal average leaf area index (LAI), as well as on total and root biomass production of two cassava varieties during a 12-month growth period in CIAT-Quilichao, Colombia

Variety	Water stress[1]	Fertilization[1]	LAI	Total biomass (kg/m^2)	Dry root yield (kg/m^2)
MCol 1684	+	+	0.96 cd	0.88 a	0.68 a
	+	–	0.74 d	0.57 b	0.48 bc
	–	+	1.36 b	0.91 a	0.70 a
	–	–	0.89 cd	0.31 c	0.26 d
CM 507-37	+	+	1.37 b	0.83 a	0.61 ab
	+	–	0.90 cd	0.53 bc	0.43 bcd
	–	+	1.70 a	0.87 a	0.62 ab
	–	–	1.06 c	0.42 bc	0.31 cd

[1] With (+) or without (-) water stress from 4 to 12 months after planting; with (+) or without (-) fertilizers
[2] Values within columns followed by the same letter(s) are not significantly different (P<0.05)
Source: De Tafur et al., 1997.

Table 11 Effect of water stress during different stages of the cassava growth cycle on the average total, shoot and root biomass at 12 MAP, as well as the nutrient use efficiency (NUE) for root production of four varieties during two crop cycles

Water regime	Dry total biomass (t/ha)	Dry root biomass (t/ha)	Dry shoot biomass (t/ha)	Nutrient use efficiency (NUE)				
				N	P	K	Ca	Mg
Control	18.9	12.9	6.0	120	660	110	300	440
Early stress	14.5	11.2	3.3	150	940	150	360	650
Mid-season stress	18.0	12.9	5.1	120	840	140	330	530
Late-season stress	16.9	11.7	5.2	110	690	120	300	520
LSD 5%	2.3	NS	1.4	12	70	9	50	65

Source: El-Sharkawy and Cadavid, 2002; El-Sharkawy, 2012.

at any stage. At the end of the early- and mid-season water stress both the root and shoot growth were markedly reduced compared with the well-watered control treatment, but plants quickly recovered once new leaves had been formed after the release of the stress. But, because water stress reduced top growth more than root growth, it also reduced nutrient uptake in the shoots, thereby enhancing the NUE for root production when the water stress occurred at the early or mid-season stage (El-Sharkawy, 2006, 2012)

It is clear that cassava is extremely tolerant of mid- and late-season water stress; this is because of partial stomatal closure at low relative humidity and reduced leaf area during water stress, as well as being able to access water at greater soil depth than most other food crops because of a deep root system. However, because early-season water stress reduces top growth more than root growth, thus increasing the HI, this early-season stress will reduce nutrient uptake in the tops and enhance NUE for root production.

Another way to obtain high yields with less nutrient supply is to plant varieties that are genetically more tolerant of low soil fertility. Since most cassava is produced by small-scale farmers, who are often living in areas where chemical fertilizers are not readily available or too expensive, it would be very useful to have available new varieties that make better use of the nutrient supply in the soil by producing more root dry matter per kg of nutrient absorbed. To facilitate breeding for this characteristic, CIAT in Colombia has screened large numbers of cassava accessions from its germplasm collection, to identify those that are not only adapted to low-P or low-K soils, but to yield well also when these nutrients are abundant, either in the soil or from fertilizer application.

From 1982 to 1989 a total of over 1,800 accessions were screened for low-P adaptation in soil that originally had only 2–3 ppm available P (Bray II) at CIAT-Quilichao. Each accession was planted in single rows in two large plots, one that received annually no P and the other with annual applications of 75 kg P/ha. During the first screening of nearly 700 accessions, both the P concentration of the indicator tissue and the mycorrhizal infection of roots were determined, but these were not related to final root yields. Thus, the low-P tolerance of each accession was based on the fresh root yield at time of harvest, usually at about 11 MAP, in the plots with 0 and 75 kg P/ha. To select germplasm that would yield well under both low and medium levels of P, a low-P adaptation index was calculated for each accession as follows:

Low-P adaptation index = (root yield at 0P x root yield at 75P)/(Avg. yield at 0P x Avg. yield at 75P)

where the Avg. yields are those of the whole trial at either 0 or 75 kg P/ha.

Accessions with the highest low-P adaptation indices were then used in the breeding programme to produce lines with better adaption to low-P soils. Table 12 shows a summary of some of these lines and varieties with high low-P adaptation indices in terms of dry root yields.

The low-P adaptation index determined on another set of 33 clones was found to be highly significantly correlated with growth and physiological characteristics such as photosynthetic rate, mesophyll conductance, dry root yield, shoot biomass, total biomass, number of storage roots, LAI and HI, indicating the validity of this index as a selectable trait for cassava improvement (El-Sharkawy, 2012).

A similar screening of 15 clones on plots receiving either 0 K or 100 kg K/ha was conducted for five consecutive years at CIAT-Quilichao. The average dry root yields of all clones during the first year were 11.1 and 13.2 t/ha for the 0K and 100K treatments. At the 5th year these yields had decreased to 3.7 and 7.3 t/ha, indicating that application of 100 kg K/ha was not enough to maintain high root yields. The very marked reduction in yield without K application indicates the serious depletion of soil K in the root harvest. Table 13 shows a summary of clones with different levels of low-K adaptation indices where the low-K adaptation index for each clone is calculated as follows:

Low-K adaptation index = (root yield at 0K x root yield at 100K)/(Avg. yield at 0K x Avg. yield at 100K)

where the Avg. yields are those of the whole trial at either 0 or 100 kg K/ha.

In this set of 15 clones, all except CM 402-11 were short- or medium-stemmed with low shoot biomass and high HI. CM 402-11 had a high adaptation index but the tallest stature,

Table 12 Dry root yield of clones with high adaptation to low-P soils as determined in a screening of 33 clones at 0P and 75 kg P/ha in CIAT-Quilichao, Colombia

	Dry root yields (t/ha)		
Clone	0P	75 kg P/ha	Low-P adaptation index
CG 996-6	13	17	2.0
CM 305-41	14	15	1.9
CG 1374-2	12	16	1.8
MBra 383	11	17	1.7
MBra 191	13	14	1.6
Av of 33 clones	**9**	**12**	**1.0**
LSD (5%) for clones	1.9	2.1	0.43

Source: El-Sharkawy, 2004.

Table 13 Dry root yields of some of the clones with high, medium and low adaptation to low-K soils as determined in a screening of 15 clones at 0K and 100 kg K/ha in CIAT-Quilichao, Colombia. Yields are the average of five consecutive cropping cycles

	Dry root yields (t/ha)		
Clone	0 kg K/ha	100 kg K/ha	Low-K adaptation index
CM4777-2	8.7	13.5	2.2 high
CM 402-11	8.7	12.6	2.0 high
CM 4729-4	7.8	9.5	1.4 medium
CG 1141-1	6.7	10.2	1.3 medium
CM 2177-2	5.2	7.3	0.7 low
CM 3299-4	3.6	5.5	0.4 low
Av. of 15 clones	**6.6**	**10.1**	**1.0**
LSD 5% for clones	1.1	1.2	

Source:El-Sharkawy, 2012.

resulting in the greatest total K uptake and the lowest K use efficiency, i.e. 158 kg dry roots/kg K at zero K and 128 kg dry roots/kg K at 100 K, compared with 210 and 148 kg dry roots/kg K for the overall trial mean. This supports the earlier observation that short- and medium-statured varieties are more efficient in the utilization of absorbed nutrients for root production than the tall ones.

8 Conclusions

1 Soil nutrient depletion and exhaustion can be prevented by application of adequate amounts of chemical fertilizers, organic manures or compost, or by incorporation of crop residues (plant tops), green manures or prunings of hedgerows.
2 Organic manures are useful sources of secondarynutrients and micronutrients; they contribute organic matter and improve soil physical conditions. For cassava, organic manures should be applied together with chemical fertilizers high in N and K.
3 Models to predict the N, P and K fertilizer requirements to obtain maximum cassava root yields per kg applied nutrient, dependent on the local indigenous soil nutrient supply and the desired target yield, are useful to maximize the efficient use of applied nutrients and minimize pollution of the environment.

9 References

Byju, G., M. Nedunchezhiyan, C. S. Ravindran, V. S. Santhosh Mithra, V. Ravi and S. K. Naskar. 2012. Modeling the response of cassava to fertilizers: A Site-Specific Nutrient Management approach for greater tuberous root yield. *Communications in Soil Science and Plant Analysis* 43: 1149–62.
Byju, G., M. Nedunchezhiyan, A. C. Hridya and Sabitha Soman. 2016a. Site-Specific Nutrient Management for cassava in southern India. *Agronomy Journal*108 (2): 1–11.

Byju, G., M. Nedunchezhiyan, James George, S. Sunitha, Sabitha Soman and A. C. Hridya. 2016b. Fertilizer best management practices by SSNM and customized fertilizers for cassava (*Manihot esculenta* Crantz) cultivation in India. *Journal of Food Agriculture and Environment* 4 (2) (accepted for publication).

Cadavid, L. F. 2011. Manual de Nutrición Vegetal – Una visión de los aspectos nutriciónales del cultivo de yuca (*Manihot esculenta* Crantz). (Plant Nutrition Manual – A vision of nutritional aspects of cassava). ISBN 978-958-44-9511-2. Colombia. 175 p.

Cadavid, L. F. 2012. Chapter 5. Soils and Fertilizers for the Cassava Crop. In B. Ospina and H. Ceballos (eds), *Cassava in the Third Millennium.* Modern Production, Processing, Use and Marketing Systems. CIAT, CLAYUCA, CTA, Cali, Colombia, pp. 113–37.

De Tafur, S. M., M. A. El-Sharkawy and L. F. Cadavid. 1997. Response of cassava (*Manihot esculenta* Crantz) to water stress and fertilization. *Photosynthetica* 34 (2): 233–9.

El-Sharkawy, M. A. 2004. Cassava biology and physiology. *Plant Molecular Biology* 56: 481–501.

El-Sharkawy, M. A. 2006. International research on cassava photosynthesis, productivity, eco-physiology, and response to environmental stresses in the tropics. *Photosynthetica* 44 (4): 481–512.

El-Sharkawy, M. A. 2012. Stress-tolerant cassava: The role of integrative ecophysiology-breeding research in crop improvement. *Open Journal of Soil Science* 2: 162–86.

El-Sharkawy, M. A., L. F. Cadavid and S. M. de Tafur. 1998. Nutrient use efficiency of cassava differs with genotype architecture. *Acta Agronómica* 48 (1/2): 23–32.

El-Sharkawy, M. A. and L. F. Cadavid. 2002. Response of cassava to prolonged water stress imposed at different stages of growth. *Experimental Agriculture* 38 (3): 333–50.

El-Sharkawy, M. A. and S. M. De Tafur. 2010. Comparative photosynthesis, growth, productivity and nutrient use efficiency among tall- and short-stemmed rain-fed cassava cultivars. *Photosynthetica* 48 (2): 173–88.

Ezui, K. S., A. C. Franke, A. Mando, B. D. K. Ahiabor, F. M. Tetteh, J. Sogbedji, B. H. Janssen and K. E. Giller. 2016. Fertiliser requirements for balanced nutrition of cassava across eight locations in West Africa. *Field Crops Research* 185: 69–78.

Guerrero, R. R. 1980. Hacia la formación de un modelo suelo-planta. In M. F. Silva (ed.), *Sociedad Colombiana de la Ciencia del Suelo (SCCS).* Bogota, DC, Colombia, pp. 1–10.

Howeler, R. H. 2002. Cassava mineral nutrition and fertilization. In R. J. Hillocks, M. J. Thresh and A. C. Bellotti (eds), *Cassava: Biology, Production and Utilization.* CABI Publishing, Wallingford, UK, pp. 115–47.

Howeler, R. H. 2004. Nutrient inputs and losses in cassava-based cropping systems – examples from Vietnam and Thailand. In R. W. Simmons, A. D. Noble and R. D. B. Lefroy (eds), *Nutrient Balances for Sustainable Agricultural Production and Natural Resource Management in SE Asia.* Proc. Intern. Workshop, held in Bangkok, Thailand, 20–22 February 2001. 30 p.

Howeler, R. H. 2007. Agronomic practices for sustainable cassava production in Asia. In R. H. Howeler (ed.), *Cassava Research and Development in Asia: Exploring New Opportunities for an Ancient Crop.* Proc. 7th Regional Workshop, held in Bangkok, Thailand, 28 October to 1 November 2002, pp. 288–314.

Howeler, R. H. 2012b. Effect of cassava production on soil fertility and the long-term fertilizer requirements to maintain high yields. In R. H. Howeler (ed.), *The Cassava Handbook.* A Reference Manual Based on the Asian Regional Cassava Training Course, held in Thailand, pp. 411–28.

Howeler, R. H. 2012e. Soil fertility maintenance: Organic solutions. In R. H. Howeler (ed.), *The Cassava Handbook.* A Reference Manual Based on the Asian Regional Cassava Training Course, held in Thailand, pp. 469–96.

Howeler, R. H. 2012f. Importance of mycorrhiza for phosphorus absorption by cassava. In R. H. Howeler (ed.), *The Cassava Handbook.* A Reference Manual Based on the Asian Regional Cassava Training Course, held in Thailand, pp. 497–523.

Howeler, R. H. 2014b. Sustainable Soil and Crop Management of Cassava in Asia – A Reference Manual. CIAT Publication No. 389. CIAT, Cali, Colombia. 280 p.

Howeler, R. and T. M. Aye. 2014. Sustainable Management of Cassava in Asia – From Research to Practice. CIAT, The Nippon Foundation. Cali, Colombia. 148 p.

Huang, Jie, Liu Zifan, Li Zhaogui, Yan Qingxiang, Lu Xiaojing and Ye Jianqiu. 2015. Cassava agronomy research at the Chinese Academy of Tropical Agricultural Sciences (CATAS) in China. In R. H. Howeler (ed.), *Sustainable Cassava Production in Asia for Multiple Uses and for Multiple Markets*. Proc. 9th Regional Workshop, held in Nanning, Guangxi, China, 27 November to 3 December 2011, pp. 172–85.

Janssen, B. H., F. C. T. Guiking, D. van der Eijk, E. M. A. Smaling, J. Wolf and H. Reuler. 1990. A system for quantitative evaluation of fertility of tropical soils (QUEFTS). *Geoderma* 46: 299–318.

Kabeerathumma, S., B. Mohankumar, C. R. Mohankumar, G. M Nair,. M. Prabhakar and N. G. Pillai. 1990. Long range effect of continuous cropping and manuring on cassava production and fertility status. In R. H Howeler (ed.), Proc. 8th Symposium International Society of Tropical Root Crops, held in Bangkok, Thailand, 30 October to 5 November 1988, pp. 259–69.

Nakviroj, C., K. Paisancharoen, O. Boonseng, C. Wongwiwatchai and S. Roongruang. 2007. Cassava long-term fertility experiments in Thailand. In R. H. Howeler (ed.), *Cassava Research and Development in Asia: Exploring New Opportunities for an Ancient Crop*. Proc. 7th Regional Workshop, held in Bangkok, Thailand, 28 October to 1 November 2002, pp. 212–23.

Pathak, H., P. K. Aggarwal, R. Roetter, N. Kalra, S. K. Bandyopadhaya, S. Prasad and H. van Keulen. 2003. Modelling the quantitative evaluation of soil nutrient supply, nutrient use efficiency, and fertilizer requirements of wheat in India. *Nutrient Cycling in Agroecosystems* 65: 105–13.

Pypers, P., W. Bimponda, J. P. Lodi-Lama, B. Lele, R. Mulumba, C. Kachaka, P. Boeckx, R. Merckx, and B. Vanlauwe. 2012. Combining mineral fertilizer and green manure for increased, profitable cassava production. *Agronomy Journal* 104 (1): 1–10.

Sattari, S. Z., M. K. van Ittersum, A. F. Bouwman, A. L. Smit and B. H. Jansen. 2014. Crop yield response to soil fertility and N, P, K inputs in different environments: testing and improving the QUEFTS model. *Field Crops Research* 157: 35–46.

Susan John, K., C. S. Ravindran and James George. 2005. Long Term Fertilizer Experiments. Three Decades Experience in Cassava. CTCRI, Thiruvananthapuram, Kerala, India. 83 p.

Utomo, W. H., Marjuki, Wargiono, Koes Hartoyo, Suharjo, E. Retnaningtyas, D. Santoso, A. Wijaya and R. Howeler. 2010. Enhancing the adoption of improved cassava production and utilization systems in Indonesia (The ACIAR Cassava Project in Indonesia). In R. H. Howeler (ed.), *A New Future for Cassava in Asia: Its Use as Food, Feed and Fuel to Benefit the Poor*. Proc. 8th Regional Workshop, held in Vientiane, Lao PDR, 20–24 October 2008, pp. 490–507.

Witt, C., A. Dobermann, S. Abdulrachman, H. C. Gines, W. Guanghuo, R. Nagarajan, S. Satawatananount, T. H. Son, P. S. Tan, L. V. Tiem, G. C. Simbahan and D. C. Olk. 1999. Internal nutrient efficiencies of irrigated lowland rice in tropical and subtropical Asia. *Field Crops Research* 63: 113–38.

Chapter 16

Intercropping and crop rotations in cassava cultivation: a production systems approach

Thomas W. Kuyper, Wageningen University, The Netherlands; and Samuel Adjei-Nsiah, International Institute of Tropical Agriculture (IITA), Ghana

1 Introduction
2 Advantages of cassava as a crop
3 Crop rotations in cassava
4 Cassava intercropping
5 Yield increases through cassava intercropping
6 Achieving yield increases in cassava intercropping
7 Soil fertility enhancement through cassava intercropping
8 Farmer experimentation during cassava cropping
9 Cassava functional traits
10 Future trends and conclusion
11 Where to look for further information
12 References

1 Introduction

Cassava is the sixth most important food crop globally (Burns et al., 2010) and it is grown in a highly diverse array of cropping systems. This diversity, which reflects the versatility of cassava, includes smallholder farming systems, but also larger-scale plantations (for animal feed, biofuels). Cassava does not have a specific and determined maturity period and it can therefore be harvested over a range of ages, depending on cropping system and socio-economic conditions. In general the cropping cycle of sweet cassava, especially on the fertile dark Amazonian soils, is much shorter than that of bitter cassava, which usually grows on oxisols and ultisols, in the same area (Fraser et al., 2012; Jakovac et al., 2016). There is a tendency in Africa towards shorter cycles of cassava in conjunction with increased commercialization of the crop (Fermont et al., 2010a).

In this chapter we will review the current knowledge on mixed cropping systems (both intercropping and crop rotations) in which cassava plays a prominent role. We will discuss yield benefits of such mixed cropping and evaluate mechanisms that have been proposed

http://dx.doi.org/10.19103/AS.2016.0014.26

to explain superior yields of such systems. We will also list the functional traits of different cassava varieties that likely explain why cassava performs well in intercropping.

2 Advantages of cassava as a crop

The cultivation of cassava is a major activity for many smallholder farmers in the tropics. Benefits of cassava cultivation for resource-constrained smallholder farmers are given by (Adjei-Nsiah et al., 2004, 2007, 2017).

2.1 High productivity

Cassava is highly productive relative to other crops, especially on soils of low fertility (Fresco, 1986; Dixon and Ssemakula, 2002). This productivity on poor soils has given rise to the faulty understanding that cassava is a major cause for soil fertility decline (Sitompul et al., 1992). The fact that cassava grows well on poor soils and that cassava cultivation area is increasing has led Hillocks (2002) to link both phenomena to declining soil fertility levels in Africa. Fermont et al. (2010a,b) argued that declining levels of soil fertility, rather than labour shortage or food shortage, drove the transformation towards cassava-based systems in Uganda.

2.2 Low fertilizer demand

Cassava seems to need little or no mineral fertilizer, except K (but see Howeler, Chapter xx); in fact, cassava does not always respond to nitrogen (N) fertilizer and sometimes shifts carbon allocation towards above-ground tissues, making it excessively leafy (Mutsaers et al., 1993) after N fertilization, with reduced yields of storage roots.

2.3 Drought tolerance

Cassava is drought tolerant, and can withstand several dry months after it has become well established in the first critical months. Cassava cultivation therefore may be part of a risk management strategy. The increasing likelihood of drier periods as a consequence of climate change could also make cassava a crop of increasing importance (Burns et al., 2010; Jarvis et al., 2012). As many other staple foods are likely to be negatively impacted by climate change, the importance of cassava as a major food will likely increase. However, higher temperatures could negatively impact cassava productivity (Adjei-Nsiah and Kermah, 2012). The increase in atmospheric CO_2 may have divergent effects on cassava – its high nutrient use efficiency may cause an increased yield (Fernandez et al., 2002), but with potentially lower nutritive value (Gleadow et al., 2009).

2.4 Low labour demand

Cassava is claimed to have a low labour demand. Before canopy closure (the first 3–6 months) intensive management is required, which makes cassava not as labour-extensive as often claimed (Fermont et al., 2010b). In the early stages labour demand can in fact even be higher than that for other crops. Harvesting can also be labour intensive.

Cassava allows a very flexible harvesting calendar, although planting needs to be done during the rainy season, or under irrigation. Staggered planting and harvesting may not only reduce constraints on labour availability, but also allow harvesting of small amounts any time it is needed over a long period, even up to two to three years.

2.5 Income generation

Income from cassava can be high. In Ghana, Adjei-Nsiah et al. (2007) showed that cassava rotation with maize had the highest return on investment. This beneficial effect was more due to the high monetary value generated by cassava than by the yield increases of subsequent maize. Also Fermont et al. (2010a) showed for Uganda that cassava generated higher income than other crops. In that study, wealthier farmers tended towards cassava mono-cropping, whereas the less well-endowed farmers maintained a larger part of their cassava fields as intercrop. Cassava also can be used as animal feed (Thailand) or biofuel (Brazil) or for local beers (Ghana). Currently, however, market linkages in Africa are underdeveloped for cassava, despite its potential as raw material for starch and ethanol (Adjei-Nsiah and Sakyi-Dawson, 2012). But with increasing urbanization, cassava may become (depending on relative food prices and food cultural preferences) a major product for the rapidly increasing urban populations (Kamau et al., 2011).

2.6 Suitability for female households

Cassava often has a prominent role in female-dominated households relative to other crops (Berry, 1993; Nweke et al., 2002); also the processing of cassava is a female-dominated activity (Chiwona-Karltun et al., 1998).

2.7 Options for intercropping

Widely spaced cassava cuttings create options for intercropping, which is intended either to suppress weeds or to grow food crops among cassava plants. If the plant stays longer in the field than the other crops, further photosynthetic carbon can be allocated below-ground until storage capacity is saturated. This factor explains why cassava, as a perennial plant, can be integrated into annual cropping systems, but can also form part of cross-year systems. It is easier to keep cassava in the field than to harvest and store the roots, especially since the roots begin to show physiological deterioration within a few days of harvest. In that sense, apart from being a source of daily food, a field of cassava is like money put in the bank – where some plants can be harvested when needs for cash increase. Other crops must be harvested at specific periods of maturity; without adequate storage there are periods of high availability (with low prices) and periods of low availability (when food reserves have run low).

By this combination of beneficial properties, even relatively poor households with not more than one or two hectares of agricultural land can achieve food security almost across the year, something that cereals are far less likely to achieve. These benefits explain why an exotic crop has now become the second-most important source of calories in Africa (after maize), with more than 500 million Africans deriving a major portion of their calorie intake from cassava (Nweke and Haggblade, 2010). Cassava has not always been popular among R&D institutions, nor a priority for national policies. Not being an export-oriented cash crop for most countries, such crops have often been neglected by policy and academia (Scott et al., 2000). However, times are changing. A book by Nweke et al. (2002) was one of the early works to reflect the increased interest in cassava as a flexible crop that could make major contributions to both food security and income generation in Africa.

However, it has been suggested that some generalizations on cassava cropping do not have sufficient empirical support. Fermont et al. (2010a) investigated five such beliefs: (1) cassava is a subsistence crop grown to avoid hunger; (2) cassava is grown by poor farmers; (3) cassava is mainly grown as intercrop, and less so as monocrop; (4) cassava cultivation is less labour demanding than most other crops; and (5) cassava requires no nutrient inputs due to its ability to grow on the poorest soils. The persistence of these beliefs, without empirical validation, could be problematical for understanding the changing role of cassava in agriculture (see Future trends and conclusion).

3 Crop rotations in cassava

Crop rotation is recommended in cassava cultivation to minimize nutrient exhaustion, weed pressure, disease build-up, and soil loss. Rotation may also result in soil organic matter build-up and promote biological soil activity. Rotation therefore maintains cassava productivity. In Colombia, Leihner and Lopez (1988) reported that within a nine-year period, fresh cassava root yield dropped from 37 to 12 t ha^{-1} when cassava was sole-cropped. In the same experiment, after five-year farmer-managed cassava sole-crop cycles, moderate fertilization did not result in increased productivity. However, when cassava was rotated with *Crotalaria juncea* (Fabaceae) as green manure, maize, cassava, maize, common bean, sorghum and cassava again, cassava yield returned to 30 t ha^{-1}. They reported that soil nutrients were not depleted under continuous cropping of cassava but that biological soil degradation resulting from continuous cropping had rendered the soil nutrients unusable by cassava. Soil organic matter under continuous cropping of cassava showed a tendency to decline. After four years of rotation, soil organic matter increased again. Spore counts of arbuscular mycorrhizal fungi were 36% higher than in continuous cassava cropping.

In rotations involving cassava it is not only the yield of cassava that is positively influenced by the system but also the yield of the crops in rotation with cassava. In a 12 years' rotational experiment in Northern Ghana, Agyare et al. (2006) reported that the best preceding crops for maize in descending order were: groundnut (3.8 t ha^{-1}), cassava (3.5 t ha^{-1}), soybean (3.4 t ha^{-1}), cowpea (3.2 t ha^{-1}), and sorghum (1.9 t ha^{-1}). They reported that a cassava-maize rotation gave the highest total energy value, indicating that cassava was the most compatible crop with maize. In the Guinea/savanna transitional agro-ecological zone of Ghana, Adjei-Nsiah et al. (2007) reported a higher maize grain yield when cassava preceded maize than when grain legumes such as pigeonpea, cowpea and groundnut preceded maize. They attributed the higher maize grain yield after cassava to the amount of nutrients returned to the soil after cassava through litter fall and green leafy biomass that was incorporated in the soil after the cassava harvest. In a rotational system involving maize with other crops such as cassava, pigeonpea, cowpea, groundnut and maize, in the semi-deciduous agro-ecological zone of Ghana, Adjei-Nsiah (2010, 2012) reported that the best preceding crops for maize in descending order were: pigeonpea, cassava, groundnut, cowpea and maize.

4 Cassava intercropping

While the literature often proposes a clear separation between crop rotations and intercropping, such a dichotomy is not always applicable in the case of cassava. Due to its long duration (12–24 months) several companion crops can be grown with cassava, either simultaneously or sequentially.

Since the review on cassava-based intercropping by Mutsaers et al. (1993) more than two decades ago, no new synthesis seems to have been published. The review concluded that in cassava intercropping the relative yield total (RYT) or land equivalent ratio (LER) is usually above 1, a phenomenon known as overyielding. However, mechanistic understanding of the benefits of cassava intercropping has made little progress.

Cassava has been intercropped in many different combinations:

- With other tuber crops (yam or sweet potato)
- With cereals (maize, upland rice)
- With pulses (cowpea, common bean, groundnut, soybean, pigeonpea)
- With maize and pulses (common bean; cowpea (but used for weed control and as green manure); groundnut; soybean)
- With commercial vegetable crops (egusi melon, pumpkin, okra, pepper, cotton, sunflower)
- With cover crops (or forage leguminous crops) as an additional contribution to weed suppression (*Crotalaria spectabilis*, *Pueraria phaseoloides*, *Canavalia ensiformis*, *Mucuna pruriens*, *Stylosanthes guianensis*, *Centrosema pubescens*)
- With fruit plants and trees (banana, pineapple, passionfruit, coffee, oil palm, coconut)
- With fruits (as above) and maize and cover crops
- With agroforestry trees (*Gliricidia sepium*, *Leucaena leucocephala*) that provide N-rich mulch in alley-cropping systems.
- With a diversity of crops as in traditional polycultures, that is intercropping arrangements without a predefined spatial arrangement
- With tree species to enhance revegetation or natural restoration on small farms in order to generate income in the first stages of afforestation (Martinotto et al., 2012).
- With any fallow plant (jachère de manioc, cassava fallow – see below).
- Cropping systems with two or more cassava varieties of different architecture also constitute a form of intercropping.

This list of intercrop combinations shows the versatility of cassava. In many, though not all cases, the intercrop is of a much shorter duration than cassava. But late-maturing crops can also be combined with cassava, and arrangements of both early- and late-maturing crops in combination with cassava have been recorded from Africa (Mutsaers et al., 1981).

5 Yield increases through cassava intercropping

Benefits or disadvantages of intercropping can be expressed in absolute or relative terms. Because of the very different nature of the harvested products, expressing benefits in relative terms is preferred. The most common parameter is RYT or LER, a parameter that expresses the sum of the relative yields of all crops when grown together compared to the summed yields when grown separately. An RYT larger than 1 is an indication of a relative yield advantage of intercropping compared to mono-cropping, but does not specify which crop benefits, or benefits more, from the intercropping arrangement.

Farmers, however, do not intercrop with the intention to increase RYT, irrespective of the crops concerned, but are specifically interested in the yield benefits or disadvantages of specific intercropping arrangements. For that reason the design of experiments by ecologists to study the underlying mechanisms of overyielding differs from the ways farmers intercrop. In ecological experiments, benefits of intercropping are usually studied in a *substitutive design*, whereby total plant density is held constant. Cassava intercropping under farmer conditions usually follows an *additive design*. In this design, cassava density is held constant at 10 000 plants ha^{-1}, with no difference in density between cassava monocrop and cassava intercrop; the other intercrop is then planted in variable densities. Farmers evaluate the extent to which their cassava harvest is reduced in this intercropping arrangement, and how this reduction is or is not compensated for by the yield of the second crop. The prevalence of additive intercropping designs with cassava follows from the fact that cassava yield in farmers' fields usually does not exceed that density of 10 000 plants ha^{-1}; at that density there is substantial unused space and light in the first months of cassava growth. Depending on the duration of the intercropping phase, the competitive interactions of cassava with the intercrop(s) and the ability of cassava to recover from initial competitive interference, the cassava yield is usually (somewhat) decreased, but the beneficial effect of the yield of the companion crop compensates for that yield loss. In exceptional cases cassava produces more as an intercrop than as sole crop. Cenukdee and Fukai (1992a) reported the intercrop combinations of soybean and small-statured cassava. The underlying mechanism was not described.

The literature abounds with data on RYT in cassava intercropping. Mutsaers et al. (1993) provided an extended table with RYT from various intercropping experiments. From their table RYT was almost always higher than 1, and reached values slightly above 2 in several cases. Modern studies have reported even higher RYTs, especially in cases of multiple-crop intercropping. The currently highest reported RYT is 3.12 for a system with cassava, maize, cowpea and melon (Udoh and Ndaeyo, 2000) in south-eastern Nigeria. In a Ghanaian intercropping system with maize, soybean, cowpea and cassava, RYTs up to 2.83 were reported (Ennin et al., 2001; Dapaah et al., 2003). In Brazil, Neto et al. (2014) described a polyculture of cassava, maize, passionfruit and pineapple, to which green manures from legumes were added, and reported RYT between 2.45 and 2.77.

A complication with the use of RYT as an indicator for intercropping benefits is the unequal period that both crops remain on the land. Considering combinations of a short-growing annual crop (cereals, legumes) with a long-duration cassava, the standard way of calculating RYT might be biased towards intercrop combinations that show large differences in phenology (temporal niche differentiation). For that reason Mutsaers et al. (1993) introduced the Area × Time Equivalency ratio (ATER). Applying ATER for cassava intercropping results in lower values than for RYT, although values still remain above unity. However, applying the correction for differential crop growth duration is not without problems, especially in cases of extended dry periods, where ATER may give low values. Subtracting the dry season can generate problems, especially in regions with a bimodal rainfall pattern. Here, including or excluding the short rainy season from ATER produces different results. With the human impact on climate, it is likely that many tropical and sub-tropical sites will receive less and less predictable rainfall; and claims on the benefits of cassava intercropping may then rely strongly on the way in which changes in dry periods are or are not discounted in ATER. Since RTY and ATER are usually correlated, we will therefore continue to use the RYT, but with the caveat that the absolute values should be interpreted with caution.

6 Achieving yield increases in cassava intercropping

While many studies have described yield increases in cassava intercropping, only very few tested specific mechanisms are responsible for that yield increase. From the literature on yield increases in intercropping (Li et al., 2014; Brooker et al., 2015) various mechanisms have been proposed. The relative importance of these mechanisms has not yet been evaluated. Our list is therefore descriptive and should be considered as an invitation for mechanistic research (as is currently taking place at IITA), in which plant traits (see below) as causal factor should be incorporated. Mechanisms for benefits with cassava intercropping include the following.

6.1 Vertical niche differentiation

Cassava has been reported to have the ability to form deep roots, thereby avoiding competition with companion (annual) crops by vertical niche differentiation. Cassava could also be able to regenerate nutrients that have leached to deeper soil layers.

6.2 Temporal niche differentiation

When cassava cuttings are planted, a certain time period is needed for adjustment of their shoot to root ratio. In this initial period light interception is low and there is substantial unused space for leaf development by companion plants. If such plants can complete their cycle before cassava forms a closed canopy, more complete light utilization can be achieved. But cassava can suffer from light limitation in intercropping. The negative relation between cassava yield and growth period of the associated soybean suggests that the temporal window to escape from competition for light becomes smaller, the longer the period of crop overlap. A major conclusion from intercropping studies is that RYT is better correlated with cassava yield than with the yield of the companion crop, implying that efficient intercropping combinations are those that have only small impacts on cassava yield. Apparently there is competition for resources as long as both crops grow together, but after harvest of the companion crop, cassava has the ability to recover. It is this ability to recover from competition that makes intercropping successful, but the mechanism of recovery and the conditions under which recovery is likely need further study. Unfortunately there are no published studies where competitive dynamics in intercropping (in terms of light capture or nutrient acquisition) have been studied, following the protocol of Trinder et al. (2013). Mutsaers et al. (1993) suggested that there is a threshold beyond which maize is too competitive to allow recovery by cassava; unfortunately, this concept of a competitive threshold has not yet been fully tested experimentally. The existence of a threshold is consistent with a hypothesis of a *temporally limited plasticity* of cassava to recover from competition for light. Because the threshold is different with legumes or cereals (Mutsaers et al., 1993), it may be possible that next to competition for light, competition for below-ground resources (including space) is important. Possibly a systematic comparison of different legumes of different statures (ranging from short-cycle cowpea to longer-cycle pigeonpea), planted prior to, simultaneously with, or one or several weeks later than cassava, will help resolve the issues of competition for light and for below-ground resources in relation to temporal niche differentiation.

6.3 Complementarity with other crops

Complementarity has been described from cropping systems with cereals and legumes. Due to their ability to fix atmospheric N_2, legumes do not compete with cereals for N, thereby leaving a larger N pool for cereals. Comparison of RYT in various intercropping combinations indicates that maize–cassava combinations are on average more productive than maize–legume combinations. That difference would suggest that below-ground niche differentiation is due rather to P or K than to N. Only two studies have suggested a cassava benefit due to N_2 fixation by a legume (Njoku et al., 2010; Umeh and Mbah, 2010). Whereas cassava takes little N and P from the soil, it takes up substantial amounts of K and cassava cropping can become K-limited, as shown by fertilizer trials (Fermont et al., 2009). A recent study by Ezui et al. (2016) indicated that for yields of up 8 ton ha^{-1} dry weight (equivalent to 20–25 ton fresh weight), the main limiting nutrient is K.

6.4 Facilitation

Facilitation describes situations whereby one plant increases the availability of a limiting nutrient in a way that benefits not only itself, but also its companion species. Facilitation has also been noted in legume–cereal interactions, with legumes producing organic anions (carboxylates) that increase phosphorus availability for the cereal; and for iron, whereby the ability of grasses to enhance iron availability in calcareous soils increases iron uptake by the legumes. The literature on cassava intercropping has not yet provided clear examples of facilitation.

6.5 Weed suppression

Intercropping may be a way to reduce weed pressure. In the initial growth stages of cassava, competition with weeds, especially rhizomatous grasses like speargrass (*Imperata cylindrica*), can result in substantial cassava yield reductions (Fermont et al., 2010a). Under these conditions cover crops can be beneficial (Ekeleme et al., 2003; Chikoye et al., 2001), although some cover crops can also result in cassava yield reductions (Chikoye et al., 2002). Weed suppression may be beneficial in cassava–pepper intercropping (Olasantan et al., 2007), resulting in higher pepper yields. A study in Brazil used cowpea to manage weeds in a cassava–cowpea intercropping system. The cowpea was not harvested, indicating the dual nature of various cowpea varieties to suppress weeds, to impact soil fertility and/or to yield grains (cf. Vissoh et al., 2008).

6.6 Reduction of erosion

Although cassava is frequently believed to contribute to erosion (Howeler, 1991), Hulugalle and Ezumah (1991) noted that cassava protected the soil against run-off due to its extensive canopy development. On the other hand a study in Nigeria by Odemerho and Avwunudiogba (1993) indicated substantial increases in erosion when polycultures with cassava were replaced by cassava monocultures. They also noted that traditional planting practices against erosion were more successful with polycultures than with monocultures. Ijima et al. (2004) noted from Indonesia that cassava mono-cropping resulted in the highest erosion rates. Ardjasa et al. (2001) found high erosion under cassava mono-cropping, especially during the first months when soil cover was still low; 90% of the annual soil loss happened during the first four months.

6.7 Enhancement of beneficial soil microbiota

Cassava is a crop that is highly dependent on and responsive to the arbuscular mycorrhizal symbiosis (Habte and Byappanahalli, 1994). Cassava could therefore build-up and maintain mycorrhizal inoculum from which the companion crop(s) benefit. Conversely, the build-up of mycorrhizal inoculum by the intercrops could also be beneficial for cassava. Enhancement of beneficial soil biota (e.g. earthworms) could also be due to microclimate improvement, such as lower soil temperatures and higher soil moisture (Olasantan et al., 1996).

In conclusion this list suggests many potential mechanisms for overyielding, but the relevance of these has in most cases not been assessed experimentally. Our interpretation from the literature suggests that more complete light use (temporal niche differentiation), weed suppression and enhanced soil biological activity likely constitute the more important mechanisms. Data from Benin (Saïdou et al., 2004) indicate that farmers mention similar mechanisms for soil fertility enhancement, including the closed canopy (less soil erosion, moderated soil temperatures, enhanced soil moisture and increased biological activity).

7 Soil fertility enhancement through cassava intercropping

Claims that cassava regenerates soil fertility (and therefore fulfils the same role as trees in bush fallow) are widespread in Africa (Saïdou et al., 2004; Adjei-Nsiah et al., 2004, 2007; Fermont et al., 2008). For that reason cassava not only is intercropped, but can also remain in the cropping cycle at the end when the land would otherwise be laid fallow. In West Africa this system is known as *cassava fallow* or, in French, *jachère de manioc*. In Benin the system was first described by Daane et al. (1997) and Totongon et al. (2000). Most descriptions of cassava fallow are from West Africa (Ghana, Benin, Nigeria), but a similar system is known from Uganda under the name *imitation fallow* (Fermont et al., 2010a). The term *cassava fallow* has also been used for indigenous agricultural systems in Honduras (House, 1997). It is likely that similar systems occur in parts of Amazonia, to judge from studies by Hecht and Posey (1990).

At first sight, the term *cassava fallow* seems an oxymoron – fallows being considered the period when no crops are grown. For that reason Fermont et al. (2010a) replaced the term *cassava fallow* with that of *imitation fallow*. We do not consider this a terminological improvement, mainly because the alternative terminology assumes an absolute distinction between the agricultural period and the fallow period. But a strict separation may be artificial – when useful crops continue to grow after the land has been abandoned and soil tillage comes to a stop before planting of new crops takes place. As cassava can remain on the land for up to 24 months, there may still be cassava after the end of what would otherwise be considered the cropping phase. And when weed pressure remains low, the cassava on that land may produce food while contributing to soil fertility recovery. In that respect cassava has partly replaced bush fallow. A benefit of cassava fallow over bush fallow is that it can shorten the fallow period; Vanlauwe et al. (2013) noted for Congo an alternation of cassava growth of 1–2 years and a subsequent fallow period of only 2–4 years. Soil data indicating recovery under two-year cassava fallow were provided by Okore et al. (2007). They noted increased labile soil organic C and N and higher microbial biomass C and N in a cassava fallow compared to a continuous cropping system. Unfortunately, the paper did not provide data on cassava planting density, nor on cassava yield.

The use of cassava fallow is not accessible to all farmers. In the areas of West Africa where cassava is commonly grown, a major influx of people from neighbouring regions and countries has taken place. Although these new immigrants are integrated in agriculture, a study by Adjei-Nsiah et al. (2008) indicated that cassava growing as part of soil fertility regeneration was much more widespread among native farmers (82%) than among immigrant farmers (44%). For bush fallow the differences are even larger, suggesting that land tenure regulations are the basis of access to such soil fertility management practices. Differences in valuation of cassava fallow have also been noted for Benin (Saïdou et al., 2004).

A question that has not been addressed is the extent to which different cassava varieties are more or less preferred by farmers for cassava fallow. Considering the high planting density (17 000 plants ha^{-1}) in cassava fallow, it is possible that certain varieties with specific traits are preferred.

8 Farmer experimentation during cassava cropping

One interesting benefit of cassava cropping systems, with their continuous harvesting times over a longer period, is that they provide opportunities for farmer experimentation that most annual cropping systems do not easily provide. For example, although rarely practised on a global basis, farmers have an opportunity to generate new genetic diversity through propagation of true seed. This will in theory allow new options for variety traits that best fit into farmers' preferred cropping systems. Basically the on-farm *breeding* system consists of farmers allowing certain plants to flower and be pollinated by natural means. (Many cassava varieties flower and set seed, but neither flowers and fruits nor dried seeds typically have value to the farmer.) When seeds set, they fall to the soil and generate new genotypes (all other cassava plants are reproduced clonally). These volunteer seedlings are then allowed to grow and are harvested at the end of the cropping season. If such volunteers are considered by farmers to have desirable traits, these new genotypes are subsequently reproduced through cuttings. By this intended (but poorly controlled process, since fertilization is random) novel genotypes are generated that allow farmers to adapt to changing conditions.

Such systems have been extensively described from the Amazon region, where farmers use an elaborate technology to balance risks of inbreeding and outbreeding (Duputié et al., 2009), but they also occur in western Africa (Manu-Aduening et al., 2005). In cassava fallow there are possibly more chances for volunteer seedlings and some of these could be tested and actually incorporated in next cycles.

9 Cassava functional traits

The benefits of intercropping for cassava yield show large differences between different cassava varieties. Classification of cassava varieties with relevant functional traits as a basis for understanding this differential performance is currently under study. Functional traits have received only little attention in the agro-ecological literature (Garnier and Navas, 2012; Martin and Isaac, 2015).

Literature on cassava in the early 1990s includes several papers that compare large numbers of cassava varieties. Descriptions of those varieties in terms of relevant functional traits are often so brief that we cannot yet link this literature to a trait-based framework. A directly selected trait like cyanogenic potential (the contrast of bitter and sweet cassava) may affect diverse characteristics such as leaf N content, leaf decomposability and N cycling (Zhao et al., 2015). They showed that linamarin (the cyanogenic glucoside from which HCN is released) is important for N resorption under drought.

From the perspective of cassava intercropping, relevant cassava traits include both those that determine the ability of cassava to yield well (exhibit only small yield reductions) in competition with intercrops, and those that determine the intercrops that yield well in competition with cassava. Clearly both sets of traits have trade-offs, for example, high competitive ability of one intercrop reduces the performance of the other. In this review we focus on potentially important cassava functional traits for intercropping.

9.1 Leaf nutrient content and leaf litter quality

Cassava leaves are rich in N, and leaf litter retains high amounts of N. According to Borin and Frankow-Lindberg (2005) cassava leaves contain 35–50 mg N g^{-1}. Cassava leaves are poor in lignin, and the combination of both traits results in high decomposition and hence high N flux rates that are likely beneficial for subsequent crops in rotation (Adjei-Nsiah et al., 2007). The actual sources for this high N content are currently unknown. Possible mechanisms include N_2 fixation by asymbiotic, free-living bacteria, the ability to prevent leaching through a deep-rooting system and/or the ability to acquire N from deep-water sources. The occurrence of N_2-fixing bacteria has been reported (Balota et al., 1999; Teixeira et al., 2007). No ^{15}N data are available to unambiguously demonstrate significant N_2 fixation by cassava. Cassava leaf litter also recycles substantial amounts of P and K (Fermont et al., 2010b).

9.2 Leaf retention and longevity

Both leaf N and leaf longevity form part of the leaf economics spectrum (Reich, 2014). The two traits (leaf N and leaf longevity) should logically be negatively correlated; however, Lenis et al. (2006) showed evidence that leaf retention was not negatively correlated with other useful agronomic traits. Cock et al. (1979), in their study of the ideal cassava phenotype, noted that higher cassava yields would be achieved with longer leaf life span. Leaf longevity determines not only the timing and the amount of N returned to the soil, but also the extent to which cassava can adapt to drought in more seasonal climates. The existence of genetic variation in that trait indicates potential for genetic improvement (Kawuki et al., 2011). The fate of N in periods of drought (e.g. resorption before senescence) seems also unknown, but it has been suggested that linamarin is important in N resorption (Zhao et al., 2015). The fact that cassava leaves are rich in N allows substantial N returns to the soil in terms of easily decomposable, N-rich litter. It is therefore not surprising that cereal crops such as maize, whose growth limitation in the West African savanna is usually N, show high productivity after a cassava fallow.

9.3 Plant architecture

Plant architecture is determined by three fundamental parameters: total plant height, height of the first branching point, and number of branching levels. Plant architecture determines the extent to which cassava can outshadow other plants, intercrops and weeds (and the extent to which it can create a suitable microclimate for more shade-tolerant intercrops and a more favourable microclimate for soil fauna). Next to variation in architecture there may be variation in the degree of plasticity. Cassava plants may change their branching after the intercrop has been harvested and light availability increases, resulting in recovery from competitive suppression. Branching habit (as well as architecture of the root system) could also be partly controlled by the arbuscular mycorrhizal symbiosis. Branching pattern and height have been mentioned as important traits by Mutsaers et al. (1993).

9.4 Timing of carbon allocation to fine and storage roots

This timing determines when the first storage roots are formed and how long storage roots continue to increase in mass. This timing is likely affected by stoichiometric relations, the relative conversion of photosynthetic carbon and nutrients into storage roots, and therefore correlated with leaf (N content, leaf longevity) and stem traits (maximum canopy height, branching architecture). A study by El-Sharkawy and De Tafur (2010) indicated that canopy height and the time after planting that cassava begins to form storage roots are positively correlated. Both are also positively correlated with total leaf area and total yield of storage roots. A corollary of their study was the suggestion that in short and in short-duration (early bulking) varieties, leaf area was less than optimal for maximum yield (although harvest index was higher in varieties of shorter stature, as also shown by Egesi et al., 2007). However, other short-duration varieties have been stated to be higher yielding than the more traditional varieties (Kamau et al., 2011).

9.5 Rooting depth

Cassava is generally considered a species with relatively superficial roots, with the largest part of the root system being in the uppermost 60 cm (Iijima et al., 2004). In a study by Subere et al. (2009) genetic variation in rooting depth was small. A study by Muhr et al. (1995a) equally indicated that, depending on variety, 58–74% of cassava root length was in the uppermost 30 cm. In intercropping there was little or no vertical niche differentiation between cassava and the intercrop (an observation that is consistent with other reports that the importance of vertical niche differentiation as an explanation for overyielding has been overestimated). A subsequent report by the same authors (Muhr et al., 1995b) indicated both the potential and the actual occurrence of severe competition during early stages of cassava growth. There are reports that cassava roots can go much deeper, up to 200 cm (El-Sharkawy et al., 1992; El-Sharkawy, 2004) or 260 cm (Connor et al., 1981). However, on acid soils, most cassava roots are likely to remain in the upper soil layers where soil organic matter levels are higher and Al toxicity is less of a problem. Deep rooting more likely occurs in coarse-textured soils than in the clayey oxisols and ultisols. Deep roots could allow cassava to acquire water from layers inaccessible to other plants, which might contribute to its higher drought tolerance (El-Sharkawy, 2007); deep roots could also be important as a safety net against N leaching. To the best of our knowledge there are no isotopic data that demonstrate that cassava has access to deep-water sources. Also the drought tolerance of cassava is not clearly linked to its deep-rooting habit. Cassava drought sensitivity (and risk of substantial yield reductions) in the first months after planting may be related to its still relatively sparse and shallow root system (Baker et al., 1989; Bakayoko et al., 2009). Its remarkable capacity to recover from drought (El-Sharkawy, 1993) is not evidently related to its ability to produce deep roots. Whether there is also hydraulic redistribution of water (as a mechanism of facilitation to explain high RYT in intercropping) is currently unknown.

9.6 Root extension and root branching

Subere et al. (2009) reported genetic variation in lateral root development and root branching, but how this variation can be harnessed for productive intercropping systems needs more attention.

9.7 Interaction with arbuscular mycorrhizal fungi

Cassava is a highly mycotrophic plant that shows large biomass increases due to arbuscular mycorrhiza (Ceballos et al., 2013; Rodriguez and Sanders, 2015). Genetic variation in mycorrhizal responsiveness of different cassava varieties has been reported by Sieverding and Galvez (1988) and Carretero et al. (2009), while genetic variation in mycorrhizal colonization was reported by Saïdou et al. (2009, 2012). Peña-Venegas (2015) reported small differences in fungal species composition between bitter and sweet cassava, when grown in the same field in the upper Amazon near Leticia, Colombia. In that area, at least 126 different species of arbuscular mycorrhizal fungi were identified with molecular methods in cassava plots. Many of these species can also form common mycorrhizal networks with other crops in intercropping systems (Cardoso and Kuyper, 2006). The mycorrhizal symbiosis also impacts on the production of strigolactones, a class of hormones that have a major impact on shoot and root branching.

9.8 Compatibility with other crops

Cassava varieties that produce most above-ground biomass in monoculture are both most successful in suppressing weeds and competitively superior to other intercropped plants. It goes without saying that the functional traits of the companion intercrop in terms of compatibility need equally to be studied – as it may ultimately be the match or mismatch of traits that determines the conditions under which intercropping results in increased relative yields of cassava and/or the intercrop. Compatibility is also relevant in cases where different cassava varieties are intercropped (Daellenbach et al., 2005). Okoli et al. (1996) compared three different cassava varieties with two different cowpea varieties. Both cowpea varieties reduced cassava height, but this resulted in different yield reductions for the different cassava varieties. Cenpukdee and Fukai (1992c) compared above-ground competition between cassava and pigeonpea, using two different cassava varieties with different growth architecture. Their study showed the importance of planting dates (as a means for temporal niche differentiation) and cassava architecture (as related to differential

Box 1 A typical cassava intercropping system within smallholder farming system in forest/savanna transitional agro-ecological zone of Ghana

Within the root-crop farming system in the forest/savanna transitional agro-ecological zone of Ghana, a typical cassava intercropping system consists of three main crops: yam, cassava and maize. The first crop to be planted after land preparation is yam. However, after preparation of the yam mounds and before the yam sets are planted, cassava cuttings are first planted in the mounds at the periphery of the field. After the yam has sprouted, maize is planted followed by cassava on the lower portions of the mound in the main part of the field. Sometimes, pigeonpea may be planted after the cassava and maize have been planted. The yam is harvested first, usually after a 9- to 10-month growing period followed by maize, and then cassava is planted at the periphery of the field. This is the first cassava crop that is harvested to fill the hunger gap before the main cassava crop is harvested later. Where pigeonpea is included in the system, it is harvested during the dry season. After the pigeonpea is harvested it is cut back to reduce competition with the cassava. Upon harvesting the cassava, the field is reverted to fallow under pigeonpea for one or two years before the field is cropped again.

ability to capture light depending on canopy height). Their study concluded that the highest RYTs were achieved by the taller cassava variety. These authors also studied seven cassava varieties in combination with soybean and pigeonpea (Cenpukdee and Fukai, 1992a). Potential height of cassava, in relation to sowing dates, was a major determinant of cassava competitive ability and hence final yield of storage roots. This suggests that breeding smaller-stature cassava may select against ability to grow in certain intercrop combinations. A third study with 18 cassava varieties and two legumes indicated that cassava canopy height was a main determinant of final cassava yield and intercrop compatibility (Cenpukdee and Fukai, 1992b). Unfortunately apart from stature, cassava traits were not analysed in detail.

10 Future trends and conclusion

This chapter documented various benefits that cassava, when grown as an intercrop or in crop rotations, can provide. However, cassava is not only grown as an intercrop in Africa. Fermont et al. (2010b) indicated that in Uganda and Kenya cassava was also grown as a monocrop and suggested that a shift towards cassava mono-cropping was taking place. This shift likely reflects the increased potential of cassava as a cash crop – either for food, or for feed or biofuel. It goes beyond the scope of this paper to analyse this shift. Shifts are likely to be driven by market conditions, where increased yields due to external inputs make mono-cropping more attractive. Also labour shortage may contribute to this shift. With shorter-duration varieties new options arise for alternative cropping arrangements, including newer forms of rotations that include cassava. New arrangements may well result in higher cassava yields. Suja and Sreekumar (2015) compared cassava yield in a rotational cycle and as intercrop (with cowpea) and noted significantly higher yields (root weight of

26 ton ha^{-1}) in a rotation than in the intercrop (20 ton ha^{-1}). A mechanism for increased cassava yield was not provided.

11 Where to look for further information

The versatility of cassava as a crop that fits in a very large diversity of agro-ecosystems has been well known. However, the knowledge has not moved much beyond the descriptive phase. The standard reference on cassava intercropping is still the review by Mutsaers et al. (1993). Many of the ideas in that paper are still highly pertinent. However, at the time of that review most studies on cassava intercropping lacked tests of specific mechanisms. Mechanistic understanding of the benefits of intercropping, through a variety of causes, is provided in the review by Li et al. (2014) and Brooker et al. (2015), although cassava is not represented in these papers. Also, understanding the specific traits of cassava (and its companion crops) was at that time in its infancy. An outline of the importance of traits in agro-ecosystems (even though the paper does not provide specific information about cassava) is provided by Martin and Isaac (2015). Next to biological studies on cassava cropping, research efforts should be geared towards understanding the socio-economic context in which cassava functions. Persistent myths about cassava were analysed and partly refuted by Fermont et al. (2010a).

12 References

Adjei-Nsiah, S. (2010). Yield and nitrogen accumulation in five cassava varieties and their subsequent effects on soil chemical properties in the forest/savanna transitional agro-ecological zone of Ghana. *Journal of Soil Science and Environmental Management* 1: 15–20.

Adjei-Nsiah, S. (2012). Evaluating sustainable cropping sequences withcassava and three grain legume crops: effects on soil fertility and maize yields in the semi-deciduous forest zone of Ghana. *Journal of Soil Science and Environmental Management* 3(2): 49–55.

Adjei-Nsiah, S., and M. Kermah (2012). Climate change and shift in cropping system: from cocoa to maize-based cropping system in Wenchi area of Ghana. *British Journal of Environment and Climate Change* 2: 137–52

Adjei-Nsiah, S., and O. Sakyi-Dawson (2012). Promoting cassava as an industrial crop in Ghana: effects on soil fertility and farming systems sustainability. *Applied and Environmental Soil Science* 2012; ID940954.

Adjei-Nsiah, S., C. Leeuwis, K. E. Giller and T. W. Kuyper (2008). Action research on alternative tenure arrangements in Wenchi, Ghana: learning from ambiguous social dynamics and self-organized institutional innovation. *Agriculture and Human Values* 25: 389–403.

Adjei-Nsiah, S., C. Leeuwis, K. E. Giller, O. Sakyi-Dawson, J. Cobbina, T. W. Kuyper, M. Abekoe and W. Van der Werf (2004). Land tenure and differential soil fertility management practices among native and migrant farmers in Wenchi, Ghana: implications for interdisciplinary action research. *NJAS – Wageningen Journal of Life Sciences* 52: 331–48.

Adjei-Nsiah, S., G. Asumugha, E. Njukwe and M. Akorada (2017). Root and Tuber Crop Farming Systems: Diversity, Complexity and High Potential Productivity. In: Dixon, J., Garrity, D. P., Boffa, J.-M., Williams, T. O., Amede, T., Auricht, C., Lott, R. and Mburathi, G. (eds), *Farming systems and food security in Africa: priorities for science and policy under global change*. Taylor & Francis, London (in press).

Adjei-Nsiah, S., T. W. Kuyper, C. Leeuwis, M. K. Abekoe and K. E. Giller (2007). Evaluating sustainable and profitable cropping sequences with cassava and four legume crops: effects on soil fertility and maize yields in the forest/savannah transitional agro-ecological zone of Ghana. *Field Crops Research* 103: 87–97.

Agyare, W. A., V. A. Clottey, H. Mercer-Quarshie and J. M. Kombiok (2006). Maize yield response in a long-term rotation and intercropping systems in the Guinea Savanna zone of northern Ghana. *Journal of Agronomy* 5: 232–8.

Ardjasa, W. S., H. Ando and M. Kimura (2001). Yield and soil erosion among cassava-based cropping patterns in South Sumatra. *Soil Science and Plant Nutrition* 47: 101–12.

Bakayoko, S., A. Tschannen, C. Nindjin, D. Dao, O. Girardin and A. Assa (2009). Impact of water stress on fresh tuber yield and dry matter content of cassava (*Manihot esculenta* Crantz) in Côte d'Ivoire. *African Journal of Agricultural Research* 4: 21–7.

Baker, G. R., S. Fukai and G. L. Wilson (1989). The response of cassava to water deficits at various stages of growth in the subtropics. *Australian Journal of Agricultural Research* 40: 517–28.

Balota, E. L., E. S. Lopes, M. Hungria and J. Dobereiner (1999). Occurrence of diazotrophic bacteria and arbuscular mycorrhizal fungi on the cassava crop. *Pesquisa Agropecuaria Brasileira* 34: 1265–76.

Berry, S. (1993). *No Condition is Permanent. The Social Dynamics of Agrarian Change in Sub-Saharan Africa.* The University of Wisconsin Press, Madison.

Borin, K., and B. E. Frankow-Lindberg (2005). Effects of legumes – cassava intercropping on cassava forage and biomass production. *Journal of sustainable Agriculture* 27: 1391–151.

Brooker R. W., A. E. Bennett, W. F. Cong, T. J. Daniell, T. S. George, P. D. Hallett, C. Hawes, P. P. M. Iannetta, H. G. Jones, A. J. Karley, L. Li, B. M. McKenzie, R. J. Pakeman, E. Paterson, C. Schob, J. B. Shen, G. Squire, C. A. Watson, C. C. Zhang, F. S. Zhang, J. L. Zhang and P. J. White (2015). Improving intercropping: a synthesis of research in agronomy, plant physiology and ecology. *New Phytologist* 206: 107–17.

Burns, A., R. Gleadow, J. Cliff, A. Zacarias and T. Cavagnaro (2010). Cassava: the drought, war and famine crop in a changing world. *Sustainability* 2: 352–3607.

Cardoso, I. M., and T. W. Kuyper (2006). Mycorrhiza and tropical soil fertility. *Agriculture, Ecosystems and Environment* 116: 72–84.

Carretero, C. L., M. Cantos, J. L. García, R. Azcón and A. Troncoso (2009). Growth responses of micropropagated cassava clones as affected by *Glomus intraradices* colonization. *Journal of Plant Nutrition* 32: 261–73.

Ceballos, I., M. Ruiz, C. Fernández, R. Peña, A. Rodriguez and I. R. Sanders (2013). The *in vitro* mass-produced model mycorrhizal fungus, *Rhizophagus irregularis*, significantly increases yield of the globally important food security crop cassava. *PLoS ONE* 8: e70633.

Cenpukdee, U., and S. Fukai (1992a). Cassava legume intercropping with contrasting cassava cultivars. 1. Competition between component crops under three intercropping conditions. *Field Crops Research* 29: 113–33.

Cenpukdee, U., and S. Fukai (1992b). Cassava legume intercropping with contrasting cassava cultivars. 2. Selection criteria for cassava genotypes in intercropping with two contrasting legume crops. *Field Crops Research* 29: 135–49.

Cenpukdee, U., and S. Fukai (1992c). Agronomic modification of competition between cassava and pigeonpea in intercropping. *Field Crops Research* 30: 131–46.

Chikoye, D., F. Ekeleme and U. E. Udensi (2001). Cogongrass suppression by intercropping cover crops in corn/cassava systems. *Weed Science* 49: 658–67.

Chikoye, D., V. M. Manyong, R. J. Crasky, F. Ekeleme, G. Gbèhounou and A. Ahanchedé (2002). Response of speargrass (*Imperata cylindrica*) to cover crops integrated with handweeding and chemical control in maize and cassava. *Crop Protection* 21: 145–56.

Chiwona-Karltun, L., J. Mkumbira, J. Saka, M. Bovin, N. M. Mahungu and H. Rosling (1998). The importance of being bitter – a qualitative study on cassava cultivar preference in Malawi. *Ecology of Food and Nutrition* 37: 219–45.

Cock, J. H., D. Franklin, G. Sandoval and P. Juri (1979). The ideal cassava plant for maximum yield. *Crop Science* 19: 271–9.

Connor, D. J., J. H. Cock and G. E. Parra (1981). Response of cassava to water shortage. I. Growth and yield. *Field Crops Research* 4: 181–200.

Daane, J., M. Breusers and E. Frederiks, eds (1997). *Dynamique paysanne sur le plateau Adja du Bénin*. Karthala, Paris.
Daellenbach, G. C., P. C. Kerridge, M. S. Wolfe, E. Frossard and M. R. Finckh (2005). Plant productivity in cassava-based mixed cropping systems in Colombian hillside farms. *Agriculture, Ecosystems and Environment* 105: 595–614.
Dapaah, H. K., J. N. Asafu-Agyei, S. A. Ennin and C. Yamoah (2003). Yield stability of cassava, maize, soya beans and cowpea intercrops. *Journal of Agricultural Science* 140: 73–82.
Dixon, A. G. O., and G. Ssemakula (2002). Prospects for cassava breeding in Sub-Saharan Africa in the next decade. *Journal of Food, Agriculture and the Environment* 6: 256–62.
Dixon, J., D. P. Garrity, J.-M. Boffa, T. O. Williams, T. Amede, C. Auricht, R. Lott and G. Mburathi, eds. *Farming systems and food security in Africa: priorities for science and policy under global change*. Taylor & Francis, London (in press).
Duputié, A., F. Massol, P. David, C. Haxaire and D. McKey (2009). Traditional Amerindian cultivators combine directional and ideotypic selection for sustainable management of cassava genetic diversity. *Journal of evolutionary Biology* 22: 1317–25.
Egesi, C. N., P. Ilona, F. O. Ogbe, M. Akoroda and A. Dixon (2007). Genetic variation and genotype x environment interaction for yield and other agronomic traits in cassava in Nigeria. *Agronomy Journal* 99: 1137–42.
Ekeleme, F., I. O. Akobundu, A. O. Inichei and D. Chikoye (2003). Cover crops reduce weed seedbanks in maize – cassava systems in southwestern Nigeria. *Weed Science* 51: 774–80.
El-Sharkawy, M. A. (1993). Drought-tolerant cassava for Africa, Asia and Latin America. *BioScience* 43: 441–51.
El-Sharkawy, M. A. (2004). International research on cassava photosynthesis, productivity, eco-physiology, and responses to environmental stresses in the tropics. *Photosynthetica* 44: 481–512.
El-Sharkawy, M. A. (2007). Physiological characteristics of cassava tolerance to prolonged drought in the tropics: implications for breeding cultivars adapted to seasonally dry and semi-arid environments. *Brazilian Journal of Plant Physiology* 19: 257–86.
El-Sharkawy, M. A., and S. M. De Tafur (2010). Comparative photosynthesis, growth, productivity, and nutrient use efficiency among tall- and short-stemmed rain-fed cassava cultivars. *Photosynthetica* 48: 173–88.
El-Sharkawy, M. A., A. D. P. Hernandez and C. Hershey (1992). Yield stability of cassava during prolonged mid-season water stress. *Experimental Agriculture* 28: 165–74.
Ennin, S. A., J. N. Asafu-Agyei, H. K. Dapaah and S. A. Ekyem (2001). Cowpea rotation with maize in cassava-maize intercropping systems. *Tropical Agriculture* 78: 218–25.
Ezui, K. S., A. C. Franke, A. Mando, B. D. K. Ahiabor, F. M. Tetteh, J. Sogbedji, B. H. Janssen and K. E. Giller (2016). Fertiliser requirements for balanced nutrition of cassava across eight locations in West Africa. *Field Crops Research* 185: 69–78.
Fermont, A. M., P. J. A. van Asten and K. E. Giller (2008). Increasing land pressure in East Africa: the changing role of cassava and consequences for sustainability of farming systems. *Agriculture, Ecosystems and Environment* 1288: 239–50.
Fermont, A. M., A. Babirye, H. M. Obiero, S. Abele and K. E. Giller (2010a). False beliefs on the socio-economic drivers of cassava cropping. *Agronomy for Sustainable Development* 30: 433–44.
Fermont, A. M., P. A. Tittonell, Y. Baguma, P. Ntawuruhunga and K. E. Giller (2010b). Towards understanding factors that govern fertilizer response in cassava: lessons from East Africa Nutrient Cycling in Agro-ecosystems 86: 133–51.
Fermont, A. M., P. J. A. Van Asten, P. Tittonell, M. T. Van Wijk and K. E. Giller (2009). Closing the cassava yield gap: an analysis from smallholder farms in East Africa. *Field Crops Research* 112: 24–36.
Fernandez, M. D., W. Tezara, E. Rengifo and A. Herrera (2002). Lack of downregulation of photosynthesis in a tropical root crop, cassava, grown under elevated CO_2 concentration. *Functional Plant Biology* 29: 805–14.
Fraser, J. A., A. Alves-Pereira, A. B. Junqueira, N. Peroni and C. R. Clement (2012). Convergent adaptations: bitter manioc cultivation systems in fertile anthropogenic dark earths and floodplain soils in Central Amazonia. *PLoS One* 7(8): e43636.

Fresco, L. O. (1986). *Cassava in Shifting Cultivation: A Systems Approach to Agricultural Technology*. Royal Tropical Institute, Amsterdam, The Netherlands.

Garnier, E., and M. L. Navas (2012). A trait-based approach to comparative functional plant ecology: concepts, methods and applications for agroecology – a review. *Agronomy for Sustainable Development* 32: 365–99.

Gleadow, R. M., J. R. Evans, S. McCaffery and T. R. Cavagnaro (2009). Growth and nutritive value of cassava (*Manihot esculenta*) are reduced when grown in elevated CO_2. *Plant Biology* 11: 76–82.

Habte, M., and M. N. Byappanahalli (1994). Dependency of cassava (*Manihot esculenta* Crantz) on vesicular-arbuscular mycorrhizal fungi. *Mycorrhiza* 4: 241–5.

Hecht, S. B., and D. A. Posey (1990). Preliminary results on soil management techniques of the Kayapó Indians. *Advances in Economic Botany* 7: 174–88.

Hillocks, R. J. (2002). Cassava in Africa. In: Hillocks, R. J., J. M. Tresh and A. C. Belotti (eds), *Cassava: Biology, Production and Utilization*. CABI Publishing, Oxon, UK, pp. 41–54.

House, P. (1997). Paysannerie forestière: étude de cas de la culture itinérante traditionelle au Honduras. *Réseau foresterie pour le Développement Rural* 21a: 1–27.

Howeler, R. H. (1991). Long-term effect of cassava cultivation on soil productivity. *Field Crops Research* 26: 1–18.

Hulugalle, N. R., and H. C. Ezumah (1991). Effects of cassava-based cropping systems on physicochemical properties of soil and earthworm casts in a tropical alfisol. *Agriculture, Ecosystems and Environment* 35: 55–63.

Iijima, M., Y. Izumi, E. Yuliadi, Sunyoto and W. S. Ardjasa (2004). Cassava-based intercropping systems on Sumatra island in Indonesia: productivity, soil erosion, and rooting zone. *Plant Production Science* 7: 347–55.

Jakovac, A. C. C., M. Peña Claros, R. C. G. Mesquita, F. Bongers and T. W. Kuyper (2016). Swiddens under transition: consequences of agricultural intensification in the Amazon. *Agriculture, Ecosystems and Environment* 218: 116–25.

Jarvis, A., J. Ramirez-Villegas, B. V. H. Campo and C. Navarro-Racines (2012). Is cassava the answer to African climate change adaptation? *Tropical Plant Biology* 5: 9–29.

Kamau, J., R. Melis, M. laing, J. Derera, P. Snanahan, C. Eliud and K. Ngugi (2011). Farmers' participatory selection for early bulking cassava genotypes in semi-arid Eastern Kenya. *Journal of Plant Breeding and Crop Science* 3(3): 44–52.

Kawuki, R. S., M. Ferguson, M. T. labuschagne, L. Herselman, J. Orone, I. Ralimanana, M. Bidiaka, S. Lukombo, M. C. Kanyanhe, G. Gashaka, G. Mkamilo, J. Gethi and H. Obiero (2011). Variation in qualitative and quantitative traits of cassava germplasm from selected national breeding programmes in sub-Saharan Africa. *Field Crops Research* 122: 151–6.

Leihner, D. E., and J. Lopez (1988). Effects of different cassava cropping patterns on soil fertility, crop yields and farm income. In: Degras, L. (ed.), *Proceedings of the International Society for Tropical Root Crops*. Gosier, Guadaloupe (FWL), 1–6 July 1985. Institut National de Recherche Agronomique (INRA), Paris, pp. 463–74.

Lenis, J. I., F. Calle, G. Jaramillo, J. C. Perez, H. Ceballos and J. H. Cock (2006). Leaf retention and cassava productivity. *Field Crops Research* 95: 126–34.

Li, L., D. Tilman, H. Lambers and F. S. Zhang (2014). Plant diversity and overyielding: insights from belowground facilitation of intercropping in agriculture. *New Phytologist* 203: 63–9.

Manu-Aduening, J. A., R. I. Lamboll, A. A. Dankyi and R. W. Gibson (2005). Cassava diversity in Ghanaian farming systems. *Euphytica* 144: 331–40.

Martin, A. R., and M. E. Isaac (2015). Plant functional traits in agroecosystems: a blueprint for research. *Journal of Applied Ecology* 52: 1425–35.

Martinotto, F., C. Martinotto, M. D. B. Coelho, R. A. B. E. Azevedo and M. C. D. E. Albuquerque (2012). Survival and initial growth of tree species native to the Cerrado intercropped with cassava. *Pesquisa Agropecuaria Brasileira* 47: 22–9.

Muhr, L., D. E. Leihner, T. H. Hilger and K. M. Mullers Amann (1995a). Intercropping of cassava with herbaceous legumes. 1. Rooting patterns and their potential importance for belowground competition. *Angewandte Botanik* 69: 17–21.

Muhr, L., D. E. Leihner, T. H. Hilger and K. M. Mullers Amann (1995b). Intercropping of cassava with herbaceous legumes. 2. Yields as affected by belowground competition. *Angewandte Botanik* 69: 22–6.
Mutsaers, H. J. W., H. C. Ezemuh and D. S. O. Osiru (1993). Cassava-based intercropping: a review. *Field Crops Research* 34: 431–57.
Mutsaers, H. J. W., P. Mbouemboue and Mouzong Boyomo (1981). Traditional food crop growing in the Yaoundé area (Cameroon). I. *AgroEcosystems* 6: 273–87.
Neto S. E. D., P. A. Campos, L. B. Tavelia, A. J. D. Solino and I. F. Da Silva (2014). Organic polyculture of passion fruit, pineapple, corn and cassava: the influene of green manure and distance between espaliers. *Ciencia e Agrotecnologia* 3: 247–55.
Njoku, D. N., S. O. Afuape and C. N. Ebeniro (2010). Growth and yield of cassava as influenced by grain cowpea population density in Southeastern Nigeria. *African Journal of Agricultural Research* 5: 278–28781.
Nweke, F. I., and S. Haggblade (2010). The cassava transformation in West and Southern Africa. In: Haggblade, S. and P. Hazell (eds), *Successes in Africal Agriculture: Lessons for the Future*. John Hopkins University Press, Baltimore, USA, pp. 29–70.
Nweke, F. I., D. S. C. Spencer and J. K. Lynam (2002). *The Cassava Transformation: Africa's Best-kept Secret*. Michigan State University Press, East Lansing, MI, USA; 273p.
Odemerho, F. O., and A. Avwunudiogba (1993). The effects of changing cassava management practices on soil loss – a Nigerian example. *Geographical Journal* 159: 63–9.
Okoli, O. O., M. A. Hossain, A. F. K. Kissidu and A. Asare Bediako (1996). Effect of planting dates and growth habits of cassava and cowpea on their yield and compatibility. *Tropical Agriculture* 73: 169–74.
Okore, I. K., H. Tijani-Eniola, A. A. Agboola and E. A Aiyelari (2007). Impact of land clearing methods and cropping systems on labile soil C and N pools in the humis zone forest of Nigeria. *Agriculture, Ecosystems and Environment* 120: 250–8.
Olasantan, F. O., A. W. Salau and E. E. Onuh (2007). Influence of cassava (*Manihot esculenta*) intercrop on growth and fruit yields of pepper (*Capsicum spp.*) in south-western Nigeria. *Experimental Agriculture* 43: 79–95.
Olasantan, F. O., H. C. Ezumah and E. O. Lucas (1996). Effects of intercropping with maize on the micro-environment, growth and yield of cassava. *Agriculture, Ecosystems and Environment* 57: 149–58.
Peña-Venegas, C. P. (2015). People, soil and manioc interactions in the upper Amazon region. PhD thesis, Wageningen University, The Netherlands.
Reich, P. B. (2014). The world-wide 'fast-slow' plant economics spectrum: a traits manifesto. *Journal of Ecology* 102: 275–301.
Rodriguez, A., and I. R. Sanders (2015). The role of community and population ecology in applying mycorrhizal fungi for improved food security. *The ISME Journal* 9: 1053–61.
Saïdou, A., A. C. Eteka, G. L. Amadji, D. G. J. M. Hougni and D. Kossou (2012). Dynamique des champignons endomycorhiziens dans les jachères manioc sur sols ferrugineux tropicaux du centre Bénin. *Annales des Sciences Agronomiques* 16: 215–28.
Saïdou, A., D. Kossou, A. Azontonde and D.-G. J. M. Hougni 2009. Effet de la nature de la jachère de la culture subséquente par les champignons endomycorhiziens: cas du système 'jachère' manioc sur sols ferrugineux tropicaux du Bénin. *International Journal of Biological and Chemical Sciences* 3: 587–97.
Saïdou, A., T. W. Kuyper, D. Kossou, R. Tossou and P. Richards (2004). Sustainable soil fertility management in Benin: learning from farmers. *NJAS – Wageningen Journal of Life Sciences* 52: 349–69.
Scott, G. J., M. W. Rosegrant and C. Ringler (2000). Root and tubers for the 21st century. Trends, projections and policy options. IFPRI Discussion paper 31; http://ebrary.ifpri.org/cdm/singleitem/collection/p15738coll2/id/125422/rec/339 (accessed 6 January 2017).
Sieverding, E., and L. Galvéz (1988). Performance of different cassava clones with various VA mycorrhizal fungi. *Angewandte Botanik* 62: 273–82.

Sitompul, M., S. Setijono, J. van der Heide and M. Van Noordwijk (1992). Crop yields and sustainability of cassava-based cropping systems on an ultisol in Lampung. *Agrivita* 15: 19–28.

Subere, J. O. Q., D. Bolatete, R. Bergantin, A. Pardales, J. J. Belmonte, A. Mariscal, R. Sebidos and A. Yamauchi (2009). Genotypic variation in responses of cassava (*Manihot esculenta* Crantz) to drought and rewatering: root system development. *Plant Production Science* 12: 462–74.

Suja, G., and J. Sreekumar (2015). Crop diversification with short-duration cassava (*Manihot esculenta*) in the humid tropics for agronomic and economic benefits. *Indian Journal of Agricultural Sciences* 85: 1619–25.

Teixeira, M. A., I. S. de Melo, R. F. Vieira, F. E. C. Costa and R. Harakava (2007). Cassava endophytic microorganisms of commercial plantings and ethnovarieties in three Brazilian states. *Pesquisa Agropecuaria Brasileira* 42: 43–9.

Totongon, J. K., B. lame and P. Zannou (2000). Quelques performances et contraintes d'adoption du pois d'Angole (*Cajanus cajan*) et du pois mascate (*Mucuna pruriens* var. *utilis*) en milieu paysan dans le Départment de l'Atlantique, République du Bénin. In: R. Carsky, A. C. Etéka, J. D. H. Keatinge and V. M. Manyong (eds), *Cover Crops for Natural Resource Management in West Africa*, pp. 109–19. IITA, Ibadan, Nigeria.

Trinder, C. J., R. W. Brooker and D. Robinson (2013). Plant ecology's guilty little secret: understanding the dynamics of plant competition. *Functional Ecology* 27: 918–29.

Udoh, A. J., and N. U. Ndaeyo (2000). Crop productivity and land use efficiency in cassava- maize system as influenced by cowpea and melon populations. *Tropical Agriculture* 77: 150–5.

Umeh, S. I., and B. N. Mbah (2010). Measuring the benefits of biological nitrogen fixation of soybean (*Glycine max* (L.) Merrill) in cassava (*Manihot esculenta* Crantz) and soybean intercrop. *African Journal of Agricultural Research* 5: 3354–9.

Vanlauwe, B., P. Mutuo, N. Mahungu and P. Pypers (2013). Boosting the productivity of cassava-based systems in DR Congo. *Best Practice* 8, Ibadan, Nigeria.

Vissoh, P. V., T. W. Kuyper, G. Gbèhounou, D. Hounkonnou, A. Ahanchédé and N. G. Röling (2008). Improving local technologies to manage speargrass (*Imperata cylindrica*) in southern Benin. *International Journal of Pest. Management* 54: 21–9.

Zhao, P. J., P. Liu, J. F. Shao, C. Q. Li, B. Wang, X. Guo, B. Yan, Y. J. Xia and M. Peng (2015). Analysis of different strategies adapted by two cassava cultivars in response to drought stress: ensuring survival or continuing growth. *Journal of Experimental Botany* 66: 1477–88.

Chapter 17

Value chain approaches to mechanization in cassava cultivation and harvesting in Africa

George Marechera and Grace Muinga, African Agricultural Technology Foundation (AATF), Kenya

1 Introduction
2 The benefits of mechanization
3 Obstacles to mechanization by smallholders
4 The AATF cassava mechanization experience
5 Mechanized production options for cassava
6 Assessing the effectiveness of mechanization
7 Business models for giving farmers access to farm machinery
8 Conclusions
9 References

1 Introduction

Cassava is a staple for 500 million people in sub-Saharan Africa (SSA), accounting for 55% of the world's cassava production. In Africa 93% of production is consumed as food (in contrast to Latin America and Asia where less than half is utilized for food consumption). Cassava is a basic staple food in SSA countries such as Nigeria, DR Congo, Angola, Ghana, Mozambique, Tanzania, Zambia and Malawi. Although Nigeria is the largest cassava producer in the world with over 40 million metric tonnes, the output per unit area is still very low (7–12 tonnes/ha) even with improved varieties as compared to 23 tonnes/ha average in Asia. Although improved cassava varieties with high genetic yield potential of 60–80 tonnes are available, their yield level on farmers' fields has remained very low due to inefficient production systems.

Recently, cassava has gained importance as a cash crop for smallholders in the region. Literally millions of African farmers are growing cassava mainly for home consumption and the local markets. The competitiveness of African cassava manufactured products in the world market has been low because cassava is produced and processed for subsistence, not as a commercial crop, and the public sector efforts to commercialize the crop have not been fully effective. The improvement of the cassava sector will benefit the food security

http://dx.doi.org/10.19103/AS.2016.0014.28

for the millions who rely on it as a staple food crop and create opportunities to produce commercialized cassava for the world market. The potential for expanding the cassava sector is exponential.

Cassava appeals to low income earners because it offers the cheapest source of calories and is an important cash crop. It can also be easily processed to produce industrial starch and is important in livestock feed production. Cassava is a high priority crop in the research and development agenda of national agricultural research efforts in Africa as a whole, due to its sustainability and guaranteed contribution towards the UN Millennium Development Goals and regional economic goals. Cassava production in SSA is mainly through manual labour which is time consuming, inefficient and not appealing to young and women farmers who perform over 75% of the work in cassava production and processing. Mechanization of cassava production is very low at the global level; nonetheless, there have been considerable advances in commercial sectors in Asia and Latin America. Cassava mechanization in Africa is almost non-existent and is one of the key constraints to the development of the sector. This chapter highlights mechanized production options in the cassava sector.

2 The benefits of mechanization

Agricultural mechanization is the application of mechanical technology and increased power to agriculture, largely as a means to enhance the productivity of human labour and often to achieve results well beyond the capacity of human labour. Agricultural mechanization embraces the use of tools, implements and machines for agricultural land development, crop production, harvesting and preparation for storage, storage and on-farm processing. The manufacture, distribution, repair, maintenance, management and utilization of agricultural tools, implements and machines is covered under this discipline with regard to how to supply mechanization inputs to the farmer in an efficient and effective manner. Mechanization is not an 'all or nothing' process, but the levels and types of improved mechanical technologies need to be appropriate, that is, compatible with local agronomic, socio-economic, environmental and industrial conditions.

Farm mechanization is an opportunity to increase yields and reduce loss in addition to lowering cost. Mechanization of farm work has three primary objectives: reduce the drudgery of farm work, increase the productivity of farm workers, and increase the timeliness and quality of farm work (Goering and Hansen, 2004). According to Goering and Hansen (2004), the hard work and low pay of many farm operations is what drives younger children off the farm. Mechanization allows for the work to be less difficult and the skill required to operate the machines oftentimes increases wages. Mechanization has also been a major contributor to the increases in hectares managed per hour of work. However, since a problem of the African countries is one of increasing the production of food, the highest priority for power and equipment should be for types that will contribute to increased yields and efficiency in operations. Mechanization can increase yield through more timely and higher quality performance of operations. There is an optimum time for performing critical farming operations such as planting and harvesting. Crop yields tend to be highest when these critical operations are done closest to the optimum time (Goering and Hansen, 2004). Additionally, mechanization improves the quality of the farm operation. For example, ploughs pulled by the farm tractor can plough deeper and at a more consistent depth than draft animal ploughing. This breaks up any underlying hardpan in the soil, allows for better mixing of the soil and permits deeper root penetration. This is particularly important for cassava as it allows for easier expansion for the cassava root underground.

Cassava mechanization implements (such as cassava planter, boom sprayer, cultivator and harvester) will not only lead to increased output, but will reduce the drudgery of farm operations making it more efficient and profitable. Experience from the African Agricultural Technology Foundation (AATF) Cassava Mechanisation and Agroprocessing Project (CAMAP) work has shown that yields can be increased from the current 7 to 10 tonnes to above 30 tonnes per ha through mechanization complemented by good planting materials, fertilizers and herbicide use and good agronomic practices. It has also been shown that the time for the field operations is greatly reduced through mechanization.

The CAMAP's practical experience in the field has shown that cassava mechanization implements are well adaptable to Nigerian conditions. The tractor service providers should move from land preparation services alone to offering a full package from land preparation operations to planting, boom spraying, weed management and harvesting services for profit maximization. Offering of a full package by the tractor service providers widens the business scope and opens the way for accessing better credit.

3 Obstacles to mechanization by smallholders

The smallholder farmers in SSA produce cassava mainly for traditional food markets, with any surplus processed on the farm or sold to local processors. Women and sometimes children are involved in the labour-intensive and non-mechanized operations along the entire cassava value chain. There are a number of constraints that affect cassava production in SSA; however, lack of mechanization in production has been one of the biggest challenges retarding growth in this sector. Cassava production is labour-intensive, with poor production techniques and no use of mechanized ploughing, planting, weeding, harvesting, loading and packaging. These activities consume a lot of time that demotivates young people to venture in cassava farming. In addition to this, cassava takes 9– 24 months to mature depending on the variety. With this in mind, very few farmers would be willing to invest their money in a crop that takes so long to return benefits. The majority of the farmers would prefer a short-term crop like maize so that they can quickly earn their profit. However, income earned from cassava has a potential to surpass that of maize by a significant margin, as confirmed by the CAMAP experience.

A comprehensive business case for technology development, transfer and the subsequent commercialization of mechanization technologies is key to address the sector constraints. Since different countries are at different levels of cassava development, these plans will need to be tailored to specific situations. Planning must be coordinated for the full value chain. For example, cassava starch mills in Nigeria have been closed mainly because of the failure of farmers to consistently supply sufficient cassava raw materials to the industries. Farmers in Ghana have failed to supply the export demand for dried cassava roots. Consumers in northern Zambia, where cassava is the main staple food, struggle to get cassava mealie meal as it has become more expensive than maize meal due to scarcity of the product. West Africa's dependence on imported maize derivatives such as the starches and alcohol shows the need to adopt the labour-saving mechanical technologies for cassava and hence expand production and save foreign exchange. The missing link has been the effective integration of production, processing and marketing. The result is that farmers around SSA have always lamented the absence of markets to give them the incentive to produce, and the processors have been operating below capacity pointing to inadequate root supply. The hypothesis we present is that mechanization of production can be a major contributor to improving the production link of the value chain. This chapter describes a range of mechanization options, their advantages and disadvantages, and the business cases for their development and adoption.

It is only through appropriate cassava mechanization and agroprocessing that smallholder farmers will be able to feed not only themselves but also the continent's burgeoning urban population. Productivity will have to increase dramatically if African farmers are able to feed their families and spare some of their production for the market. Thus there is need to regard cassava as a commercial crop that can contribute to industrial development and the achievements of the MDGs and CADAAP objectives.

Adoption of cassava mechanization has long been eyed as an important component that could propel SSA to increased levels of productivity. Trends in mechanization worldwide show that there are strong correlations between economic growth and mechanization (Clarke and Bishop, 2002). Historically, while there have been several attempts at bringing mechanization to SSA, it is widely accepted that most of these attempts have failed and the situation of the smallholder seems to have improved little. A disjointed approach in cassava mechanization and to agriculture in general is the reason that these attempts have failed over the years. In the past, mainly from the 1960s to 1980s, agricultural mechanization was seen to cause rural unemployment and had other negative social impact on the communities and this led to less investment in the sector. During the same period, tractors were donated generously to developing countries by various donors with an aim to boost the sector. In addition, government agencies targeted at providing mechanization services often failed due to intrinsic inefficiencies of government-run businesses (Houmy, Kienzle, and Ashburner, 2016).

According to Kienzle et al. (2016) tractor hire schemes operated by the Ugandan government have deteriorated and have been abandoned owing to numerous inefficiencies. In the 1950s and early 1980s, farm machinery like tractors and ploughs were donated or made available at subsidized rates to developing countries in the form of aid (Benin, 2014). The payment plan to the buyers was staggered and allowed payment over time with an intention to attract more investors in mechanization. Unfortunately, training for use of these machines in most of the African countries was not provided in a timely fashion and therefore the efforts failed to meet their intended purpose. In addition, mechanization was only aimed at ploughing operations, without other operations like harrowing, planting, herbicide application and harvesting. This led to the mechanization business not being viable economically for the service providers, as the minimum number of hectares per year needed to break even was not attainable.

With recent improvements in Africa's economic performance, many SSA countries have focused on agricultural mechanization by seeking assistance from FAO to develop strategies. As a consequence of this resurgence in tractor-based mechanization and the availability of low-cost tractors, a number of African governments have proceeded to purchase large quantities of tractors directly from manufacturers without having previously formulated a strategy to support their distribution, economic operation and maintenance. This is a worrying trend as it may lead governments to return to the direct provision of tractor-based services. As already indicated, such public sector tractor hire schemes did not prove successful when previously introduced in the 1960s (Ashburner and Kienzle, 2009).

These efforts by governments and donors to make mechanization affordable to the developing country citizens made the machines artificially cheap in comparison to labour and animal-drawn implements. This, coupled with the lack of training in the use of the machines, and social side effects of mechanization, motivated the communities to dis-adopt mechanization, causing donors to also pull out of funding of the same. On the other hand, success stories for mechanization have been reported elsewhere, examples being the privately owned shallow tube wells for irrigation in South Asia, axial flow threshers in Southeast Asia, single-axle tractors in Thailand, cassava mechanization in Brazil and various forms of farm mechanization in many parts of China

A systems approach to cassava mechanization involving all value chain players can be very transformational for most African countries where cassava is a major crop. The introduction of agricultural technology, including mechanization, is a complex process. Thus, cassava mechanization will very much depend on country-specific characteristics of the economy, its level of development, the agriculture sector and the cassava industry. Within the levels of mechanizing technology, degrees of sophistication must be distinguished. When evaluating the agricultural sector as a whole in a particular country, different options for enhancing land and labour productivity, as well as their economic and financial implications will be considered. Thus, the size and design of the machines will take into consideration the field size and layout, the soil types, the level of affordability by the farmers and the repair and maintenance access and costs. The basic principles to keep in mind when exploring cassava production mechanization are as follows:

- History in Africa has shown that the process of cassava mechanization is driven by the need to increase operational efficiency and productivity.
- The reasons for cassava mechanization should be economic as it reduces drudgery, increases efficiency and creates wealth. The driving force behind mechanization is the farmer's effort to increase or maintain net income leading to improved livelihoods.
- Cassava mechanization should be demand-driven: ultimately it is the farmer who will decide what machine to buy, from whom to buy and how to use it given the options.
- Experience has shown that cassava mechanization must be left to the private sector as much as possible and other players like international organizations should provide the enabling environment, capacity building and market linkages to ensure that it is sustainable. Experience tends to support the premise that governments should not actively get involved in the manufacture, import, distribution and repair of agricultural machinery and its operation, but provide the incentives for private sector response.
- The advantages of cassava mechanization are that it reduces work drudgery, save time and money, facilitates high precision and quality in farm operations and contributes to higher yields through good land preparation and doing all operations in a timely manner.
- Mechanized planting allows for the stake's length and depth to be uniform, along with uniformity in the distance between rows and stakes, leading to the desired plant populations. The application of fertilizer is also uniformly distributed by augers.

Promoters of cassava mechanization have begun to look holistically at the process, focusing on the whole value chain and system. This means that the selection of implements for mechanization is done across the whole production cycle, for example, from ploughing, to harvesting and eventually to processing of the cassava. With this concept in place and best agronomic practices employed, high yields per unit area are achievable. A solid example of this approach in operation is CAMAP, implemented in Nigeria, Zambia and Uganda, in partnership with the private sector, farmers and government departments. This project run by AATF employs mechanization from ploughing of cassava fields, to harrowing, planting, weeding and harvesting. This full system approach has made it possible for the project to realize yields of up to 30 tonnes per hectare on farmer fields in Nigeria from the usual 7–12 tonnes per hectare (AATF, 2014). The project emphasizes good agronomy such as selection of high-yielding improved cassava cuttings of disease-resistant varieties, timely planting, use of fertilizer, use of herbicides and regular weeding of the crop. The increase in yield can be attributed to both mechanization and good agronomic practices. This is a positive step in promoting cassava as a food security crop within the increasing population in Africa. Efforts are currently underway to replicate this approach in several African countries to bring it to scale in order for it to have sustainable impact in the continent.

This approach by AATF contrasted with approaches of previous researchers who disregarded mechanization as unsustainable. For example, Houssou et al. (2013) concentrated only on primary tillage and concluded that it was not sustainable. Although this is likely correct when applied in isolation from other production practices, mechanization across the whole system can be profitable and sustainable. This emphasizes the need to employ the approach across the whole chain to get better returns and ensure efficiency of the process. This whole-system approach needs to be considered by implementers as they proceed with cassava mechanization.

4 The AATF cassava mechanization experience

The AATF is a not-for-profit organization that facilitates and promotes public/private partnerships (PPP) to access, develop, adapt and deliver appropriate agricultural technologies for use by resource-poor smallholder farmers in SSA. AATF works through PPP to solve the constraints of agriculture production for smallholder farmers in SSA. Through the partnerships, it identifies productivity challenges and does technology scouting to identify technologies that can address the challenges.

In 2012, AATF undertook a critical analysis of the cassava sector in SSA to assess the constraints affecting the sector and come up with the technology needs. It was noted that mechanization of labour-intensive cassava production activities is one of the major interventions that can rapidly be applied to enhance field operations and improve production systems. The analysis pointed to the inefficient cassava productive systems coupled with the high levels of drudgery. On realizing the technology gap in the cassava sector, AATF conducted some cassava production technology assessments across the world. The studies showed that there was cassava equipment for planting, cultivation and harvesting in Brazil, China and Thailand. This equipment has supported the industrialization of the cassava sector in the respective countries as they have made cassava production efficient, profitable and timely. This equipment could be accessed and used in SSA but needed to be tested for efficacy in the respective countries.

To facilitate testing of the equipment, a technology needs assessment within cassava producing countries was done. AATF developed selection criteria to choose the countries for piloting the technologies. The criteria highlighted that AATF needed to target countries with a government strategy on cassava production, cassava as a major crop, improved seed systems and other initiatives on transforming the cassava sector. There was need to choose countries with different levels of cassava production to include the highest producer in Africa (Nigeria), a mid-level producer (Uganda) and another at the lower end but with the minimum conditions to ensure that AATF can achieve progress in the pilots (Zambia).

AATF has been working with smallholder farmers in Nigeria, Zambia and Uganda under CAMAP since 2013 to offer solutions that would boost productivity for cassava farming. The overall goal of the project is to improve cassava production through mechanization for sustainable food security and commercialization towards enhanced incomes and livelihoods of farmers in SSA. This is being achieved through the upgrading and expanding of traditional planting, harvesting and processing technologies. These changes are contributing to the development of competitive cassava commodity value chains resulting in a reliable supply of processed products for food and non-food industrial use. The project objectives are to negotiate access and transfer of cassava mechanization and agroprocessing technologies for use by smallholder farmers; to increase cassava production through mechanization across the entire value chain and thus reduce post-harvest losses and demand for intensive labour; to ensure economical and sustainable provision of mechanization services to smallholder farmers; to support the cassava industry through value addition and the creation of market linkages between smallholder farmers and agroprocessing centres; to build the capacity of local entrepreneurs to design prototype machines, manufacture, maintain and repair equipment for planting, harvesting and processing cassava; and to expand the utilization of safe, quality, diversified, value-added cassava products and derivatives.

AATF identified cassava-specific field machinery and implements of interest, including cassava planters, root diggers, cultivators and boom sprayers from Planti-Center in Brazil. AATF partnered with these design engineers in Brazil in 2013, to avail these state-of-the-art technologies for cassava production and harvesting. To ensure that the cassava mechanization equipment was ideal for African conditions, AATF accessed three sets of planters, cassava cultivators and root harvesters for testing in Nigeria, Zambia and Uganda.

CAMAP's business case involved a two-pronged approach of upgrading the existing mechanization activities and small-to-medium enterprises to produce more efficiently and at larger capacity, while expanding the processing capacity with larger processing plants (Fig. 1).

This business case looks at a systems approach to cassava mechanization, starting with participatory constraint mapping to identify the needs of the smallholder farmers. Once the needs are identified, there is technology scouting and identification to ensure that the right type of technology for the right beneficiaries is identified. As most technologies are patented or have intellectual property restrictions, there is need for negotiation and access to the technology, which is a key mandate for the AATF to ensure that smallholder farmers can also access technologies that are useful and would otherwise not be accessible without this intervention.

To ensure that the accessed equipment is useful to the farmers, AATF accessed three sets of planters, cassava cultivators and cassava root diggers, for piloting and testing in two states of Osun and Kwara in Nigeria. In partnership with the National Center for Agricultural Mechanisation (NCAM), pilot farmers and sites were selected and cassava mechanization work commenced in 100 farms each measuring 1 hectare. It was through

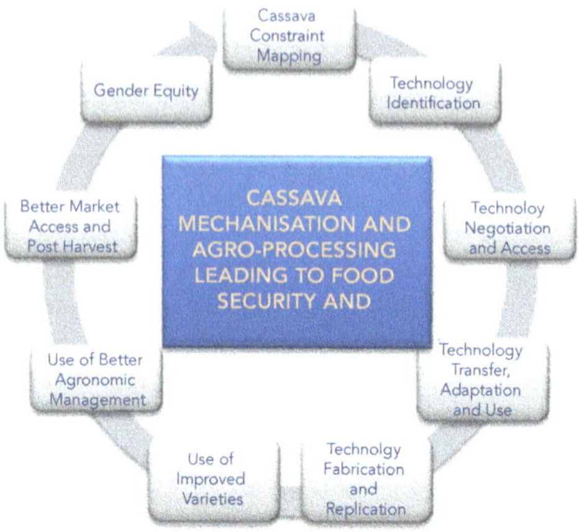

Figure 1 Step-wise approach to cassava mechanization and agro-processing.

these pilots that many farmers and project partners realized that cassava can be fully mechanized and this attracted thousands of farmers to get involved into the project.

During the pilot phase, AATF supported all the farmers with high-yielding disease-resistant cassava stems, four bags of fertilizer per ha, herbicides (glyphosate and Diuron) and provided all the mechanization services for land preparation (ploughing and harrowing), herbicide application and planting. The farmers provided the land and the labour for weeding. Results from the crop performance during the first month of planting was overwhelming as the crop vigour was high compared to that which was planted manually. During harvest, AATF provided the cassava root harvesters for piloting and farmers managed to harvest on average more than 30 tonnes per hectare compared to less than nine tonnes they were harvesting before. Thus more farmers started joining in the project which was then expanded to other states like Kogi, Ogun, Ekiti, and Nasarawa and Benue. This was then replicated in Zambia and Uganda and the same good results were achieved.

However, for sustainability of the project, a business model based on private service providers has now been instituted. As most smallholder farmers have land which is less than 5 ha, it does not make any business sense for each farmer to own the equipment. Thus farmers are organized into clusters where an entrepreneur can provide mechanization services as a business and farmers pay for the services. These service providers need to be trained in business management and financial management.

Through the project farmers are now being linked to mechanization services on full cost recovery in which they pay a commercial rate for the mechanization services in line with farming as a business. Examples of these service providers are the NCAM and Plantgate serving Kwara and Ogun states in Nigeria. Some of these service providers have their own tractors, ploughs and harrows to assist in land preparation. AATF comes in with cassava-specific implements to aid the mechanization. To date more than 6500 farmers with more than 15 000 hectares have benefited in the three countries. In Nigeria, CAMAP is spread across Osun, Ogun, Kwara, Kogi, Edo and Oyo states. In Zambia, CAMAP is implemented is Mansa, Kawambwa, Samfya and Kaoma districts, whereas in Uganda it is in Apac, Kiryandogo and Nwoya districts. .

For sustainability of mechanization, there was need to build local expertise in the fabrication of the spare parts to avoid continuous importation. This means capacity building of local fabricators for the mechanization to be sustainable. A number of local fabricators were identified for integration into the project.

As pointed out earlier, mechanization alone will not increase yields. Thus, the approach is to integrate cassava mechanization with improved cassava varieties. These are high-yielding, disease-resistant varieties being promoted by the governments and research institutions in the respective countries. For example TM419 in Nigeria and Mweru in Zambia. Most farmers have been planting the landraces; however, these will not give them the yield that will make cassava mechanization economically viable. By introducing new improved varieties, farmers are able to get yields of up to 30 tonnes, making the business profitable.

Along with mechanization and varieties, there is need to bring in the best agronomic practices, which include the right planting time, optimized plant populations, appropriate fertilizers and application rates, effective herbicides to control weeds, best timing for weeding and the harvesting time when the starch content in the roots is high. This systematic approach has been instrumental for success of the CAMAP.

Additionally, most farmers complain that there are not sufficient markets for cassava, thus they do not have the incentive to produce more. However, when you talk to processors, they lament that they do not have enough cassava for processing. This means that there is a missing link between farmers and processors. Thus through the CAMAP, AATF identified processors like the Allied Atlantic Distilleries Limited (AADL) and negotiated for good prices for the farmers. With the market linkages, farmers are incentivized to produce more cassava as they are assured of a market for their cassava. The AADL is benefiting from the relationship as they have a steady supply of roots to process.

As most tasks in the cassava sector are done by women and youth, the approach is to look at how the cassava technologies can assist these target groups. During the selection of farmers, concerted effort is made to ensure that we have more female-led households. A minimum of 30% of the beneficiaries should be women and youth and the youth are also trained and employed in service provision, repairs and maintenance.

5 Mechanized production options for cassava

Traditionally, cassava production is done manually using the hoe and the axe. This work is physically difficult and very inefficient, and sometimes leads to low yields. Several options exist to make cassava production easier. Some producers in Asia and Latin America have long used machines for their cassava production; however, Africa is yet to fully embrace mechanization.

To facilitate cassava mechanization and maximize yields, the right type of land must be identified. This has to be land that is flat, without rocks, without trees, stumps and anthills. Using tractor-drawn implements, cassava production can be mechanized from ploughing to harvesting. The land preparation is done by a tractor-driven disc plough, an operation that normally takes an average of 1 hour for a hectare, depending on the accessibility and tractability of the land. This is followed by harrowing to break the soil crumbs and equally distribute the soil uniformly on the bed. This operation takes about 45 minutes to complete. The ploughs and harrows are locally available in most African countries. The following section looks at the equipment accessed by AATF through CAMAP.

Once the land is ploughed, planting of cassava can be done by a cassava planter which can plant from one to six rows, depending on the size of the farm to be planted. The bigger the farm the bigger the type of planter used. The two-row planter shown in Fig. 2 has been tested to be the one suitable to smallholder farmers with less than 10 ha, as it is easy to use in small fields. However, as mechanization becomes popular and farmers start increasing land to more than 20 ha of cassava, the four-row planter will became technically and economically viable. With the two-row planter, you can plant at least 5–7 ha per day using a 60 horsepower tractor. Planting manually would take farmers more than eight days which means the plants in the same fields will sprout at different days and are not uniform in growth pattern.

The cassava planter, a tractor-pulled implement, will have people feeding stems into it. The planter concurrently or sequentially opens a furrow in the soil, applies fertilizers, cuts the stems into equal pieces of 18 cm and drops the stems into the opened furrow. The stems are cut through a rotor, drop to the soil and are pressed by a rubber wheel to achieve uniform depth placement. A depth control wheel adjusts the depth and finishing with the closing discs. A four-row planter can plant between 13 and 15 hectares per day and it will require at least a 110 horsepower tractor.

Advantages of mechanical planting are that stem sizes are the same size and uniformly planted, leading to good plant population, which is key to yields. The work is done on time. Few people are involved which means that other available labour can provide other economic activity to generate income for the household. Some mechanical planters are able to apply fertilizer along the row at the time of planting, a key part of good early root establishment.

Once the field has been planted, there will be the need to apply herbicides to control weeds. Weeding is one of the most demanding operations in cassava, demanding up to 12 person days per hectare. Using a boom sprayer which can hold between 200 and 500 L of herbicides, and extending to 10 m across the field, pre- and post-emergence herbicides like glyphosate and Diuron are applied at least 5– 7 days after planting as shown in Fig. 3. This ensures that there will be minimum weed pressure affecting the establishment of the

Figure 2 Two-row planter.

Figure 3 Boom sprayer.

cassava in the early days, a critical time for both foliage development and root system establishment. Machine application of herbicides takes an average of 30 minutes per ha and the applicator can cover at least 12–15 hectares per day on small farms, or much more in a single large field.

In traditional management systems farmers spent most of the cassava production time doing weeding. Traditionally, due to the fact that cassava takes more than 12 months to harvest, most farmers would plant cassava after all other crops have been planted. They would also not weed the cassava until dry periods when the farm work demands are low. This meant that the weeds would have competed for nutrients, sunlight and water with the cassava plant, resulting in the poor growth of cassava crop, and leading to poor harvests. Through the CAMAP, a cassava cultivator was identified and tested. Once the crop has sprouted and grown up to the 8th to the 12th weeks, a cassava cultivator can be used for effective weeding. Fig. 4 shows a two-row cultivator which can cultivate up to 9 hectares a day, which equates to about 45 minutes to weed 1 hectare. The cultivator requires a tractor of about 45 horsepower or more. This cultivator also loosens the soil to permit aeration and good water retention

For bigger fields, a four-row cultivator shown on Fig. 5 can cover 18 hectares per day.

The same cultivator can have a 200–500 L tank for herbicide application in which case it applies herbicides concurrently whilst weeding to ensure that the weeds that are not destroyed by the cultivator can be destroyed by the herbicides. This is shown in Fig. 6. The herbicides are applied just below the leaves so that the chemicals do not affect the cassava plants.

When the cassava is ready for harvesting, a root digger is used (Fig. 7). This root digger is pulled by an 80 horsepower tractor as it loosens the soil, ensuring that part of the roots are brought to the soil surface. This greatly facilitates pulling the roots out of the soil

Figure 4 Cassava cultivator.

Figure 5 Four-row cultivator.

Figure 6 Cultivator with herbicide application.

Figure 7 Cassava uprooter/harvester.

completely (manually), and heaping the roots ready for loading into trucks. A full manual harvest requires more than 45 person days to harvest 1 hectare. By comparison, with this machine harvest, it can take less than 3 hours for five people to finish harvesting of 1 hectare.

It can harvest an average of 5 hectares per day with average yield per hectare being 30 tonnes. This equipment permits less drudgery and reduces post-harvest losses as harvesting is done on time, efficiently and with minimum root damage. Itodo and Daudu (2013) supported this as they reported breakages of the same cultivar tested after hand harvesting reduced from 32.0 to 2.6%.

All this equipment (planter, cultivator and root harvester) are made in such a way that the spacing can be synchronized for all operations.

As farmers increase their yields due to best agronomic practices coupled with mechanization, there is also need to fast track the loading of the cassava into bigger trucks so that the cassava can be sent to the market quickly. To aid in this there is the cassava crane loader with which the cassava is loaded into big bags and into trucks. The crane is attached to a 75 horsepower tractor and can load big bags of up to 1200 kg. Figure 8 shows the cassava crane loader. This crane can load a 30-ton truck in less than 30 minutes.

> This effectively means cassava can be harvested and sent to the market within 5 hours. Through the efficient, timely and lower-cost operations, the mechanization of the cassava sector is poised for growth. Farmers are already reaping high income from cassava.

The CAMAP initiative has proven that a paradigm shift can be introduced in the cassava value chain in SSA. To date, AATF has accessed, delivered and tested mechanization equipment for land preparation, planting, weeding and harvesting. The project has thirty cassava planters, thirty cassava root diggers, twenty cultivators, ten boom sprayers, eight tractors, eight ploughs and eight harrows.

We have also observed that youth are getting interested in going into cassava production because of the mechanization opportunities. However, the continuous availability of appropriate machines and associated service required to enable farmers

Figure 8 Cassava crane loader.

take full advantage of mechanization remain a big challenge in Nigeria. If the current projection by the government of Nigeria to double cassava production in smallholder fields in Nigeria is to be attained, a sustainable agribusiness that provides mechanization services to smallholder farmers in Nigeria would as a necessity be developed to create an enabled and functional production system for the country. Unless this gap is urgently addressed, and fully supported by the establishment of a strong network of service providers to provide mechanization services, it is unlikely that cassava farmers will benefit from the huge potential global market opportunities that require increased cassava productivity.

6 Assessing the effectiveness of mechanization

Adewoyin and Ajav (2013) noted that selection of machines and their correct use has substantial impact on land and labour productivity. This increases farming profits from one's land and in turn, the livelihoods of the community. A success story of the Green Revolution in Asia is that mechanization increased land productivity more quickly than labour availability. With this in place, tractor mechanization in Asia in the 1960s and 1970s accelerated due to increased demand for timely land preparation caused largely by the introduction of high-yielding varieties of wheat and rice, and the opportunities arose for intensifying agricultural production (Monk 2015). All these developments led to an increase in the land cultivated, increased cropping intensity as well as timely land preparation (Kienzle, 2016). These practical experiences in Asia indicate that mechanization has the potential to increase productivity as well as reduce poverty (Takeshima et al., 2013). Mrema et al. (2008) indicated that countries that have had exceptional economic growth have also transformed agriculture through mechanization.

Results from CAMAP activities have shown that mechanization of farm operations has the capability to address challenges in cassava production through timely and efficient execution of operations. Under CAMAP, cassava fresh root yields on farmers' field in Nigeria were over 30 tonnes/ha, almost three times higher than the 7–10 tonnes/ha current national average. The project has also contributed significantly to decreasing the cost of production, reducing drudgery and manual labour in farming activities, increasing cassava productivity and improving the overall operational efficiency of production, especially for women who provide the bulk of labour in cassava operations in Nigeria.

Through CAMAP, the plant distances of 0.8 m in between rows and 1.0 m between rows to get a plant population of over 12 500 plants per hectare are used. The land preparation and planting is done just at the onset of the rainy season to allow for good time for the cassava to grow. Four bags of 50 kg each of fertilizers (NPK 15 17 15) are used per hectare at planting. Some farmers apply split fertilizers application in which case they apply two bags at planting and another two bags after the crop is between 6 and 8 weeks old. Efforts are currently being made to do the testing of the soils to ensure that the appropriate amounts of fertilizers are applied depending on the soil types and the nutrient needs of the crop.

The mechanization of cassava was complimented by supporting the farmers with technical assistance in best agronomic management practices regarding timing for planting, weeding and herbicides applications. In Nigeria and Uganda, the cassava varieties grown there are ready for harvesting in 12 months. However for Zambia, the varieties take longer and are normaly ready for harvesting after 18 months, farmers got yields of over 35 tonnes per ha with one farmer, Mr. Mushili of Nsonga Cluster in Mansa getting up to 56 tonnes per ha in 2015 from the average of 8 tonnes per hectare they were getting before embarking on mechanization.

The increased yields and the access to markets meant that farmers are generating more income. Table 1 illustrates gross margins. Assuming the price for cassava roots is $0.08 per kilogram and total yield per hectare is 30 tonnes, net income is expected to be more than $1500.00 per hectare after 15 months.

Further, farmers pocketed higher profits with prices of US$88 per tonne instead of the usual US$62 per tonne resulting from the market linkages that were created through the project. In Zambia, a farmer who was getting US$620 from his 1 hectare farm managed to get US$2640 from use of improved high-yielding disease-resistant varieties, good land preparation, right planting time, use of fertilizers and herbicides, right plant density and reduced damage at harvesting due to the use of the right implements.

To better quantify the economic benefits of cassava mechanization, efforts are currently underway to compare detailed manual and mechanical operations under well-designed, side-by-side experiments, under different agronomic practices and inputs (time of planting, varieties, fertilizer and herbicide application) and these will be compared with plots which will be done mechanically. This will give a clear picture of the contribution of each of the operations to the yields and efficiency in cassava production.

Cassava mechanization has increased efficiency in operation and reduced drudgery in the farm operations across the cassava value chain. These technologies have provided a paradigm shift in which smallholder cassava farmers in SSA who are used to using rudimentary farming implements for ploughing, harvesting and processing cassava, which are tiring, are exposed to efficient and effective tools. This has made cassava production and processing attractive and profitable to the youth and women who constitute 75% of their farming time working on cassava. The implementation of CAMAP has led to land and labour optimization and created efficiencies in the cassava sector using the cassava planter and root digger which is illustrated in Table 2. Using the specialized cassava mechanization

Table 1 Gross margins for cassava (traditional vs mechanical) per hectare

Data	Nigeria		Zambia	
	Traditional	CAMAP	Traditional	CAMAP
Average price of cassava in the market ($/kg)	0.08	0.08	0.08	0.08
Average quantity of cassava produced (kg/ha)	8,000	30,000	8,000	30,000
Total revenue ($/ha)	640.00	2,400.00	640.00	2,400.00
Average cost of cassava stems used for ($/ha)	30.00	80.00	20.00	50.00
Average cost of fertilizer used in planting ($/ha)	0	100.00	0	180.00
Average cost of agrochemicals used in planting ($/ha)	0	40.00	0	30.00
Total input costs ($/ha)	30.00	220.00	20.00	260.00
Average cost of labour ($/ha)				
Land preparation ($/ha)	150.00	100.00	280.00	200.00
Planting ($/ha)	50.00	40.00	80.00	50.00
Weeding ($/ha)	200.00	80.00	180.00	60.00
Harvesting ($/ha)	150.00	100.00	140.00	120.00
Transportation ($)	200.00	400.00	250.00	600.00
Total variable costs ($)	750.00	720.00	930.00	1,030.00
Total costs ($)	780.00	940.00	950	1,290.00
Gross profit ($/ha)	(140.00)	1,4600.00	(310.00)	1,110.00

Assumptions: The cost of cassava roots is $0.08 per kg. All farmers are buying planting materials based on traditional, local varieties while the CAMAP uses improved varieties. Based on 2013 market prices for inputs.
Source: CAMAP.

equipment from Brazil, time taken for field operations have been reduced by more than 400%. AATF and partners have managed to reduce the time for land preparation from 240 man-hours to 3 man-hours, planting from 64 man-hours to 1 man-hour, harvesting from 320 man-hours to 8 man-hours. The project has contributed significantly to decreasing the cost of production, reducing drudgery in farming activities, including providing employment, as well as decreasing manual labour and increasing land productivity and overall operational efficiency, especially for women who provide the bulk of labour in cassava operations.

The experience has shown that tractor-mounted machines (plough, harrow, planter) take 3 hours to till the land and plant one hectare compared to 30 person days if done manually. As mechanization reduces drudgery, increases efficiency in operations and contributes to increased production, it has been noted that women and youth will now spend less time in the field and have the opportunity to do other income-generating activities like market gardening. They will also have the time to rest and to attend social gatherings and thereby improve their livelihoods.

The inefficient traditional production and processing methods of cassava in Africa have led to cassava products failing to compete in the local and global markets. However, through CAMAP, farmers produce high-quality cassava after linkage with processors

Table 2 Land and labour efficiency using planter and root harvester (per hectare) under CAMAP in Nigeria

Process	Manual	Mechanized
Land preparation per ha	30 person days (240 hours)	1 hour
Stem preparation and planting per ha	8 person days (64 hours)	45 minutes
Weeding per ha	12 person days (96 hours)	30 minutes
Weeding cost (US$)	300	50 – Use of herbicides
Harvesting per ha	45 person days (320 hours)	5 hours
Cost reduction per ha		60%
Yields per ha	7–9 ton	30–45 tonnes
Income per ha	US$700–900	US$3,000–4,500
Market linkages	Traditional (brokers and middlemen)	Linked to processors

who buy their products in bulk. The processors make significant large investments in production, processing and delivery of high-quality cassava products. Thus, mechanization technologies in the cassava industry will boost the competitiveness of the cassava industry. Access to markets will eventually be transformed and value-added cassava products will be produced and efficiently marketed.

Throughout the entire cassava productive process, women and youth typically carry out 75% of the work. Much of this work is labour-intensive. However, despite the predominance of women in the agricultural sector, they remain the most vulnerable segment of the population in the cassava growing regions. Women are the caregivers and are responsible for safeguarding the health and nutrition of their children. Agriculture is a food and income source for them. Without addressing the limited or lack of access to the necessary tools for women farmers, the battle of alleviating food insecurity and poverty in the region cannot be achieved. The goal of this innovation is also to give women and youth access to the tools that help them farm more successfully with greater rewards and comfort and thus create employment opportunities. Thus, the project is targeting at least 30% of the beneficiaries to be women and youth. Women and youth are now employed to put the stems into the planters, drive the tractors, service the equipment and cultivate the fields mechanically as compared to doing it manually.

As part of the implementation structure of the project, there has been the identification of local fabricators to manufacture, maintain and repair the necessary equipment for cassava planting, harvesting and processing which is necessary for ensuring timely availability. However, a number of these local fabricators are not well trained in producing quality equipment. In some cases, they do not have the right machines to manufacture spare parts. Thus, the provision of continued technical support will be critical to the success of this project to ensure that the local market is not flooded with imports. Building this capacity is also central for the sustainability of mechanization as a whole. Already a number of private sector partners are working in the project. Continued use of improved tools and equipment for production and processing calls for local artisan support in fabrication, maintenance and repair services. This means that in case of breakdowns, routine servicing and replacement of worn-out parts, local sources will be invoked.

Adoption support for continued upscaling from simpler to more sophisticated machinery, tools and equipment will be developed among producers, researchers, extension agents and policy makers.

As service providers start to work with farmers, there is also a need to build the capacity of local entrepreneurs to design prototype machines, manufacture, maintain and repair the necessary equipment for cassava planting, harvesting and processing for sustainability of the service provision. This will ensure that there will not be any continuous importation of the equipment as they will be locally fabricated.

7 Business models for giving famers access to farm machinery

There are a number of approaches to ensure that cassava mechanization is economically viable to farmers, service providers and the processors. This section will look at some of the approaches, their strengths and their weaknesses.

For cassava mechanization to be successful, there should be the implementation of inclusive business models in the specific context of each of the different countries and regions within countries. These model options include sole private business ownership, contract farming, outgrower schemes, individual ownership, joint ventures, cooperatives, PPP, licensing and franchising. The applicability of the respective models is based on their advantages and disadvantages, the conditions under which they could evolve and be sustainable, the expected roles of other stakeholders and their inclusiveness. Based on the analysis and understanding of the respective countries' conditions, a number of options will be implemented.

7.1 Private business model

There is overwhelming evidence that private ownership (which includes several possible structures as described below) is often the most economically feasible and sustainable way to provide the core capacity for efficient cassava mechanization, supported as well by public sector services such as extension. Private contractors are most appropriate to provide mechanization technology to small farmers. There needs to be identification of private service providers for land preparation, planting and harvesting and to build capacity to ensure that they will work with the farmers on a commercial basis for the project to be sustainable. The service providers may be entrepreneurs who are already investing in providing these services along with those who will be able to fabricate parts for the machines.

The service providers will be identified and linked with the farmers to provide the services on a contract basis. In this model farmers will ideally also get guaranteed markets and the processors will get a guaranteed supply of roots, at pre-established prices. It is anticipated that this arrangement will be done on purely commercial lines. For sustainability of this arrangement, the farmers have to pay for ongoing services after a period of start-up assistance.

This approach has been successfully adopted by AATF under the CAMAP. After two years of implementation, proof of concept and lessons learnt, a payment mechanism by farmers has now been instituted. Farmers will now pay for the mechanization services offered by the project. Each country is unique and payment mechanisms will differ from country to country as well as from farmer to farmer. Two scenarios will apply:

- If a private company is available to offer mechanization services, for example, ploughing and disking, then farmers will pay for these services directly to the service provider. Using the project equipment, CAMAP will only provide services in those areas for which no private sector provider yet exists, and could be for planting, spraying, cultivation and harvesting, for example. CAMAP will charge a fee for the services offered, at a cost-recovery rate. CAMAP will enter into a contract with the farmer and agree on the mode of payment for the services.
- If no private company offers mechanization services in the area, then CAMAP will provide all services, that is, ploughing, disking, planting, spraying, cultivating and harvesting. These services are all chargeable and a mode of payment agreed upon with the farmer. In both scenarios, the farmer will provide their own herbicide, fertilizer and weeding services.

7.2 Cooperative business model

An analysis done in most targeted countries has shown that cooperative ownership of agricultural machinery has usually failed because of management problems. However, where certain conditions are in place and management capacity building can be done, it is recommended that community-based co-ops will be used to provide needed machinery or services not currently available. This will increase competition within a market if it is implemented alongside other business models. Starting a business often requires greater capital and management capacity than one person can provide. Banding together to form a co-op with other like-minded people can be the answer. Cooperatives aimed at cassava mechanization will work best when there are people with a strong common interest, who want to maintain a broad base of control, and the benefits of owning and controlling the business extend beyond receiving company profits and increasing share values. The cooperatives will then be developed into either machinery syndicates or rings.

- Machinery syndicates
  These are mechanisms of joint ownership of agricultural machinery and may fall into three categories: informal groups where two or more farmers agree to cooperate; formal groups where a group of farmers reduce individual capital outlay by jointly investing in machinery; and cooperatives, where farmers will be stakeholders in a business run by paid employees. In cooperatives management is formal and it is vital to establish operating rules at the time of establishing the syndicate.
- Machinery rings
  These are groups of farmers, some or all of whom periodically suffer imbalances of machinery capacity. The ring cooperates to use excess capacity on one farm to satisfy the demand on another. Rings need to be administered by a manager and are financed by subscriptions and levies on all transactions.

7.3 Individual ownership model

The project will also facilitate those farmers who can own the machinery and equipment on their own. It is anticipated that as the project progresses, more farmers will generate enough income to buy machinery. Under individual ownership, sole use of machines improves timeliness, equipment matching and maintenance assurance, there are also no costs for management or travelling, which may be incurred in the sharing possibilities.

There are other options which offer lower-cost solutions. These options includes hiring for a fixed term which allows costs to be known in advance. Repair and maintenance will usually be the responsibility of the owner and so the fixed costs of ownership are avoided. There is also contracting, which is a traditional way for farmers to deal with work peaks. One drawback could be that timeliness of operations may suffer when demand outstrips supply. However, the service provider may offer up-to-date equipment with skilled operators and thus work quality can be very high.

7.4 Contract farming model

This model links farmers to processors through contract farming in which cassava production is carried out according to an agreement between a buyer and farmers. The contract establishes conditions for the production and marketing of a farm product or products. Thus, the farmer agrees to provide agreed quantities of cassava. These should meet the quality standards of the purchaser and be supplied at the time determined by the purchaser. In turn, the buyer commits to purchase the product and, in some cases, to support production through, for example, the supply of farm inputs, land preparation and the provision of technical advice.

Both partners engaged in this contract farming can benefit. Farmers have a guaranteed market outlet, reduce their uncertainty regarding prices and often are supplied with loans in kind, through the provision of farming inputs such as seeds and fertilizers. Purchasing firms benefit from having a guaranteed supply of agricultural products that meet their specifications regarding quality, quantity and timing of delivery.

However, as with any other form of contractual relationship, there are also potential disadvantages and risks associated with contract farming. If the terms of the contract are not respected by one of the contracting parties, then the affected party stands to lose. Common contractual problems include farmer sales to a different buyer (side selling or extra-contractual marketing), a company's refusal to buy products at the agreed prices, or the downgrading of produce quality by the company. A frequent criticism of contract farming arrangements is the uneven nature of the business relationship between farmers and their buyers. Buying firms, who are invariably more powerful than farmers, may use their bargaining clout to their short-term financial advantage, although in the long run this would be counterproductive as farmers would cease to supply them. These problems notwithstanding, the balance between advantages and disadvantages for both firms and farmers seems to be on the positive side: contractual arrangements are more and more frequently being used in agriculture worldwide.

7.5 Outgrower scheme model

An outgrower scheme for cassava is where farmers are clustered around a company. This is an innovative approach in which small farmers are organized around selected local commercial farmers (nucleus farmers). The commercial farmer provides support to the smallholder farmers (mechanization, credit, seed, chemicals and transport). The commercial farmer is directly involved in the production, processing and marketing of the cassava (vertical integration). The smallholder farmers are supported throughout the value chain by the commercial farmer, who then buys the produce from the smallholder farmers. The business case is that there will be enhanced knowledge, adapted technology, management skills and secure markets that enable farmers to improve their productivity in a sustainable manner.

The biggest challenge with outgrower schemes is that when the prices are low, most smallholder farmers will end up doing side marketing, in which case they will sell to other processors at the expense of the commercial farmer who will have given them inputs.

7.6 Structural linkages

The guiding principles for the successful implementation of mechanization will be based on careful and insightful industry analysis of production, processing and market systems which are demand- and market-driven and include both the private and public sectors. More emphasis is on adding value to cassava products to achieve greater returns to both producers and processors. There is need to build innovative PPP between international organizations, producers, processors, marketing people, machine fabricators, financiers and government leaders, with an overall goal of agricultural transformation leading to increased food security and poverty reduction in Africa. Importantly, there is need to acknowledge the country and regional differences and identify innovations particular to each country (no one-fit-for-all).

7.7 Transportation of cassava roots to the processors

Transportation of goods and services in the cassava sector depends strongly on the road network, and this is frequently a challenge. Many farms have been established along major paved roads but there are also many hinterland farms that still have difficulty to bring roots to nearby markets. Each kilogram transported over 1 km adds a cost to the root price, thereby making the roots even more un-competitive for these isolated farmers. A model will be utilized whereby the processor engages third-party transporters or owns his own trucks to transport roots to the factory, and the producer is charged the cost of diesel or transport. This model is also the predominant way of getting fresh roots to the factory in large starch processing plants in Brazil.

Transportation costs can also be managed through a pre-processing to reduce water content of the raw material, that is, converting raw roots to pressed mash using hydraulic presses. This reduces the water content by up to 75%. In this way, the whole roots are not transported but collecting and intermediate processing is undertaken near the farms in smaller batches and the pressed mash is more easily transported to a central processing point where the final drying and milling takes place. The mash can be processed into gari, starch, HQCF and other related products.

Alternatively, mobile teams of partial processors engage in the quality assurance of processing the roots on-farm and moving the mash to the central flour processing plant. Companies such as DADTCO have successfully employed their Automobile Processing Unit (AMPU) technology to do just this in Nigeria, Mozambique and Ghana. DADTCO, a private company, developed a patented innovative mobile processing technology that has changed dramatically the way cassava is perceived, grown and processed in Africa. They process fresh cassava on-farm or nearby, instead of transporting the perishable roots over long distances to a central factory. The mobile concept essentially overcomes three main obstacles to the commercialization of cassava across Africa which are the high water content of the roots (65–75%), the rapid perishability of the roots once harvested (must be used within 48 hours, preferably 24) and its bulky and irregular shape.

Coupled with the wide scattering of smallholder farmers, these factors contribute to the combined barriers of high transportation costs and challenging logistics, in effect often making the crop uneconomic for commercial exploitation. The AMPU processes fresh cassava roots (<24 hours after harvesting) into cassava cake that in its turn is refined in a mobile refinery unit to produce cassava starch cake. Cassava starch cake is an intermediate product of dough-like consistency that has a proven shelf life, under normal ambient conditions, of over six months. The cake is highly compact and therefore easily transportable by truck to a central location.

The throughput of the AMPU is 4–5 MT cassava roots per hour, but the annual capacity depends on daily operating hours and shift configurations. At maximum, one AMPU can process up to 100 MT of roots per day (on the basis of 5 MT per hour of roots on 3 × 8 hour shifts totalling 20 hours per day net production time on 300 working days per year). This translates into an annual roots processing capacity of 30 000 MT, in turn producing close on 15 000 MT pa of cassava starch cake.

8 Conclusions

Cassava mechanization in SSA is technically and economically feasible and the experience from the AATF CAMAP can provide the basis for building a more robust and systematic pathway for commercialization of cassava mechanization equipment. The focus of cassava mechanization should be mainly on smallholder farmers, small-to-medium farms and commercial farms in communities where cassava is a major contributor to family livelihoods. The farmers need to be aggregated into clusters so that each cluster has dedicated mechanization service providers who provide the services along commercial business lines.

To ensure that cassava mechanization is sustainable, there is need to engage all the cassava actors with the intention of boosting the entire value chain by:

- Transforming agriculture from a subsistence to a commercial activity;
- Fostering demand for high-quality inputs such as seeds, fertilizers and agrochemicals, as well as appropriate forms of farm equipment;
- Fostering the demand for cassava mechanization technology and services by exploring ways of mechanizing more operations such as planting, spraying, harvesting, transportation, processing and packaging;
- Fostering the financial sector, including rural finance networks to recognize the immense need for medium-term finance schemes, especially for production and processing equipment;
- Farmer empowerment and increased numbers of commercially active farmers and agriculture-related rural entrepreneurs (female farmers, youth and entrepreneurs are included);
- Development of a viable agricultural machinery industry, including manufacture of a range of agricultural implements.

There are great opportunities for deployment of cassava mechanization across SSA.<b >Based on the above achievements, AATF would like to outscale this effort to reach out to millions of farmers, processors and other cassava value chain players. AATF and partners therefore intend to:

- Support the private sector to introduce mechanical planters, boom sprayers, cultivators, stem cutters, harvesters and loaders
- Increase cassava production for smallholder and large-scale farmers through linking them with mechanization equipment, and thereby reduce post-harvest losses and high labour cost.
- Introduce intensive cassava production models in partnership with large-scale producers
- Promote mechanical best practices (use of quality planting materials, fertilizer, herbicides, weeding, harvesting) including new varieties and improved cassava agronomy to increase yields
- Build capacity of service providers in the efficient use and management of agricultural machinery
- Strengthen capacity of local entrepreneurs in the maintenance and repairs of farm and processing machinery and in prototype design of machines including the manufacture of low-cost machines.

To ensure sustainability and commercial viability of mechanization in the cassava production chain, AATF will continue working with mechanization service providers to support farmers in commercial business models. The proposed model for scaling up is to design a programme of support to recruit, train and sustain a network of such mechanization service providers, including helping them access the equipment and finance. Through this business model and the designed programme, the investment will create a pool of well-trained, profit-making service providers with full access to equipment for the benefit of the smallholder farmers. There will be the involvement of local equipment importers and dealers and local fabricators for machine repairs and fabrication of spare parts for the equipment.

AATF and partners foresee outscaling the project to more states and districts in Nigeria, Zambia and Uganda and to other countries including DR Congo, Ghana, Mozambique, Tanzania, Malawi, Sierra Leone and Angola. This project is expected to cost US$35 million to reach out to more than 3.5 million beneficiaries across SSA in the next five years.

9 References

Abdulai, S., P. K. Nkegbe and S. A. Donkoh. 2013. 'Technical efficiency of maize production in Northern Ghana.' *African Journal of Agricultural Research* 8(43): 5251–9.

Adewoyin, A. O. and E. A. Ajav. 2013. 'Field and economic performance of three tractor models for ploughing operations in sandy-clay soil of Nigeria.' *American Society of Agricultural and Biological Engineers*. doi:10.13031/aim.20131578419.

Ashburner, J. and J. Kienzle. 2009. A review of some public sector driven mechanization schemes and cases of private sector models in Africa. In: Investment in agricultural mechanization in Africa. A Round Table Meeting, Arusha, Tanzania, 3–5 June 2009. Part II Summary of the papers. FAO, Rome, 2010 (in press).

Benin, S. 2014. *Impact of Ghana's Agricultural Mechanization Services Center Program*: IFPRI discussion paper 1330. International Food Policy Research Institute (IFPRI). https://ideas.repec.org/p/fpr/ifprid/1330.html.

Clarke, L. and C. Bishop. 2002. 'Farm power-present and future availability in developing countries.' *Agricultural Engineering International: CIGR Journal.* http://www.cigrjournal.org/index.php/Ejounral/article/viewFile/335/329.

De Janvry, A. and E. Sadoulet. 2010. 'Agricultural growth and poverty reduction: Additional evidence.' *The World Bank Research Observer* 25(1): 1–20. doi:10.1093/wbro/lkp015.

FAO. 2016. 'FAOSTAT.' Accessed 11 April. http://faostat3.fao.org/browse/Q/QC/E.

Goering, C. E. and A. C. Hansen. 2004. 'Chapter 1 – Introduction.' In C. E. Goering and A. C. Hansen (Eds), *Engine & Tractor Power, 4th Edition*, pp. 1–16. St. Joseph, MI: American Society of Agricultural and Biological Engineers. http://elibrary.asabe.org/abstract.asp?JID=4&AID=24129&CID=etp2004&T=1.

Houmy, K., J. Kienzle and J. Ashburner. 2012. 'The Current Situation of the Intervention of Private Sector in Agricultural Mechanization Development in African Countries.' In C – 1105. CIGR-EurAgEng.

Houssou, N., X. Diao, F. Cossar, S. Kolavalli, K. Jimah and P. O. Aboagye. 2013. 'Agricultural mechanization in Ghana: Is specialized agricultural mechanization service provision a viable business model?' *American Journal of Agricultural Economics* 95(5): 1237–44. doi:10.1093/ajae/aat026.

Itodo, I. N. and J. O. Daudu. 2013. 'Performance evaluation of a tractor mounted yam [Dioscorea Spp] harvester fitted with a collection unit.' *American Society of Agricultural and Biological Engineers.* doi:10.13031/aim.20131592710.

Kienzle, J., W. Hancox and J. Ashburner. 2016. 'Tractor Based Services and Related Support Policies in Developing Countries.' Accessed 16 April. http://www.csbe-scgab.ca/docs/meetings/2010/CSBE100700.pdf.

Monk, K. 2015. 'An Economic Analysis of a Mechanized Systems Approach to Cassava Production in Sub-Saharan Africa.' University of Illinois at Urbana-Champaign. https://www.ideals.illinois.edu/handle/2142/88085.

Mrema, G. C., D. C. Baker and D. Kahan. 2008. *Agricultural Mechanization in Sub-Saharan Africa: Time for a New Look.* Rome: Food and Agriculture Organization on the United Nations.

Rijk, A. G. 1999. 'Agricultural Mechanization Strategy,' CIGR Handbook of Agricultural Engineering. *Plant Production Engineering* 3: 536–53.

Takeshima, H., A. Nin–Pratt and X. Diao. 2013. 'Mechanization and agricultural technology evolution, agricultural intensification in Sub-Saharan Africa: Typology of agricultural mechanization in Nigeria.' *American Journal of Agricultural Economics* 95(5): 1230–6. doi:10.1093/ajae/aat045.

Index

AATF. *see* African Agricultural Technology Foundation (AATF)
adhesives, cassava starch for 85
African Agricultural Technology Foundation (AATF) 380–383
agbelima 40
age, of planting material 273
animal feed
  cassava chips as 62
  and cassava cultivation in sub-Saharan Africa (SSA) 142
anti-nutrients, in cassava root products 46–47
arbuscular mycorrhizal fungi 367
Asia
  cassava cultivation in
    area and yield in 104
    climate and -growing soils 103–104
    improvement in crop management 110
    overview 101–102
    pest, disease and weed incidence 112–113
    planting and harvesting time 110–111
    production systems 102–103
    soil fertility maintenance 111–112
    varietal improvement 104–109
  soil and nutrient management in 317–323
attieke 40

Basic Research to Enable Agricultural Development programme (BREAD) 216
bâton de manioc 38–39
beverages, cassava starch for 84
bioethanol production, cassava chips for 62–67
biofortification, of cassava cultivation 145
biofortified cassava root products 47–50
Brazilian cassava flour (farinha) 164–165
BREAD. *see* Basic Research to Enable Agricultural Development programme (BREAD)
business models, for mechanization
  contract farming model 394
  cooperative business model 393
  individual ownership model 394
  outgrower scheme model 394–395
  private business model 392–393
  structural linkages 395
  transportation of cassava roots 395–396

calcium deficiency 307
Cambodia, cassava cultivation in
  area and yield in 114–115
  climate and -growing soils 114
  production systems 113–114
  soil fertility management 117–118
  varietal improvement 115–117
carbon allocation 366
cassava
  in Africa and Madagascar 17–19
  boiled and fried 165–171
    cooking 167–169
    research 166–167
    varietal improvement 169–171
  chips
    for animal feed 62
    for bioethanol production 62–67
    cassava pellets 62
    for citric acid 67–68
    production 60–61
    quality 61–62
  and climate change 192–195
  crop health
    pest and disease control 244
    soil erosion control 235–240
    soil fertility and soil health 235
    weed control 240–243
  cultural domestication 20–21
  current and developing production and use 89–92
  -derived foods
    cassava leaves 135–136
    dried roots 134
    fresh root 134
    granulated products 135
    pasty products 134–135
  development
    as crop 177–179
    role 179–180
  diversity in traditional products 35
  domestication 6–8
    and cyanogenesis 11–15
  high quality cassava flour (HQCF) 92–96
  for industrial uses
    overview 57–58
    root composition 58–59
  mixed clonal/sexual reproductive system of 8–11, 21–24
  overview 3–4
  pre-harvesting
    intercropping 233–234
    land preparation 229–231
    overview 227–228
    plant density and planting arrangement 232–233
    planting methods 231–232
    selection and production of planting materials 228–229
  research paradigm for 191–192
  starch
    for adhesives and glue 85

for beverages 84
chemical modification 79–80
chemicals 86
for confectionery use 85
for food products 84
hydrolysates/sweeteners and derivatives 81–84
for paper 85
pearls 76–77
for pharmacy and cosmetics 85–86
physical modification 77–79
production 68–72
properties 72–76
for textiles 85
in Tropical Asia and Oceania 19
wild ancestors of 4–6
Cassava: Adding Value for Africa (C:AVA) 207–209
cassava bacterial blight (CBB) 260
cassava brown streak disease (CBSD) 260
cassava cooking 167–169
cassava cultivation
abiotic and biotic constraints 180–181
as crop 177–179
crop and soil management 185–187
genetic diversity and breeding 182–185
mechanization
and African Agricultural Technology Foundation (AATF) 380–383
benefits of 376–377
business models for 392–396
effectiveness of 388–392
obstacles to 377–380
overview 375–376
production options 383–388
overview 175–177
policy
on agriculture in fragile ecosystems 190
on biofuels 190
on credit and crop insurance 191
on food security and equity 189
on market development 189
on models for research and extension support 190–191
trade 189–190
processing
market and gender issues 187–189
and market constraints 181–182
role 179–180
and soil productivity
nutrient losses in eroded sediments and runoff 293–294
nutrient removal 287–290
overview 285–287
soil analyses 295–297
soil fertility and yields 290–292
symptoms of deficiencies and toxicities 295
tissue analysis 297–298
cassava cultivation in Asia
area and yield in 104
climate and -growing soils 103–104
improvement in crop management 110
overview 101–102
pest, disease and weed incidence 112–113
planting and harvesting time 110–111
production systems 102–103
soil fertility maintenance 111–112
varietal improvement 104–109
cassava cultivation in Cambodia
area and yield in 114–115
climate and -growing soils 114
production systems 113–114
soil fertility management 117–118
varietal improvement 115–117
cassava cultivation in Latin America
boiled and fried cassava 165–171
cassava cooking 167–169
research 166–167
varietal improvement 169–171
Brazilian cassava flour (farinha) 164–165
competitiveness
flour for human consumption 157–158
overview 154–155
starch 155–157
improvement opportunities for 161–162
mechanization of planting and harvesting 160–161
overview 149–150
production systems 150–154
propagation of 159
rapid multiplication 160
starch industry
fermented starch and modified products 163
overview 162–163
technologies and types 163
tissue culture 159–160
varietal improvement 158
cassava cultivation in sub-Saharan Africa (SSA)
area and yield in 127
biofortification of 145
cassava-derived foods
cassava leaves 135–136
dried roots 134
fresh root 134
granulated products 135
pasty products 134–135
and climate change 144
commercial production of 143
contribution to diets 136–139
expanded use of processing machinery 140–141
gender in 129–130
industrial uses

animal feed 142
ethanol 142
high-quality cassava flour (HQCF) 142–143
starch 141
pest, disease and weed incidence 127–129
and population dynamics 143–144
production to consumption system 123–125
as robust crop for future 133
science, technology and innovation 144–145
as source of foreign exchange 133
as source of household cash income 130–132
trade in food products 143
trends in production 125–127
varietal improvement 140
cassava frog skin disease (CFSD) 261
cassava intercropping 233–234
advantages of
drought tolerance 356
high productivity 356
income generation 357
low fertilizer demand 356
low labour demand 356–357
suitability for female households 357
complementarity 362
crop rotations 358
description 359
facilitation 362
farmer experimentation 364
functional traits
arbuscular mycorrhizal fungi 367
compatibility 367–368
leaf nutrient content and leaf litter quality 365
leaf retention and longevity 365–366
plant architecture 366
root extension and root branching 367
rooting depth 366–367
timing of carbon allocation 366
options for 357–358
overview 355–356
reduction of erosion 362–363
soil fertility 363–364
soil microbiota 363
temporal niche differentiation 361–362
vertical niche differentiation 361
weed suppression 362
yield growth 359–361
cassava leave products
anti-nutrients in 46–47
chemical composition of
macronutrients 44–45
minerals and vitamins 45
general process 40–41
recipes 41–43
cassava leaves 135–136
cassava mosaic disease (CMD) 259–260
cassava pearls 76–77
cassava pellets 62
cassava post-harvesting 244–245
cassava production technologies
Cassava: Adding Value for Africa (C:AVA) 207–209
fufu project 205–206
goat keeping systems in Nigeria 206–207
overview 199–201
research on improving production and yield 201–202
smallholder decision making 202
smallholders' support systems 203–205
understanding smallholder farmers 202–203
cassava root products
anti-nutrients in 46–47
biofortified 47–50
chemical composition of
macronutrients 43–44
minerals and vitamins 44
fermentation by submersion
bâton de manioc 38–39
chikwangue 38–39
fufu 37–38
lafun 38
non-fermented
industrial products 36–37
in single-step process 36
overview 33–36
solid-state fermented products
agbelima 40
attieke 40
gari and kpokpogari 39–40
cassava root transportation 395–396
C:AVA. *see* Cassava: Adding Value for Africa (C:AVA)
CBB. *see* cassava bacterial blight (CBB)
CBSD. *see* cassava brown streak disease (CBSD)
CFSD. *see* cassava frog skin disease (CFSD)
chemical composition
of cassava leave products 44–45
of cassava root products 43–44
chemical-defence polymorphism 20–21
chemical energy, and cassava starch 86
chemical fertilizers 335–336
chemical modification, of cassava starch 79–80
chikwangue 38–39
chips, cassava
for animal feed 62
for bioethanol production 62–67
cassava pellets 62
for citric acid 67–68
production 60–61
quality 61–62
citric acid, cassava chips for 67–68
climate change
and cassava 192–195
cultivation 144

clonal cleaning 161
CMD. *see* cassava mosaic disease (CMD)
compatibility, of cassava
intercropping 367–368
complementarity, of cassava intercropping 362
confectionery, cassava starch for 85
contract farming model 394
cooperative business model 393
cosmetics, cassava starch in 85–86
crop rotations 358
cultural domestication, of cassava 20–21
cyanogenesis, and domestication 11–15

degradation, of planting material 258
derivatives, and cassava starch 81–84
diet contribution, and cassava
cultivation 136–139
domestication, of cassava 6–8
and cyanogenesis 11–15
dried cassava roots 134

erosion, and cassava intercropping 362–363
ethanol, and cassava cultivation 142

facilitation, of cassava intercropping 362
farinha (Brazilian cassava flour) 164–165
fermentation, of cassava root products
bâton de manioc 38–39
chikwangue 38–39
fufu 37–38
lafun 38
fermented starch and modified products 163
food products, cassava starch for 84
formal seed systems management 257–258
classes of 272
quality standards of 273–274
stages of 273
fresh cassava roots 134
fufu 37–38
project 205–206
functional traits, of cassava intercropping
arbuscular mycorrhizal fungi 367
compatibility 367–368
leaf nutrient content and leaf litter
quality 365
leaf retention and longevity 365–366
plant architecture 366
root extension and root branching 367
rooting depth 366–367
timing of carbon allocation 366

gari and kpokpogari 39–40
GCP21 (global cassava partnership for
the 21st century)
collaboration among cassava
community 215–216
description 214–215
flow and dissemination of
information 219–220
gaps in R&D 217–218
importing and exporting technologies 219
overview 213–214
'whole system' approach 216–217
gender, in cassava cultivation 129–130
glue, cassava starch for 85
goat keeping systems, in Nigeria 206–207
good agricultural practices (GAP)
for cassava crop health
pest and disease control 244
soil erosion control 235–240
soil fertility and soil health 235
weed control 240–243
in cassava pre-harvesting
intercropping 233–234
land preparation 229–231
overview 227–228
plant density and planting
arrangement 232–233
planting methods 231–232
selection and production of planting
materials 228–229
granulated cassava products 135

high-quality cassava flour (HQCF) 92–96,
142–143
HQCF. *see* high-quality cassava flour (HQCF);
High Quality Cassava Flour (HQCF)
hydrolysates/sweeteners, and cassava
starch 81–84

ICAR-CTCRI mini-stake system 274–280
harvest 280
nursery management 278–279
nursery preparation 278
overview 274
planting in shade-net house 277
planting mini-stakes in nursery 278
preparation of mini-stakes 274, 277
transplanting mini-stakes 279–280
individual ownership model 394
industrial uses
cassava
overview 57–58
root composition 58–59
cassava cultivation in sub-Saharan Africa
(SSA)
animal feed 142
ethanol 142
high-quality cassava flour (HQCF) 142–143
starch 141
informal seed systems management 257–258
inorganic nutrient sources 331–335
integrated crop management
adoption of GAP by farmers 245–249
in cassava post-harvesting 244–245
GAP for cassava crop health
pest and disease control 244
soil erosion control 235–240

soil fertility and soil health 235
weed control 240–243
GAP in cassava pre-harvesting
intercropping 233–234
land preparation 229–231
overview 227–228
plant density and planting arrangement 232–233
planting methods 231–232
selection and production of planting materials 228–229
overview 225–226
in vitro systems 270–272

kpokpogari and gari 39–40

lafun 38
land preparation
and cassava pre-harvesting 229–231
and planting for seeds 264
Latin America
cassava cultivation in
boiled and fried cassava 165–171
Brazilian cassava flour (farinha) 164–165
competitiveness 154–158
improvement opportunities for 161–162
mechanization of planting and harvesting 160–161
overview 149–150
production systems 150–154
propagation of 159
rapid multiplication 160
starch industry 162–163
tissue culture 159–160
varietal improvement 158
soil and nutrient management in 311–315
leaf litter quality 365
leaf nutrient content 365
leaf retention and longevity 365–366

macronutrients
in cassava leave products 44–45
in cassava root products 43–44
magnesium deficiency 307–308
manures and soil amendments 335–336
mechanization
and African Agricultural Technology Foundation (AATF) 380–383
benefits of 376–377
business models for
contract farming model 394
cooperative business model 393
individual ownership model 394
outgrower scheme model 394–395
private business model 392–393
structural linkages 395
transportation of cassava roots 395–396
effectiveness of 388–392
obstacles to 377–380
overview 375–376
production options 383–388
micronutrient deficiencies and toxicities 308–309
minerals and vitamins
in cassava leave products 45
in cassava root products 44
mini-stake system 269–270
ICAR-CTCRI case study 274–280
harvest 280
nursery management 278–279
nursery preparation 278
overview 274
planting in shade-net house 277
planting mini-stakes in nursery 278
preparation of mini-stakes 274, 277
transplanting mini-stakes 279–280
mixed clonal/sexual reproductive system, of cassava 8–11, 21–24

NextGen project 215
nitrogen, phosphorus and potassium (NPK) 311, 336–338
nitrogen deficiency 302–304
non-fermented cassava root products
industrial products 36–37
in single-step process 36
NPK. *see* nitrogen, phosphorus and potassium (NPK)
NUE. *see* nutrient use efficiency (NUE)
nutrient removal, and cassava cultivation 287–290
nutrient sources
chemical fertilizers 335–336
fertilizer requirements for balanced nutrition 342–347
manures and soil amendments 335–336
nitrogen, phosphorus and potassium (NPK) balance 336–338
nutrient use efficiency (NUE) 347–352
organic and inorganic 331–335
overview 331
site-specific nutrient management (SSNM) 338–341
nutrient use efficiency (NUE) 347–352
nutritional disorders
calcium deficiency 307
magnesium deficiency 307–308
micronutrient deficiencies and toxicities 308–309
nitrogen, phosphorus and potassium (NPK) requirements 311
nitrogen deficiency 302–304
overview 301–302
phosphorus deficiency 304–306
potassium deficiency 306
soil acidity and alkalinity 309–310
soil and nutrient management
in Asia 317–323

in Latin America 311–315
in sub-Saharan Africa 315–317
sulphur deficiency 308

organic nutrient sources 331–335
outgrower scheme model 394–395

papermaking, cassava starch for 85
pasty cassava products 134–135
pest and disease control 244
pests and diseases
cassava bacterial blight (CBB) 260
cassava brown streak disease (CBSD) 260
cassava frog skin disease (CFSD) 261
cassava mosaic disease (CMD) 259–260
overview 258–259
root rots 261
super-elongation disease (SED) 260–261
pharmaceutical industries, cassava starch in 85–86
phosphorus deficiency 304–306
physical modification, of cassava starch 77–79
physiological quality, of planting material 261–262
pith thickness 273
plant architecture, and cassava intercropping 366
plant density and planting arrangement 232–233
planting material, and seeds
age of 273
degradation of 258
number of nodes 273
physiological quality of 261–262
pith thickness 273
sprouting capacity 274
visual inspection of stems/stakes 273
planting materials, selection and production of 228–229
planting methods, and cassava pre-harvesting 231–232
potassium deficiency 306
pre-harvesting, of cassava
intercropping 233–234
land preparation 229–231
overview 227–228
plant density and planting arrangement 232–233
planting methods 231–232
selection and production of planting materials 228–229
private business model 392–393
processing machinery, and cassava cultivation 140–141

quality planting materials, and seeds
field maintenance and inspection 264–265
land preparation and planting 264
ratoon cropping 269
selection of 263–264
stem harvesting and collection 265–266
storage of 267–268
transport of 268–269
types of varieties 263

rapid multiplication techniques
mini-stake system 269–270
rooted shoots 270
somatic embryogenesis 272
in vitro systems 270–272
ratoon cropping 269
recipes, of cassava leave products 41–43
rooted shoots 270
root extension and branching 367
rooting depth 366–367
root rots 261

SED. *see* super-elongation disease (SED)
seed systems management
description of 256–257
formal seed system 257–258
classes of 272
quality standards of 273–274
stages of 273
ICAR-CTCRI case study 274–280
harvest 280
nursery management 278–279
nursery preparation 278
overview 274
planting in shade-net house 277
planting mini-stakes in nursery 278
preparation of mini-stakes 274, 277
transplanting mini-stakes 279–280
informal 257–258
overview 255–256
pests and diseases
cassava bacterial blight (CBB) 260
cassava brown streak disease (CBSD) 260
cassava frog skin disease (CFSD) 261
cassava mosaic disease (CMD) 259–260
overview 258–259
root rots 261
super-elongation disease (SED) 260–261
planting material
age of 273
degradation of 258
number of nodes 273
physiological quality of 261–262
pith thickness 273
sprouting capacity 274
visual inspection of stems/stakes 273
quality planting material
field maintenance and inspection 264–265
land preparation and planting 264
ratoon cropping 269

selection of 263–264
stem harvesting and collection 265–266
storage of 267–268
transport of 268–269
types of varieties 263
rapid multiplication techniques
mini-stake system 269–270
rooted shoots 270
somatic embryogenesis 272
in vitro systems 270–272
site-specific nutrient management (SSNM) 338–341
smallholder farmers
decision making 202
support systems 203–205
understanding 202–203
soil acidity and alkalinity 309–310
soil amendments 335–336
soil analysis 295–297
soil and nutrient management
in Asia 317–323
in Latin America 311–315
in sub-Saharan Africa 315–317
soil erosion control 235–240
soil fertility
cassava intercropping 363–364
and soil health 235
and yields 290–292
soil microbiota 363
soil productivity, and cassava cultivation
nutrient losses in eroded sediments and runoff 293–294
nutrient removal 287–290
overview 285–287
soil analysis 295–297
soil fertility and yields 290–292
symptoms of deficiencies and toxicities 295
tissue analysis 297–298
solid-state fermented cassava root products
agbelima 40
attieke 40
gari and kpokpogari 39–40
somatic embryogenesis 272
sprouting capacity, of planting material 274
SSNM. *see* site-specific nutrient management (SSNM)
starch, cassava
for adhesives and glue 85
for beverages 84
chemical modification 79–80
chemicals 86
for confectionery use 85
cultivation in sub-Saharan Africa (SSA) 141
for food products 84
hydrolysates/sweeteners and derivatives 81–84
for paper 85
pearls 76–77
for pharmacy and cosmetics 85–86
physical modification 77–79
production 68–72
properties 72–76
for textiles 85
starch industry, in Latin America
fermented starch and modified products 163
overview 162–163
technologies and types 163
stem harvesting, and collection 265–266
structural linkages, for mechanization 395
sub-Saharan Africa (SSA)
cassava cultivation in
area and yield in 127
biofortification of 145
cassava-derived foods 134–136
and climate change 144
commercial production of 143
contribution to diets 136–139
expanded use of processing machinery 140–141
gender in 129–130
industrial uses 141–143
pest, disease and weed incidence 127–129
and population dynamics 143–144
production to consumption system 123–125
as robust crop for future 133
science, technology and innovation 144–145
as source of foreign exchange 133
as source of household cash income 130–132
trade in food products 143
trends in production 125–127
varietal improvement 140
soil and nutrient management in 315–317
sulphur deficiency 308
super-elongation disease (SED) 260–261

temporal niche differentiation 361–362
textile industry, cassava starch in 85
tissue analysis 297–298
transportation, of cassava roots 395–396

vertical niche differentiation 361
visual inspection, of stems/stakes 273

weed control 240–243
weed suppression 362
'whole system' approach 216–217

yield growth, and cassava intercropping 359–361

Lightning Source UK Ltd.
Milton Keynes UK
UKHW020646310719
347143UK00001B/10/P

9 781786 760005